Essays on Coding Theory

Critical coding techniques have developed over the past few decades for data storage, retrieval and transmission systems, significantly mitigating costs for governments and corporations that maintain server systems containing large amounts of data. This book surveys the basic ideas of these coding techniques, which tend not to be covered in the graduate curricula, including pointers to further reading. Written in an informal style, it avoids detailed coverage of proofs, making it an ideal refresher or brief introduction for students and researchers in academia and industry who may not have the time to commit to understanding them deeply. Topics covered include fountain codes designed for large file downloads; LDPC and polar codes for error correction; network, rank-metric and subspace codes for the transmission of data through networks; post-quantum computing; and quantum error correction. Readers are assumed to have taken basic courses on algebraic coding and information theory.

IAN F. BLAKE is Honorary Professor in the Department of Electrical and Computer Engineering at the University of British Columbia, Vancouver. He is a fellow of the Royal Society of Canada, the Institute for Combinatorics and its Applications, the Canadian Academy of Engineers and a Life Fellow of the IEEE. In 2000, he was awarded an IEEE Millennium Medal. He received his undergraduate degree at Queen's University, Canada and doctorate degree at Princeton University in 1967. He also worked in industry, spending sabbatical leaves with IBM and M/A-Com Linkabit, and working with the Hewlett-Packard Labs from 1996 to 1999. His research interests include cryptograph and algebraic coding theory, and he has written several books in these areas.

"This book is an essential resource for graduate students, researchers and professionals delving into contemporary topics in coding theory not always covered in textbooks. Each expertly-crafted essay offers a clear explanation of the fundamental concepts, summarizing key results with a consistent notation and providing valuable references for further exploration."

Frank R. Kschischang, *University of Toronto*

"This volume lives up to its title: it explores many modern research directions in coding theory without always insisting on complete proofs. Professor Blake nevertheless manages to explain not only the results themselves, but also why things work the way they do. This volume will be a wonderful supplement to in-depth presentations of the topics that it covers."

Alexander Barg, *University of Maryland*

"This book provides an excellent and comprehensive presentation of 16 major topics in coding theory. The author brings the highly mathematical subjects down to a level that can be understood with basic knowledge in combinatorial mathematics, modern algebra, and coding and information theory. It can be used as a textbook for a graduate course in electrical engineering, computer sciences and applied mathematics. It is also an invaluable reference for researchers and practitioners in the areas of communications and computer sciences."

Shu Lin, *retired from University of California, Davis*

"This very unique contribution by Professor Blake consists of a collection of essays on coding theory that can be read independently and yet are coherently written. It covers a comprehensive list of topics of interest and is an excellent reference for anyone who is not an expert on all of these topics."

Raymond W. Yeung, *The Chinese University of Hong Kong*

Essays on Coding Theory

Ian F. Blake
University of British Columbia

CAMBRIDGE
UNIVERSITY PRESS

CAMBRIDGE
UNIVERSITY PRESS

Shaftesbury Road, Cambridge CB2 8EA, United Kingdom

One Liberty Plaza, 20th Floor, New York, NY 10006, USA

477 Williamstown Road, Port Melbourne, VIC 3207, Australia

314–321, 3rd Floor, Plot 3, Splendor Forum, Jasola District Centre, New Delhi – 110025, India

103 Penang Road, #05–06/07, Visioncrest Commercial, Singapore 238467

Cambridge University Press is part of Cambridge University Press & Assessment, a department of the University of Cambridge.

We share the University's mission to contribute to society through the pursuit of education, learning and research at the highest international levels of excellence.

www.cambridge.org
Information on this title: www.cambridge.org/9781009283373

DOI: 10.1017/9781009283403

First published 2024

A catalogue record for this publication is available from the British Library

A Cataloging-in-Publication data record for this book is available from the Library of Congress

ISBN 978-1-009-28337-3 Hardback

To Betty, always

Contents

Preface

The subject of algebraic coding theory arose in response to Shannon's remarkable work on information theory and the notion of capacity of communication channels, which showed how the structured introduction of redundancy into a message can be used to improve the error performance on the channel. The first example of an error-correcting code was a Hamming code in Shannon's 1948 paper. This was followed by the Peterson book on error-correcting codes in 1961. The books of Berlekamp in 1968 and the joint volume of Peterson and Weldon in 1972 significantly expanded access to the developing subject. An impressive feature of these books was a beautiful treatment of the theory of finite fields – at a time when most engineers had virtually no training in such algebraic concepts.

Coding theory developed through the 1960s, although there were concerns that the encoding and decoding algorithms were of such complexity that their use in practice might be limited, given the state of electronic circuits at that time. This was dispelled during the 1970s with the increasing capabilities of microelectronics. They are now included in many communications and storage applications and standards and form a critical part of such systems.

Coding theory has expanded beyond the original algebraic coding theory with very significant achievements in systems, such as LDPC coding, polar coding and fountain codes, and these systems are capable of achieving capacity on their respective channels that have been incorporated into numerous standards and applications. It has also embraced new avenues of interest such as locally decodable codes, network codes, list decoding and codes for distributed storage, among many others.

While topics such as LDPC coding and polar coding are of great importance, only a few graduate departments will be able to devote entire courses to them or even partially cover them in more general courses. Given the pressure departments face, many of the other topics considered here may not fare so well.

The idea of this book was to create a series of presentations of modest length and depth in these topics to facilitate access to them by graduate students and researchers who may have an interest in them but defer from making the commitment of time and effort for a deeper understanding. Each chapter is designed to acquaint the reader with an introduction to the main results and possibilities without many of the proofs and details. They can be read independently and a prerequisite is a basic course on algebraic coding and information theory, although some of the topics present technical challenges.

There are as many reasons not to write such a book as to write it. A few of the areas have either excellent monographs or tutorials available on the web. Also, it might be argued that an edited book on these topics with chapters written by acknowledged experts would be of more value. Indeed, such a volume is *A Concise Encyclopedia of Coding Theory*, W.C. Huffman, J.-L. Kim and P. Solé, eds., 2021, CRC Press. However, the entries in such a volume, as excellent as they usually are, are often of an advanced nature, designed for researchers in the area to bring them abreast of current research directions. It was felt that a series of chapters, written from a fairly consistent point of view and designed to introduce readers to the areas covered, rather than provide a deep coverage, might be of interest. I hope some readers of the volume will agree. For many of the areas covered, the influence of the seminal papers on the subjects is impressive. The attempt here is to explain and put into context these important works, but for a serious researcher in an area, it does not replace the need to read the original papers.

Choosing the level of the presentation was an interesting challenge. On the one hand it was desired to achieve as good an appreciation of the results and implications of an area as possible. The emphasis is on describing and explaining contributions rather than proving and deriving, as well as providing a few examples drawn mainly from the literature. On the other hand the inclusion of too much detail and depth might discourage reading altogether. It is hoped the compromise reached is satisfactory. While efforts were made to render readable accounts for the topics, many readers might still find some of the topics difficult.

Another problem was to choose a consistent notation when describing results from different authors. Since one of the goals of the work was to provide an entrée to the main papers of an area, an effort was made to use the notation of the seminal works. Across the chapters, compromises in notation had to be made and the hope is that these were reasonable. Generally there was a bias toward describing code construction techniques for the areas which tended to make some sections technically challenging.

I would like to thank the many colleagues around the world who provided helpful and useful comments on many parts of the manuscript. First among these are Shu Lin and Frank Kschischang, who read virtually all of the work and consistently supported the effort. I cannot thank them enough for their comments and suggestions. I would also like to thank Raymond Yeung, Vijay Kumar, Eitan Yaakobi, Rob Calderbank, Amir Tasbihi and Lele Wang, who read several of the chapters and provided expert guidance on many issues.

I would also like to thank the Department of Electrical and Computer Engineering at the University of British Columbia and my colleagues there, Vijay Bhargava, Lutz Lampe and Lele Wang, for providing such a hospitable environment in which to pursue this work.

Finally, I would like to thank my wife Betty without whose love, patience and understanding this book would not have been written.

1
Introduction

Since the early 2000s we have seen the basic notions of coding theory expand beyond the role of error correction and algebraic coding theory. The purpose of this volume is to provide a brief introduction to a few of the directions that have been taken as a platform for further reading. Although the approach is to be descriptive with few proofs, there are parts which are unavoidably technical and more challenging.

It was mentioned in the Preface that the prerequisite for this work is a basic course on algebraic coding theory and information theory. In fact only a few aspects of finite fields, particularly certain properties of polynomials over finite fields, Reed–Solomon codes and Reed–Muller codes and their generalizations are considered to provide a common basis and establish the notation to be used. The trace function on finite fields makes a few appearances in the chapters and its basic properties are noted. Most of the information will be familiar and stated informally without proof. A few of the chapters use notions of information theory and discrete memoryless channels and the background required for these topics is also briefly reviewed in Section 1.2. The final Section 1.3 gives a brief description of the chapters that follow.

1.1 Notes on Finite Fields and Coding Theory

Elements of Finite Fields

A few basic notions from integers and polynomials will be useful in several of the chapters as well as considering properties of finite fields. The *greatest common divisor* (gcd) of two integers or polynomials over a field will be a staple of many computations needed in several of the chapters. Abstractly, an integral domain is a commutative ring in which the product of two nonzero elements is nonzero, sometimes stated as a commutative ring with identity which has no zero divisors (i.e., two nonzero elements a, b such that $ab = 0$). A *Euclidean domain* is an integral domain which is furnished with a norm function, in which the division of an element by another with a remainder of lower degree can be formulated. Equivalently the Euclidean algorithm (EA) can be formulated in a Euclidean domain.

Recall that the gcd of two integers $a, b \in \mathbb{Z}$ is the largest integer d that divides both a and b. Let $\mathbb{F}$ be a field and denote by $\mathbb{F}[x]$ the ring of polynomials over $\mathbb{F}$ with coefficients from $\mathbb{F}$. The gcd of two polynomials $a(x), b(x) \in \mathbb{F}[x]$ is the monic polynomial (coefficient of the highest power of x is unity) of the greatest degree, $d(x)$, that divides both polynomials. The EA for polynomials is an algorithm that produces the gcd of polynomials $a(x)$ and $b(x)$ (the one for integers is similar) by finding polynomials $u(x)$ and $v(x)$ such that

$$d(x) = u(x)a(x) + v(x)b(x). \tag{1.1}$$

It is briefly described as follows. Suppose without loss of generality that $\deg b(x) < \deg a(x)$ and consider the sequence of polynomial division steps producing quotient and remainder polynomials:

$$\begin{aligned}
a(x) &= q_1(x)b(x) + r_1(x), & \deg r_1 &< \deg b \\
b(x) &= q_2(x)r_1(x) + r_2(x), & \deg r_2 &< \deg r_1 \\
r_1(x) &= q_3(x)r_2(x) + r_3(x), & \deg r_3 &< \deg r_2 \\
&\ \vdots \qquad \vdots & & \\
r_k(x) &= q_{k+2}(x)r_{k+1}(x) + r_{k+2}(x), & \deg r_{k+2} &< \deg r_{k+1} \\
r_{k+1}(x) &= q_{k+3}(x)r_{k+2}(x), & d(x) &= r_{k+2}(x).
\end{aligned}$$

That $d(x)$, the last nonzero remainder, is the required gcd is established by tracing back divisibility conditions. Furthermore, tracing back shows how two polynomials $u(x)$ and $v(x)$ are found so that Equation 1.1 holds.

A similar argument holds for integers. The gcd is denoted (a, b) or $(a(x), b(x))$ for integers and polynomials, respectively. If the gcd of two integers or polynomials is unity, they are referred to as being *relatively prime* and denoted $(a, b) = 1$ or $(a(x), b(x)) = 1$.

If the prime factorization of n is

$$n = p_1^{e_1} p_2^{e_2} \cdots p_k^{e_k}, \quad p_1, p_2, \ldots, p_k \text{ distinct primes,}$$

then the number of integers less than n that are relatively prime to n is given by the *Euler Totient function* $\phi(n)$ where

$$\phi(n) = \prod_{i=1}^{k} p_i^{e_i - 1}(p_i - 1). \tag{1.2}$$

A *field* is a commutative ring with identity in which elements have additive inverses (0 denotes the additive identity) and nonzero elements have multiplicative inverses (1 denotes the multiplicative identity). It may also be viewed as an integral domain in which the nonzero elements form a multiplicative group.

A finite field is a field with a finite number of elements. For a finite field, there is a smallest integer c such that each nonzero element of the field added to itself a total of c times yields 0. Such an integer is called the *characteristic of the field.* If c is not finite, the field is said to have characteristic 0. Notice that in a finite field of characteristic 2, addition and subtraction are identical in that $1 + 1 = 0$. Denote the set of nonzero elements of the field $\mathbb{F}$ by $\mathbb{F}^*$.

Denote by $\mathbb{Z}_n$ the set of integers modulo n, $\mathbb{Z}_n = \{0, 1, 2, \ldots, n-1\}$. It is a finite field iff n is a prime p, since if $n = ab$, $a, b \in \mathbb{Z}$ is composite, then it has zero divisors and hence is not a field. Thus the characteristic of any finite field is a prime and the symbol p is reserved for some arbitrary prime integer. In a finite field $\mathbb{Z}_p$, arithmetic is modulo p. If $a \in \mathbb{Z}_p, a \neq 0$, the inverse of a can be found by applying the EA to $a < p$ and p which yields two integers u, $v \in \mathbb{Z}$ such that

$$ua + vp = 1 \quad \text{in } \mathbb{Z}$$

and so $ua + vp \pmod{p} \equiv ua \equiv 1 \pmod{p}$ and $a^{-1} \equiv u \pmod{p}$. The field will be denoted $\mathbb{F}_p$. In any finite field there is a smallest subfield, a set of elements containing and generated by the unit element 1, referred to as the *prime subfield*, which will be $\mathbb{F}_p$ for some prime p.

Central to the notion of finite fields and their applications is the role of polynomials over the field. Denote the ring of polynomials in the indeterminate x over a field $\mathbb{F}$ by $\mathbb{F}[x]$ and note that it is a Euclidean domain (although the ring of polynomials with two variables $\mathbb{F}[x, y]$ is not). A polynomial $f(x) = f_n x^n + f_{n-1}x^{n-1} + \cdots + f_1 x + f_0 \in \mathbb{F}[x], f_i \in \mathbb{F}$ is monic if the leading coefficient f_n is unity.

A polynomial $f(x) \in \mathbb{F}[x]$ is called reducible if it can be expressed as the product of two nonconstant polynomials and irreducible if it is not the product of two nonconstant polynomials, i.e., there do not exist two nonconstant polynomials $a(x), b(x) \in \mathbb{F}[x]$ such that $f(x) = a(x)b(x)$. Let $f(x)$ be a monic irreducible polynomial over the finite field $\mathbb{F}_p$ and consider the set of p^n polynomials taken modulo $f(x)$ which will be denoted

$$\mathbb{F}_p[x]/\langle f(x)\rangle = \left\{a_{n-1}x^{n-1} + a_{n-2}x^{n-2} + \cdots + a_1 x + a_0, a_i \in \mathbb{F}_p\right\}$$

where $\langle f(x)\rangle$ is the ideal in $\mathbb{F}_p$ generated by $f(x)$. Addition of two polynomials is obvious and multiplication of two polynomials is taken modulo the irreducible polynomial $f(x)$, i.e., the remainder after division by $f(x)$. The inverse of a nonzero polynomial $a(x) \in \mathbb{F}_p[x]/\langle f(x)\rangle$ is found via the EA as before. That is since by definition $(a(x), f(x)) = 1$ there exist polynomials $u(x), v(x)$ such that

$$u(x)a(x) + v(x)f(x) = 1$$

and the inverse of $a(x) \in \mathbb{F}_p[x]/\langle f(x)\rangle$ is $u(x)$. Algebraically this structure might be described as the factor field of the ring $\mathbb{F}_p[x]$ modulo the maximal ideal $\langle f(x)\rangle$.

It follows the set $\mathbb{F}_p[x]/\langle f(x)\rangle$ forms a finite field with p^n elements. It is conventional to denote $q = p^n$ and the field of p^n elements as either $\mathbb{F}_{p^n}$ or $\mathbb{F}_q$. Every finite field can be shown to have a number of elements of the form $q = p^n$ for some prime p and positive integer n and that any two finite fields of the same order are isomorphic. It will be noted that an irreducible polynomial of degree n will always exist (see Equation 1.4) and so all finite fields can be constructed in this manner.

In general, suppose $q = p^m$ and let $f(x)$ be a monic irreducible polynomial over $\mathbb{F}_q$ of degree m (which will be shown to always exist). The set of q^m polynomials over $\mathbb{F}_q$ of degree less than m with multiplication modulo $f(x)$ will then be a finite field with q^m elements and designated $\mathbb{F}_{q^m}$. For future reference denote the set of polynomials of degree less than m by $\mathbb{F}_q{}^{<m}[x]$ and those less than or equal by $\mathbb{F}_q{}^{\leq m}[x]$. Since it involves no more effort, this general finite field $\mathbb{F}_{q^m}$ will be examined for basic properties. The subset $\mathbb{F}_q \subseteq \mathbb{F}_{q^m}$ is a field, i.e., a subset that has all the properties of a field, a subfield of $\mathbb{F}_{q^m}$.

The remainder of the subsection contains a brief discussion of the structure of finite fields and polynomials usually found in a first course of coding theory.

It is straightforward to show that over any field $\mathbb{F}$ $(x^m - 1)$ divides $(x^n - 1)$ iff m divides n, written as

$$(x^m - 1)\big|(x^n - 1) \text{ iff } m \mid n.$$

Further, for any prime p,

$$(p^m - 1)\big|(p^n - 1) \text{ iff } m \mid n.$$

The multiplicative group of a finite field, $\mathbb{F}^*_{q^m}$, can be shown to be cyclic (generated by a single element). The order of a nonzero element α in a field $\mathbb{F}$ is the smallest positive integer ℓ such that $\alpha^\ell = 1$, denoted as ord $(\alpha) = \ell$ and referred to as the order of α. Similarly if ℓ is the smallest integer such that the polynomial $f(x) \mid \left(x^\ell - 1\right)$, the polynomial is said to have order ℓ over the understood field. The order of an irreducible polynomial is also the order of its zeros.

If β has order ℓ, then β^i has order $\ell/(i,\ell)$. Similarly if β has order ℓ and γ has order κ and $(\ell,\kappa) = 1$, then the order of $\beta\gamma$ is $\ell\kappa$.

An element $\alpha \in \mathbb{F}_{q^m}$ of maximum order $q^m - 1$ is called a *primitive element.* If α is primitive, then α^i is also primitive iff $(i, q^m - 1) = 1$ and there are $\phi(q^m - 1)$ primitive elements in $\mathbb{F}_{q^m}$.

A note on the representation of finite fields is in order. The order of an irreducible polynomial $f(x)$ over $\mathbb{F}_q$ of degree k can be determined by successively dividing the polynomial $(x^n - 1)$ by $f(x)$ over $\mathbb{F}_q$ as n increases. If the smallest such n is $q^k - 1$, the polynomial is primitive. To effect the division, arithmetic in the field $\mathbb{F}_q$ is needed. If $f(x)$ is primitive of degree k over $\mathbb{F}_q$, one could then take the field as the elements

$$\mathbb{F}_{q^k} = \left\{0, 1, x, x^2, \ldots, x^{q^k-2}\right\}.$$

By definition the elements are distinct. Each of these elements could be taken modulo $f(x)$ (which is zero in the field) which would result in the field elements being all polynomials over $\mathbb{F}_q$ of degree less than k. Multiplication in this field would be polynomials taken modulo $f(x)$. The field element x is a primitive element. While this is a valid presentation, it is also common to identify the element x by an element α with the statement "let α be a zero of the primitive polynomial $f(x)$ of degree k over $\mathbb{F}_q$." The two views are equivalent.

There are $\phi(q^k - 1)$ primitive elements in $\mathbb{F}_{q^k}$ and since the degree of an irreducible polynomial with one of these primitive elements as a zero is necessarily k, there are exactly $\phi(q^k - 1)/k$ primitive polynomials of degree k over $\mathbb{F}_q$.

Suppose $f(x)$ is an irreducible nonprimitive polynomial of degree k over $\mathbb{F}_q$. Suppose it is of order $n < q^k - 1$, i.e., $f(x) \mid (x^n - 1)$. One can define the field $\mathbb{F}_{q^k}$ as the set of polynomials of degree less than k

$$\mathbb{F}_{q^k} = \left\{a_{k-1}x^{k-1} + a_{k-2}x^{k-2} + \cdots + a_1 x + a_0,\ a_i \mathbb{F}_q\right\}$$

with multiplication modulo $f(x)$. The element x is not primitive if $n < (q^k - 1)$ but is an element of order $n, n \mid q^k - 1$ (although there are still $\phi(q^k - 1)$ primitive elements in the field).

Let $\alpha \in \mathbb{F}_{q^m}$ be an element of maximum order $(q^m - 1)$ (i.e., primitive) and denote the multiplicative group of nonzero elements as

$$\mathbb{F}^*_{q^m} = \langle \alpha \rangle = \left\{1, \alpha, \alpha^2, \ldots, \alpha^{q^m-2}\right\}.$$

Let $\beta \in \mathbb{F}_{q^m}^*$ be an element of order ℓ which generates a cyclic multiplicative subgroup of $\mathbb{F}_{q^m}^*$ of order ℓ and for such a subgroup $\ell \mid (q^m - 1)$. The order of any nonzero element in $\mathbb{F}_{q^m}$ divides $(q^m - 1)$. Thus

$$x^{q^m} - x = \prod_{\beta \in \mathbb{F}_{q^m}} (x - \beta), \quad x^{q^m - 1} - 1 = \prod_{\beta \in \mathbb{F}_{q^m}^*} (x - \beta) \qquad (1.3)$$

is a convenient factorization (over $\mathbb{F}_{q^m}$).

Suppose $\mathbb{F}_{q^m}$ has a subfield $\mathbb{F}_{q^k}$ – a subset of elements which is itself a field. The number of nonzero elements in $\mathbb{F}_{q^k}$ is $(q^k - 1)$ and this set must form a multiplicative subgroup of $\mathbb{F}_{q^m}^*$ and hence $(q^k - 1) \mid (q^m - 1)$ and this implies that $k \mid m$ and that $\mathbb{F}_{q^k}$ is a subfield of $\mathbb{F}_{q^m}$ iff $k \mid m$. Suppose $\mathbb{F}_{q^k}$ is a subfield of $\mathbb{F}_{q^m}$. Then

$$\beta \in \mathbb{F}_{q^m} \text{ is in } \mathbb{F}_{q^k} \text{ iff } \beta^{q^k} = \beta$$

and $\beta = \alpha^j \in \mathbb{F}_{q^m}$ is a zero of the monic irreducible polynomial $f(x)$ of degree k over $\mathbb{F}_q$. Thus

$$f(x) = x^k + f_{k-1}x^{k-1} + \cdots + f_1 x + f_0, \quad f_i \in \mathbb{F}_q,\ i = 0, 1, 2, \ldots, k-1$$

and $f(\alpha^j) = 0$. Notice that

$$\begin{aligned} f(x)^q &= \left(x^k + f_{k-1}x^{k-1} + \cdots + f_1 x + f_0\right)^q \\ &= x^{kq} + f_{k-1}^q x^{q(k-1)} + \cdots + f_1^q x^q + f_0^q \\ &= x^{kq} + f_{k-1}x^{q(k-1)} + \cdots + f_1 x^q + f_0, \text{ as } f_i^q = f_I \text{ for } f_i \in \mathbb{F}_q \\ &= f(x^q) \end{aligned}$$

and since $\beta = \alpha^j$ is a zero of $f(x)$ so is β^q. Suppose ℓ is the smallest integer such that $\beta^{q^\ell} = \beta$ (since the field is finite there must be such an ℓ) and let

$$C_j = \left\{\alpha^j = \beta, \beta^q, \beta^{q^2}, \ldots, \beta^{q^{\ell-1}}\right\}$$

referred to as the conjugacy class of β. Consider the polynomial

$$g(x) = \prod_{i=0}^{\ell-1} \left(x - \beta^{q^j}\right)$$

and note that

$$g(x)^q = \prod_{i=0}^{\ell-1} \left(x - \beta^{q^j}\right)^q = \prod_{i=0}^{\ell-1} \left(x^q - \beta^{q^{j+1}}\right) = \prod_{i=0}^{\ell-1} \left(x^q - \beta^{q^j}\right) = g(x^q)$$

and, as above, $g(x)$ has coefficients in $\mathbb{F}_q$, i.e., $g(x) \in \mathbb{F}_q[x]$. It follows that $g(x)$ must divide $f(x)$ and since $f(x)$ was assumed monic and irreducible it must be that $g(x) = f(x)$. Thus if one zero of the irreducible $f(x)$ is in $\mathbb{F}_{q^m}$, all are. Each conjugacy class of the finite field corresponds to an irreducible polynomial over $\mathbb{F}_q$.

By similar reasoning it can be shown that if $f(x)$ is irreducible of degree k over $\mathbb{F}_q$, then $f(x) \mid (x^{q^m} - x)$ iff $k \mid m$. It follows that the polynomial $x^{q^m} - x$ is the product of all monic irreducible polynomials whose degrees divide m. Thus

$$x^{q^m} - x = \prod_{\substack{f(x) \text{ irreducible} \\ \text{over } \mathbb{F}_q \\ \text{degree } f(x)=k|m}} f(x).$$

This allows a convenient enumeration of the polynomials. If $N_q(m)$ is the number of monic irreducible polynomials of degree m over $\mathbb{F}_q$, then by the above equation

$$q^m = \sum_{k|m} k N_q(k)$$

which can be inverted using standard combinatorial techniques as

$$N_q(m) = \frac{1}{m} \sum_{k|m} \mu\left(\frac{m}{k}\right) q^k \tag{1.4}$$

where $\mu(n)$ is the Möbius function equal to 1 if $n = 1$, $(-1)^s$ if n is the product of s distinct primes and zero otherwise. It can be shown that $N_q(k)$ is at least one for all prime powers q and all positive integers k. Thus irreducible polynomials of degree k over a field of order q exist for all allowable parameters and hence finite fields exist for all allowable parameter sets.

Consider the following example.

Example 1.1 Consider the field extension $\mathbb{F}_{2^6}$ over the base field $\mathbb{F}_2$. The polynomial $x^{2^6} - x$ factors into all irreducible polynomials of degree dividing 6, i.e., those of degrees 1, 2, 3 and 6. From the previous formula

$$N_2(1) = 2,\ N_2(2) = 1,\ N_2(3) = 2,\ N_2(6) = 9.$$

For a primitive element α the conjugacy classes of $\mathbb{F}_{2^6}$ over $\mathbb{F}_2$ are (obtained by raising elements by successive powers of 2 mod 63, with tentative polynomials associated with the classes designated):

$$\begin{array}{l}
\alpha^1, \alpha^2, \alpha^4, \alpha^8, \alpha^{16}, \alpha^{32} \approx f_1(x) \\
\alpha^3, \alpha^6, \alpha^{12}, \alpha^{24}, \alpha^{48}, \alpha^{33} \approx f_2(x) \\
\alpha^5, \alpha^{10}, \alpha^{20}, \alpha^{40}, \alpha^{17}, \alpha^{34} \approx f_3(x) \\
\alpha^7, \alpha^{14}, \alpha^{28}, \alpha^{56}, \alpha^{49}, \alpha^{35} \approx f_4(x) \\
\alpha^9, \alpha^{18}, \alpha^{36} \approx f_5(x) \\
\alpha^{11}, \alpha^{22}, \alpha^{44}, \alpha^{25}, \alpha^{50}, \alpha^{37} \approx f_6(x)
\end{array}$$

$$\alpha^{13},\alpha^{26},\alpha^{52},\alpha^{41},\alpha^{19},\alpha^{38} \approx f_7(x)$$
$$\alpha^{15},\alpha^{30},\alpha^{60},\alpha^{57},\alpha^{51},\alpha^{39} \approx f_8(x)$$
$$\alpha^{21},\alpha^{42} \approx f_9(x)$$
$$\alpha^{23},\alpha^{46},\alpha^{29},\alpha^{58},\alpha^{53},\alpha^{43} \approx f_{10}(x)$$
$$\alpha^{27},\alpha^{54},\alpha^{45} \approx f_{11}(x)$$
$$\alpha^{31},\alpha^{62},\alpha^{61},\alpha^{59},\alpha^{55},\alpha^{47} \approx f_{12}(x).$$

By the above discussion a set with ℓ integers corresponds to an irreducible polynomial over $\mathbb{F}_2$ of degree ℓ. Further, the order of the polynomial is the order of the conjugates in the corresponding conjugacy class.

Notice there are $\phi(63) = \phi(9 \cdot 7) = 3 \cdot 2 \cdot 6 = 36$ primitive elements in $\mathbb{F}_{2^6}$ and hence there are $36/6 = 6$ primitive polynomials of degree 6 over $\mathbb{F}_2$. If α is chosen as a zero of the primitive polynomial $f_1(x) = x^6 + x + 1$, then the correspondence of the above conjugacy classes with irreducible polynomials is

Poly. No.	Polynomial	Order
$f_1(x)$	$x^6 + x + 1$	63
$f_2(x)$	$x^6 + x^4 + x^3 + x^2 + x + 1$	21
$f_3(x)$	$x^6 + x^5 + x^2 + x + 1$	63
$f_4(x)$	$x^6 + x^3 + 1$	9
$f_5(x)$	$x^3 + x^2 + 1$	7
$f_6(x)$	$x^6 + x^5 + x^3 + x^2 + 1$	63
$f_7(x)$	$x^6 + x^4 + x^3 + x + 1$	63
$f_8(x)$	$x^6 + x^5 + x^4 + x^2 + 1$	21
$f_9(x)$	$x^2 + x + 1$	3
$f_{10}(x)$	$x^6 + x^5 + x^4 + x + 1$	63
$f_{11}(x)$	$x^3 + x + 1$	7
$f_{12}(x)$	$x^6 + x^5 + 1$	63

The other three irreducible polynomials of degree 6 are of orders 21 (two of them, $f_2(x)$ and $f_8(x)$) and nine ($f_4(x)$). The primitive element α in the above conjugacy classes could have been chosen as a zero of any of the primitive polynomials. The choice determines arithmetic in $\mathbb{F}_{2^6}$ but all choices will lead to isomorphic representations. Different choices would have resulted in different associations between conjugacy classes and polynomials.

Not included in the above table is the conjugacy class $\{\alpha^{63}\}$ which corresponds to the polynomial $x + 1$ and the class $\{0\}$ which corresponds to the polynomial x. The product of all these polynomials is $x^{2^6} - x$.

The notion of a *minimal polynomial* of a field element is of importance for coding. The *minimal polynomial* $m_\beta(x)$ of an element $\beta \in \mathbb{F}_{q^n}$ over $\mathbb{F}_q$ is that the monic irreducible polynomial of least degree that has β as a zero. From the above discussion, every element in a conjugacy class has the same minimal polynomial.

Further notions of finite fields that will be required include that of a *polynomial basis* of $\mathbb{F}_{q^n}$ over $\mathbb{F}_q$ which is one of a form $\{1,\alpha,\alpha^2,\ldots,\alpha^{n-1}\}$ for some $\alpha \in \mathbb{F}_{q^n}$ for which the elements are linearly independent over $\mathbb{F}_q$. A basis of $\mathbb{F}_{q^n}$ over $\mathbb{F}_q$ of the form $\{\alpha,\alpha^q,\alpha^{q^2},\ldots,\alpha^{q^{n-1}}\}$ is called a *normal basis* and such bases always exist. In the case that $\alpha \in \mathbb{F}_{q^n}$ is primitive (of order $q^n - 1$) it is called a primitive normal basis.

The Trace Function of Finite Fields

Further properties of the trace function that are usually discussed in a first course on coding will prove useful at several points in the chapters. Let $\mathbb{F}_{q^n}$ be an extension field of order n over $\mathbb{F}_q$. For an element $\alpha \in \mathbb{F}_{q^n}$ the *trace function* of $\mathbb{F}_{q^n}$ over $\mathbb{F}_q$ is defined as

$$\mathrm{Tr}_{q^n|q}(\alpha) = \sum_{i=0}^{n-1} \alpha^{q^i}.$$

The function enjoys many properties, most notably that [8]

(i) $\mathrm{Tr}_{q^n|q}(\alpha+\beta) = \mathrm{Tr}_{q^n|q}(\alpha) + \mathrm{Tr}_{q^n|q}(\beta), \quad \alpha,\beta \in \mathbb{F}_{q^n}$
(ii) $\mathrm{Tr}_{q^n|q}(a\alpha) = a\mathrm{Tr}_{q^n|q}(\alpha),\ a \in \mathbb{F}_q, \alpha \in \mathbb{F}_{q^n}$
(iii) $\mathrm{Tr}_{q^n|q}(a) = na,\ a \in \mathbb{F}_q$
(iv) $\mathrm{Tr}_{q^n|q}$ is an onto map.

To show property (iv), which the trace map is onto (i.e., codomain is $\mathbb{F}_q$), it is sufficient to show that there exists an element α of $\mathbb{F}_{q^n}$ for which $\mathrm{Tr}_{q^n|q}(\alpha) \neq 0$ since if $\mathrm{Tr}_{q^n|q}(\alpha) = b \neq 0, b \in \mathbb{F}_q$, then (property ii) $\mathrm{Tr}_{q^n|q}(b^{-1}\alpha) = 1$ and hence all elements of $\mathbb{F}_q$ are mapped onto. Consider the polynomial equation

$$x^{q^{n-1}} + x^{q^{n-2}} + \cdots + x = 0$$

that can have at most q^{n-1} solutions in $\mathbb{F}_{q^n}$. Hence there must exist elements of $\beta \in \mathbb{F}_{q^n}$ for which $\mathrm{Tr}_{q^n|q}(\beta) \neq 0$. An easy argument shows that in fact exactly q^{n-1} elements of $\mathbb{F}_{q^n}$ have a trace of $a \in \mathbb{F}_q$ for each element of $\mathbb{F}_q$.

Notice that it also follows from these observations that

$$x^{q^n} - x = \prod_{a \in \mathbb{F}_q} \left(x^{q^{n-1}} + x^{q^{n-2}} + \cdots + x - a \right)$$

since each element of $\mathbb{F}_{q^n}$ is a zero of the LHS and exactly one term of the RHS.

Also, suppose [8] $L(\cdot)$ is a linear function from $\mathbb{F}_{q^n}$ to $\mathbb{F}_q$ in the sense that for all $a_1, a_2 \in \mathbb{F}_q$ and all $\alpha_1, \alpha_2 \in \mathbb{F}_{q^n}$

$$L(a_1\alpha_1 + a_2\alpha_2) = a_1 L(\alpha_1) + a_2 L(\alpha_2).$$

Then $L(\cdot)$ must be of the form

$$L(\alpha) = \mathrm{Tr}_{q^n|q}(\beta\alpha) \overset{\Delta}{=} L_\beta(\alpha)$$

for some β. Thus the set of such linear functions is precisely the set

$$L_\beta(\cdot),\ \beta \in \mathbb{F}_{q^n}$$

and these are distinct functions for distinct β.

A useful property of the trace function ([8], lemma 3.51, [11], lemma 9.3) is that if $u_1, u_2, \ldots, u_n$ is a basis of $\mathbb{F}_{q^n}$ over $\mathbb{F}_q$ and if

$$\mathrm{Tr}_{q^n|q}(\alpha u_i) = 0 \text{ for } i = 1, 2, \ldots, n, \quad \alpha \in \mathbb{F}_{q^n},$$

then $\alpha = 0$. Equivalently if for $\alpha \in \mathbb{F}_{q^n}$

$$\mathrm{Tr}_{q^n|q}(\alpha u) = 0 \quad \forall u \in \mathbb{F}_{q^n}, \tag{1.5}$$

then $\alpha = 0$. This follows from the trace map being onto. It will prove a useful property in the sequel. It also follows from the fact that

$$\begin{bmatrix} u_1 & u_1^q & \cdots & u_1^{q^{n-1}} \\ u_2 & u_2^q & \cdots & u_2^{q^{n-1}} \\ \vdots & \vdots & \vdots & \vdots \\ u_n & u_n^q & \cdots & u_n^{q^{n-1}} \end{bmatrix}$$

is nonsingular iff $u_1, u_2, \ldots, u_n \in \mathbb{F}_{q^n}$ are linearly independent over $\mathbb{F}_q$. A formula for the determinant of this matrix is given in [8].

If $\mu = \{\mu_1, \mu_2, \ldots, \mu_n\}$ is a basis of $\mathbb{F}_{q^n}$ over $\mathbb{F}_q$, then a basis $\nu = \{\nu_1, \nu_2, \ldots, \nu_n\}$ is called a *trace dual basis* if

$$\mathrm{Tr}_{q^n|q}(\mu_i \nu_j) = \delta_{i,j} = \begin{cases} 1 & \text{if } i = j \\ 0 & \text{if } i \neq j \end{cases}$$

and for a given basis a unique dual basis exists. It is noted that if $\mu = \{\mu_1, \ldots, \mu_n\}$ is a dual basis for the basis $\{\nu_1, \ldots, \nu_n\}$, then given

$$y = \sum_{i=1}^{n} a_i \,\mu_i \quad \text{then} \quad y = \sum_{i=1}^{n} \mathrm{Tr}_{q^n|q}(y\nu_i)\,\mu_i, \quad a_i \in \mathbb{F}_q. \tag{1.6}$$

Thus an element $y \in \mathbb{F}_{q^n}$ can be represented in the basis μ, by the traces $\mathrm{Tr}_{q^n|q}(y\nu_j),\ j = 1, 2, \ldots, n$.

It can be shown that for a given normal basis, the dual basis is also normal. A convenient reference for such material is [8, 12].

Elements of Coding Theory

A few comments on BCH, Reed–Solomon (RS), Generalized Reed–Solomon (GRS), Reed–Muller (RM) and Generalized Reed–Muller (GRM) codes are noted. Recall that a cyclic code of length n and dimension k and minimum distance d over $\mathbb{F}_q$, designated as an $(n,k,d)_q$ code, is defined by a polynomial $g(x) \in \mathbb{F}_q[x]$ of degree $(n-k)$, $g(x) \mid (x^n - 1)$ or alternatively as a principal ideal $\langle g(x) \rangle$ in the factor ring $\mathcal{R} = \mathbb{F}_q[x]/\langle x^n - 1 \rangle$.

Consider a BCH code of length $n \mid (q^m - 1)$ over $\mathbb{F}_q$. Let β be a primitive n-th root of unity (an element of order exactly n). Let

$$g(x) = \operatorname{lcm}\left\{m_\beta(x), m_{\beta^2}(x), \ldots, m_{\beta^{2t}}(x)\right\}$$

be the minimum degree monic polynomial with the sequence $\beta, \beta^2, \ldots, \beta^{2t}$ of $2t$ elements as zeros (among other elements as zeros). Define the BCH code with length n designed distance $2t+1$ over $\mathbb{F}_q$ as the cyclic code $C = \langle g(x) \rangle$ or equivalently as the code with null space over $\mathbb{F}_q$ of the parity-check matrix

$$H = \begin{bmatrix} 1 & \beta & \beta^2 & \cdots & \beta^{(n-1)} \\ 1 & \beta^2 & \beta^4 & \cdots & \beta^{2(n-1)} \\ \vdots & \vdots & \vdots & \vdots & \vdots \\ 1 & \beta^{2t} & \beta^{2(2t)} & \cdots & \beta^{2t(n-1)} \end{bmatrix}.$$

That the minimum distance bound of this code, $d = 2t+1$, follows since any $2t \times 2t$ submatrix of H is a Vandermonde matrix and is nonsingular since the elements of the first row are distinct.

A cyclic Reed–Solomon $(n, k, d = n-k+1)_q$ code can be generated by choosing a generator polynomial of the form

$$g(x) = \prod_{i=1}^{n-k} (x - \alpha^i), \ \alpha \in \mathbb{F}_q, \ \alpha \text{ primitive of order } n.$$

That the code has a minimum distance $d = n-k+1$ follows easily from the above discussion.

A standard simple construction of Reed–Solomon codes over a finite field $\mathbb{F}_q$ of length n that will be of use in this volume is as follows. Let $\boldsymbol{u} = \{u_1, u_2, \ldots, u_n\}$ be a set, referred to as the *evaluation set* (and viewed as a set rather than a vector – we use boldface lowercase letters for both sets and vectors) of $n \leq q$ distinct evaluation elements of $\mathbb{F}_q$. As noted, $\mathbb{F}_q^{<k}[x]$ is the set of polynomials over $\mathbb{F}_q$ of degree less than k. Then another incarnation of a Reed–Solomon code can be taken as

$$RS_{n,k}(\boldsymbol{u},q) = \left\{ \boldsymbol{c}_f = (f(u_1), f(u_2), \ldots, f(u_n)), f \in \mathbb{F}_q^{<k}[x] \right\}$$

where $\boldsymbol{c_f}$ is the codeword associated with the polynomial $\boldsymbol{f}$. That this is an $(n,k,d = n-k+1)_q$ code follows readily from the fact that a polynomial of degree less than k over $\mathbb{F}_q$ can have at most $k-1$ zeros. As the code satisfies the Singleton bound $d \leq n-k+1$ with equality it is referred to as *maximum distance separable* (MDS) code and the dual of such a code is also MDS. Of course the construction is valid for any finite field, e.g., $\mathbb{F}_{q^\ell}$.

The dual of an RS code is generally not an RS code.

A slight but useful generalization of this code is the *Generalized Reed–Solomon* (GRS) code denoted as $GRS_{n,k}(\boldsymbol{u},\boldsymbol{v},q)$, where $\boldsymbol{u}$ is the evaluation set of distinct field nonzero elements as above and $\boldsymbol{v} = \{v_1, v_2, \ldots, v_n\}$, $v_i \in \mathbb{F}_q^*$ (referred to as the *multiplier set*) is a set of not necessarily distinct nonzero elements of $\mathbb{F}_q$. Then $GRS_{n,k}(\boldsymbol{u},\boldsymbol{v},q)$ is the (linear) set of codewords:

$$GRS_{n,k}(\boldsymbol{u},\boldsymbol{v},q) = \left\{ \boldsymbol{c}_f = (v_1 f(u_1), v_2 f(u_2), \ldots, v_n f(u_n)),\ f \in \mathbb{F}_q^{<k}[x] \right\}.$$

Since the minimum distance of this linear set of codewords is $n-k+1$ the code is MDS, for the same reason noted above. Clearly an RS code is a $GRS_{n,k}(\boldsymbol{u},\boldsymbol{v},q)$ code with $\boldsymbol{v} = (1, 1, \ldots, 1)$.

The dual of any MDS code is MDS. It is also true [7, 9, 10] that the dual of a GRS code is also a GRS code. In particular, given $GRS_{n,k}(\boldsymbol{u},\boldsymbol{v},q)$ there exists a set $\boldsymbol{w} \in (\mathbb{F}_q^*)^n$ such that

$$\begin{aligned} GRS_{n,k}^{\perp}(\boldsymbol{u},\boldsymbol{v},q) &= GRS_{n,n-k}(\boldsymbol{u},\boldsymbol{w},q) \\ &= \left\{ w_1 g(u_1), w_2 g(u_2), \ldots, w_n g(u_n),\ g \in \mathbb{F}_q^{<n-k}[x] \right\}. \end{aligned} \tag{1.7}$$

In other words, for any $f(x) \in \mathbb{F}_q^{<k}[x]$ and $g(x) \in \mathbb{F}_q^{<n-k}[x]$ for a given evaluation set $\boldsymbol{u} = \{u_1, \ldots, u_n\}$ and multiplier set $\boldsymbol{v} = \{v_1, \ldots, v_n\}$ there is a multiplier set $\boldsymbol{w} = \{w_1, \ldots, w_n\}$ such that the associated codewords $\boldsymbol{c}_f \in GRS_{n,k}(\boldsymbol{u},\boldsymbol{v},q)$ and $\boldsymbol{c}_g \in GRS_{n,n-k}(\boldsymbol{u},\boldsymbol{w},q)$ are such that

$$(\boldsymbol{c}_f, \boldsymbol{c}_g) = v_1 f(u_1) w_1 g(u_1) + \cdots + v_n f(u_n) w_n g(u_n) = 0.$$

Indeed the multiplier set vector $\boldsymbol{w}$ can be computed as

$$w_i = \left(v_i \prod_{j \neq i} (u_i - u_j) \right)^{-1}. \tag{1.8}$$

To see this, for a given evaluation set $\boldsymbol{u}$ (distinct elements), denote

$$e(x) = \prod_{i=1}^{n}(x - u_i) \quad \text{and} \quad e_i(x) = e(x)/(x - u_i) = \prod_{k \neq i}(x - u_k),$$

a monic polynomial of degree $(n - 1)$. It is clear that

$$\frac{e_i(u_j)}{e_i(u_i)} = \begin{cases} 1 \text{ if } & j = i \\ 0 \text{ if } & j \neq i. \end{cases}$$

It follows that for any polynomial $h(x) \in \mathbb{F}_q[x]$ of degree less than n that takes on values $h(u_i)$ on the evaluation set $\boldsymbol{u} = \{u_1, u_2, \ldots, u_n\}$ can be expressed as

$$h(x) = \sum_{i=1}^{n} h(u_i)\frac{e_i(x)}{e_i(u_i)}.$$

To verify Equation 1.7 consider applying this interpolation formula to $f(x)g(x)$ where $f(x)$ is a codeword polynomial $f(x) \in \mathbb{F}_q^{<k}[x]$ (in $GRS_{n,k}(\boldsymbol{u},\boldsymbol{v},q)$) and $g(x) \in \mathbb{F}_q^{<n-k}[x]$ (in $GRS_{n,k}(\boldsymbol{u},\boldsymbol{v},q)^{\perp} = GRS_{n,n-k}(\boldsymbol{u},\boldsymbol{w},q)$) where it is claimed that the two multiplier sets $v = \{v_1, v_2, \ldots, v_n\}$ and $w = \{w_1, w_2, \ldots, w_n\}$ are related as in Equation 1.8.

Using the above interpolation formula on the product $f(x)g(x)$ (of degree at most $(n - 2)$) gives

$$f(x)g(x) = \sum_{k=1}^{n} f(u_k)g(u_k)\frac{e_k(x)}{e_k(u_k)}.$$

The coefficient of x^{n-1} on the left side is 0 while on the right side is 1 (as $e_k(x)$ is monic of degree $(n - 1)$) and hence

$$\begin{aligned} 0 &= \sum_{k=1}^{n} \frac{1}{e_k(u_k)} f(u_k)g(u_k) = \sum_{k=1}^{n}(v_k f(u_k))\left(\frac{v_k^{-1}}{e_k(u_k)}g(u_k)\right) \\ &= \sum_{k=1}^{n}(v_k f(u_k))(w_k g(u_k)) \quad \text{(by Equation 1.8)} \\ &= (\boldsymbol{c_f}, \boldsymbol{c_g}) = 0. \end{aligned}$$

It is noted in particular that

$$RS_{n,k}^{\perp}(\boldsymbol{u},q) = GRS_{n,n-k}(\boldsymbol{u},\boldsymbol{w},q)$$

for the multiplier set $w_i = \prod_{j \neq i}(u_i - u_j)$.

Reed–Muller Codes

Reed–Muller (RM) codes are discussed in some depth in most books on coding (e.g., [3, 4]) with perhaps the most comprehensive being [2] which considers their relationship to Euclidean geometries and combinatorial designs. The properties of RM codes are most easily developed for the binary field but the general case will be considered here – the *Generalized Reed–Muller* (GRM) codes (generalized in a different sense than the GRS codes). The codes are of most interest in this work for the construction of locally decodable codes (Chapter 8) and their relationship to multiplicity codes introduced there.

Consider m variables $x_1, x_2, \ldots, x_m$ and the ring $\mathbb{F}_q[x_1, x_2, \ldots, x_m] = \mathbb{F}_q[\boldsymbol{x}]$ of multivariate polynomials over $\mathbb{F}_q$ (see also Appendix B). The set of all monomials of the m variables and their degree is of the form

$$\left\{ \boldsymbol{x}^{\boldsymbol{i}} = x_1^{i_1} x_2^{i_2} \cdot x_m^{i_m},\ \boldsymbol{i} \sim (i_1, i_2, \ldots, i_m),\ \text{degree } = \sum_j i_j \right\}. \tag{1.9}$$

A multivariate polynomial $f(\boldsymbol{x}) \in \mathbb{F}_q[\boldsymbol{x}]$ is the sum of monomials over $\mathbb{F}_q$ and the degree of f is largest of the degrees of any of its monomials. Notice that over the finite field $\mathbb{F}_q, x_i^q = x_i$ and so only degrees of any variable less than q are of interest.

In the discussion of these codes we will have the need for two simple enumerations: (i) the number of monomials on m variables of degree *exactly* d and (ii) the number of monomials of degree *at most* d. These problems are equivalent to the problems of the number of partitions of the integer d into at most m parts and the number of partitions of all integers *at most* d into at most m parts. These problems are easily addressed as "balls in cells" problems as follows.

For the first problem, place d balls in a row and add a further m balls. There are $d+m-1$ spaces between the $d+m$ balls. Choose $m-1$ of these spaces in which to place markers (in $\binom{d+m-1}{m-1}$ ways). Add markers to the left of the row and to the right of the row. Place the balls between two markers into a "bin" – there are m such bins. Subtract a ball from each bin. If the number of balls in bin j is i_j, then the process determines a partition of d in the sense that $i_1 + i_2 + \cdots + i_m = d$ and all such partitions arise in this manner. Thus the number of monomials on m variables of degree equal to d is given by

$$\binom{d+m-1}{m-1} = \left|\{i_1 + i_2 + \cdots + i_m = d,\ i_j \in \mathbb{Z}_{\geq 0}\}\right|. \tag{1.10}$$

To determine the number of monomials on (at most) m variables of total degree *at most* d, consider the setup as above except now add another ball to the row to have $d+m+1$ balls and choose m of the spaces between the balls in $\binom{d+m}{m}$ ways in which to place markers corresponding to $m+1$ bins. As before subtract a ball from each bin. The contents of the last cell are regarded as superfluous and discarded to take into account the "at most" part of the enumeration. The contents of the first m cells correspond to a partition and the number of monomials on m variables of total degree at most d is

$$\binom{d+m}{m} = \big|\{i_1 + i_2 + \cdots + i_m \leq d,\, i_j \in \mathbb{Z}_{\geq 0}\}\big|. \tag{1.11}$$

Note that it follows that

$$\sum_{j=1}^{d} \binom{j+m-1}{m-1} = \binom{d+m}{m},$$

(i.e., the number of monomials of degree at most d is the number of monomials of degree exactly j for $j = 1, 2, \ldots, d$) as is easily shown by induction.

Consider the code of length q^m denoted $GRM_q(d,m)$ generated by monomials of degree at most d on m variables for $d < q-1$, i.e., let $f(\boldsymbol{x}) \in \mathbb{F}_q[\boldsymbol{x}]$ be an m-variate polynomial of degree at most $d \leq q-1$ (the degree of polynomial is the largest degree of its monomials and no variable is of degree greater than $q-1$). The corresponding codeword is denoted

$$\boldsymbol{c}_f = \Big(f(\boldsymbol{a}),\ \boldsymbol{a} \in \mathbb{F}_q^m,\ f \in \mathbb{F}_q[\boldsymbol{x}],\ f \text{ of degree at most } d \Big),$$

i.e., a codeword of length q^m with coordinate positions labeled with all elements of $\mathbb{F}_q^m$ and coordinate labeled $\boldsymbol{a} \in \mathbb{F}_q^m$ with a value of $f(\boldsymbol{a})$. It is straightforward to show that the codewords corresponding to the monomials are linearly independent over $\mathbb{F}_q$ and hence the code has length and dimension

$$\text{code length } n = q^m \quad \text{and} \quad \text{code dimension } k = \binom{m+d}{d}.$$

To determine a bound on the minimum distance of the code the theorem ([8], theorem 6.13) is used that states the maximum number of zeros of a multivariate polynomial of m variables of degree d over $\mathbb{F}_q$ is at most dq^{m-1}. Thus the maximum fraction of a codeword that can have zero coordinates is d/q and hence the normalized minimum distance of the code (code distance divided by length) is bounded by

$$1 - d/q, \quad d < q-1.$$

(Recall d here is the maximum degree of the monomials used, not code distance.) The normalized (sometimes referred to as fractional or relative) distance of a code will be designated as $\Delta = 1 - d/q$. (Many works use δ to denote this, used for the erasure probability on the BEC here.) Thus, e.g., for

$m = 2$ (bivariate polynomials) this subclass of GRM codes has the parameters $\left(q^2, \binom{d+2}{d}, q^2 - dq\right)_q$. Note that the rate of the code is

$$\binom{d+2}{2} \Big/ q^2 \approx d^2/2q^2 = (1-\Delta)^2/2.$$

Thus the code can have rate at most 1/2.

For a more complete analysis of the GRM codes the reader should consult ([2], section 5.4). Properties of GRS and GRM codes will be of interest in several of the chapters.

1.2 Notes on Information Theory

The probability distribution of the discrete random variable X, $Pr(X = x)$, will be denoted $P_X(x)$ or as $P(x)$ when the random variable is understood. Similarly a joint discrete random variable $X \times Y$ (or XY) is denoted $Pr(X = x, Y = y) = P_{XY}(x,y)$. The conditional probability distribution is denoted $Pr(Y = y \mid X = x) = P(y \mid x)$. A probability density function (pdf) for a continuous random variable will be designated similarly as $p_X(x)$ or a similar lowercase function.

Certain notions from information theory are required. The entropy of a discrete ensemble $\{P(x_i), i = 1, 2, \dots\}$ is given by

$$H(X) = -\sum_i P(x_i) \log P(x_i)$$

and unless otherwise specified all logs will be to the base 2. It represents the amount of uncertainty in the outcome of a realization of the random variable.

A special case will be important for later use, that of a binary ensemble $\{p, (1-p)\}$ which has an entropy of

$$H_2(p) = -p \log_2 p - (1-p)\log_2(1-p) \tag{1.12}$$

referred to as the binary entropy function. It is convenient to introduce the q-ary entropy function here, defined as

$$H_q(x) = \begin{cases} x\log_q(q-1) - x\log_q x - (1-x)\log_q(1-x), & 0 < x \le \theta = (q-1)/q \\ 0, & x = 0 \end{cases} \tag{1.13}$$

an extension of the binary entropy function. Notice that $H_q(p)$ is the entropy associated with the q-ary discrete symmetric channel and also the entropy of the probability ensemble $\{1-p, p/(q-1), \dots, p/(q-1)\}$ (total of q values). The binary entropy function of Equation 1.12 is obtained with $q = 2$.

Similarly the entropy of a joint ensemble $\{P(x_i, y_i), i = 1, 2, \ldots\}$ is given by

$$H(X,Y) = -\sum_{x_i, y_i} P(x_i, y_i) \log P(x_i, y_i)$$

and the conditional entropy of X given Y is given by

$$H(X \mid Y) = -\sum_{x_i, y_i} P(x_i, y_i) \log P(x_i \mid y_i) = H(X,Y) - H(Y)$$

which has the interpretation of being the expected amount of uncertainty remaining about X after observing Y, sometimes referred to as equivocation.

The mutual information between ensembles X and Y is given by

$$I(X;Y) = \sum_{x_i, y_i} P(x_i, y_j) \log \frac{P(x_i, y_j)}{P(x_i)P(y_j)} \tag{1.14}$$

and measures the amount of information knowledge that one of the variables gives on the other. The notation $\{X;Y\}$ is viewed as a joint ensemble. It will often be the case that X will represent the input to a discrete memoryless channel (to be discussed) and Y the output of the channel and this notion of mutual information has played a pivotal role in the development of communication systems over the past several decades.

Similarly for three ensembles it follows that

$$I(X;Y,Z) = \sum_{i,j,k} P(x_i, y_j, z_k) \log \frac{P(x_i, y_j, z_k)}{P(x_i)P(y_j, z_k)}.$$

The conditional information of the ensemble $\{X;Y\}$ given Z is

$$I(X;Y \mid Z) = \sum_{i,j,k} P(x_i, y_j, z_k) \log \frac{P(x_i, y_j \mid z_k)}{P(x_i \mid z_k)P(y_j \mid z_k)}$$

or alternatively

$$I(X;Y \mid Z) = \sum_{i,j,k} P(x_i, y_j, z_k) \log \frac{P(x_i, y_j, z_k)P(z_k)}{P(x_i, z_k)P(y_j, z_k)}.$$

The process of conditioning observations of X, Y on a third random variable Z may increase or decrease the mutual information between X and Z but it can be shown the conditional information is always positive. There are numerous relationships between these information-theoretic quantities. Thus

$$I(X;Y) = H(X) - H(X \mid Y) = H(X) + H(Y) - H(X,Y).$$

Our interest in these notions is to define the notion of capacity of certain channels.

A *discrete memoryless channel* (DMC) is a set of finite inputs $\mathcal{X}$ and a discrete set of outputs $\mathcal{Y}$ such that at each instance of time, the channel accepts an input $x \in \mathcal{X}$ and with probability $W(y \mid x)$ outputs $y \in \mathcal{Y}$ and each use of the channel is independent of other uses and

$$\sum_{y \in \mathcal{Y}} W(y \mid x) = 1 \text{ for each } x \in \mathcal{X}.$$

Thus if a vector $\boldsymbol{x} = (x_1, x_2, \ldots, x_n)$, $x_i \in \mathcal{X}$ is transmitted in n uses of the channel, the probability of receiving the vector $\boldsymbol{y} = (y_1, y_2, \ldots, y_n)$ is given by

$$P(\boldsymbol{y} \mid \boldsymbol{x}) = \prod_{i=1}^{n} W(y_i \mid x_i).$$

At times the DMC might be designated simply by the set of transition probabilities $W = \{W(y_i \mid x_i)\}$.

For the remainder of this chapter it will be assumed the channel input is binary and referred to as a binary-input DMC or BDMC where $\mathcal{X} = \{0, 1\}$ and that $\mathcal{Y}$ is finite. Important examples of such channels include the *binary symmetric channel* (BSC), the *binary erasure channel* (BEC) and a general BDMC, as shown in Figure 1.1 (a) and (b) while (c) represents the more general case.

Often, rather than a general BDMC, the additional constraint of symmetry is imposed, i.e., a *binary-input discrete memoryless symmetric channel* by which is meant a binary-input $\mathcal{X} = \{0, 1\}$ channel with a channel transition probability $\{W(y \mid x), x \in \mathcal{X}, y \in \mathcal{Y}\}$ which satisfies a symmetry condition, noted later.

The notion of mutual information introduced above is applied to a DMC with the $\mathcal{X}$ ensemble representing the channel input and the output ensemble $\mathcal{Y}$ to the output. The function $I(X; Y)$ then represents the amount of information the output gives about the input. In the communication context it would be desirable to maximize this function. Since the channel, represented by the channel transition matrix $W(y \mid x)$, is fixed, the only variable that can be adjusted is the set of input probabilities $P(x_i), x_i \in \mathcal{X}$. Thus the maximum

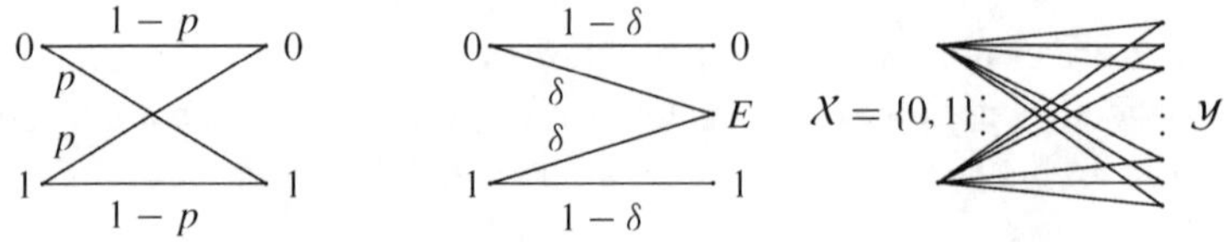

BSC, C = $1 - H(p)$ BEC, C = $1 - \delta$ BDMC $W(y \mid x)$

Figure 1.1 Binary-input DMCs

amount of information that on average can be transmitted through the channel in each channel use is found by determining the set of input probabilities that maximizes the mutual information between the channel input and output.

It is intuitive to define the *channel capacity* of a DMC as the maximum rate at which it is possible to transmit information through the channel, per channel use, with an arbitrarily low error probability:

$$\begin{aligned}\text{channel capacity} &= C \stackrel{\Delta}{=} \max_{P(x), x \in \mathcal{X}} I(X,Y) = I(W) \\ &= \max_{P(x), x \in \mathcal{X}} \sum_{x \in \mathcal{X}, y \in \mathcal{Y}} W(y \mid x) P(x) \log \frac{W(y \mid x) P(x)}{P(x)(P(y)}\end{aligned}$$

where $P(y) = \sum_{x \in \mathcal{X}} W(y|x)P(x)$. For general channels, determining channel capacity can be a challenging optimization problem. When the channels exhibit a certain symmetry, however, the optimization is achieved with equally likely inputs:

Definition 1.2 ([6]) A DMC is *symmetric* if the set of outputs can be partitioned into subsets in such a way that for each subset, the matrix of transition probabilities (with rows as inputs and columns as outputs) has the property each row is a permutation of each other row and each column of a partition (if more than one) is a permutation of each other column in the partition.

A consequence of this definition is that for a symmetric DMC, it can be shown that the channel capacity is achieved with equally probable inputs ([6], theorem 4.5.2). It is simple to show that both the BSC and BEC channels are symmetric by this definition. The capacities of the BSC (with crossover probability p) and BEC (with erasure probability δ) are

$$C_{BSC} = 1 + p \log_2 p + (1-p)\log_2(1-p) \quad \text{and} \quad C_{BEC} = 1 - \delta. \tag{1.15}$$

This first relation is often written

$$C_{BSC} = 1 - H_2(p)$$

and $H_2(p)$ is the binary entropy function of Equation 1.12.

For channels with continuous inputs and/or outputs the mutual information between channel input and output is given by

$$I(X,Y) = \int_x \int_y p(x,y) \log\left(\frac{p(x,y)}{p(x)p(y)}\right) dxdy$$

for probability density functions $p(\cdot,\cdot)$ and $p(\cdot)$.

Versions of the Gaussian channel where Gaussian-distributed noise is added to the signal in transmission are among the few such channels that offer

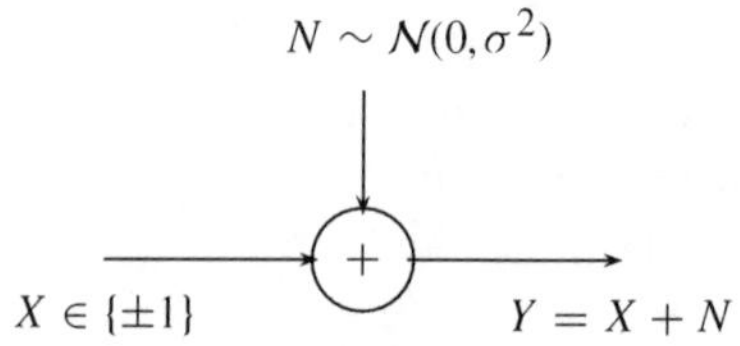

Figure 1.2 The binary-input additive white Gaussian noise channel

tractable solutions and are designated *additive white Gaussian noise* (AWGN) channels. The term "white" here refers to a flat power spectral density function of the noise with frequency. The binary-input AWGN (BIAWGN) channel, where one of two continuous-time signals is chosen for transmission during a given time interval $(0, T)$ and experiences AWGN in transmission, can be represented as in Figure 1.2:

$$Y_i = X_i + N_i, \text{ and } X_i \in \{\pm 1\}, \qquad \text{BIAWGN},$$

where N_i is a Gaussian random variable with zero mean and variance σ^2, denoted $N_i \sim \mathcal{N}(0, \sigma^2)$. The joint distribution of (X, Y) is a mixture of discrete and continuous and with $P(X = +1) = P(X = -1) = 1/2$ (which achieves capacity on this channel) and with $p(x) \sim \mathcal{N}(0, \sigma^2)$. The pdf $p(y)$ of the channel output is

$$\begin{aligned} p(y) &= \frac{1}{2} \cdot \frac{1}{\sqrt{2\pi\sigma^2}} e^{-\frac{(y+1)^2}{2\sigma^2}} + \frac{1}{2} \cdot \frac{1}{\sqrt{2\pi\sigma^2}} e^{-\frac{(y-1)^2}{2\sigma^2}} \\ &= \frac{1}{\sqrt{8\pi\sigma^2}} \left(\exp\left(-\frac{(y+1)^2}{2\sigma^2} \right) + \exp\left(-\frac{(y-1)^2}{2\sigma^2} \right) \right) \end{aligned} \tag{1.16}$$

and maximizing the expression for mutual information of the channel (equally likely inputs) reduces to

$$C_{BIAWGN} = -\int_y p(y) \log_2 p(y) dy - \frac{1}{2} \log_2(2\pi e \sigma^2). \tag{1.17}$$

The general shape of these capacity functions is shown in Figure 1.3 where SNR denotes signal-to-noise ratio.

Another Gaussian channel of fundamental importance in practice is that of the band-limited AWGN channel (BLAWGN). In this model a band-limited signal $x(t), t \in (0, T)$ with signal power $\leq S$ is transmitted on a channel band-limited to W Hz, i.e., $(-W, W)$ and white Gaussian noise with two-sided power spectral density level $N_o/2$ is added to the signal in transmission. This channel can be discretized via orthogonal basis signals and the celebrated and much-used formula for the capacity of it is

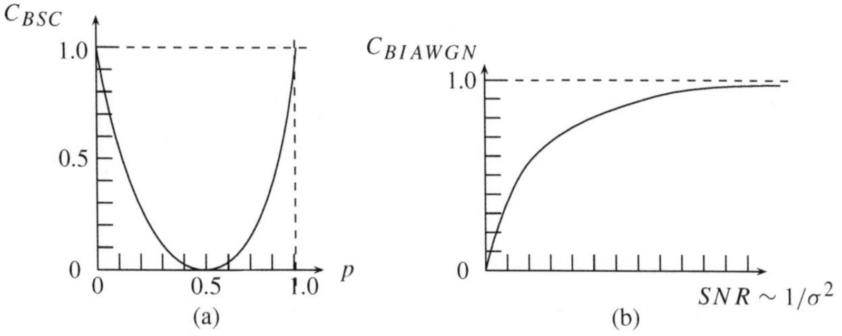

Figure 1.3 Shape of capacity curves for (a) BSC and (b) BIAWGN

$$C_{BLAWGN} = W \log_2(1 + S/N_o W) \text{ bits per second.} \tag{1.18}$$

The importance of the notion of capacity and perhaps the crowning achievement of information theory is the following coding theorem, informally stated, that says k information bits can be encoded into n coded bits, code rate $R = k/n$, such that the bit error probability P_e of the decoded coded bits at the output of a DMC of capacity C can be upper bounded by

$$P_e \leq e^{-nE(R)}, \; R < C \tag{1.19}$$

where $E(R)$, the error rate function, is > 0 for all $R < C$. The result implies that for any code rate $R < C$ there will exist a code of some length n capable of transmitting information with negligible error probability. Thus reliable communication is possible even though the channel is noisy as long as one does not transmit at too high a rate.

The information-theoretic results discussed here have driven research into finding efficient codes, encoding and decoding algorithms for the channels noted over many decades. The references [5, 15] present a more comprehensive discussion of these and related issues.

1.3 An Overview of the Chapters

A brief description of the following chapters is given.

The chapter on coding for erasures is focused on the search for erasure-correcting algorithms that achieve linear decoding complexity. It starts with a discussion of Tornado codes. Although these codes did not figure prominently in subsequent work, they led to the notion of codes from random graphs with irregular edge distributions that led to very efficient decoding algorithms. In turn these can be viewed as leading to the notion of fountain codes which are not erasure-correcting codes. Rather they are codes that can efficiently recreate a file from several random combinations of subfiles. Such codes led to the important concept of Raptor codes which have been incorporated into numerous standards for the download of large files from servers in a multicast network while not requiring requests for retransmissions of portions of a file that a receiver may be missing, a crucial feature in practice.

Certain aspects of low-density parity-check (LDPC) codes are then discussed. These codes, which derive from the work of Gallager from the early 1960s, have more recently assumed great importance for applications as diverse as coding for flash memories as well as a wide variety of communication systems. Numerous books have been written on various aspects of the construction and analysis of these codes. This chapter focuses largely on the paper of [13] which proved crucial for more recent progress for the analytical techniques it introduced.

The chapter on polar codes arose out of the seminal paper [1]. In a deep study of information-theoretic and analytical technique it produced the first codes that provably achieved rates approaching capacity. From a binary-input channel with capacity $C \leq 1$, through iterative transformations, it derived a channel with $N = 2^n$ inputs and outputs and produced a series of NC sub-channels that are capable of transmitting data with arbitrarily small error probability, thus achieving capacity. The chapter discusses the central notions to assist with a deeper reading of the paper.

The chapter on network coding is devoted to the somewhat surprising idea that allowing nodes (servers) in a packet network to process and combine packets as they traverse the network can substantially improve throughput of the network. This raises the question of the capacity of such a network and how to code the packets in order to achieve the capacity. This chapter looks at a few of the techniques that have been developed for multicast channels.

With the wide availability of the high-speed internet, access to information stored on widely distributed databases became more commonplace. Huge amounts of information stored in distributed databases made up of standard computing and storage elements became ubiquitous. Such elements fail with some regularity and methods to efficiently restore the contents of a failed server are required. Many of these early distributed storage systems simply replicated data on several servers and this turned out to be an inefficient method of achieving restoration of a failed server, both in terms of storage and transmission costs. Coding the stored information greatly improved the efficiency with which a failed server could be restored and Chapter 6 reviews the coding techniques involved. The concepts involved are closely related to locally repairable codes considered in Chapter 7 where an erased coordinate in a codeword can be restored by contacting a few other coordinate positions.

Chapter 8 considers coding techniques which allow a small amount of information to be recovered from errors in a codeword without decoding the entire codeword, termed locally decodable codes. Such codes might find application where very long codewords are used and users make frequent requests for modest amounts of information. The research led to numerous other variations such as locally testable codes where one examines a small portion of data and is asked to determine if it is a portion of a codeword of some code, with some probability.

Private information retrieval considers techniques for users to access information on servers in such a manner that the servers are unaware of which information is being sought. The most common scenario is one where the servers contain the same information and users query information from individual servers and perform computations on the responses to arrive at the desired information. More recent contributions have shown how coded information stored on the servers can achieve the same ends with greatly improved storage efficiency. Observations on this problem are given in Chapter 9.

The notion of a batch code addresses the problem of storing information on servers in such a way that no matter which information is being sought no single server has to download more than a specified amount of information. It is a technique to ensure load balancing between servers. Some techniques to achieve this are discussed in Chapter 10.

Properties of expander graphs find wide application in several areas of computer science and mathematics and the notion has been applied to the construction of error-correcting codes with efficient decoding algorithms. Chapter 11 introduces this topic of considerable current interest.

Algebraic coding theory is based on the notion of packing spheres in a space of n-tuples over a finite field $\mathbb{F}_q$ according to the Hamming metric. Rank-metric codes consider the vector space of matrices of a given shape over a finite field with a different metric, namely the distance between two such matrices is given by the rank of the difference of the matrices which can be shown to be a metric on such a space. A somewhat related (although quite distinct) notion is to consider a set of subspaces of a vector space over a finite field with a metric defined between such subspaces based on their size and intersection. Such codes of subspaces have been shown to be of value in the network coding problem. The rank-metric codes and subspace codes are introduced in Chapter 12.

A problem that was introduced in the early days of information theory was the notion of list decoding where, rather than the decoding algorithm producing a unique closest codeword to the received word, it was acceptable to produce a list of closest words. Decoding was then viewed as successful if the transmitted codeword was on the list. The work of Sudan [14] introduced innovative techniques for this problem which influenced many aspects of coding theory. This new approach led to numerous other applications and results to achieve capacity on such a channel. These are overviewed in Chapter 13.

Shift register sequences have found important applications in numerous synchronization and ranging systems as well as code-division multiple access (CDMA) communication systems. Chapter 14 discusses their basic properties.

The advent of quantum computing is likely to have a dramatic effect on many storage, computing and transmission technologies. While still in its infancy it has already altered the practice of cryptography in that the US government has mandated that future deployment of crypto algorithms should be quantum-resistant, giving rise to the subject of "postquantum cryptography." A brief discussion of this area is given in Chapter 15. While experts in quantum computing may differ in their estimates of the time frame in which it will become significant, there seems little doubt that it will have a major impact.

An aspect of current quantum computing systems is their inherent instability as the quantum states interact with their environment causing errors in the computation. The systems currently implemented or planned will likely rely on some form of quantum error-correcting codes to achieve sufficient system stability for their efficient operation. The subject is introduced in Chapter 16.

The final chapter considers a variety of other coding scenarios in an effort to display the width of the areas embraced by the term "coding" and to further illustrate the scope of coding research that has been ongoing for the past few decades beyond the few topics covered in the chapters.

The two appendices cover some useful background material on finite geometries and multivariable polynomials over finite fields.

References

[1] Arıkan, E. 2009. Channel polarization: a method for constructing capacity-achieving codes for symmetric binary-input memoryless channels. *IEEE Trans. Inform. Theory*, **55**(7), 3051–3073.

[2] Assmus, Jr., E.F., and Key, J.D. 1992. *Designs and their codes*. Cambridge Tracts in Mathematics, vol. 103. Cambridge University Press, Cambridge.

[3] Blahut, R.E. 1983. *Theory and practice of error control codes*. Advanced Book Program. Addison-Wesley, Reading, MA.

[4] Blake, I.F., and Mullin, R.C. 1975. *The mathematical theory of coding*. Academic Press, New York/London.

[5] Forney, G.D., and Ungerboeck, G. 1998. Modulation and coding for linear Gaussian channels. *IEEE Trans. Inform. Theory*, **44**(6), 2384–2415.

[6] Gallager, R.G. 1968. *Information theory and reliable communication*. John Wiley & Sons, New York.

[7] Huffman, W.C., and Pless, V. 2003. *Fundamentals of error-correcting codes*. Cambridge University Press, Cambridge.

[8] Lidl, R., and Niederreiter, H. 1997. *Finite fields*, 2nd ed. Encyclopedia of Mathematics and Its Applications, vol. 20. Cambridge University Press, Cambridge.

[9] Ling, S., and Xing, C. 2004. *Coding theory*. Cambridge University Press, Cambridge.

[10] MacWilliams, F.J., and Sloane, N.J.A. 1977. *The theory of error-correcting codes: I and II*. North-Holland Mathematical Library, vol. 16. North-Holland, Amsterdam/New York/Oxford.

[11] McEliece, R.J. 1987. *Finite fields for computer scientists and engineers*. The Kluwer International Series in Engineering and Computer Science, vol. 23. Kluwer Academic, Boston, MA.

[12] Menezes, A.J., Blake, I.F., Gao, X.H., Mullin, R.C., Vanstone, S.A., and Yaghoobian, T. 1993. *Applications of finite fields*. The Kluwer International Series in Engineering and Computer Science, vol. 199. Kluwer Academic, Boston, MA.

[13] Richardson, T.J., and Urbanke, R.L. 2001. The capacity of low-density parity-check codes under message-passing decoding. *IEEE Trans. Inform. Theory*, **47**(2), 599–618.

[14] Sudan, M. 1997. Decoding of Reed Solomon codes beyond the error-correction bound. *J. Complexity*, **13**(1), 180–193.

[15] Ungerboeck, G. 1982. Channel coding with multilevel/phase signals. *IEEE Trans. Inform. Theory*, **28**(1), 55–67.

2

Coding for Erasures and Fountain Codes

A coordinate position in a received word is said to be an erasure if the receiver is using a detection algorithm that is unable to decide which symbol was transmitted in that position and outputs an erasure symbol such as E rather than risk making an error, i.e., outputting an incorrect symbol. One might describe an erasure as an error whose position is known. For binary information symbols the two most common discrete memoryless channels are shown in Figure 1.1, the binary erasure channel (BEC) and the binary symmetric channel (BSC), introduced in Chapter 1. The symbols p and δ will generally refer to the channel crossover probability for the BSC and erasure probability for the BEC, respectively. Both symbols sometimes occur with other meanings as will be noted. Each such channel has a capacity associated with it which is the maximum rate at which information (per channel use) can be sent through the channel error-free, as discussed in Chapter 1. The subject of error-correcting codes arose to meet the challenge of realizing such performance.

It is to be emphasized that two quite different channel error models are used in this chapter. The BEC will be the channel of interest in the first part of this chapter. Thus a codeword (typically of a linear code) is received which contains a mix of correct received symbols and erased symbols. The job of the code design and decoder algorithm is then to "fill in" or interpolate the erased positions with original transmitted symbols, noting that in such a model the unerased positions are assumed correct.

Codes derived for the BEC led to the notion of irregular distribution codes where the degrees of variable and check nodes of the code Tanner graph, to be introduced shortly, are governed by a probability distribution. These in turn led to fountain codes, which is the interest of the last section of the chapter. In such channels each transmitted packet is typically a linear combination of information packets over some fixed finite field. The receiver gathers a sufficient number of transmitted packets (assumed without errors) until it is able to retrieve the information packets by, e.g., some type of matrix inversion algorithm on the set of packets received. The retriever then does not care which particular packets are received, just that they receive a sufficient number of them to allow the decoding algorithm to decode successfully. This is often described as a "packet loss" channel, in that coded packets transmitted may be lost in transmission due to a variety of network imperfections such as buffer overflow or failed server nodes, etc. Such a packet loss situation is not modeled by the DMC models considered.

While the two channel models examined in this chapter are quite different, it is their common heritage that suggested their discussion in the same chapter.

2.1 Preliminaries

It is convenient to note a few basic results on coding and DMCs for future reference. Suppose $C = (n,k,d)_q$ is a linear block code over the finite field of q elements $\mathbb{F}_q$, designating a linear code that has k *information symbols* (dimension k) and $(n - k)$ *parity-check symbols* and minimum distance d. Much of this volume is concerned with binary-input channels and $q = 2$.

Suppose a codeword $\boldsymbol{c} = (c_1, c_2, \ldots, c_n)$ is transmitted on a BEC and the received word is $\boldsymbol{r} = (r_1, r_2, \ldots, r_n)$ which has e erasures in positions $\mathcal{E} \subset \{1,2,\ldots,n\}$, $|\mathcal{E}| = e$. Then $r_i = E$ for $i \in \mathcal{E}$ for E the erasure symbol. The unerased symbols received are assumed correct.

A parity-check matrix of the code is an $(n - k) \times n$ matrix H over $\mathbb{F}_q$ such that

$$\boldsymbol{H} \cdot \boldsymbol{c}^t = \boldsymbol{0}^t_{n-k}$$

for any codeword $\boldsymbol{c}$ where $\boldsymbol{0}_{n-k}$ is the all-zero $(n - k)$-tuple, a row vector over $\mathbb{F}_q$. If the columns of $\boldsymbol{H}$ are denoted by $\boldsymbol{h}^{(i)}$, $i = 1,2,\ldots,n$, then $\boldsymbol{H} \cdot \boldsymbol{c}^t$ is the sum of columns of $\boldsymbol{H}$ multiplied by the corresponding coordinates of the codeword, adding to the all-zero column $(n - k)$-vector $\boldsymbol{0}^t$. Similarly let $\boldsymbol{H}_e$ be the $(n-k) \times e$ submatrix of columns of $\boldsymbol{H}$ corresponding to the erased positions and $\boldsymbol{c}_e$ be the e-tuple of the transmitted codeword on the erased positions. Then

$$\boldsymbol{H}_e \cdot \boldsymbol{c}^t_e = -\boldsymbol{y}^t_e$$

where $\boldsymbol{y}^t_e$ is the $(n - k)$-tuple corresponding to the weighted sum of columns of $\boldsymbol{H}$ in the nonerased positions, i.e., columns of $\boldsymbol{H}$ multiplied by the known (unerased) positions of the received codeword.

As long as $e \leq d - 1$ the above matrix equation can be solved uniquely for the erased word positions, i.e., $\boldsymbol{c}_e$. However, this is generally a task of cubic complexity in codeword length, i.e., $O(n^3)$. The work of this chapter will show how a linear complexity with codeword length can be achieved with high probability.

This chapter will deal exclusively with binary codes and the only arithmetic operation used will be that of XOR (exclusive or), either of elements of $\mathbb{F}_2$ or of packets of n bits in $\mathbb{F}_2^n$. Thus virtually all of the chapter will refer to packets or binary symbols (bits) equally, the context being clear from the problem of interest.

It is emphasized that there are two different types of coding considered in this chapter. The first is the use of linear block codes for erasure correction while the second involves the use of *fountain codes* on a packet loss channel.

Virtually all of this chapter will use the notion of a bipartite graph to represent the various linear codes considered, a concept used by Tanner in his prescient works [37, 38, 39]. A bipartite graph is one with two sets of vertices, say U and V and an edge set E, with no edges between vertices in the same set. The graph will be called (c,d)-regular bipartite if the vertices in U have degree c and those of V have degree d. Since $\mid U \mid = n$, then $\mid V \mid = (c/d)n$. The U set of vertices will be referred to as the left vertices and V the right vertices. Bipartite graphs with irregular degrees will also be of interest later in the chapter.

There is a natural connection between a binary linear code and an $(n-k)\times n$ parity-check matrix and a bipartite graph. Often the left vertices of the bipartite code graph are associated with the entire n codeword coordinate positions and referred to as the *variable* or *information* nodes or vertices. Equivalently they represent the columns of the parity-check matrix. Similarly the $(n-k)$ right nodes or vertices are the *constraint* or *check* nodes which represent the rows of the parity-check matrix. The edges of the graph correspond to the ones in the check matrix in the corresponding rows and columns. Such a graphical representation of the code is referred to as the *Tanner graph* of the code, a notion that will feature prominently in many of the chapters.

The binary parity-check matrix of the code is an alternate view of the incidence matrix of the bipartite graph. The following illustrates the Tanner graph associated with the parity-check matrix for a Hamming $(8,4,4)_2$ code which is used in Example 2.3 shown also in Figures 2.1 and 2.3.

As a second graph representation of a binary linear code, it is equally possible to have the left nodes of the graph as the k information nodes and the $(n-k)$ right nodes as the check nodes and this is the view for most of the next section. As a matter of convenience this representation is referred to as the

$$H = \begin{bmatrix} 0 & 0 & 0 & 1 & 1 & 1 & 1 & 0 \\ 0 & 1 & 1 & 0 & 0 & 1 & 1 & 0 \\ 1 & 0 & 1 & 0 & 1 & 0 & 1 & 0 \\ 1 & 1 & 1 & 1 & 1 & 1 & 1 & 1 \end{bmatrix}$$

(a)

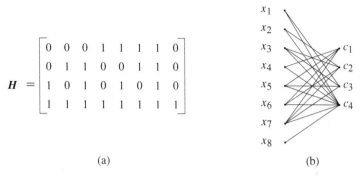

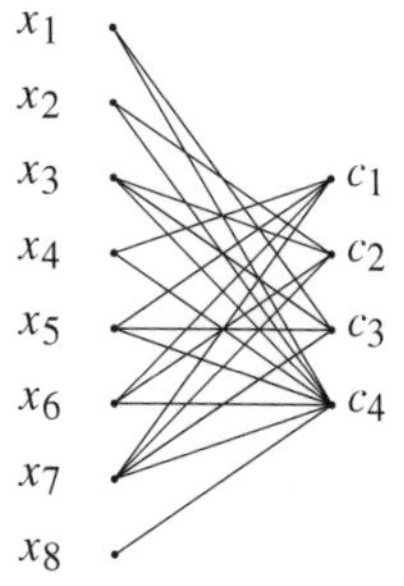

(b)

Figure 2.1 (a) The Hamming $(8,4,4)_2$ code and (b) its Tanner graph

normal graph representation of a code in this work, although some literature on coding has a different meaning for the term "normal." The Tanner graph representation seems more common in current research literature.

The next section describes a class of linear binary codes, the *Tornado codes*, which use a cascade of (normal) bipartite graphs and a very simple decoding algorithm for correcting erasures. To ensure the effectiveness of decoding it is shown how the graphs in the cascade can be chosen probabilistically and this development introduced the notion of irregular distributions of vertex/edge degrees of the left and right vertices of each graph in the cascade. This notion has proved important in other coding contexts, e.g., in the construction of LDPC codes to be considered in Chapter 3.

Section 2.3 introduces the notion of *LT codes*, standing for *Luby transform*, the first incarnation of the important notion of a *fountain code* where coded packets are produced at random by linearly XORing a number of information packets, according to a probability law designed to ensure efficient decodability. That section also considers *Raptor* codes, a small but important modification of LT codes that has been standardized as the most effective way to achieve large downloads over the Internet.

The notion of Tornado codes introduced the idea of choosing random bipartite graphs to effect erasure decoding. Such a notion led to decoding algorithms of fountain codes where a file is comprised of randomly linearly encoded pieces of the file. These decoding algorithms achieve linear complexity rather than the normally cubic complexity associated with Gaussian elimination with a certain probability of failure. As noted, there is no notion of "erasure" with fountain codes as there is with Tornado codes. A significant feature of these fountain codes is that they do not require requests for retransmission of missing packets. This can be a crucial feature in some systems since such requests could overwhelm the source trying to satisfy requests from a large number of receivers, a condition referred to as *feedback implosion*. This is the *multicast* situation where a transmitter has a fixed number of packets (binary n-tuples) which it wishes to transmit to a set of receivers through a network. It is assumed receivers are not able to contact the transmitter to make requests for retransmissions of missing packets. The receiver is able to construct the complete set of transmitted information packets from the reception of *any* sufficiently large set of coded packets, not a specific set. Typically the size of the set of received packets is just slightly larger than the set of information packets itself, to ensure successful decoding with high probability, leading to a very efficient transmission and decoding process.

Most of the algorithms in the chapter will have linear complexity (in codeword length or the number of information symbols), making them very attractive for implementation.

2.2 Tornado Codes and Capacity-Achieving Sequences

The notion of Tornado codes was first noted in [7] and further commented on in [2] with a more complete account in [10] (an updated version of [7]). While not much cited in recent works they introduced novel and important ideas that have become of value in LDPC coding and in the formulation of fountain codes. At the very least they are an interesting chapter in coding theory and worthy of some note.

For this section it is assumed transmission is on the BEC. Since only the binary case is of interest the only arithmetic operation will be the XOR between code symbols and thus the code symbols (coordinate positions) can be assumed to be either bits or sequences of bits (packets). Any received packet is assumed correct – no errors in it. Packets that are erased will be designated with a special symbol, e.g., E (either a bit or packet) when needed.

Tornado codes can be described in three components: a cascade of a sequence of bipartite graphs; a (very simple) decoding algorithm for each stage, as decoding proceeds from the right to the left and a probabilistic design algorithm for each of the bipartite graphs involved. As mentioned, the design algorithm has proven influential in other coding contexts.

Consider the first bipartite graph B_0 of Figure 2.2 with k left vertices, associated with the k information packets and βk, $\beta < 1$, right nodes, the *check* nodes (so each code in the cascade is a normal graph – one of the few places in these chapters using normal graphs). It is assumed the codes are binary and, as noted, whether bits or packets are used for code symbols is immaterial.

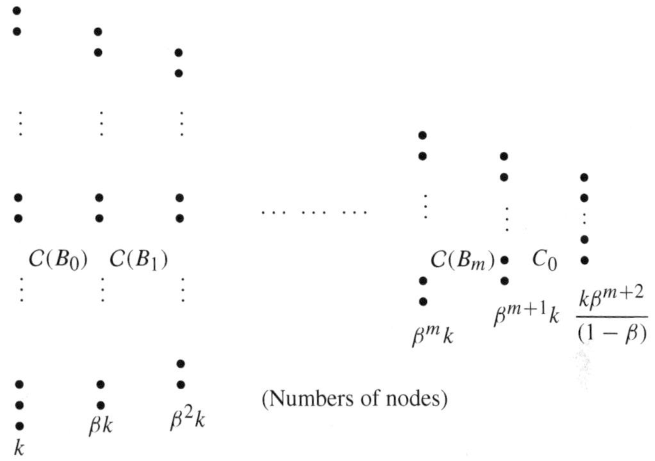

Figure 2.2 The cascade construction of the Tornado code $C(B_0, B_1, \ldots, B_m, C_0)$ – code rate = $1 - \beta$, $\beta \in (0, 1)$, code length $k/(1-\beta)$. Graph edges are not shown for clarity

For the sake of concreteness packets will be assumed. The parity-check matrix could have been used here but the equivalent graph is more convenient. The terms vertices and nodes are used interchangeably. The term packets will often be used to emphasize the application of the ideas to general networks. For a given set of information packets the check packets are a simple XOR of the connected information packets. Note that the edges have not been included in the figure of this cascade of graphs as it tended to obscure other aspects of importance.

To show how decoding could take place on such a graph, consider the first graph on the right of the cascade, C_0, and assume for the moment that all check packets (the furthest right nodes) have been received without erasures and consider the following simple decoding algorithm [7, 10].

Algorithm 2.1 Given the value of a check packet and all but one of its neighbor information packets, set the missing information packet to the XOR of the check packet and its neighbors.

Thus to decode, one searches for a check node at each stage of the decoding satisfying the criterion – namely that only one of its neighbor variable nodes is erased, and decodes that packet, finding the value of the missing information packet. That information packet is then XORed with its neighbor check packets. Each stage proceeds similarly. For the decoding to be successful, there must exist suitable check nodes of degree one at each stage of the decoding until all variable (information) packets for that stage are found – which are then the check nodes for the next (left) stage of the graph.

This first code in the cascade (on the left) is designated as $C(B_0)$ and consists of the k left information packets and βk right check packets. The next bipartite graph in the cascade will consist of the βk left nodes (the check packets of $C(B_0)$ and the "information packets" of $C(B_1)$) and $\beta^2 k$ check packets on these for some fixed β. This stage of the "code" will be designated as $C(B_1)$. The process of constructing the cascade continues, each graph consisting of the previous check nodes and constructing a new set of check nodes of size β times that of the set of left nodes. The cascade components are designated $C(B_i), i = 0, 1, \ldots, m$. Each stage of the cascade has $k\beta^i$ left nodes and $k\beta^{i+1}, \quad i = 0, 1, 2, \ldots, m$ check nodes. The cascade is concluded with a conventional block code C_0 which is assumed to be a simple conventional erasure-correcting code with $k\beta^{m+1}$ left nodes and $k\beta^{m+2}/(1 - \beta)$ check nodes, capable of correcting any fraction of β erasures with high probability. This code is not specified further – any of a number of standard codes will suffice. The whole cascade is designated as $C(B_0, B_1, \ldots, B_m, C_0)$ (see Figure 2.2). As mentioned, the edges of the graph have been omitted as they would tend to obscure the simple process involved. The code has k information packets and

$$\sum_{i=1}^{m+1} k\beta^i + k\beta^{m+2}/(1 - \beta) = k\beta/(1 - \beta), \quad \beta \in (0, 1)$$

check packets and hence the overall code has rate $1 - \beta$.

To decode this code one starts at the extreme right graph/code C_0. It is assumed this ordinary erasure-correcting code is strong enough to correct all erasures in its parity checks. At each stage of decoding, as it progresses from right to left, it will be assumed the check nodes are all determined. Applying the above decoding algorithm, the variable node values of C_0 can be determined as long as there is a check node satisfying the criterion noted for the decoding algorithm, i.e., that all but one of its neighbor variables are determined. If all such variable nodes of C_0 are determined, one proceeds to decoding $C(B_m)$, all of whose check nodes (variable nodes of C_0) are now known. Applying the decoding algorithm to these determines its variable nodes – assuming the supply of suitable check nodes is not exhausted prior to completion. The process continues to the left. The decoding is successful only if each stage is successful, i.e., at each stage a suitable check node is always available with high probability until all information nodes for that stage are determined.

The problem remains as to how to choose the component graphs in this cascade (i.e., the variable/check node connections) so that the algorithm will complete with high probability.

The above assumed a graph representation for each stage of the code in what has been designated as the normal form with information nodes on the left and check nodes on the right. It is clear that there are equivalent versions of the decoding algorithm for a Tanner graph representation of a code, with the n codeword symbols (information and check symbols) on the left and $(n - k)$ check symbols associated with the nodes on the right (a single stage only), with erasures randomly occurring possibly in both variable and check nodes. Much of the recent work on coding uses the Tanner graph representation and a modification of the above discussion for such graphs might be as follows:

Algorithm 2.2 Identify the received codeword symbols (or packets) (with erasures) with the n variable (left) nodes and associate a register with each of the $(n - k)$ check nodes on the right, each initially set to the all-zero packet. XOR the value of each nonerased variable node to its check neighbors and delete those variable nodes and edges emanating from them – thus only variable nodes that have been erased remain at this stage. If there is a check node of degree one, transfer its contents to its variable node neighbor, say v, and then XOR this value to the check neighbors of v and remove all associated edges. Continue in this manner until there are no check nodes of degree one. If decoding completes, the values of the variable nodes are the decoded word. If at any stage before completion, there are no check nodes of degree one the algorithm fails (there are still edges left in the graph).

This algorithm can be adapted for use with the cascade structure of a Tornado code. In both of the algorithms, decoding will be successful only if there is a sufficient number of suitable right nodes of degree one at each stage until completion. Thus the problem of designing each stage of the decoding algorithm to ensure this condition with high probability is of importance. Henceforth only one stage of the graph is of interest and the Tanner graph description of the binary linear code is assumed. The connection to the normal graph representation of interest above is immediate.

Example 2.3 Consider a parity-check matrix of a $(8, 4, 4)_2$ extended Hamming code:

$$\boldsymbol{H} = \begin{bmatrix} 0 & 0 & 0 & 1 & 1 & 1 & 1 & 0 \\ 0 & 1 & 1 & 0 & 0 & 1 & 1 & 0 \\ 1 & 0 & 1 & 0 & 1 & 0 & 1 & 0 \\ 1 & 1 & 1 & 1 & 1 & 1 & 1 & 1 \end{bmatrix}$$

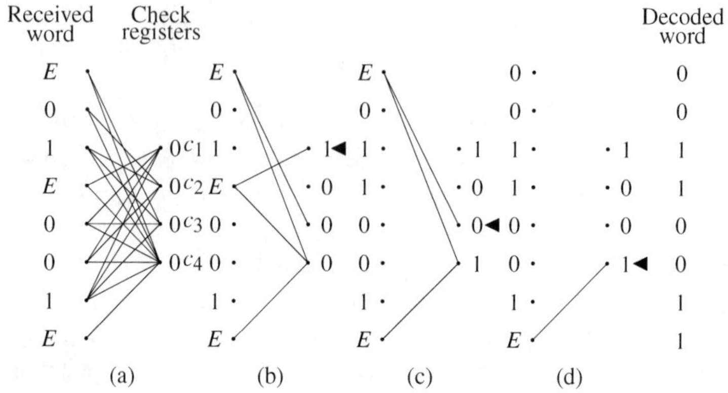

Figure 2.3 Decoding the triple erasure-correcting extended Hamming $(8,4,4)_2$ code

As a code of distance 4 it is capable of correcting three erasures. Suppose the codeword $c = (0,0,1,1,0,0,1,1)$ is transmitted on a BEC and the received word is $y = (E,0,1,E,0,0,1,E)$ is received. The decoding process is shown in Figure 2.3.

Algorithms where messages are passed on graph edges are termed *message-passing algorithms*. When the messages reflect a belief on the bit values involved they are termed *belief propagation* (BP) algorithms. A *Gaussian elimination* (GE) decoding algorithm is where decoding is achieved by solving a matrix equation via matrix reduction, involving received symbols and information symbols (packets). Such decoding algorithms are often optimum achieving a minimum probability of error – hence a *maximum-likelihood* (ML) algorithm – but suffer from a much higher complexity, often $O(n^3)$, than BP. In BP algorithms it is quite possible that the system of equations to be solved is of full rank – hence possesses a unique solution, yet the BP decoding algorithm fails because there is no appropriate check node. Hence BP algorithms are generally not ML.

Decoding algorithms that achieve linear decoding complexity are sought. For successful decoding of the Tornado codes it is necessary for there to be at least one suitable check node at each iteration and the analysis of the algorithm to compute the probability of complete decoding involves computing the probability of there being such a check node at each stage of decoding until completion. The construction of graphs that allow for such an analysis is needed – an overview of the following innovative approach of [11] that achieves this goal is discussed.

For a graph edge, define its *left (resp. right) degree* to be i if it is connected to a variable (resp. check) node of degree i. Thus all edges incident with a degree i vertex (either variable or check) are of edge degree i. Define a pair of degree sequences, $(\lambda_1, \ldots, \lambda_m)$ for the left (variable) nodes and right $(\rho_1, \ldots, \rho_m)$ for the right check nodes, with λ_i the fraction of edges with left degree i and ρ_i the fraction of edges with right degree i. It is convenient for the analysis to follow to define two polynomials

$$\lambda(x) = \sum_i \lambda_i x^{i-1} \quad \text{and} \quad \rho(x) = \sum_i \rho_i x^{i-1}.$$

The exponents of $(i-1)$ rather than i in these polynomials are an artifact of the analysis. Graphs with these edge degree fractions will be denoted (λ, ρ) distribution graphs or equivalently $C^n(\lambda, \rho)$. Thus by a $C^n(\lambda, \rho)$ graph is meant some incarnation of a bipartite graph with some n variable nodes with these edge distributions. One can think of choosing a graph uniformly at random from this ensemble of graphs for analysis.

The case of (d_v, d_c) biregular bipartite discussed previously corresponds to $\lambda(x) = x^{d_v-1}$, $\rho(x) = x^{d_c-1}$.

If E is the number of edges in the bipartite graph (previously used to indicate a coordinate position containing an erasure) then the number of left or variable nodes of degree i is given by $E\lambda_i/i$ and the total number of variable nodes is then $E\sum_i \lambda_i/i$, with similar comments for check nodes. The average left degree of nodes in the graph is

$$a_L = 1\Big/\Big(\sum_i \lambda_i/i\Big) = 1\Big/\Big(\int_0^1 \lambda(x)dx\Big)$$

and the average right degree is

$$a_R = 1\Big/\Big(\sum_i \rho_i/i\Big) = 1\Big/\Big(\int_0^1 \rho(x)dx\Big).$$

The code rate is given by

$$R = \frac{k}{n} = 1 - \frac{(n-k)}{n} \geq 1 - \frac{E\sum_i \rho_i/i}{E\sum_i \lambda_i/i} = 1 - \frac{a_L}{a_R} = 1 - \frac{\int_0^1 \rho(x)dx}{\int_0^1 \lambda(x)dx}$$

with equality only if the check equations are linearly independent.

Our goal will be to determine conditions on these two-degree distributions that ensure, with high probability, the availability of degree one check nodes at each stage of decoding to allow the decoding algorithm to complete. Before considering this problem, it is noted that given such degree distributions it is not difficult to give a random graph construction with such distributions, as follows.

To construct a Tanner graph corresponding to a code of length $n = k + m, m = n - k$ and dimension k the number of left variable nodes is n and the number of check right nodes is m. Decide on an appropriate number of graph edges E which is the total number of 1's in the parity-check matrix and is proportional to the complexity of the proposed decoding algorithm. For a given $(\lambda(x), \rho(x))$ distribution the number of left (resp. right) nodes of degree i is $E\lambda_i/i$ (resp. $E\rho_i/i$). For the sake of argument assume all such quantities are integers. To construct a bipartite graph with the given edge distributions, imagine an array consisting of four columns of nodes, the first column is the set of n variable (codeword nodes) and the fourth column corresponding to the m check nodes. The second and third columns each have E nodes. From each left variable node of degree i, designate, for each i, $E\lambda_i/i$ variable (first column) nodes to be of degree i and generate i edges emanating from each of them terminating in a total of $E\lambda_i$ (disjoint) nodes in the second column. Similarly for each right check node of degree i, designate, for each i, $E\rho_i/i$ nodes to be of degree i and generate i edges from each of them terminating in a total of $E\rho_i$ (disjoint) nodes in the third column. The second and third columns each contain E nodes and so far are of degree one. A random permutation of order E is generated to join nodes in the second and third columns. The nodes in the first and fourth columns are now joined (in the sense of edges between them) and the nodes in the second and third columns can be deleted. There is a small probability that this procedure might generate a bipartite graph with multiple edges between nodes in the first and fourth columns. These may be removed and the effect on the probabilistic analysis is minimal. Similarly there is a probability the associated parity-check matrix will not be of full rank.

It is noted again that for such graphs with edge distributions, the central question is how to choose the distributions $\lambda(x)$ and $\rho(x)$ in order to have the previous algorithm complete the decoding with high probability? Equivalently how to choose the distributions so that, with high probability, there is at least one degree one check node at each stage of the decoding. A formal analysis involving the use of martingales and differential equations is given in [7, 10] and will be commented on later. The following informal argument is instructive and gives an idea as to why the condition on the degree distributions developed arises [9].

Consider an edge e joining variable node v and check node c_{i-1} with left edge degree i and right edge degree j. Consider the graph generated by variable node v and all paths emanating from v with the first edge e for ℓ iterations, with one iteration being one traverse from a variable node to a check node and back again to another variable node. If the girth of the graph is greater than 4ℓ, this graph will be a tree and all variable nodes encountered when traversing from v to the variable nodes (the leaves of the graph) at level 0 will have been erased on the channel independently, each with probability δ.

Let p_ℓ be the probability that a variable node is not resolved (its value is not known) by level ℓ. A recursion is developed for $p_{\ell+1}$ and a condition found on the two edge distributions that ensures these probabilities are decreasing as the algorithm continues. The implication is that if the conditions stated remain satisfied as the number of iterations increases, the probability will vanish implying successful decoding. Suppose all variable node neighbors of check node c_{i-1} are resolved. Since each such node has the same probability of being resolved at level ℓ, the probability they are resolved is $(1 - p_\ell)^{j-1}$. The probability the edge e has right degree j is ρ_j and the probability that at least one such neighbor of c_{i-1} is unresolved is

$$1 - \sum_j \rho_j (1 - p_\ell)^{(j-1)} = 1 - \rho(1 - p_\ell)$$

(hence the reason for the $(i - 1)$ exponent in the polynomial definition). Now the variable node v at iteration $\ell + 1$ will be unresolved only if all $i - 1$ check nodes of v at lower level in the tree have at least one lower-level variable node unresolved and since each edge joining v to these lower-level check nodes has left degree i with probability λ_i, the probability the variable node v remains unresolved at level $\ell + 1$ is

$$p_{\ell+1} = \delta \sum_i \lambda_i (1 - \rho(1 - p_\ell))^{(i-1)} = \delta\lambda(1 - \rho(1 - p_\ell)) \tag{2.1}$$

where $p_0 = \delta$, the erasure probability on the channel, the initial condition that variable node v was unresolved (received as an erasure). Thus [28] successful decoding gives the condition that the probability a variable node being unresolved decreases with the level:

$$p_{\ell+1} = \delta\lambda(1 - \rho(1 - p_\ell)) < p_\ell. \tag{2.2}$$

Thus as the number of iterations ℓ increases the probabilities will tend to 0 and decoding will be successful if the condition holds. In other words, finding edge distributions $\lambda(x)$ and $\rho(x)$ that satisfy this condition will, with some probability, assure the completion of the decoding algorithm.

Alternatively, replacing p_ℓ by a variable x, if neighborhoods of a variable node of depth ℓ are trees (which assures the independent erasure condition holds), and if

$$\delta\lambda(1 - \rho(1 - x)) < (1 - \epsilon)x, \quad 0 < x < \delta,$$

then after ℓ iterations the probability a variable node is unresolved is at most $(1-\epsilon)^{\ell}\delta$. Thus edge distributions $\rho(x)$ and $\lambda(x)$ are sought that, for a given erasure channel δ, satisfy the condition

$$\delta\lambda(1-\rho(1-x)) < x, \quad 0 < x < \delta \tag{2.3}$$

so that with high probability, decoding on a BEC channel with erasure probability at most δ will be successful, i.e., the probabilities the nodes are unresolved decrease with iterations.

Notice that as a polynomial with positive coefficients $\lambda(x)$ is an increasing function on $(0,1)$ and hence is invertible and letting $x = \delta\lambda(1-y)$ the condition can be written as

$$\rho(1-\delta\lambda(1-y)) \geq 1-\lambda^{-1}(\lambda(1-y)) = y \quad \text{on} \quad (0,1) \tag{2.4}$$

and the equivalent (dual) condition to Equation 2.3 is [29]

$$\rho(1-\delta\lambda(1-y)) > y, \quad 0 < y < 1.$$

Equivalently by letting $y = \rho^{-1}(1-x)$ this equation can be written as

$$\delta\lambda(1-\rho(y)) < 1-y, \quad y \in [0,1) \tag{2.5}$$

(which is also obtained by substituting $y = 1-x$ in Equation 2.3). Degree distributions that satisfy these conditions for as large a value of δ as possible are sought, where δ is the probability of erasure on the channel. Thus with high probability, for $(\lambda(x), \rho(x))$ distributions that satisfy these conditions, the decoding will complete for as high a channel erasure probability as possible. Indeed, distributions that result in a δ satisfying a code rate $R = 1-\delta$ achieve capacity on a BEC.

The informal development for the condition in Equation 2.3 can be made formal. The approach in [7, 8, 10, 23, 24] is outlined. Let $\ell_t^{(i)}$ and $r_t^{(i)}$ be the *fraction* of left (resp. right) edges at stage (time) t of the algorithm of degree i, where by fraction is meant the actual number of left (resp. right) divided by the original total number of edges E. A discrete time differential equation is developed in terms of the degree distributions and solved. In particular it is shown ([10], proposition 1) that the fraction of degree one right edges is

$$r_1(x) = \delta\lambda(x)[x-1+\rho(1-\delta\lambda(x))]$$

where x is defined via $dx/d\tau = -x/e(\tau)$ where $e(\tau)$ is the fraction of edges remaining at time τ. Thus the decoding continues as long as $r_1(\tau) > 0$ which leads to the proposition:

Proposition 2.4 ([10], proposition 2) *Let B be a bipartite graph that is chosen at random with edge degree distributions* $\lambda(x)$ *and* $\rho(x)$. *Let* δ *be fixed so that*

$$\rho(1 - \delta\lambda(x)) > 1 - x, \quad x \in (0, 1]. \tag{2.6}$$

For all $\eta > 0$ *there is some* k_0 *such that for all* $k \geq k_0$, *if the k (left) message bits of* $C(B)$ *are erased independently with probability* δ, *then with probability at least* $1 - k^{2/3} \exp(-\sqrt[3]{k/2})$ *the recovery algorithm terminates with at most* ηk *message bits erased.*

Notice the proposition only gives an upper bound on the number of erasures remaining after the algorithm terminates.

The result can be used to determine suitable distributions to ensure there is at least one check node available of degree one for the completion of the algorithm. Returning to Tornado codes, the following conclusion can be shown as a condition on the degree distributions:

Theorem 2.5 ([10], theorem 2) *Let k be an integer and suppose that*

$$C = C(B_0, \ldots, B_m, C_0)$$

is a cascade of bipartite graphs where B_0 *has k variable nodes. Suppose each* B_i *is chosen at random with edge degrees specified by* $\lambda(x)$ *and* $\rho(x)$ *such that* $\lambda_1 = \lambda_2 = 0$ *and suppose that* δ *is such that*

$$\rho(1 - \delta\lambda(x)) > 1 - x, \quad 0 < x < 1.$$

Then, if at most a δ*-fraction of the coordinates of an encoded word in C are erased independently at random, the erasure-decoding algorithm terminates successfully with probability* $1 - O(k^{-3/4})$ *and does so in* $O(k)$ *steps.*

The $O(k)$ complexity follows from the fact the average node degree is a constant. Recall that if the probability of an erasure on the BEC is δ, the capacity of the channel, the maximum rate at which information can be transmitted through the channel reliably, is $R = 1 - \delta$, or conversely, the maximum-erasure rate that a code of rate R may be used reliably on a BEC is $\delta = 1 - R$. The following result shows this:

Theorem 2.6 ([10], theorem 3) *For any rate R with* $0 < R < 1$, *any* ϵ *with* $0 < \epsilon < 1$ *and sufficiently large block length n, there is a linear code and a decoding algorithm that, with probability* $1 - O(n^{-3/4})$, *is able to correct a random* $(1 - R)(1 - \epsilon)$*-fraction of erasures in time proportional to* $n \ln(1/\epsilon)$.

Note that this result achieves the goal of a linear-time (in code length) decoding algorithm. Thus the challenge is to devise a distribution pair

$(\lambda(x), \rho(x))$ such that the rate R of the corresponding code (the related bipartite graph) is such that the code is able to correct a fraction of $1 - R$ erasures, asymptotically on average. Distribution pairs that achieve this relationship are referred to as *capacity-achieving sequences.*

The average node degrees on the left and right, a_L and a_R, were shown to be

$$a_L = \left(\sum_i \lambda_i / i\right)^{-1} \quad \text{and} \quad a_R = \left(\sum_i \rho_i / i\right)^{-1}.$$

The following theorem is of interest:

Theorem 2.7 ([27], theorem 1) *Let G be a bipartite graph with distributions* $(\lambda(x), \rho(x))$ *and let* δ *be a positive number such that*

$$\delta\lambda(1 - \rho(1 - x)) \leq x, \quad 0 < x \leq \delta.$$

Then

$$\delta \leq \frac{a_L}{a_R}(1 - (1 - \delta)^{a_R}).$$

It is also of interest to note ([27], lemma 2) that if λ and ρ are polynomials satisfying the above with the above notation, then $\delta \leq \rho'(1)/\lambda'(1)$. Thus the distribution pair determines the rate of the code and this result determines a bound on how close it will be to achieving capacity in the sense it gives a bound on the erasure-correcting capability of the code. In addition it is of interest to find good degree distributions so as to yield as large a value of erasure probability δ as possible.

We return to the notion of a Tornado code which is a cascade of graphs each of whose distributions satisfy the above conditions. They are referred to as Tornado [2] as in practice it often occurs that as the substitution process progresses (recovery of variable nodes by check nodes of degree one), the decoding process typically proceeds slowly until the resolution of one more variable node results in the whole set of variable nodes being resolved quickly.

To summarize, the arguments that led to the conditions Equations 2.3 and 2.4 say that under certain conditions on the edge distributions, the probability of a node not being resolved by a certain iteration decreases with the number of iterations and hence tends to zero and to complete decoding. The results depend only on edge distributions and hence can apply also to the Tanner graph representation of a code. The decoding algorithm is equivalent to that of Tornado codes and, as noted, the crucial property is to have check nodes of degree one with high probability at each iteration, which the stated edge degree distributions tend to fulfill.

The fact that the differential equation/martingale approach to the decoding problem and this rather different approach leads to the same conditions is an interesting confirmation of the results.

Examples of Capacity-Achieving Distributions

Numerous works give examples of pairs of distributions $(\lambda(x), \rho(x))$ that satisfy the above conditions. Recall that the average left and right degrees of a graph with edge distributions $(\lambda(x), \rho(x))$ are $1/\left(\sum_i \lambda_i/i\right)$ and $1/\left(\sum_i \rho_i/i\right)$ $\left(\text{or } 1/\int_0^1 \lambda(x)dx \text{ and } 1/\int_0^1 \rho(x)dx\right)$. The rate R of the code is at least $1 - \left(\int_0^1 \rho(x)dx / \int_0^1 \lambda(x)dx\right)$ (depending on the corresponding matrix having linearly independent rows). It can be shown [28] that for given $(\lambda(x), \rho(x))$ distributions satisfying condition of Equation 2.3, δ is always less than or equal to $1 - R$ for R the rate of the resulting code derived from the edge distributions. For the formal definition of a capacity-achieving sequence we use the following:

Definition 2.8 ([28], section 3.4) An edge distribution sequence $(\lambda(x), \rho(x))$ is called *capacity achieving* of rate R if

(i) the corresponding graphs give rise to codes of rate at least R;
(ii) for all $\epsilon > 0$ there exists an η_0 such that for all $\eta > \eta_0$ we have

$$\lambda(1 - \rho(1 - x)) < x \text{ for } x \in (0, (1 - R)(1 - \epsilon))$$

where η is the length of the probability distributions λ and ρ.

Example 2.9 The first example of such distributions, cited in numerous works (e.g., [2, 10, 27, 28, 29]) is referred to as *heavy-tailed/Poisson* sequences for reasons that will become clear.

Let D be a positive integer (that will be an indicator of how close δ can be made to $1 - R$ for the sequences obtained). Let $H(D) = \sum_{i=1}^{D} 1/i$ denote the harmonic sum truncated at D and note that for large D, $H(D) \approx \ln(D)$. The two distributions parameterized by the positive integer D are

$$\lambda_D(x) = \frac{1}{H(D)} \sum_{i=1}^{D} x^i/i \quad \text{and} \quad \rho_D(x) = e^{\mu(x-1)} \ (D \text{ terms}).$$

Here μ is the solution to the equation

$$\frac{1}{\mu}(1 - e^{-\mu}) = \frac{1 - R}{H(D)}\left(1 - \frac{1}{D+1}\right)$$

and such edge distributions give a code of rate at least R and note the average left degree is $a_L = H(D)(D+1)/D$ and $\int_0^1 \lambda_D(x)dx = (1/H(D))(1 - 1/(D+1))$. The right degree edge distribution is the truncated Poisson distribution. It can also be established that

$$\begin{aligned} \delta\lambda_D(1-\rho_D(1-x)) &= \delta\lambda_D(1-e^{-\mu x}) \\ &\leq \frac{-\delta}{H(D)} \ln(e^{-\mu x}) \\ &= \frac{\delta\mu x}{H(D)}. \end{aligned}$$

For the right-hand side of the last equation to be at most x it is required that

$$\delta \leq H(D)/\mu$$

(δ is the largest erasure probability possible on the channel possible for the conditions) which can be shown to be equal to

$$(1-R)(1-1/(D+1))/(1-e^{-\mu}).$$

That this pair of distributions satisfies condition

$$(1-R)(1-1/D)\lambda_D(1-\rho_D(1-x)) < x, \quad 0 < x < (1-R)(1-1/D),$$

is shown in [28]. (Note: Such codes are referred to as Tornado codes there in contrast to the definition of such codes used here.)

Example 2.10 Another example from [27, 28] is the following: Recall the general binomial theorem

$$(x+y)^\alpha = \sum_{j=0}^{\infty} \binom{\alpha}{j} x^j y^{\alpha-j}$$

for α any real or complex number and the fractional binomial coefficients are given by

$$\binom{\alpha}{n} = \alpha(\alpha-1)\cdots(\alpha-(n-1))/n!\ .$$

Consider the right regular graphs with the check nodes all of degree a, for a positive integer $a \geq 2$ and the distributions

$$\lambda_{a,n}(x) = \frac{\sum_{k=1}^{n-1} \binom{\alpha}{k}(-1)^{k+1}x^k}{1-n\binom{\alpha}{k}(-1)^{k+1}} \quad \text{and} \quad \rho_a(x) = x^{a-1}$$

where $\alpha = 1/(a-1)$ is real, positive and noninteger. Notice that the fractional binomial coefficients alternate in sign for $\alpha < 1$ and hence the coefficients of the distribution $\lambda_{a,n}$ are positive. It is convenient to introduce a parameter

ν such that $0 < \nu < 1$ by $n = \nu^{-1/\alpha}$ (ignoring integer constraints). It is shown in ([27], proposition 2 and theorem 2), that for the code rate defined as $R = 1 - a_L/a_R$ and $a_R = a = (\alpha + 1)/\alpha$

$$\frac{\delta}{1-R} \geq 1 - \nu^{\alpha+1}/\alpha = 1 - \nu^{a_R}.$$

As $\nu < 1$ this suggests $\delta \approx 1 - R$ for large a_R, approaching capacity.

The properties of these distributions and the sense in which they are asymptotically optimal and satisfy the conditions such as Equation 2.3 or 2.4 is discussed in [27]. Notice in this case the graphs are right regular – all right check nodes of the same degree.

Example 2.11 A variety of techniques have been developed to determine capacity-achieving sequences, including a linear programming approach and density evolution, a method used effectively in the analysis of LDPC codes (e.g., see [10, 34] and Chapter 3).

As an example, the following distribution pair given in [34] was found using density evolution:

$$\begin{aligned}\lambda(x) &= 0.29730x + 0.17495x^2 + 0.24419x^5 + 0.28353x^{19}\\ \rho(x) &= 0.33181x^6 + 0.66818x^7\end{aligned}$$

The Example 2.10 is interesting in that one is able to construct good sequence pairs with one of the pair being a monomial, i.e., all right nodes have the same degree (right regular). However, it is shown in [7] that sequence pairs which are both monomials (hence biregular graphs) perform poorly.

The relationship of these results on the BEC to those obtained for other channels, most notably the BIAWGN using belief propagation as discussed in Chapter 3, is of interest. To briefly note the approach taken there, the pdf p_ℓ of the log likelihood ratios (LLRs) (not the same p_ℓ used in the previous analysis) passed during the decoding process with the code graph described by the distribution pair $(\lambda(x), \rho(x))$ satisfied the recursion (Equation 3.28 of Chapter 3) is shown there to satisfy

$$p_\ell = p_0 \otimes \lambda(\Gamma^{-1}(\rho\Gamma(p_{\ell-1}))), \quad \ell \geq 1$$

where $\otimes$ indicates convolution and Γ is an operator introduced in Chapter 3 that gives the pdf of it argument. It should be noted the arguments in that chapter will be likelihood ratios as messages passed on the decoding graph and these are random since they depend on received random messages.

Applying this equation to the erasure channel, the pdf [28] is a two-point mass function with a mass of p_ℓ at 0 and mass $1 - p_\ell$ at ∞. Performing the convolutions with such mass functions in the above equation can be shown to yield the result

$$p_\ell = \delta\lambda(1 - \rho(1 - p_{\ell-1}))$$

where, as before, δ is the channel erasure probability. This is essentially the Equation 2.2 obtained by very different means, an interesting confirmation.

2.3 LT and Raptor Codes

The formulation of the Tornado codes of the previous section introduced (at least) two novel ideas: the notion of deriving edge probability distributions to generate bipartite code graphs where certain simple decoding algorithms could prove effective. In a sense to be discussed these ideas led to the notion of fountain codes.

To initiate the discussion, consider the following simple situation. A server has k information packets of data of some fixed number of bits to be downloaded over a network to a large number of users. The users are interested in receiving a complete set of the k packets. Packets can be XORed. One possibility to achieve this download is to simply forward the packets on the Internet with each user obtaining the packets in some order. However, due to imperfections in the Internet such as buffer overflow or node servers failing, it is likely some users will miss one or several of the packets leading to the requirement of some feedback mechanism to the source where each user is able to request retransmission of their missing packets. This can lead to inefficient operation and congestion (implosion) on the network.

A more interesting possibility is for the server to first *code*, each coded packet consisting of the XOR of a random selection of information packets. This notion will be the basis of fountain codes.

Consider first a (not very efficient) case of random fountain coding where each of the k information packets is included in the formation of a coded packet with a probability of $1/2$ independently and transmitted on the Internet, i.e., the coder chooses uniformly at random a selection of information packets, XORs them together to form a coded packet which is transmitted on the network.

A user now must gather a sufficient number of *any* of the coded packets to allow the solution of the set for the original information packets. In effect the user must gather a set of $(k + m)$ coded packets for m a small positive integer to allow for a full-rank random $k \times (k+m)$ matrix equation to be formed which is necessary to guarantee solution. The fact that any of the coded packets can be used is a positive feature of the scheme. However, solving for the original

k information packets involves the solution of a $k \times (k+m)$ binary random matrix, say by Gaussian elimination, an operation that is $O(k^3)$ in complexity which, since k might easily on the order of several thousand be far higher than is typically of interest. By Equation A.8 of Appendix A the probability such a matrix is of full rank, $Q_{k,k+m}$ (hence solvable by Gaussian elimination), is

$$0.999 < Q_{k,k+10} < 0.999 + \epsilon, \quad \epsilon < 10^{-6}.$$

Such codes are generally referred to as *fountain codes* for the obvious reason. They are also referred to as *rateless codes* as their design involves no particular rate in the sense of the design of block codes.

A possible remedy to the large complexity of the Gaussian elimination decoding argument might be as with the Tornado codes and form a coding graph in the obvious manner. The headers of the coded packets contain the information as to which information packets were XORed to form the coded packet allowing the code graph to be formed consisting of information nodes on the left corresponding to information packets (nodes) and code packets (nodes) on the right. There is an edge between an information packet and a code packet if the information packet is involved with the formation of the code packet. The following decoding algorithm is as for Tornado codes.

Algorithm 2.12 If there is a coded node of degree 1, XOR the contents of the coded node to the neighbor information node thus resolving that information node. XOR this information node to its other neighbor coded nodes and delete all edges involved (decreasing the number of unresolved information nodes by one and the number of coded nodes not yet used by at least one).

This algorithm is simple with linear decoding complexity. Notice that if the $k \times (k+m)$ matrix formed, corresponding to the graph, is of full rank (k), Gaussian elimination is guaranteed to provide a solution for the k information packets. On the other hand the decoding algorithm above, while very simple with linear complexity, is unlikely to run to completion and will almost certainly fail.

The remedy for this situation is not to choose the packets for XORing to form a coded packet uniformly at random but rather formulate a probability distribution for the number of information packets to be XORed, much as for a single section of the Tornado code algorithm, so that with high probability, the previous simple decoding algorithm will be successful, i.e., at each stage of decoding find a check node of degree one all the way to completion. With such a distribution the simple (linear complexity) decoding algorithm can be used rather than the computationally expensive Gaussian elimination. This is the genius behind the LT (Luby transform) [6] fountain codes discussed below.

LT Codes

The approach of Tornado codes and capacity-achieving sequences of the previous section suggests the following possibility [6, 31]. As above, assume there are k information packets to be coded (to give coded packets). A discrete probability distribution (the literature typically uses $\Omega(x)$ for this distribution, a convention which is followed here)

$$\Omega_i,\ \ i = 1,2,\ldots,k, \quad \sum_{i=1}^{k} \Omega_i = 1, \quad \Omega(x) = \sum_{i=1}^{k} \Omega_i x^i$$

is to be chosen and used in the following manner to form coded packets: for each coded packet, the distribution is first sampled – if d is obtained (with probability $\Omega_d,\ d \in [k]$) – then d distinct information packets are chosen uniformly at random from among the k and XORed together to form a coded packet to be transmitted on the network. The choice of packets is stored in the coded packet header. The process is repeated as required. As noted, such a coding process is referred to as a fountain code or as a rateless code since there is no specific rate associated with the process of forming coded words. It is the form of the distribution that restores the complexity of decoding to linear and makes the above decoding algorithm effective.

The receiver gathers slightly more than k, say $k + \epsilon(k)$ (to be discussed), coded packets from the network and forms the bipartite graph with k information packet nodes (considered as initially empty registers), on the left, denoted $\boldsymbol{x}_1,\boldsymbol{x}_2,\ldots,\boldsymbol{x}_k$ (to be determined by the decoding algorithm) and $k + \epsilon(k)$ received coded packets on the right, denoted $\boldsymbol{y}_1,\boldsymbol{y}_2,\ldots,\boldsymbol{y}_{k+\epsilon(k)}$.

The problem is to choose the distribution $\Omega(x)$ to have the probability the decoding algorithm is successful – runs to completion. Clearly at least one check node of degree one is required at each stage of decoding until all information symbols are found. If this is not the case the decoding algorithm fails. Clearly a balance is needed in the sense a sufficient supply of degree one coded nodes is required at each stage to make the decoding algorithm robust and successful while too many degree one nodes will make it inefficient and give poor performance. The analysis is somewhat intricate and the reader is referred to [6, 31, 33] for details. The essence of the results is described here without proofs.

To discuss the situation further, the behavior of the algorithm as it iterates – at each stage finding a coded node of degree one, transferring the coded packet associated with such a node to the connecting information node and then XORing that packet to all neighbor check nodes and removing the information node and all its edges. The terms "node," "symbol" and "packet" are used interchangeably.

Some definitions are needed. The number of coded symbols gathered for decoding, beyond the number of information symbols, is termed the *overhead* of the decoding algorithm. Thus if $k(1 + \epsilon k)$ coded packets are used the overhead is $k\epsilon$. A simple argument ([33], proposition 2.1) shows that if an ML decoder is to succeed with high probability for an overhead fraction ϵ that tends to zero with k, then Ω_1 has to converge to zero. At stage i of the algorithm define the *output ripple* (or simply ripple) as the set of coded nodes of degree one. At this stage, since one information symbol (packet) is released at each stage of the algorithm, there are $k - i$ undecoded (or unresolved) information nodes remaining at stage i. A coded node is said to be *released* at stage $i + 1$ if its degree is greater than one at stage i and equal to one at step $i + 1$. To calculate the probability [30, 31, 33] a coded node of initial degree d is released at stage $i + 1$ it is assumed the original neighbor nodes of a coded node are chosen with replacement – convenient for the analysis and having little effect on the code performance. The probability a coded node, originally of degree d, has only one neighbor at stage $i + 1$ among the $k - (i + 1)$ information nodes not yet recovered and that all of its other $(d - 1)$ information node neighbors are in the set of recovered information nodes is

$$d\left(\frac{k-(i+1)}{k}\right)\left(\frac{i+1}{k}\right)^{d-1}$$

(this is where the choosing with replacement assumption appears).

Another way of viewing this formula is to reverse the situation and suppose the set of variable (information) nodes resolved at the i-th iteration is V_i, $\mid V_i \mid = i$ and at the $(i + 1)$-st iteration is $V_{i+1} \supset V_i$, $\mid V_{i+1} \mid = i + 1$. We ask what is the probability the edges of a constraint (coded) node c of original degree d are chosen so it is released at the $(i + 1)$-st iteration with these sets of resolved variable nodes. Thus at iteration i there are at least two variable (information) neighbors of c unresolved and at iteration $i + 1$ there is only one unresolved – c is in the output ripple at that stage. There are d ways of choosing the edge from c that will be unresolved at the $(i + 1)$-st iteration and the probability it is unresolved then is $\left(1 - \frac{i+1}{k}\right)$. For the constraint node c to be released at the i-th iteration it has $(d - 1)$ edges attached to resolved nodes at the $(i + 1)$-st iteration but not all of these were resolved at the i-th iteration. The probability of this is

$$\left(\left(\frac{i+1}{k}\right)^{d-1} - \left(\frac{i}{k}\right)^{d-1}\right).$$

Thus, as in the above formula, the probability a coded node, originally of degree d, is released at stage $i + 1$ of the decoding is approximately

$$d\left(1-\left(\frac{i+1}{k}\right)\right)\left(\left(\frac{i+1}{k}\right)^{d-1}-\left(\frac{i}{k}\right)^{d-1}\right).$$

Multiplying this expression by Ω_d and summing yields

$$\Pr(\text{output symbol is released at stage } i+1)$$
$$=\left(1-\frac{i+1}{k}\right)\left(\Omega'\left(\frac{i+1}{k}\right)-\Omega'\left(\frac{i}{k}\right)\right).$$

Recalling the approximation for a smooth continuous function (like polynomials) $f''(x) \approx \big(f'(x+\Delta)-f'(x)\big)/\Delta$ for small Δ it is clear that

$$\Omega'\left(\frac{i+1}{k}\right)-\Omega'\left(\frac{i}{k}\right)\approx\frac{1}{k}\Omega''\left(\frac{1}{k}\right)$$

and the expected number of coded nodes released at step $i+1$ is n times this amount ($n = k+\epsilon(k)$, the number of coded packets collected)

$$\frac{n}{k}\left(1-\frac{i+1}{k}\right)\Omega''\left(\frac{i}{k}\right). \tag{2.7}$$

If one sets $x = i/k$, if the probability the last coded symbol is released at stage $k+\epsilon(k) = n$ is one, this is equivalent to the equation

$$(1-x)\Omega''(x) = 1, \quad 0 < x < 1$$

which, with the required condition that $\Omega(1) = 1$, has the solution

$$\Omega(x) = \sum_{i\geq 2}\frac{x^i}{i(i+1)}.$$

This is referred to as the *limited degree distribution* [33]. Clearly this distribution is useless for our purpose since it produces no coded nodes of degree one and decoding could not start. In addition, the distribution is infinite – the needed distribution should produce no coded nodes of degree greater than k, the number of information nodes assumed. The analysis, however, is instructive. The following modified distribution is suggested [6]:

Definition 2.13 The *(ideal) soliton distribution* is defined as

$$\Omega_i^{sol} = \begin{cases} \frac{1}{k}, & i = 1 \\ \frac{1}{i(i-1)}, & i = 2,3,\ldots,k. \end{cases} \tag{2.8}$$

The term soliton arises in a refraction problem in physics with similar requirements. The distribution is associated with the polynomial

$$\Omega^{sol}(x) = \frac{x}{k} + \sum_{i=2}^{k} \frac{x^i}{(i-1)i}.$$

That this is a distribution (probabilities sum to unity) follows readily by an induction argument or by observing that $1/(i(i-1)) = 1/(i-1) - 1/i$.

Notice that this distribution has a coded node of expected degree

$$\sum_{i=1}^{k} i\Omega_i^{sol} = \frac{1}{k} + \sum_{i=2}^{k} i\frac{1}{i(i-1)} = \sum_{i=1}^{k} \frac{1}{i} \approx \ln(k)$$

which is the first k terms of the harmonic series which is well approximated by $\ln(k)$ for large k. This is the minimum possible to give a reasonable possibility of each information node being involved with at least one coded node in the following sense. The problem is equivalent to throwing n balls (the coded nodes) into k cells (the information nodes) and asking for the probability there is no empty cell (although with replacement). This is just one less the probability at most $(k-1)$ cells are occupied which is

$$1 - \binom{k}{k-1}\left(\frac{k-1}{k}\right)^n.$$

If it is wished to have a probability of $1 - \alpha$ to have no cell empty (all information nodes covered) is

$$1 - \alpha = 1 - k\left(1 - \frac{1}{k}\right)^n \quad \text{or} \quad \left(1 - \frac{1}{k}\right)^n = \alpha$$

and in the limit

$$\lim_{k \longrightarrow \infty} \left(1 - \frac{1}{k}\right)^{k\cdot(n/k)} \longrightarrow \exp(-n/k) \approx \alpha/k \quad \text{or} \quad n \approx k\log(k/\alpha).$$

Thus for the encoding process to cover each information symbol at least once with probability at least $1-\alpha$, the average degree of the approximately k coded nodes must be at least $\log(k/\alpha)$.

As discussed below, the ideal soliton distribution will prove to be unsatisfactory for several reasons. One problem is that the variance of the number of coded nodes in the ripple at each stage of the decoding is so large that the probability there is no node of degree one at each stage is too high. Nonetheless it does have some interesting properties. If the probability that a coded node of initial degree i is released when there are L information nodes remaining unrecovered is denoted $r(i, L)$ and $r(L)$ is the overall probability of release [6],

$$r(L) = \sum_i r(i, L).$$

Then for the ideal soliton distribution it is shown ([6], proposition 10) that $r(L) = 1/k$ and the probability a coded node is released at each stage of decoding is $r(L) = 1/k$, i.e., the probability of release is the same at each stage of decoding.

For the following let η (δ in [6], used for erasure probability here) be the target probability the decoding algorithm fails to complete decoding. In an effort to improve the performance of the ideal soliton distribution the *robust soliton distribution* was proposed, defined as follows ([6], definition 11):

Definition 2.14 The *robust soliton distribution* denoted Ω^{rs} is defined using the ideal Soliton distribution and the function τ_i as follows. Let $R = c\sqrt{k}\ln(k/\eta)$ for a constant $c > 0$ and define

$$\tau_i = \begin{cases} R/ik & \text{for } i = 1, 2, \ldots, k/R - 1 \\ R\ln(R/\eta)/k & \text{for } i = k/R \\ 0 & \text{for } i = k/R + 1, \ldots, k. \end{cases}$$

Then the robust soliton distribution is

$$\Omega_i^{rs} = (\Omega_i^{sol} + \tau_i)/\beta, \quad i = 1, 2, \ldots, k \tag{2.9}$$

for the normalizing constant $\beta = \sum_{i=1}^{k} (\Omega_i^{sol} + \tau_i)$.

The rationale for this distribution is that it tends to maintain the size of the ripple, the number of coded nodes of degree one at each stage of decoding, preventing the size to fall to zero at any stage before the end – which would lead to decoding failure.

The distribution chosen for the formation of the coded packets is critical for performance of the decoding algorithm. The relationship between the distribution and the number of extra coded packets (beyond the number of information packets k) required to achieve a given error probability (the probability the decoding fails to complete) is complex. For the Robust Soliton distribution it is shown ([6], theorems 12 and 17) that for a decoder failure probability of η an overhead of $K = k + O(\sqrt{k} \cdot \ln^2(k/\eta))$ is required and that the average degree of a coded node is $O(\ln(k/\eta))$ and that the release probability when L information nodes have not been covered is of the form $r(L) \geq L/((L - \theta R)K), L \geq R$ for some constant θ and R as in the above definition.

It has been observed above the average degree of a coded node under the Robust Soliton distribution is $O(\log(k))$ (for constant probability of decoding error η) and hence each coded symbol requires $O(\log(k))$ operations to produce it and the decoding operation is of complexity $O(k\log(k))$. For very large k a decoder with linear complexity would be more desirable and this will be possible with the Raptor codes discussed in the next section.

The literature on the analysis of aspects of decoding LT codes is extensive. The references [4, 13] are insightful. A detailed analysis of the decoding error probability is given in ([31], section 3).

Shokrollahi [30] defines a *reliable decoding algorithm* as one that can recover the k information symbols from any set of n coded symbols and errs with a probability that is at most inversely polynomial in k, i.e., a probability of the form $1/k^c$. Recall the overhead of a decoding algorithm is the number of coded symbols beyond k needed to achieve the target probability of error. A *random LT code* is one with a uniform distribution on the choice of information symbols to combine for a coded symbol. This corresponds to the choice of the distribution

$$\Omega(x) = (1+x)^k \cdot (1/2^k)$$

and corresponds to the random analysis given earlier in this section. It can be shown ([30], propositions 1 and 2) that for *any* LT code with k information symbols there is a constant c such that the graph associated with the decoder has at least $ck\log(k)$ edges and that any random LT code with k information symbols has an encoding complexity of $k/2$ and ML decoding is a reliable decoder with an overhead of $O(\log(k)/k)$. ML is Gaussian elimination for the $k \times (k + O(\log(k)/k))$ matrix formed at the decoder with complexity $O(k^3)$. As noted previously in general BP (the algorithm described above) is not ML – indeed it is possible that if K coded symbols are gathered that the $k \times K$ matrix can be of full rank k (for which ML decoding would be successful) and yet the BP algorithm fail to completely decode (run out of coded nodes of degree one before completing). However, the low computational requirements of the BP decoding algorithm and its error performance are impressive.

The implication of these comments is that it will not be possible to have a reliable LT decoder and achieve linear time encoding and decoding. The development of the Raptor codes of the next section shows how this goal can be achieved by a slight modification of LT codes.

Raptor Codes

The idea behind Raptor codes (the term is derived from RAPid TORnado) is simple. It was observed that decoding failures for LT codes tend to occur toward the end of the decoding process when there are few information symbols left unresolved. This suggests introducing a simple linear block erasure-correcting code to correct the few remaining erasures after the LT decoding fails.

For Raptor codes the k information symbols/packets are first *precoded* with an efficient simple (easy to decode) linear block erasure-correcting $C_n = (n,k,d)_2$ code and then these n packets are LT coded. Note this introduces $(n-k)$ parity packets – i.e., the parities of the code C_n are formed across the k information packets. For decoding, some $n(1+\epsilon)$ of the LT coded packets are gathered for LT decoding, as before. Using the LT decoding on the precoded bits, the decoding might (typically) stall with a few uncoded packets left undecoded (no more coded nodes of degree one in the ripple). These can then be decoded with the linear erasure-correcting code C_n. It is also desired to have linear coding and decoding complexity and because of the reduced requirements on the LT decoding, this will be possible. Recall from the above discussion that a reliable decoding algorithm for LT codes must have at least $O(\log k)$ information node degrees and an overhead of $O(\log(k)/k)$.

Much of the development and commercial deployment of Raptor codes is due largely to the work of Michael Luby and Amin Shokrollahi and their colleagues although the key idea behind them was independently found in the work of Maymounkov [15, 16]. Both of the papers [30] and [15] are important reading.

A Raptor code then has parameters $(k, C_n, \Omega(x))$ where C_n is a binary linear $(n,k,d)_2$ erasure-correcting code and $\Omega(x)$ is a distribution on n letters. For simplicity Ω is used in the remainder of this chapter or Ω_D when the parameter D is needed. The coordinates of the code C_n will be an $(n,k,d)_2$ code and will produce n *intermediate packets* from the k information packets and $(n-k)$ parity packets. This code is not specified further – as noted its requirements are minimal in that a variety of simple codes, capable of correcting a few erasures with linear complexity, can be used. It is used essentially to correct a few erasures if the LT decoding leaves a few nodes unresolved (viewed as erasures) toward the end of its decoding process. That is, if the LT decoding does not quite complete to the end, leaving a few unresolved packets, the code C_n completes the decoding, treating the unresolved packets as erasures. The output of the LT code will be referred to as the coded packets and $(n, \Omega_D(x))$ LT coding, as before.

It will be important to recognize that while LT codes cannot have linear decoding complexity, using the LT decoding process with an appropriate distribution to decode up to a few remaining erasures can have linear complexity.

As before, a reliable decoding algorithm for the Raptor code will have a probability of error of inverse polynomial form, $1/k^c$ for some positive constant c. Such codes will be analyzed with respect to their space requirements, overhead and cost or complexity. The two extremes of such codes are the LT codes where C is the trivial $(k,k,1)_2$ code and the *precode only* (PCO) code where there is no LT code. The *decoding cost* of such a code is the expected number of arithmetic operations per information symbol. The *overhead* of the code is the number of coded symbols *beyond* k needed to recover all information symbols. The Raptor code $(k, C_n, \Omega_D(x))$ is thought of in terms of three columns of nodes or symbols (the left part of the previous figure):

- a left column of k information packets (associated with information packets);
- a middle column of *intermediate* packets consisting of the k information packets and $(n-k)$ check packets computed using the linear erasure-correcting code $C_n = (n,k,d)_2$;
- a right column of $N = n(1+\epsilon)$ coded packets consisting of LT coding of the intermediate packets using the distribution $\Omega_D(x)$.

A representation of the Raptor encoder/decoder is shown in Figure 2.4.

In analyzing Raptor codes it will be assumed ([30], proposition 3) the precode C_n can be encoded with complexity βk for some constant $\beta > 0$ and that there is an $\epsilon > 0$ such that C_n can be decoded over a BEC with erasure probability $1 - R(1+\epsilon)$ with high probability with cost/complexity γk. With such assumptions Raptor codes will be designed with constant encoding and decoding cost per symbol (hence both of overall complexity $O(k)$), and their space consumption is close to one and their overhead close to zero.

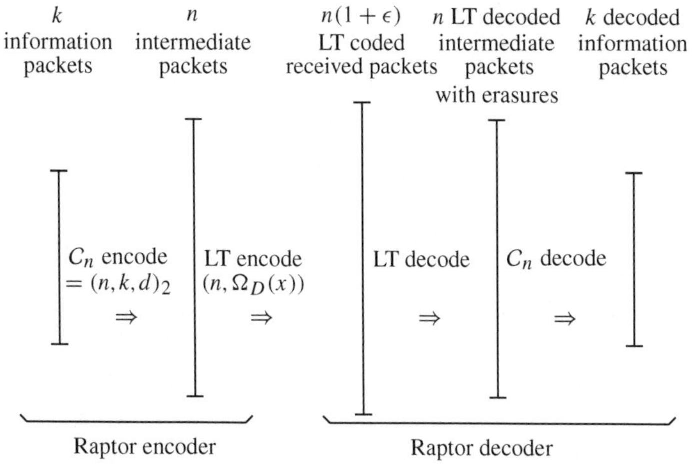

Figure 2.4 Basic structure of a Raptor code $(k, C_n, \Omega_D(x))$ system – encoder/decoder – graph edges not shown for clarity

We discuss the section VI of [30] on Raptor codes with good asymptotic performance. The arguments needed to achieve linear encoding and decoding require a delicate balancing of parameters.

The distribution used for the LT coding ([30], section VI) is a slightly modified ideal soliton:

$$\Omega_D(x) = \frac{1}{\mu+1}\left(\mu x + \sum_{i=2}^{D} \frac{x^i}{i(i-1)} + \frac{x^{D+1}}{D}\right) = \sum_{i=1}^{D} \Omega_i x^i \qquad (2.10)$$

where $\mu = (\epsilon/2) + (\epsilon/2)^2$ and $D = \lceil 4(1+\epsilon)/\epsilon \rceil$. That this is in fact a distribution is seen by observing

$$\sum_{i=2}^{D} \frac{1}{i(i-1)} = \sum_{i=2}^{D}\left(\frac{1}{i-1} - \frac{1}{i}\right) = 1 - \frac{1}{D}$$

and hence it is a distribution for any positive number μ and positive integer $D > 2$. It will be shown later that for constant D this distribution will yield coded nodes with constant average degree (as opposed to being a function of k) which allows the overall algorithm to have a linear complexity. With this distribution we have:

Lemma 2.15 ([30], lemma 4) *There exists a positive real number c (depending on ϵ) such that with error probability at most e^{-cn} any set of $(1+\epsilon/2)n+1$ LT coded symbols with parameters $(n, \Omega_D(x))$ are sufficient to recover at least $(1-\delta)n$ intermediate symbols via BP decoding where $\delta = (\epsilon/4)(1+\epsilon)$.*

The proof of this important lemma is straightforward although it requires some interesting manipulations. Some discussions are given to assist – the lemma itself is not proved.

The right-degree node (packet) distribution is by assumption $\Omega_D(x)$ and hence the right edge degree distribution is

$$\omega(x) = \frac{\Omega'_D(x)}{\Omega'_D(1)}.$$

Recall there are n input (intermediate – or left nodes for this argument) nodes to the LT coding process and $N = (1+\epsilon/2)n + 1$ coded (right) nodes for the decoding. To compute the left edge degree distribution $\iota(x)$, let α be the average degree of an output node (LT coded) node which is $\alpha = \Omega'_D(1)$. Let u be a left node and w an output (right) node. The probability that u is a neighbor of w, since on average w has α input neighbors, is α/n. Assuming each output node has this same probability of having u as a neighbor independently, on average the degree of u will have a binomial distribution, i.e., the probability u has degree ℓ is given by

$$\binom{N}{\ell}\left(\frac{\alpha}{n}\right)^{\ell}\left(1-\frac{\alpha}{n}\right)^{N-\ell}, \quad N = n(1+\epsilon/2)+1$$

and the generating function of the degree distribution of the left input nodes is thus

$$L(x) = \sum_{\ell=0}^{N}\binom{N}{\ell}\left(\frac{\alpha}{n}\right)^{\ell}\left(1-\frac{\alpha}{n}\right)^{N-\ell} x^{\ell} = \left(1-\frac{\alpha(1-x)}{n}\right)^{N}. \tag{2.11}$$

The edge distribution is $\iota(x)$ where $\iota(x) = L'(x)/L'(1)$ and

$$\iota(x) = \left(1-\frac{\alpha(1-x)}{n}\right)^{(1+\epsilon/2)n}.$$

For these edge distributions Equation 2.5 is equivalent to the condition that

$$\iota(1-\omega(1-x)) < 1-x \quad \text{for} \quad x \in [0, 1-\delta].$$

Since $\lim_{n\to\infty}(1-b/n)^n \longrightarrow \exp(-b)$ and that for $0 < b < m$, $(1-b/m)^m < \exp(-b)$ then

$$\begin{aligned}\iota(1-\omega(1-x)) &= \left(1-\frac{\alpha}{n}\frac{\Omega_D'(1-x)}{\Omega_D'(1)}\right)^{(1+\epsilon/2)n}\\ &< \exp\left(-\frac{\alpha}{n}\frac{\Omega_D'(1-x)}{\Omega_D'(1)}(1+\epsilon/2)n\right)\\ &< \exp(-(1+\epsilon/2)\Omega_D'(1-x)), \quad \text{since } \alpha = \Omega_D'(1).\end{aligned}$$

The above condition then reduces to showing that

$$\exp\Big(-(1+\epsilon/2)\Omega_D'(x)\Big) < 1-x, \quad x \in [0, 1-\delta]. \tag{2.12}$$

To establish this inequality using the form of $\Omega_D'(x)$ of Equation 2.10 is a technical development [10], which completes the proof of Lemma 2.15.

The analysis implies ([31], p. 79) that, asymptotically, the input ripple (the expected fraction of input symbols connected to LT coded symbols of degree one when a fraction of input symbols that have been recovered is x), is given by

$$1-x-\exp\Big(-(1+\epsilon)\Omega_D'(x)\Big). \tag{2.13}$$

The analysis of [9] further gives the fact that if x_0 is the smallest root of Equation 2.13 in $[0,1)$, then the expected fraction of input symbols not recovered at the termination of the decoding process is $1 - x_0$. Thus if distributions are designed so that x_0 is maximized then [30] the associated Raptor codes will have an average coded node degree of $O(\log(1/\epsilon))$ a linear decoding complexity of $O(k\log(1/\epsilon))$ and a decoding error probability which decreases inversely polynomial in k.

The conditions on the choice of erasure-correcting code C_n are simple:

(i) The rate of C_n is at least $(1+\epsilon/2)(1+\epsilon)$.
(ii) The BP decoder for C_n used on a BEC can decode to an erasure probability of $\delta = (\epsilon/4)/(1+\epsilon)$ (which is half the capacity) and has linear decoding complexity.

The details are omitted. The key theorem for Raptor codes is the following:

Theorem 2.16 ([30], theorem 5) *Let ϵ be a positive real number, k an integer, $D = \lceil 4(1+\epsilon)/\epsilon \rceil$, $R = (1+\epsilon/2)/(1+\epsilon)$, $n = k/R$ and let C_n be a block erasure code satisfying the conditions above. The Raptor code $(k, C_n, \Omega_D(x))$ has space consumption $1/R$, overhead ϵ and cost $O(\log(1/\epsilon))$ with respect to BP decoding of both the precode C_n and LT code.*

Lemma 2.15 shows the LT code with these parameters is able to recover at least a fraction of $(1-\delta)$ of the intermediate symbols and the erasure code is, by assumption and design, to correct this fraction of erasures to recover the k information symbols. It remains to show the cost is linear in k. The average degree of coded nodes is (from Equation 2.10)

$$\begin{aligned}\Omega'_D(1) &= \frac{1}{\mu+1}\left(\mu + \sum_{i=1}^{D-1}\frac{1}{i} + \frac{D+1}{D}\right)\\ &= 1 + H(D)/(1+\mu) = 1 + \ln\left(\frac{1}{\epsilon}\cdot 4(1+\epsilon)\right) + 1 = \ln\left(\frac{1}{\epsilon}\right) + O(\epsilon)\end{aligned}$$

where the standard upper bound on the truncated Harmonic series $H(D) \leq \ln(D)+1$ has been used. Thus the cost of encoding the LT code is $O(\ln(1/\epsilon))$ per coded symbol which is also the cost of LT decoding and also the cost of decoding the code C_n by assumption. Notice the overhead of the LT code is approximately $\epsilon/2$.

It is noted again that the linear encoding/decoding here is achieved by relaxing the condition that the LT have a high probability of complete decoding – it is sufficient to achieve almost complete decoding and complete the task with the linear code. This allows the average graph degree to be constant rather than linear in $\log(k)$.

It has been observed [11] that if the graph that remains of the LT decoding algorithm above when it stalls (all constraint node degrees greater than 1) is a good enough expander graph, then a hard decision decoding algorithm can proceed and with high probability complete the decoding process – thus all errors can be corrected. A brief informal explanation of this "expander-based completion" argument is given as it will also apply to certain LDPC arguments.

A bipartite graph has expansion (α, β) if for all variable node subsets S of size at most α, the set of neighbor constraint nodes is at least $\beta \mid S \mid$. For such codes/graphs the following two simple decoding algorithms are proposed [36]:

Algorithm 2.17 (Sequential decoding) If there is a variable node that has more unsatisfied constraint neighbors than satisfied, flip the value of the variable node. Continue until no such variable exists.

Algorithm 2.18 (Parallel decoding) In parallel, flip each variable node that is in more unsatisfied than satisfied constraints and repeat until no such nodes exist.

It can be shown ([11], lemma 2, [36], theorems 10 and 22) that if the graph remaining after LT decoding as discussed above is an (α, β) and is a good enough expander ($\beta > 3/4 + \epsilon$), then the above parallel and sequential decoding algorithms will correct up to αn errors in linear time, n the number of variable nodes. Comments on the probability the remaining graph will be such an expander are given in [11] – and hence this switching of decoding algorithms after the LT decoding will lead to complete decoding.

A brief interesting heuristic discussion of the analysis of the decoding performance of Raptor codes from ([30], section VII) is given. The intermediate and coded edge degree distributions have been noted as $\iota(x)$ and $\omega(x)$, respectively. Let p_{i+1} be the probability an edge in the decoding graph is of right degree one at iteration i. The analysis of Section 2.2 shows that

$$p_{i+1} = \omega(1 - \iota(1 - p_i)) \tag{2.14}$$

and it is desired to design the distribution $\Omega(x)$ such that this quantity decreases. Equation 2.11 giving the binomial moment generating function (mgf), for large n, can be approximated with the standard Poisson mgf

$$\iota(x) = \exp(\alpha(x - 1))$$

where α is the average degree of an input node. This function is also the input node degree distribution since it is easily verified that

$$\iota(x) = \frac{\iota'(x)}{\iota'(1)}.$$

This analysis required the tree involved to have no cycles and the variables involved being independent. Let u_i denote the probability that an input symbol is recovered at step i of the LT decoding algorithm. Then given an unresolved input node of degree d (note input node degrees only change when the node is resolved), then

$$u_i = 1 - (1 - p_i)^d$$

and averaging over the input degree distribution gives

$$u_i = 1 - \iota(1 - p_i) = 1 - \exp(-\alpha p_i).$$

Hence $p_i = -\ln(1 - u_i)/\alpha$.
From this relation and Equation 2.14 write

$$\begin{aligned} p_{i+1} &= \omega(1 - \exp(1 - (1 - p_i))) = \omega(1 - \exp(-\alpha p_i)) \\ &= \omega(1 - \exp(-\alpha\{-\ln(1 - u_i)\}/\alpha)) = \omega(1 - (1 - u_i)) \\ &= \omega(u_i) \end{aligned}$$

and so, since $\alpha\omega(x) = (1 + \epsilon)\Omega_D^{'}(x)$ it is argued [30, 31, 33] that, for $k(1 + \epsilon)$ coded (output) nodes the expected fraction of symbols in the input ripple is given, when the expected fraction of input nodes that have already been recovered is x, by

$$1 - x - \exp\Big(- (1 + \epsilon)\Omega_D^{'}(x)\Big).$$

It is further argued this fraction should be large enough to ensure continued operation of the decoding algorithm to completion and it is suggested as a heuristic that this fraction satisfy

$$1 - x - \exp\Big(- (1 + \epsilon)\Omega_D^{'}(x)\Big) \geq c\sqrt{\frac{1-x}{k}}, \quad x \in [0, 1 - \delta],\ \delta > c/\sqrt{k}.$$

Probability distributions that satisfy this criterion are found to be similar to the Soliton distribution for small values of d and give good performance.

This has been a heuristic overview of aspects of the performance of Raptor codes. The reader is referred to the literature for more substantial treatments on both the performance analysis and code design using such techniques as density evolution and linear programming, in [4, 5, 7, 8, 10, 11, 13, 14, 14, 18, 19, 21, 22, 24, 27, 29, 30, 31, 33, 34, 40].

The remainder of the section considers aspects of Raptor codes that prove important in their application, including the notion of inactivation decoding, the construction of systematic Raptor codes and their standardization.

Inactivation Decoding of LT Codes

It has been noted that BP decoding of LT codes is very efficient but is clearly not ML since it is possible to construct a situation where the decoding process will not complete due to the lack of degree one nodes, yet the decoding matrix formed from the information symbols and received coded symbols is of full rank – and hence ML, which is Gaussian elimination, would yield the unique solution. For a $k \times (k + m)$ matrix, ML would have complexity $O(k^3)$

which for the file and symbol sizes considered here, would be far too large. Inactivation decoding seeks to combine the two techniques in a manner that can significantly decrease the overhead and probability of failing to decode for a slight increase of decoding complexity. It takes advantage of the efficiency of the BP decoding and the optimality of ML decoding to yield an efficient version of an ML algorithm.

The following description of the inactivation decoding procedure is slightly unconventional but intuitive. Consider a first phase of BP decoding on the bipartite graph representing n coded symbols (received, known) and k information symbols whose values are sought. To start with, the bipartite graph of the set of k unknown information nodes (registers) is on the left and $n = k(1+\epsilon)$ received coded nodes (registers) is on the right. A coded node of degree one is sought and if found, its value is transferred to the attached information node and its value is then added (XORed) to all its right neighbors. The associated edges involved are all deleted. The process continues until no coded node of degree one is available. In this case suppose a coded symbol of degree two is available (in a well-designed LT code this will often be the case). Referring to Figure 2.5, suppose the register associated with the node contains the symbol r_{i_3}. If the information nodes that are neighbors are u and v, then create a variable x_1 (an inactivation variable) and assign a variable $r_{i_3} \oplus x_1$ to u and x_1 to v. The variable value $r_{i_3} \oplus x_1$ is then added symbolically to all coded neighbors of u and x_1 to all neighbors of v and the BP process is continued as though the value of the variables are known. The process is shown in Figure 2.5. The process continues and each time no node of degree one is available, a

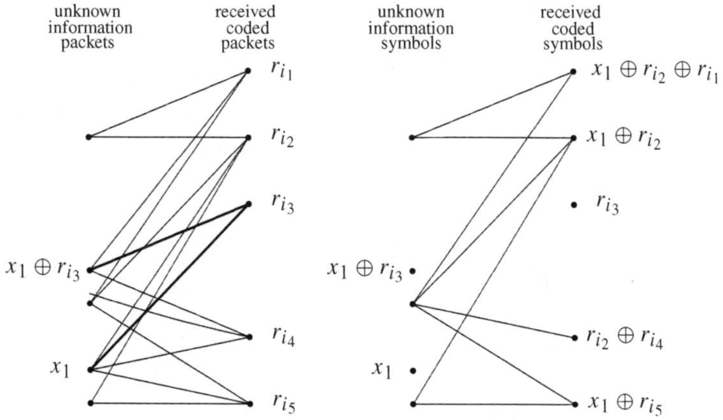

Figure 2.5 One stage of inactivation decoding

sufficient number of variables are introduced to allow the process to continue. At the end of the process each information node is "resolved" – although many may contain unknown variables. Typically the number of unknown variables is small. Suppose z inactivation variables have been introduced. Using the original received values of the right nodes and the values assigned to the information nodes, it is possible then to set up a linear relationship between the nodes on the left containing inactivation variables and the known received values on the right. These equations can then be solved, a process equivalent to solving a $z \times z$ matrix equation of complexity $O(z^3)$. If the original received nodes are equivalent to a full-rank system, the process is guaranteed to be successful and is ML. However, since the number of inactivation variables is typically quite small, this use of a combination of BP and ML is typically far more efficient than performing ML on the original received variables.

The above discussion gives an intuitive view of inactivation decoding. More efficient versions are found in the literature, including [1, 12, 20, 33] as well as the original patent description in [26].

Systematic Encoding

There are many situations where a systematic Raptor code is beneficial, i.e. one that with the usual Raptor coding produces the original information symbols among the coded output symbols. A natural technique might be to simply transmit all the information symbols first and then the usual Raptor coded symbols. This turns out not to be a good idea as it gives very poor error performance and requires large overheads to make it work at all. A convincing mathematical proof that this is so is given in [31].

A brief description of the more successful approach of [30] is noted. The terminology of that work is preserved except we retain our notation of information, intermediate and coded symbols/packets for consistency. As before a packet is viewed as a binary word of some fixed length and the only arithmetic operations employed will be the XOR of words or packets of the same length. The approach is to determine a preprocessing of the information packets in such a way as to ensure the original information packets are among the Raptor encoded packets. Thus the Raptor encoding method, and the impact of the distribution derived, is not disturbed ensuring that its promised efficiency is retained. The process is somewhat involved.

First consider the usual Raptor encoding. Let $\boldsymbol{x} = (\boldsymbol{x}_1, \boldsymbol{x}_2, \ldots, \boldsymbol{x}_k)$ denote the k information packets and consider the Raptor code $(k, C_n, \Omega_D(x))$. Let $\boldsymbol{G}$

be a binary (fixed) $n \times k$ generator matrix of the precode C_n assumed to be of full rank.

To initiate the process $k(1+\epsilon)$ LT encoded vectors are generated by sampling $\Omega_D(x)$ this number of times to combine packets (n-tuples chosen randomly from $\mathbb{F}_2{}^n$) to generate the $(k,\Omega_D(x))$ coded vectors $\boldsymbol{v}_1,\boldsymbol{v}_2,\ldots,\boldsymbol{v}_{k(1+\epsilon)}$. It will be assumed the n-tuples $\boldsymbol{v}_{i_1},\boldsymbol{v}_{i_2},\ldots,\boldsymbol{v}_{i_k}$ are linearly independent over $\mathbb{F}_2$. Let $\boldsymbol{A}$ be the $k \times n$ matrix with these independent n-tuples as rows and let $\boldsymbol{R}=\boldsymbol{AG}$, an invertible $k \times k$ matrix.

To encode the k information packets $\boldsymbol{x} = (\boldsymbol{x}_1,\boldsymbol{x}_2,\ldots,\boldsymbol{x}_k)$, a vector whose i-th component ([30], algorithm 11) is the information packet $\boldsymbol{x}_i$. Compute the vector $\boldsymbol{y}^t = (\boldsymbol{y}_1,\boldsymbol{y}_2,\ldots,\boldsymbol{y}_k)^t = \boldsymbol{R}^{-1}\boldsymbol{x}^t$ and encode these packets via the C_n code as $\boldsymbol{u} = (\boldsymbol{u}_1,\boldsymbol{u}_2,\ldots,\boldsymbol{u}_n)$ where $\boldsymbol{u}^t = \boldsymbol{G}\cdot\boldsymbol{y}^t$. From the packets $\boldsymbol{v}_j$ (vectors) introduced above compute the inner products

$$z_i = \boldsymbol{v}_i \cdot \boldsymbol{u}^t = (v_{i,1},v_{i,2},\ldots,v_{i,k}) \cdot \begin{bmatrix} \boldsymbol{u}_1 \\ \vdots \\ \boldsymbol{u}_k \end{bmatrix}, \quad i = 1,2,\ldots,k(1+\epsilon)$$

which are the output packets. Notice that from the generating process for the packets $\boldsymbol{v}_j$ this last operation is a $(k,\Omega_D(x))$ LT coding of the coded packets $\boldsymbol{u}_j$. Further packets $z_{k(1+\epsilon)+1}, z_{k(1+\epsilon)+2}\cdots$ are formed by $(k,\Omega_D(x))$ coding of the $\boldsymbol{u}_j$ packets. Thus these output packets are LT codings of the $\boldsymbol{y}_j$ packets.

The point of the process is to notice that the output packet $z_{i_j} = x_j, j = 1,2,\ldots,k$. To see this ([30], proposition 12) note that since $\boldsymbol{R} = \boldsymbol{A}\cdot\boldsymbol{G}$ and $\boldsymbol{R}\boldsymbol{y}^t = \boldsymbol{x}^t$. Then $\boldsymbol{A}\cdot\boldsymbol{G}\boldsymbol{y}^t = \boldsymbol{A}\cdot\boldsymbol{u}^t = \boldsymbol{x}^t$. By construction of the matrix $\boldsymbol{A}$, the j-th row is $\boldsymbol{v}_{i_j}$ and hence this last equation gives the inner product

$$\boldsymbol{v}_{i_j} \cdot \boldsymbol{x}^t = \boldsymbol{x}_j, \quad j = 1,2,\ldots,k.$$

Thus the Raptor code is systematic.

To decode the $z_j, j = 1,2,\ldots,k(1+\epsilon)$ one uses the usual decoding procedure for the $(k,C_n,\Omega_D(x))$ code to yield the packets $\boldsymbol{y}_1,\boldsymbol{y}_2,\ldots,\boldsymbol{y}_k$ and then compute $\boldsymbol{x}^t = \boldsymbol{R}\cdot\boldsymbol{y}^t$ for the original information packets.

It is important to note that the point of this process is to retain the optimality of the distribution of the LT process, yet have certain coded outputs as the input information packets. Of course the process requires the preprocessing of information packets regarding the matrix $\boldsymbol{R}$ and systematic row indices to be known to receivers. The process can be made efficient and is incorporated into many of the standards of Raptor codes. More details on the process are given in [30] which also contains a complexity analysis. Additional information is in the patents regarding these techniques [26, 32, 35].

Standardized Raptor Codes

Raptor codes have been adopted in a large number of mobile and broadcast and multicast standards. These are discussed extensively in the monograph [33]. Two versions of Raptor codes are available commercially, the Raptor version 10, or R10 code and the RaptorQ code. Both are systematic and use versions of inactivation decoding. The R10 code supports blocks of up to 8,192 source symbols per block and up to 65,536 coded blocks. It achieves a probability of failure of 10^{-6} with only a few blocks of overhead across the entire range of block sizes. It is described in detail in [25].

The RaptorQ code is designed for a much wider range of applications and achieves faster encoding and decoding with good failure probability curves. It supports blocks of up to 56,403 source symbols and up to 16,777,216 coded blocks. It uses the observation that the rank properties of random matrices are superior in the sense of achieving full rank for much lower overheads. The field of 256 elements, $\mathbb{F}_{256}$ is used for this code which is described in [17], although the reference [33] gives an excellent description of both of these codes.

Other Aspects of Fountain Codes

The term "online codes" has also been used [15] for the term "fountain codes." Additionally [3] it has been used to describe decoders for fountain codes that adapt to changing conditions of the channel and decoder and are thus able to achieve better performance. It requires some method to measure states and transmission back to the encoder, an important consideration in implementation.

The performance of similar techniques on other channels such as the BSC for error correction remains an interesting possibility.

Comments

The chapter has traced the progress in the remarkable development of erasure-correcting and lost packet codes/fountain codes with linear encoding and decoding complexity. The notion of choosing the parity checks according to a probability distribution in order to ensure, with high probability, a particular decoding algorithm completes, is novel and surprising. The success of both the theoretical and practical developments techniques is impressive.

References

[1] Blasco, F.L. 2017. Fountain codes under maximum likelihood decoding. *CoRR*, abs/1706.08739v1. arXiv:1706.08739v1,2017.

[2] Byers, J.W., Luby, M., Mitzenmacher, M., and Rege, A. 1998. A digital fountain approach to reliable distribution of bulk data. Pages 56–67 of: *Proceedings of the ACM SIGCOMM '98*. ACM, New York.

[3] Cassuto, Y., and Shokrollahi, A. 2015. Online fountain codes with low overhead. *IEEE Trans. Inform. Theory*, **61**(6), 3137–3149.

[4] Karp, R., Luby, M., and Shokrollahi, A. 2004 (June). Finite length analysis of LT codes. Page 39: *Proceedings of the IEEE International Symposium on Information Theory, June 2004*. ISIT.

[5] Kim, S., Lee, S., and Chung, S. 2008. An efficient algorithm for ML decoding of raptor codes over the binary erasure channel. *IEEE Commun. Letters*, **12**(8), 578–580.

[6] Luby, M.G. 2002. LT codes. Pages 271–280 of: *Proceedings of the 43rd Annual IEEE Symposium on Foundations of Computer Science.*

[7] Luby, M.G., Mitzenmacher, M., Shokrollahi, M.A., Speileman, D., and Stenman, V. 1997. Practical loss-resilient codes. Pages 150–159 of: *Proceedings of the Twenty-Ninth Annual ACM Symposium on the Theory of Computing*. ACM, New York.

[8] Luby, M.G., Mitzenmacher, M., Shokrollahi, M.A., and Spielman, D. 1998. Analysis of low density codes and improved designs using irregular graphs. Pages 249–258 of: *Proceedings of the Thirtieth Annual ACM Symposium on the Theory of Computing*. ACM, New York.

[9] Luby, M.G., Mitzenmacher, M., and Shokrollahi, M.A. 1998. Analysis of random processes via And-Or tree evaluation. Pages 364–373 of: *Proceedings of the Ninth Annual ACM-SIAM Symposium on Discrete Algorithms (San Francisco, CA, 1998)*. ACM, New York.

[10] Luby, M.G., Mitzenmacher, M., Shokrollahi, M.A., and Spielman, D.A. 2001. Efficient erasure correcting codes. *IEEE Trans. Inform. Theory*, **47**(2), 569–584.

[11] Luby, M.G., Mitzenmacher, M., Shokrollahi, M.A., and Spielman, D.A. 2001. Improved low-density parity-check codes using irregular graphs. *IEEE Trans. Inform. Theory*, **47**(2), 585–598.

[12] Lzaro, F., Liva, G., and Bauch, G. 2017. Inactivation decoding of LT and Raptor codes: analysis and code design. *IEEE Trans. Commun.*, **65**(10), 4114–4127.

[13] Maatouk, G., and Shokrollahi, A. 2009. Analysis of the second moment of the LT decoder. *CoRR*, abs/0902.3114.

[14] Maneva, E., and Shokrollahi, A. 2006. New model for rigorous analysis of LT-codes. Pages 2677–2679 of: *2006 IEEE International Symposium on Information Theory.*

[15] Maymounkov, P. 2002. Online codes. Technical report. New York University.

[16] Maymounkov, P., and Mazières, D. 2003. Rateless codes and big downloads. Pages 247–255 of: Kaashoek, M.F., and Stoica, I. (eds.), *Peer-to-peer systems II*. Springer, Berlin, Heidelberg.

[17] Minder, L., Shokrollahi, M.A., Watson, M., Luby, M., and Stockhammer, T. 2011 (August). *RaptorQ forward error correction scheme for object delivery*. RFC 6330.

[18] Oswald, P., and Shokrollahi, A. 2002. Capacity-achieving sequences for the erasure channel. *IEEE Trans. Inform. Theory*, **48**(12), 3017–3028.

[19] Pakzad, P., and Shokrollahi, A. 2006 (March). Design principles for Raptor codes. Pages 165–169 of: *2006 IEEE Information Theory Workshop – ITW '06 Punta del Este*.

[20] Paolini, E., Liva, G., Matuz, B., and Chiani, M. 2012. Maximum likelihood erasure decoding of LDPC codes: Pivoting algorithms and code design. *IEEE Trans. Commun.*, **60**(11), 3209–3220.

[21] Pfister, H.D., Sason, I., and Urbanke, R.L. 2005. Capacity-achieving ensembles for the binary erasure channel with bounded complexity. *IEEE Trans. Inform. Theory*, **51**(7), 2352–2379.

[22] Rensen, J.H.S., Popovski, P., and Ostergaard, J. 2012. Design and analysis of LT codes with decreasing ripple size. *IEEE Trans. Commun.*, **60**(11), 3191–3197.

[23] Richardson, T.J., and Urbanke, R.L. 2001. The capacity of low-density parity-check codes under message-passing decoding. *IEEE Trans. Inform. Theory*, **47**(2), 599–618.

[24] Richardson, T.J., Shokrollahi, M.A., and Urbanke, R.L. 2001. Design of capacity-approaching irregular low-density parity-check codes. *IEEE Trans. Inform. Theory*, **47**(2), 619–637.

[25] Shokrollahi, A., Stockhammer, T., Luby, M.G., and Watson, M. 2007 (October). *Raptor forward error correction scheme for object delivery*. RFC 5053.

[26] Shokrollahi, A.M., Lassen, S., and Karp, R. 2005 (February). Systems and processes for decoding chain reaction codes through inactivation. US Patent 6856263. www.freepatentsonline.com/6856263.html.

[27] Shokrollahi, M.A. 1999. New sequences of linear time erasure codes approaching the channel capacity. Pages 65–76 of: *Applied algebra, algebraic algorithms and error-correcting codes (Honolulu, HI, 1999)*. Lecture Notes in Computer Science, vol. 1719. Springer, Berlin.

[28] Shokrollahi, M.A. 2000. Codes and graphs. Pages 1–12 of: *In STACS 2000 (invited talk), LNCS No. 1770*.

[29] Shokrollahi, M.A. 2001. Capacity-achieving sequences. Pages 153–166 of: *Codes, systems, and graphical models (Minneapolis, MN, 1999)*. The IMA Volumes in Mathematics and Its Applications, vol. 123. Springer, New York.

[30] Shokrollahi, M.A. 2006. Raptor codes. *IEEE Trans. Inform. Theory*, **52**(6), 2551–2567.

[31] Shokrollahi, M.A. 2009. Theory and applications of Raptor codes. Pages 59–89 of: *MATHKNOW – Mathematics, applied sciences and real life*. Modeling, Simulation & Analysis (MS&A), vol. 3. Springer, Milan.

[32] Shokrollahi, M.A., and Luby, M. 2004 (April). Systematic encoding and decoding of chain reaction codes. US Patent WO 2004/034589 A2. www.freepatentsonline.com/y2005/0206537.html

[33] Shokrollahi, M.A., and Luby, M. 2009. Raptor codes. Pages 213 – 322 of: *Foundations and trends in communications and information theory*, vol. 6. NOW Publishers.

[34] Shokrollahi, M.A., and Storn, R. 2005. *Design of efficient erasure codes with differential evolution*. Springer, Berlin Heidelberg.
[35] Shokrollahi, M.A., Lassen, S., and Karp, R. 2005 (February). Systems and processes for decoding chain reaction codes through inactivation. US Patent 20050206537. www.freepatentsonline.com/y2005/0206537.html.
[36] Sipser, M., and Spielman, D.A. 1996. Expander codes. *IEEE Trans. Inform. Theory*, **42**(6, part 1), 1710–1722.
[37] Tanner, R.M. 1981. A recursive approach to low complexity codes. *IEEE Trans. Inform. Theory*, **27**(5), 533–547.
[38] Tanner, R.M. 1984. Explicit concentrators from generalized N-gons. *SIAM J. Algebraic Discrete Methods*, **5**(3), 287–293.
[39] Tanner, R.M. 2001. Minimum-distance bounds by graph analysis. *IEEE Trans. Inform. Theory*, **47**(2), 808–821.
[40] Xu, L., and Chen, H. 2018. New constant-dimension subspace codes from maximum rank distance codes. *IEEE Trans. Inform. Theory*, **64**(9), 6315–6319.

3
Low-Density Parity-Check Codes

The inspiration for a large amount of work of the past two decades on low-density parity-check (LDPC) codes and for their enormous importance in application is the thesis of Gallager [8]. Although largely ignored for three decades after its appearance, it remains a major focus of coding activity and the genesis of a large number of new directions in coding theory. It has been shown that such codes can achieve very low error probabilities and low error floors making them suitable for a wide variety of applications. Many of the ideas explored in this chapter have overlap with Chapter 2.

The term *low density* stems from the work of Gallager whose interest was in finding algorithms that approximated maximum-likelihood (ML) decoding for a code defined by a parity-check matrix. Such algorithms often have complexity that is exponential in codeword length and often too computationally expensive to implement in practice. His approach used message-passing techniques and while the work does not mention graphs, considering the equivalence (essentially) of a binary parity-check matrix and a bipartite Tanner graph (see Chapter 1), it has become common to interpret his work as message passing on edges of graphs. Note that message-passing algorithms include the subset of *belief propagation* algorithms where the messages being passed on edges represent a belief as to the value of the associated variable nodes. The computational complexity of such algorithms is often a function of the number of edges in the graph (also the number of 1s in the parity-check matrix). It can be shown that in some circumstances the performance of these algorithms can approach ML. One interpretation of the terms *sparse* or *low density* is a code with parity-check matrix with a number of ones that is linear in code length – hence a fairly sparse matrix. As noted it might also be taken to mean coding graphs with a number of edges that render the decoding algorithm feasible.

Note also that the two methods noted for constructing LDPC codes, by constructing a parity-check matrix at random and constructing a graph by the method noted in the previous chapter using random permutations are not quite identical in that the latter method could result in multiple edges between two nodes while with the former method this could not happen. In either case one imposes a uniform probability distribution on the ensembles. For large code lengths the two techniques are "essentially" equivalent. While Gallager [8] used the parity-check matrix method, Richardson and Urbanke [20] and Luby et al. [17] make a strong case that the graph ensembles are more convenient to use for analysis and this method is used here.

This chapter is at odds with the stated purpose of this volume in that important work on these codes is available in numerous published books (see [11, 12, 13, 19, 22]) as well as at least two excellent survey articles [10, 25]. One purpose of including such a chapter is for the relationship the material has with the graph algorithms of Chapter 2 and other chapters. This chapter largely ignores many of the important design and performance issues for LDPC codes, covered in the books mentioned. This includes omitting a discussion of important structural features such as trapping sets, absorbing sets and small cycle distribution that to a large extent determine the performance of them. The February 2001 issue of the *IEEE Transactions on Information Theory* (volume 47, no. 2), a special issue on Codes on Graphs and Iterative Algorithms, contains numerous seminal papers that have shaped the subject. In particular the three papers [17, 20, 21] in that volume are of fundamental importance and this chapter attempts to convey the sense of some of the material in these three papers with less detail. The papers are, however, essential reading for any researcher in the area.

The graph-theoretical background needed for this work is modest, consisting mainly of the essential (especially for large code lengths) equivalence between a binary matrix and a bipartite graph (Tanner graph) noted in Chapters 1 and 2 and the construction and a few of the properties of the irregular graphs with $(\lambda(x), \rho(x))$ edge distributions, which as noted previously will also be referred to as in the class $C^n(\lambda, \rho)$, the ensemble of graphs with n variable nodes and edge distributions $\lambda(x)$ and $\rho(x)$.

The codes are described either by a binary parity-check matrix $\boldsymbol{H}$ where n columns represent codeword positions and $(n-k)$ rows parity checks or, equivalently, by a bipartite (Tanner) graph with n variable or left nodes and $(n-k)$ check, constraint or right nodes.

This chapter will restrict its attention to binary-input symmetric channels and deal with the performance of belief propagation (BP) algorithms and the analysis of LDPC codes on such channels. While the analysis and techniques developed can be applied to general binary-input symmetric channels, the channels of particular interest are the binary symmetric channel (BSC) and the

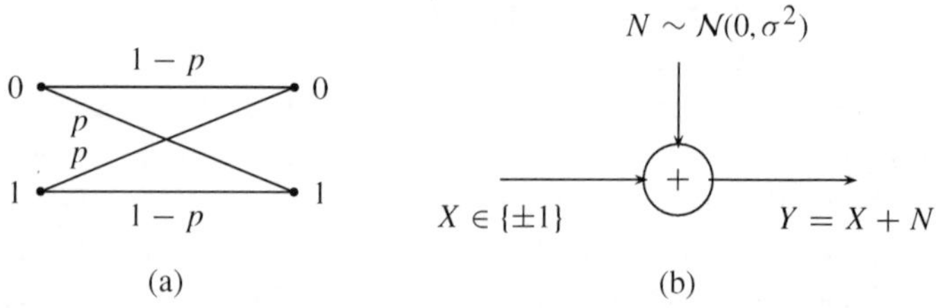

Figure 3.1 Binary-input channels: (a) BSC and (b) BIAWGN

binary-input additive white Gaussian noise (BIAWGN) channel whose figures are repeated here for convenience in Figure 3.1. The binary erasure channel (BEC), of interest in Chapter 2, will be mentioned occasionally. The BEC tends to be a simple case where many formulae can be stated exactly rather than approximated. The same situation occurs with polar codes whose properties on the BEC can be analyzed more successfully and explicitly than on other channels.

The capacity formulae for these channels were noted in Chapter 1. For the BSC with crossover probability p it is

$$C_{BSC} = 1 - H_2(p), \quad H_2(x) = -x\log_2(x) - (1-x)\log_2(1-x), \tag{3.1}$$

$H_2(p)$ the binary entropy function. For the BIAWGN channel the joint distribution of (X,Y), input and output, is a mixture of discrete and continuous and with $P(X=+1) = P(X=-1) = 1/2$ (which achieves capacity on this channel) and pdf $p(x) \sim \mathcal{N}(0,\sigma^2)$. The pdf $p(y)$ of the output is

$$\begin{aligned} p(y) &= \frac{1}{2}\cdot\frac{1}{\sqrt{2\pi\sigma^2}}e^{-\frac{(y+1)^2}{2\sigma^2}} + \frac{1}{2}\cdot\frac{1}{\sqrt{2\pi\sigma^2}}e^{-\frac{(y-1)^2}{2\sigma^2}} \\ &= \frac{1}{\sqrt{8\pi\sigma^2}}\left(\exp\left(-\frac{(y+1)^2}{2\sigma^2}\right) + \exp\left(-\frac{(y-1)^2}{2\sigma^2}\right)\right) \end{aligned}$$

and the expression for capacity reduces to

$$C_{BIAWGN} = -\int_y p(y)\log_2 p(y)dy - \frac{1}{2}\log_2(2\pi e\sigma^2). \tag{3.2}$$

The general shape of these capacity functions was shown in Figure 1.3.

Much of the analysis in the literature is for the case of (c,d)-biregular bipartite codes where the Tanner graph has variable nodes of degree c and the check nodes of degree d. It will be convenient to refer to these degree parameters as (d_v, d_c). The notion of irregular graphs considered in Chapter 2 is used here as well for both the BSC and BIAWGN case. Recall that such irregular graphs are assumed to have an edge distribution pair $(\lambda(x), \rho(x))$ where

$$\lambda(x) = \sum_i \lambda_i x^{i-1}, \quad \rho(x) = \sum_i \rho_i x^{i-1}.$$

The construction of such graphs has been previously described. Note that the (d_v, d_c)-biregular case corresponds to $\lambda(x) = x^{d_v-1}$ and $\rho(x) = x^{d_c-1}$. The equivalent distributions for node degrees are easily derived from these edge distributions (see Chapter 2). As shown there such irregular distributions play a crucial role in determining effective erasure-correcting codes. Indeed Chapter 2 shows that such irregular edge distributions are able to achieve capacity on the BEC. The results for the binary-input channels of interest in this chapter, the BSC and BIAWGN, are far less complete and more complex and challenging. Nonetheless the progress for these channels in the past few decades has been impressive.

It will be of value for this chapter to briefly review a result of Section 2.2 for the BEC. Suppose in a Tanner graph of a code with edge distributions $(\lambda(x), \rho(x))$, channel symbols are erased with probability δ and let p_ℓ be the probability a variable node is unresolved at the ℓ-th iteration. It was shown in Chapter 2 how a recursion can be developed for the $(\ell + 1)$-st iteration of the decoding Algorithm 2.2. A simple argument yields that

$$p_{\ell+1} = \delta\lambda(1 - \rho(1 - p_\ell))$$

and a condition for successful decoding is that $p_{\ell+1} < p_\ell$, i.e., with each iteration the probability a variable node remains unresolved decreases. This translates into the condition that for a given BEC with erasure probability δ, the edge distributions should be chosen so that (Equation 2.3 of Chapter 2)

$$\delta\lambda(1 - \rho(x)) < x, \; 0 < x < \delta. \tag{3.3}$$

If the code graph is chosen with such edge distributions, then with high probability the BP decoding algorithm will decode successfully with up to a fraction of symbols δ being erased. Furthermore the rate of the code, determined by the distributions $\lambda(x)$ and $\rho(x)$, should approach the capacity of the channel $1 - \delta$. That such distributions can be found, and hence capacity achieved on the BEC is a major result of Chapter 2.

The notion of graphs with irregular edge distributions $(\lambda(x), \rho(x))$ seems to have first appeared in [14] for the BEC. These notions will also be of importance to codes on more general channels as considered in this chapter.

It is assumed all graphs will have no self-loops or multiple edges. All channels considered here will have binary inputs, be memoryless and will be parameterized by a single parameter, say γ which often allows the notion of a threshold to be formulated as discussed below.

The notion of *thresholds* already present in the work of Gallager [8] is of importance. Consider the ensemble of codes $C^n(\lambda, \rho)$ for fixed distributions and channels that can be characterized with a single parameter, say γ. For the BEC, BSC and BIAWGN these would be δ (erasure probability on the BEC), p (crossover probability on the BSC) and σ (noise standard deviation on the BIAWGN). The parameter γ is generic for this purpose. Then Richardson [21] shows there exists a *threshold* γ^* which is a maximum channel parameter such that for any $\epsilon > 0$ and $\gamma < \gamma^*$ there exists a code length $n(\epsilon, \gamma)$ and a number $\ell(\epsilon, \gamma)$ such that almost every code in the ensemble $C^n(\lambda, \rho)$ with $n > n(\epsilon, \gamma)$ has a bit error smaller than ϵ assuming that the transmission is over a channel with parameter γ and the decoding algorithm performs $\ell(\epsilon, \gamma)$ iterations. Conversely, if the channel parameter $\gamma > \gamma^*$, then for almost all codes in the $C^n(\lambda, \rho)$ ensemble for a fixed number of decoding algorithm iterations, the probability of bit error will exceed some constant η which depends on channel parameter γ but not on the number of iterations.

Finally it is noted that although the case of the BEC was not treated in the work of Gallager [8], the other notions of importance here, namely message passing on a graph and the analysis involving iterations and convergence were clearly present in that work, although augmented since by the work of many others. As in Chapter 2 the input and output alphabets are binary $\{0, 1\}$ for the hard decision algorithms of the next section on the BSC. For Section 3.2 on the BIAWGN channel, the input alphabet will be $\{\pm 1\}$ and the output and message-passing alphabets will be the real numbers. Most algorithms considered here will be BP algorithms,

As noted, the results for channels such as the BSC and BIAWGN channels treated here are not as complete as those for the BEC and the algorithms involved are more complex than for the BEC. The BSC is considered in the next section where Gallager algorithms A and B are given and then recast into message-passing algorithms in preparation for the application of similar notions to the BIAWGN channel. The case of the BIAWGN is then treated in Section 3.2 where the work of [20] is considered. The final section considers the notions related to the performance of the algorithms of Section 3.2 such as an *extrinsic information transfer* (EXIT) chart as an aid in tracing the progress of the decoding algorithm as the number of iterations increases and the notions of concentration of densities and thresholds.

3.1 Gallager Decoding Algorithms A and B for the BSC

Gallager [8], in his influential thesis, considered the performance of LDPC codes on several classes of binary-input channels. He gave two algorithms for decoding on the BSC that have come to be known as Gallager algorithms A and B. The analysis of these two algorithms for decoding on the BSC follows. Although, as noted, Gallager did not use graph terminology, they are the first true message-passing algorithms. The algorithms differ in the manner in which they formulate the messages to be passed. These algorithms were originally determined for the case of (d_v, d_c)-biregular Tanner graphs (in Gallager's work, matrices with constant row and column sums) and were generalized in [17] to the case of ensembles of irregular $(\lambda(x), \rho(x))$ graphs, $C^n(\lambda, \rho)$, which are also considered here. The section concludes by recasting the original description of these algorithms as message-passing algorithms, as described in [20], in preparation for the discussion of these ideas on the BIAWGN in Section 3.2.

Consider the Tanner graph of an $(n,k)_2$ code with n variable nodes of degree d_v and $m = (n - k)$ check nodes of degree d_c. Thus the edge distributions for such biregular graphs are $\lambda(x) = x^{d_v - 1}$ for variable or information nodes and $\rho(x) = x^{d_c - 1}$ for check or constraint nodes.

In this section, all messages on the graph edges are binary from the alphabet $\{0, 1\}$ and the channel a BSC.

Gallager Decoding Algorithm A

The algorithm is described as follows.

Algorithm 3.1 In the first round (round 0) the binary received word $r = (r_1, r_2, \ldots, r_n), r_i \in \{0, 1\}$, is associated with the n variable nodes and these values are retained by the variable nodes for the entire algorithm. It will be convenient to label the received bit for the variable node v as $r_v, v = 1, 2, \ldots, n$. The algorithm proceeds in rounds or iterations. One iteration will consist of messages from variable nodes to check nodes (first phase) and from the check nodes back to variable nodes (second phase).

Assuming an edge from variable node v to check node c, at round 0, node v transmits its received value r_v to its neighboring check nodes. In this and each succeeding iteration, each check node c sends to an attached variable node v the parity (mod 2 sum) of the messages it received from all its variable neighbors **except node v immediately previously**. Thus c sends to v the value it believes node v should have. In each succeeding iteration, variable node v sends to neighboring check node c the following: if all the messages in the previous iteration to node v, **except that from** c, were b, then v sends b to c; otherwise, it sends its original received value r_v.

Note again that all messages are either 0 or 1. Assuming there is an edge between nodes v and c, denote the messages sent in the ℓ-th iteration as $m_{vc}^{(\ell)}$ and $m_{cv}^{(\ell)}$, respectively. The maximum number of iterations that can be sustained and still have the variables involved be statistically independent is $g/4$ where g is the girth of the bipartite graph. In essence this means that for a fixed node v as one traces edges out from v out to a depth of less than $g/2$ edges the evolved graph would be a tree and hence all the variable nodes visited are distinct and their associated values are statistically independent, a requirement for the analysis, at times referred to as the *cycle-free* case. Denote the set of neighbors of a variable node v (respectively check node c) by N_v (respectively N_c). Further define $N_{v,c}$ (respectively $N_{c,v}$) as the set of neighbors of v (respectively c) except c (respectively v). This implies that messages transmitted use only *extrinsic* information from previous rounds, i.e., not messages received from that vertex in the previous round, in order to preserve independence of the information. Independence is of course crucial for the validity of the analysis.

Note that the variable N denotes white Gaussian noise leading to overuse of this variable. The distinction of this usage from neighbor sets will be by subscripts and should cause no confusion.

For each edge (v,c) of the graph the message passed is denoted $m_{vc}^{(\ell)}$, the message passed from variable node v to check node c at iteration ℓ, as

$$m_{vc}^{(\ell)} = \begin{cases} r_v \ \text{if } \ell = 0 \\ b \ \text{if } m_{c'v}^{(\ell-1)} = b \quad \forall c' \in N_{v,c}, \ \ell \geq 1 \\ r_v \ \text{otherwise} \end{cases} \tag{3.4}$$

and similarly for $m_{cv}^{(\ell)}$, the message passed from check node c to variable node v at iteration ℓ:

$$m_{cv}^{(\ell)} = \bigoplus_{v' \in N_{c,v}} m_{v',c}^{(\ell)} \quad \ell \geq 0. \tag{3.5}$$

The algorithm is both a *message-passing algorithm* and a *belief propagation* (BP) algorithm.

The last equation is the parity of the bits seen by the check node, exclusive of the bit received from the variable node being sent to. Thus the check node would be satisfied if the variable node receiving the message had the same value. If the input alphabet $\{\pm 1\}$ is used, as for the BIAWGN, the message sent from the check node could be expressed as

$$m_{cv}^{(\ell)} = \prod_{v' \in N_{c,v}} m_{v',c}^{(\ell)} \quad \ell \geq 0.$$

A further comment on this will be given later. Until then assume all messages transmitted are binary $\{0,1\}$ as for the BSC.

The analysis of the algorithm simply assesses the various probabilities of the transmitted messages being correct. For the analysis of the algorithm it is useful to verify that if a random variable has a binomial distribution with parameters k, p, then

$$\sum_{j \text{ even}} \binom{k}{j} p^j (1-p)^{k-j} = \frac{1+(1-2p)^k}{2},$$

a result achieved by considering the sums

$$(a+b)^k = \sum_{j=0}^{k} \binom{k}{j} a^j b^{k-j} \quad \text{and} \quad (-a+b)^k = \sum_{j=0}^{k} \binom{k}{j} (-a)^j b^{k-j}$$

and taking $a = -p$ and $b = 1-p$. Summing these terms with the values indicated, the odd terms cancel and hence

$$\sum_{j \text{ even}} \binom{k}{j} p^j (1-p)^{k-j} = \frac{1+(1-2p)^k}{2}$$

and similarly

$$\sum_{j \text{ odd}} \binom{k}{j} p^j (1-p)^{k-j} = \frac{1-(1-2p)^k}{2}.$$

Assuming an edge (v,c) in the graph, let $p_{\ell+1}$ be the probability that v sends an incorrect value to c in the $(\ell+1)$-st iteration. The probability that a variable node was originally received correctly is $1-p_0$, $p_0 = p$, the crossover probability on the BSC. Conditioned on this event, the probability the node passes on an error is the probability that all neighbor vertices of v, except c, will indicate an error which is the probability that each neighbor c' in $N_{v,c}$, passes an error to v (i.e., each $c' \in N_{v,c}$ receives and odd number of errors), which is

$$\frac{1-(1-2p_\ell)^{d_c-1}}{2} \tag{3.6}$$

and hence the probability r_v was received correctly but sent incorrectly at the $(\ell+1)$-st iteration is

$$(1-p_0)\left(\frac{1-(1-2p_\ell)^{d_c-1}}{2}\right)^{d_v-1}.$$

In this event the node v will pass an error to its $N_{v,c}$ neighbors.

Similarly suppose node v was originally received in error with probability p_0. The probability that node v receives no error indication from each of its $d_v - 1$ neighbors (i.e., its neighbors except c) is

$$\left(\frac{1+(1-2p_\ell)^{d_c-1}}{2}\right)^{d_v-1}.$$

The probability that at least one neighbor of v (other than c) passes an error indication to v is 1 minus this quantity. Putting the two statements together gives

$$p_{\ell+1} = p_0\left(1-\left(\frac{1+(1-2p_\ell)^{d_c-1}}{2}\right)^{d_v-1}\right)+(1-p_0)\left(\frac{1-(1-2p_\ell)^{d_c-1}}{2}\right)^{d_v-1}, \tag{3.7}$$

which is the sum of the probabilities the received symbol r_v was received in error and passed on as correct at the $(\ell+1)$-st iteration plus the probability r_v was received correctly and passed on as incorrect at that iteration.

The algorithm continues until values at all variable nodes satisfy the check equations or until some preset number of iterations at which point failure may be declared or the probabilities may be consulted and decisions made.

The idea of Gallager [8] is then to find a threshold for the BSC crossover probability p^*, such that for crossover probabilities less than this, the p_ℓ decreases monotonically. Recall the analysis is only accurate if the independence assumption holds, i.e., if the graph has a sufficiently large girth or if the probabilities converge sufficiently fast. Typically, randomly chosen graphs do not have large girth so constructions of good graphs are required.

Some exact thresholds for Gallager's algorithm A are discussed in ([3], table 1). For example, for the standard $d_v = 3, d_c = 6$ rate $1/2$ code, the threshold using the Gallager A decoding algorithm is shown to be $\tau(3,6) = (1-\sqrt{\sigma})/2 \sim 0.039$) where σ is the (single) positive root of the equation $r(x) = -1 - x - x^2 + x^3 + x^4$. Thus for a code used on a BSC with crossover probability below this threshold, decoded with Gallager A, asymptotically with length, will decode successfully with high probability and the error probability will be above a constant if the crossover probability is above this threshold. Further developments on this problem are given for code ensembles (and Gallager A decoding) for code ensembles $C^n(\lambda,\rho)$ discussed later in the section.

Gallager Decoding Algorithm B

Gallager [8] suggests a modification to the above decoding algorithm to improve performance. For the ℓ-th iteration a threshold b_ℓ is proposed, depending on the iteration number, such that if at least b_ℓ neighbors of v (except c) sent the same bit in the previous round ℓ, then v sends this bit to c in the $(\ell + 1)$-st round. Otherwise it will send the original received bit r_v. The thought here is that rather than insisting all extrinsic information is the same, as in Algorithm A above, if a large fraction of them are the same it is likely to be the correct value and that performance of the algorithm is likely to improve if this is taken into account. Note the threshold b_ℓ is a function of the iteration number.

The above formulae are modified in a straightforward manner to:

$$\begin{aligned} p_{\ell+1} = p_0 - p_0 \sum_{j=b_\ell}^{d_v-1} \binom{d_v-1}{j} \left(\frac{1+(1-2p_\ell)^{d_c-1}}{2} \right)^j \cdot \left(\frac{1-(1-2p_\ell)^{d_c-1}}{2} \right)^{d_v-1-j} \\ + (1-p_0) \sum_{j=b_\ell}^{d_v-1} \binom{d_v-1}{j} \left(\frac{1-(1-2p_\ell)^{d_c-1}}{2} \right)^j \cdot \left(\frac{1+(1-2p_\ell)^{d_c-1-j}}{2} \right)^{d_v-1-j}. \end{aligned} \tag{3.8}$$

It is desired to choose the threshold b_ℓ to minimize $p_{\ell+1}$. This can be achieved as follows. Parameterize the probabilities with the threshold b_ℓ, the threshold at the ℓ-th iteration, as $p_{\ell+1,b_\ell}$ and consider the smallest value of b_ℓ such that

$$p_{\ell+1,b_i} - p_{\ell+1,b_{i-1}} < 0,$$

i.e., at the $(\ell + 1)$-st iteration choose the smallest threshold that ensures the probabilities get smaller. This is equivalent to ([8], equation 4.16) choosing b_ℓ as the smallest integer that satisfies the equation

$$\frac{1-p_0}{p_0} \leq \left[\frac{1+(1-2p_\ell)^{d_c-1}}{1-(1-2p_\ell)^{d_c-1}} \right]^{(2b_\ell - d_v + 1)}.$$

Note that b_ℓ decreases as p_ℓ decreases. Typically, as before, the algorithm runs until the variable values satisfy all the check equations or until a fixed number of iterations is reached. The following quantifies the decoding performance somewhat:

Theorem 3.2 ([17], Theorem 1) *Let $\ell > 0$ be an integer constant and let Z_ℓ be the random variable describing the fraction of edges containing incorrect messages after round ℓ of the above algorithm. Further, let p_ℓ be as given in Equation 3.8. Then there is a constant η (which depends on the maximum of (d_v, d_c)) such that for any $\epsilon > 0$ and sufficiently large n we have*

$$Pr(| Z_\ell - p_\ell |> \epsilon) < \exp(-c\epsilon^2 n).$$

Recall that n is the number of bits in a codeword (the number of variable nodes in the graph). Thus the probabilities of the above formulae accurately track the number of errors at each stage of the algorithm. The following corollary further supports this view:

Corollary 3.3 ([17], corollary 1) *Given a random (d_v, d_c)-biregular code with p_ℓ defined as above, if the sequence p_ℓ converges to 0, then for any $\eta > 0$ there is a sufficiently large message size such that Gallager's hard decision decoding algorithm decodes all but at most ηn bits in some constant number ℓ_η rounds with high probability.*

Thus the Gallager algorithm is capable of correcting almost all errors, but typically not all. The situation is reminiscent of that of LT codes that in moving to Raptor codes employed a simple erasure-correcting code to correct the residual errors. Here, in order to correct *all* the errors it is suggested that an expander-based graph argument be used to complete the decoding, as noted in an earlier chapter. See [17] for further details.

The above discussion is for the biregular graph/code case. The extension to irregular graphs is contained in [15, 17]. The results are simply stated below. Suppose the code graph is a member of the ensemble of graphs $C^n(\lambda, \rho)$. As before let $p_{\ell+1}$ denote the probability a variable node sends an incorrect message at the $(\ell+1)$-st iteration. The probability a check node in the irregular code receives an even number of error messages, using an argument identical to that leading to Equation 3.6, conditioned on the check degree being d_c, then is

$$\frac{1 + (1 - 2p_\ell)^{d_c - 1}}{2}.$$

If the degree, however, is chosen according to the distribution $\rho(x) = \sum_i \rho_i x^{i-1}$, then with probability ρ_i the right edge degree will be i–1. Averaged over all possibilities, the probability the check node receives an even number of errors is

$$\frac{1 + \rho(1 - 2p_\ell)}{2}.$$

It is then straightforward to generalize Equation 3.8 to the recursion going from round i to $i + 1$, here d_ℓ is the number of terms in $\rho(x)$:

$$\begin{aligned} p_{i+1} &= f(p_i) \quad \text{where} \\ f(x) &= p_0 - \sum_{j=1}^{d_\ell} \lambda_j \left[p_0 \sum_{t=b_{i,j}}^{j-1} \binom{j-1}{t} \left[\frac{1+\rho(1-2x)}{2}\right]^t \cdot \left[\frac{1-\rho(1-2x)}{2}\right]^{j-1-t} \right. \\ &\quad \left. +(1-p_0) \sum_{t=b_{i,j}}^{j-1} \binom{j-1}{t} \left[\frac{1-\rho(1-2x)}{2}\right]^t \cdot \left[\frac{1+\rho(1-2x)}{2}\right]^{j-1-t} \right]. \end{aligned} \tag{3.9}$$

The threshold [17] $b_{i,j}$ is chosen as the smallest integer that satisfies the equation

$$\frac{1-p_0}{p_0} \le \left[\frac{1+\rho(1-2p_i)}{1-\rho(1-2p_i)}\right]^{2b_{i,j}-j+1}.$$

Notice that the exponent in the above equation can be expressed as

$$2b_{i,j} - j + 1 = b_{i,j} - (j - 1 - b_{i,j}).$$

There remains the problem of choosing the distributions $\lambda(x)$ and $\rho(x)$ in order to yield the largest possible value of p_0 (BSC crossover probability) such that the sequence $\{p_\ell\}$ decreases to 0. The following procedure does not find optimum sequences but does give sequences that perform better than biregular ones. For a fixed p_0 and a given sequence $\rho_1, \rho_2, \ldots, \rho_{d_c}$. Let

$$b_{i,j} = \left\lceil \left(j - 1 + \frac{\log(1-p_0)/p_0}{\log(1+\rho(1-2x))/(1-\rho(1-2x))}\right) \Big/ 2 \right\rceil.$$

The criterion then is, for a given p_0 and right sequence $\rho(x)$, to find the left degree sequence $\lambda(x)$ such that $f(p_\ell) = p_{\ell+1}$ and $f(x) < x$ on the open interval $(0, p_0)$ for $f(x)$ given by Equation 3.9. For details see [17].

As a final comment, it is noted that the above analysis does not guarantee the decoding process always completes successfully. It can be shown, using graph expansion arguments and modified decoding, that the type of random graph constructed here will typically be a good expander and that with these notions the decoding can be completed with high probability. In the context of bipartite graphs the notion of graph expansion will hold if (informally), for a given subset of variable nodes, the number of its check node neighbors is large. Such notions can be used to show that if the variable subset is identified with undecoded bits, the existence of a sufficiently large set of neighbors can be used to complete the decoding (by switching to a different decoding algorithm).

3.2 Performance of LDPC Codes on the BIAWGN Channel

The performance of belief propagation decoding for a BIAWGN is considered. Most of the analysis here applies to other binary input symmetric channels but our interest will be in the BIAWGN.

The capacity of the BIAWGN channel is given by Equation 3.2 and the general shape of its capacity curve is shown in Figure 1.3(b). A binary code will be assumed and the inputs will be from the alphabet $\{\pm 1\}$ for transmission on the AWGN channel. As noted previously, these are often derived from a binary $\{0,1\}$ alphabet via the transformation $0 \mapsto 1,\ 1 \mapsto -1$, i.e., via $(-1)^c, c \in \{0,1\}$. The messages on the Tanner graph of the code will be log likelihood ratios, hence real numbers representing beliefs (probabilities) as to the values of the various variables.

The decoding algorithm will consist of passing messages back and forth on the Tanner graph of the code with the thought that as the number of iterations progresses the log likelihoods polarize to either minus infinity or infinity indicating a bit decision. This section will give an overview of the construction of the messages, and the performance of the algorithm. The details are contained in [20] and [21], and are required reading for researchers in the area. The aim here is to give background and discussion to complement reading of the original papers.

For notation the *cumulative distribution function* (cdf) of a random variable X is denoted by capital $F_X(x)$ and its corresponding *probability density function* (pdf) by $f_X(x)$ although the subscripts will be dropped when the random variable is understood. In the case of smooth continuous functions they are related by differentiation. From the previous section recall the definitions of N_v, $N_{v,c}$, N_c and $N_{c,v}$ as sets of neighbor vertices.

The decoding algorithm will proceed in iterations of passing messages on the graph edges, one iteration will consist of forming messages at each of the variable nodes, passing the messages from variable nodes to check nodes (first phase of an iteration), forming new messages at the check nodes and then passing the messages back from check nodes to the variable nodes (second phase). The method of forming the messages for each type of node will be described. Since they will be log likelihoods they will represent a belief as to the values associated with the variable nodes, given past messages, and the algorithm is thus a belief propagation (BP) algorithm as well as a message-passing algorithm.

To determine the performance of the algorithm the probability densities of the messages passed on the edges have to be determined and this is the technical part of the development. Finally some properties of the pdfs that aid in determining decoding error probabilities are commented on.

The messages passed on the graph at the ℓ-th iteration from a variable node v to a check node c, assuming (v,c) is an edge in the graph will be designated $m_{vc}^{(\ell)}$ and those from c to v $m_{cv}^{(\ell)}$.

Various notions of symmetry are important to the discussion, although perhaps not entirely intuitive, assuming inputs $\{\pm 1\}$:

Definition 3.4 ([20, 21])

(i) *Channel symmetry:* The channel is *output symmetric* if

$$p(y = q \mid x = +1) = p(y = -q \mid x = -1)$$

where $p(\cdot \mid \cdot)$ is the pdf of the (continuous) channel output, given the input of the channel.

(ii) *Check node symmetry:* The signs factor out of the check node message maps:

$$m_{cv}^{(\ell)}(b_1 m_1, \dots, b_{d_c-1} m_{d_c-1}) = m_{cv}^{(\ell)}(m_1, \dots, m_{d_c-1}) \left(\prod_i b_i \right)$$

for any ± 1 input sequence $\{b_i\}$ and where the arguments of the functions $m_{cv}^{(\ell)}$ are the inputs to the check nodes except that from node v.

(iii) *Variable node symmetry:* Inverting the signs of all $\{\pm 1\}$ incoming messages to a variable node v will invert the sign of the messages sent from v.

(iv) *Message pdf and cdf symmetry:* A pdf f is symmetric if $f(x) = e^x f(-x)$, $x \in \mathbb{R}$ and equivalently for the cdf.

The notion of message symmetry is slightly unusual but arises out of the analysis. For an example that will be of use later, consider a Gaussian (normal) pdf with mean μ and variance σ^2 denoted by $\mathcal{N}(\mu, \sigma^2)$. It is readily verified that for such a pdf to satisfy the symmetry condition it is necessary and sufficient that $\sigma^2 = 2\mu$, i.e., if

$$\frac{1}{\sqrt{2\pi}\sigma} \exp\left(-(x-\mu)^2/2\sigma^2\right) = \frac{1}{\sqrt{2\pi}\sigma} \exp\left(-(-x-\mu)^2/2\sigma^2\right) \cdot \exp(x). \tag{3.10}$$

The messages passed on the code graph edges represent belief as to the transmitted value in the following manner. Consider a variable node v and connected check node c. The message passed from v to c in a given iteration is the belief of the value of the variable node v given all the messages (beliefs) passed to v from all neighbor check nodes except c in the previous phase. Similarly the message passed from c to v is the belief as to v's value, given all the messages passed to c in the previous phase except that from v. It is assumed that all messages are independent meaning that at the ℓ-th iteration there are not cycles of length 2ℓ or less in the graph. The term *belief* in this will be the likelihood ratio or log likelihood ratio in the following sense.

Suppose a binary $\{\pm 1\}$ value X is transmitted and a continuous value Y is received and assuming the BIAWGN, $Y = X + N$ where N is a Gaussian random variable with mean zero and variance σ^2, $Y \sim \mathcal{N}(\pm 1, \sigma^2)$. The *conditional likelihood* or *likelihood ratio* (LR) value is defined as

$$L(x \mid y) = P(X = 1 \mid Y)/P(X = -1 \mid Y) \tag{3.11}$$

which is the likelihood (probability) the transmitted value was $+1$ given the received value Y, compared to the probability it was -1, given that Y was received. Note that the LR is a random variable and the values of $L(x \mid y)$ range from 0 to $+\infty$ – a value near 0 indicates a belief the transmitted value was -1 while a large value represents a belief it was 1. For the BIAWGN the *likelihood ratio (LR)* is

$$L(x \mid y) = \frac{P(X = +1 \mid Y)}{P(X = -1 \mid Y)} = \frac{\exp(-(y-1)^2/2\sigma^2)}{\exp(-(y+1)^2/2\sigma^2)} = \exp\left(\frac{2}{\sigma^2}y\right) \tag{3.12}$$

More often the log likelihood ratio (LLR) $\ln L(x \mid y)$ is used. Notice that $\ln L(x \mid y) = \frac{2}{\sigma^2}y$ for the case above is a normal random variable (recall the received signal is of the form $Y = X + N$) and for $Y > 0$ its mean is $2/\sigma^2$ and variance is $4/\sigma^2$, i.e., $\sim \mathcal{N}(2/\sigma^2, 4/\sigma^2)$, and, as can be checked, the density for this LLR is symmetric by the above definition.

As above, a high positive value for an LLR message represents a belief the variable value was $+1$ and a strong negative value that it was -1.

In Equation 3.12 let $p_1 = P(X = +1 \mid Y)$ and $p_{-1} = P(X = -1 \mid Y)$ and let $L(x \mid y) = p_1/p_{-1}$ and $\ell\ell(x \mid y) = \ln L(x \mid y)$ and

$$m = \ln \frac{p_1}{p_{-1}} = \ln L(x \mid y) \stackrel{\Delta}{=} \ell\ell(x \mid y).$$

The hyperbolic tan function is defined as

$$\tanh(x) = \frac{e^x - e^{-x}}{e^x + e^{-x}} = \frac{e^{2x} - 1}{e^{2x} + 1}.$$

For any probability pair (p_1, p_{-1}), $p_1 + p_{-1} = 1$, $m = \ln L = \ln\left(\frac{p_1}{p_{-1}}\right)$,

$$\tanh\left(\frac{m}{2}\right) = \frac{e^m - 1}{e^m + 1} = \frac{\frac{p_1}{p_{-1}} - 1}{\frac{p_1}{p_{-1}} + 1} = \frac{L-1}{L+1} = p_1 - p_{-1}, \tag{3.13}$$

which will be a useful relationship. Equivalently

$$L(x \mid y) = \frac{1 + (p_1 - p_{-1})}{1 - (p_1 - p_{-1})} == \frac{1 + \tanh(m/2)}{1 - \tanh(m/2)}. \tag{3.14}$$

It is assumed the channel input variables $X \in \pm 1$ are equally likely. It follows that $L(x \mid y) = L(y \mid x)$ since by Bayes theorem, informally,

$$L(x \mid y) = \frac{P(X=+1,Y)/p(Y)}{P(X=-1,Y)/p(Y)} = \frac{P(X=+1,Y)}{P(X=-1,Y)}$$

and

$$L(y \mid x) = \frac{P(X=+1,Y)/P(X=+1)}{P(X=-1,Y)/P(X=-1)} = \frac{P(X=+1,Y)}{P(X=-1,Y)}$$

where $P(X=i,Y)$ is a joint discrete and continuous probability density which requires greater care.

Consider a check node c with a set of variable node neighbors N_c. It will be convenient, when the code parity properties are under consideration, to view the codeword chosen $\boldsymbol{c} = (c_1,\dots,c_n)$ as binary $c_i \in \{0,1\}$ and view the word transmitted on the channel as $\boldsymbol{x} = (x_1,\dots,x_n)$ as binary $x_i \in \{\pm 1\}$. As noted, the two are related as $x_i = (-1)^{c_i}$ or $0 \longrightarrow 1$ and $1 \longrightarrow -1$. It will be clear from the context which is under consideration and the equivalence will be assumed without comment. The related probabilities are denoted $\{p_0, p_1\}$ and $\{p_1, p_{-1}\}$, with superscripts indicating iteration number if needed. The received channel output at variable node v will be denoted $Y_v = X_v + N$. Note the double duty of the notation N – without any subscript it will denote a Gaussian noise variable.

In the absence of channel noise the value of a variable node $v \in N_c$ could be obtained by noting that the check values are all zero for a valid codeword input. Thus the bit associated with any check node would be 0 and the belief of check node c as to the value associated with variable node v, x_v would be

$$0 = \bigoplus_{v_i \in N_c} X_{v_i} \quad \text{or} \quad X_v = \bigoplus_{v_i \in N_{c,v}} X_{v_i}, \quad N_{c,v} = N_c \backslash \{v\} = \{v_1, v_2, \dots, v_f\}.$$

The message to be passed from check node c to variable node v at the ℓ-th iteration would be

$$\begin{aligned} m_{c.v}^{(\ell)} &= \ell\ell\big(x_v \mid y_{v_1},\dots,y_{v_f}\big),\ v_i \in N_{c.v} \\ &= \ln L\big(x \mid y_{v_1},\dots,y_{v_f}\big) \\ &= \ln \frac{p\big(X_v = +1 \mid Y_{v_1},\dots,Y_{v_f}\big)}{p\big(X_v = -1 \mid Y_{v_1},\dots,Y_{v_f}\big)} \end{aligned} \tag{3.15}$$

with $N_{c,v}$ as above with corresponding variable node values Y_{v_i}, i.e., the messages transmitted to c from the variable nodes in the previous phase. To compute these expressions note the following. Since the y_i initial received variables are assumed independent the joint density factors:

$$p(y_{v_1}, y_{v_2}, \ldots, y_{v_f} \mid x_v) = \prod_{i=1}^{f} p(y_{v_i} \mid x_v).$$

The assumption of independence requires the path generated from an original node to be cycle-free. Then

$$L(y_{v_1}, y_{v_2}, \ldots, y_{v_f} \mid x_v) = \prod_{i=1}^{f} p(y_{v_i} \mid x_{v_i} = +1)/p(y_{v_i} \mid x_{v_i} = -1) = \prod_{i=1}^{f} L(y_{v_i} \mid x_{v_i})$$

and by the above observation that $L(y_{v_i} \mid x_{v_i}) = L(x_{v_i} \mid y_{v_i})$ for equally likely data variables and

$$L(x_v \mid y_{v_1}, \ldots, y_{v_f}) = \prod_{i=1}^{f} L(x_v \mid y_{v_i}). \tag{3.16}$$

Suppose $X_v = X_{v_1} \oplus X_{v_2} \oplus \cdots \oplus X_{v_f}$ (where $Y_{v_i} = X_{v_i} + N$) and Y_i depends only on X_i and assuming binary $\{0, 1\}$ code symbols. Define

$$L(x_{v_i} \mid y_{v_i}) = \frac{P(X_{v_i} = 1 \mid Y_{v_i})}{P(X_{v_i} = 0 \mid Y_{v_i})} = \frac{p_1^{(i)}}{p_0^{(i)}}, \quad i = 1, 2, \ldots, f.$$

Let

$$\begin{aligned} \rho_0 &= P(X_{v_1} \oplus \cdots \oplus X_{v_f} = 0 \mid Y_{v_1}, \ldots Y_{v_f}) \\ \text{and} \quad \rho_1 &= P(X_{v_1} \oplus \cdots \oplus X_{v_f} = 1 \mid Y_{v_1}, \ldots Y_{v_f}). \end{aligned}$$

Under the above conditions it is claimed that

$$\rho_0 - \rho_1 = \prod_{i=1}^{f} \left(p_0^{(i)} - p_1^{(i)}\right). \tag{3.17}$$

This can be shown by induction. Suppose the relation is true for $f-1$ variables and let $Z_{f-1} = X_{v_1} \oplus \cdots \oplus X_{v_{f-1}}$. Then

$$\begin{aligned} \rho_0 &= P(X_{v_1} \oplus \cdots \oplus X_{v_f} = 0 \mid Y_{v_1}, \ldots Y_{v_f}) \\ &= P(Z_{f-1} = 0 \mid Y_{v_1}, \ldots Y_{v_{f-1}})P(X_{v_f} = 0 \mid Y_{v_f}) \\ &\quad + P(Z_{f-1} = 1 \mid Y_{v_1}, \ldots Y_{v_{f-1}})P(X_{v_f} = 1 \mid Y_{v_f}) \end{aligned}$$

and similarly

$$\begin{aligned} \rho_1 &= P(X_{v_1} \oplus \cdots \oplus X_{v_f} = 1 \mid Y_{v_1}, \ldots Y_{v_f}) \\ &= P(Z_{f-1} = 1 \mid Y_{v_1}, \ldots Y_{v_{f-1}})P(X_{v_f} = 0 \mid Y_{v_f}) \\ &\quad + P(Z_{f-1} = 0 \mid Y_{v_1}, \ldots Y_{v_{f-1}})P(X_{v_f} = 1 \mid Y_{v_f}) \end{aligned}$$

Subtracting the expressions gives

$$P(Z_{f-1} = 0 \mid Y_{v_1}, \dots Y_{v_{f-1}}) \left(P(X_{v_f} = 0 \mid Y_{v_f}) - P(X_{v_f} = 1 \mid Y_{v_f})\right)$$
$$- P(Z_{f-1} = 1 \mid Y_{v_1}, \dots Y_{v_{f-1}}) \left(P(X_{v_f} = 0 \mid Y_{v_f}) - P(X_{v_f} = 1 \mid Y_{v_f})\right)$$

which leads to

$$\left(p_0^{(f)} - p_1^{(f)}\right) \prod_{i=1}^{f-1} \left(p_0^{(i)} - p_1^{(i)}\right) = \rho_0 - \rho_1 \tag{3.18}$$

as claimed.

Interpreting the above developments for the message $m_{cv}^{(\ell)}$ from check node c to variable node $v \in N_c$ at the (second phase of the) ℓ-th iteration it follows from Equation 3.15 and Equations 3.13, 3.14 and 3.17 that

$$\begin{aligned} m_{cv}^{(\ell)} &= \ln \frac{1 + \left(\prod_{i=1}^{f} \left(p_1^{(i)} - p_{-1}^{(i)}\right)\right)}{1 - \left(\prod_{i=1}^{f} \left(p_1^{(i)} - p_{-1}^{(i)}\right)\right)} \\ &= \ln \frac{\left(1 + \prod_{i=1}^{f} \tanh(\ell\ell_i/2)\right)}{\left(1 - \prod_{i=1}^{f} \tanh(\ell\ell_i/2)\right)} \qquad \ell \geq 0, \quad \ell\ell_i = \ln \frac{p_1^{(i)}}{p_{-1}^{(i)}}, i = 1,2,\dots,f \\ &= \ln \frac{\left(1 + \prod_{v' \in N_{c,v}} \tanh\left(m_{v'c}^{(\ell)}/2\right)\right)}{\left(1 - \prod_{v' \in N_{c,v}} \tanh\left(m_{v'c}^{(\ell)}/2\right)\right)} \qquad \ell \geq 0. \end{aligned} \tag{3.19}$$

For the messages passed from the variable nodes to the check nodes, consider the following. For the 0-th iteration at variable node v, $m_{vc}^{(0)}$, the message passed to check node $c \in N_v$ will be $m_{vc}^{(0)} = \ln P(X_v = +1 \mid Y_v)/P(X_v = -1 \mid Y_v)$ given by Equation 3.16. For later iterations it is given by

$$m_{vc}^{(\ell)} = \begin{cases} m_{vc}^{(0)}, & \ell = 0 \\ m_{vc}^{(0)} + \sum_{c' \in N_{v,c}} m_{c'v}^{(\ell-1)}, & \ell \geq 1 \end{cases} \tag{3.20}$$

which is the LLR $\ln(P(X_v = +1 \mid Y_1, \dots, y_f)/P(X_v = -1 \mid Y_1, \dots, Y_f))$ where as before the Y variables correspond to nodes in $N_{v,c}$.

The message sent from v to c in the ℓ-th iteration is the variable node v's estimate of its (i.e., v's) value via LLRs which is the simple sum of the LLRs it received in the second phase of the previous iteration. The additive form of this equation is convenient. Since the tanh and ln tanh functions are monotonic on $[-\infty, +\infty]$ the messages can be represented in a (sign, absolute value) representation as follows:

$$\begin{aligned}\gamma : &\longrightarrow \mathbb{F}_2 \times [0, +\infty] \\ x &\longmapsto \gamma(x) = (\gamma_1(x), \gamma_2(x)) = \left(\operatorname{sgn}(x), \ln \tanh |\tfrac{x}{2}|\right)\end{aligned} \tag{3.21}$$

where the sgn function is defined as

$$\operatorname{sgn}(x) = \begin{cases} 0, & x > 0 \\ 0, & \text{with probability} 1/2 \text{ if } x = 0 \\ 1, & \text{with probability} 1/2 \text{ if } x = 0 \\ 1, & \text{if } x < 0. \end{cases}$$

The function $\gamma(\cdot)$ has a well-defined inverse γ^{-1}. The message representation used then is

$$m_{cv}^{(\ell)} = \gamma^{-1}\left(\sum_{v' \in \mathcal{V}_{cv}} \gamma\left(m_{v'c}^{(\ell)}\right)\right). \tag{3.22}$$

The advantages of this additive version of the message passing are clear, compared to Equation 3.19 .

To recap, the message formats that have been developed are:

$$m_{vc}^{(\ell)} = \begin{cases} m_v, & \ell = 0 \\ m_v + \sum_{c' \in C_{v,c}} m_{c'v}^{(\ell-1)}, & \ell \geq 1 \end{cases} \tag{3.23}$$

and

$$m_{cv}^{(\ell)} = \gamma^{-1}\left(\sum_{v' \in \mathcal{V}_{cv}} \gamma\left(m_{v'c}^{(\ell)}\right)\right) \tag{3.24}$$

and recall that the pdf of the initial LLR message from variable nodes is the pdf of the signal $Y_v = X_v + N,\ N \sim \mathcal{N}(0, \sigma^2)$ received from the channel at the variable node v is (from Equation 3.11) $\mathcal{N}(2/\sigma^2, 4/\sigma^4)$, i.e., the LLR received from the channel in the 0-th iteration is $m_{vc}^{(0)} = 2y/\sigma^2$ and

$$p_{vc}^{(0)}(z) = \frac{\sigma}{\sqrt{2\pi 2}} \exp\left(-\left(z - \frac{2}{\sigma^2}\right)^2 \Big/ (8/\sigma^2)\right).$$

While the pdfs cease to be Gaussian as the iterations progress, they remain symmetric for the original Gaussian channel noise, i.e., this pdf is symmetric by Definition 3.4 (part iv).

Having determined the form of the messages passed on the edges of the code Tanner graph, it remains to examine the performance of this message-passing algorithm, i.e., to determine the probability of error one might expect. This will be discussed in the remainder of this section. Section 3.3 will comment on various aspects and properties of the decoding algorithm.

Denote by $p_{vc}^{(\ell)}(x)$ the pdf of the message $m_{vc}^{(\ell)}$ and $P_{vc}^{(\ell)}$ the corresponding cdf. Let $P_e^{(\ell)}$ denote the error probability at the ℓ-th iteration. As noted, the message pdfs ($p_{cv}^{(\ell)}$ and $p_{vc}^{(\ell)}$) are symmetric functions as the message passage algorithm decoder iterates. A consequence of this is that ([20], lemma 1) the error probability is independent of the codeword chosen for transmission. Thus without loss of generality it can be assumed the all $+1$ codeword (in the $\{\pm 1\}$ code) was transmitted to compute the error probability.

The message $m_{vc}^{(\ell)}$ at the ℓ-th iteration is the log likelihood of node v's belief it represents a transmitted $+1$ over that of -1. Since it is assumed the all $+1$ codeword was transmitted, if the process was stopped at this iteration an error would be made if this message was negative. In the following analysis it is assumed (without further comment) the messages are independent random variables, i.e., the cycle-free case, together with the fact that each node uses only extrinsic information in forming messages. Thus the error probability, given the all-ones codeword was transmitted, is given by

$$P_e^{(\ell)} = \int_{-\infty}^{0} p_{vc}^{(\ell)}(x)dx \tag{3.25}$$

or in the case the densities have discrete probability components, in terms of the cdfs,

$$P_e^{(\ell)} = \frac{1}{2}\left(P_v^{(\ell)}(0)^+ + P_v^{(\ell)}(0)^-\right) \tag{3.26}$$

where half of the probability at the origin is added.

Note again that while the decoder message pdfs are symmetric they are not Gaussian – except for the initial pdf from the channel $p_{vc}^{(0)}$ as shown in Equation 3.12 although for a large number of variable nodes the pdf of $m_{vc}^{(\ell)}$ might be modeled as such since, as a sum of a large number of independent random variables, the Central Limit Theorem might be applied. However, the form of the messages $m_{cv}^{(\ell)}$ as in Equation 3.20 with log tanh functions precludes normality of the associated pdfs. The issue will be discussed further in the next section.

An iterative process will be derived to determine the pdfs of the messages passed as the iterations progress which will allow expressions for the probability of error of the algorithm. Before addressing this problem two comments are in order.

It was noted earlier that typically, the decoding process will continue until either some fixed number of iterations has been reached or the LLRs passed to the variable nodes exceed some predetermined threshold to indicate some confidence the decision has been reached. Note that BP often represents a feasible (achievable) approximation to ML decoding. Again, the cycle-free case is assumed although in some cases in practice, the number of iterations exceeds this limit with satisfactory results.

The pdfs of the messages transmitted are required as the iterations proceed. The tracking of the pdfs to be described will be an iterative process and is termed *density evolution*. It is required to determine the pdfs of the messages of the form in Equations 3.23 and 3.24. The situation is complicated somewhat by the fact that the messages $m_{cv}^{(\ell)}$ are represented on a product space $\mathbb{F}_2 \times [0, +\infty]$ with one part of the probability involving discrete components (probability mass functions or weighted impulse functions) and the other part continuous. Nonetheless the basic tool is convolution and an overview of the technique and results will be described. The pdf of a sum of d independent random variables will be the d-fold convolution of their respective pdfs.

The pdfs of the message Equations 3.23 and 3.24 for the messages $m_{vc}^{(\ell)}$ and $m_{cv}^{(\ell)}$ at the ℓ-th iteration, denoted $p_{vc}^{(\ell)}$ and $p_{cv}^{(\ell)}$, respectively, are to be found. The sum of independent random variables can be computed as convolutions of their pdfs or products of their Fourier transforms. The following analysis is technical and only a sketch of the ideas of the developments in [21] (with minor changes in notation) is given. The messages $m_{cv}^{(\ell)}$ involve γ^{-1} of a sum of random variables, each involving the function γ and defined over $\mathbb{F}_2 \times [0, +\infty]$.

The equation for the pdfs of the messages $m_{vc}^{(\ell)}$ is straightforward. Assume for the moment the code graph is (d_v, d_c)-biregular: then

$$p_{vc}^{(\ell+1)} = p_{vc}^{(0)} \otimes \left(p_{cv}^{(\ell)}\right)^{\otimes(d_v-1)}$$

where $\left(p_{cv}^{(\ell-1)}\right)^{\otimes(d_v-1)}$ is the convolution of $p_{cv}^{(\ell)}$ with itself applied $d_c - 1$ times, as the pdfs are the same for each check node. For the irregular $(\lambda(x), \rho(x))$ graph, this argument is conditioned on the edge having a certain degree and averaged over the distribution of edge degrees and hence for the irregular graph with these edge distributions it follows that

$$\begin{gathered} p_{vc}^{(\ell+1)} = p_{vc}^{(0)} \otimes \left(\sum_i \lambda_i \left(p_{cv}^{(\ell)}\right)^{\otimes(i-1)}\right) = p_{vc}^{(0)} \otimes \lambda\left(p_{cv}^{(\ell)}\right), \\ \text{where } \lambda\left(p_{cv}^{(\ell)}\right) = \sum_i \lambda_i \cdot \left(p_{cv}^{(\ell)}\right)^{\otimes(i-1)}. \end{gathered} \tag{3.27}$$

Finding the pdfs for the messages $m_{cv}^{(\ell)}$ is more difficult due to their form as 2-tuples over the product of fields $\mathbb{F}_2 \times (0, \infty)$ and that this form combines continuous and discrete functions. Hence the convolutions of such probability expressions are more subtle.

Let $\Gamma(p_{vc}^{(\ell)})$ be the pdf of the random variable $\gamma(m_{vc}^{(\ell)})$ (a 2-tuple) and the pdf of the addition of the random variables is found by convolution of these expressions. It is noted this map has a well-defined inverse Γ^{-1}. Using the argument above, the pdf of the message $m_{cv}^{(\ell)}$ on an irregular graph with (λ, ρ) edge distribution, as given in Equation 3.24, is given by ([21], equation 7)

$$p_{cv}^{(\ell+1)} = \Gamma^{-1}\left(\rho\left(\Gamma\left(p_{vc}^{(\ell)}\right)\right)\right) \text{ where } \rho\left(\Gamma\left(p_{vc}^{(\ell)}\right)\right) = \sum_{i\geq 2} \rho_i \left(\Gamma\left(p_{vc}^{(\ell)}\right)\right)^{\otimes(i-1)}. \tag{3.28}$$

Substituting the expression for $p_{cv}^{(\ell)}$ into Equation 3.27 yields ([21], theorem 2)

$$p_{vc}^{(\ell+1)} = p_{vc}^{(0)} \otimes \lambda\left(\Gamma^{-1}\left(\rho\left(\Gamma\left(p_{vc}^{(\ell)}\right)\right)\right)\right), \tag{3.29}$$

which, as noted, is a symmetric function.

It was noted in Chapter 2 that this development is entirely consistent with the result of erasure coding on (the BEC) irregular graphs in that from $p_{vc}^{(\ell)}$ above, one identifies the variable p_ℓ of that chapter and the above relation can be shown to be equivalent to the relationship for BECs

$$p_{\ell+1} = p_0\lambda(1 - \rho(1-_{\ell})), \quad p_0 = \delta$$

(where now p_ℓ has the interpretation of the probability a variable node is unresolved at iteration ℓ) which is the important result of the equation in Theorem 2.7 in Chapter 2 achieved by different means.

Several important properties of this result are developed in [20]. As noted in Equation 3.26 the expression for the probability of error is given by

$$P_e^{(\ell)} = \frac{1}{2}\left(P_v^{(\ell)}(0)^- + P_v^{(\ell)}(0)^+\right) \tag{3.30}$$

and ([20], corollary 1) this function converges to 0 as $\ell \longrightarrow \infty$ iff $p_v^{(\ell)}$ converges to δ_∞, the delta function at infinity (probability mass of 1).

Further interesting and important results on the probability of error are noted, namely that under symmetry and stability conditions, the probability of error is a nonincreasing function of the number of iterations ℓ and will decrease to zero (see [20], theorems 7 and 8 and corollary 2).

3.3 Thresholds, Concentration, Gaussian Approximation, EXIT Charts

There are numerous issues around the design and performance of LDPC codes with BP decoding that are not addressed in this brief treatment. A major one is the design of edge distributions $\lambda(x)$ and $\rho(x)$ to minimize the probability of error. This issue was briefly touched on previously and in the previous chapter for erasure codes but is not considered here.

The topics of this section include indications of the notions of concentration of performance of code, the establishment of thresholds, the use of Gaussian approximation to simplify the analysis of density evolution and the decoding algorithm and the notion of an EXIT chart as a visual aid in viewing the performance of the decoding algorithm.

Thresholds and Concentration

The class of *binary memoryless symmetric* (BMS) channels is considered and in particular the subclass that is parameterized by a real parameter γ not to be confused with the mapping of Equation 3.21. Thus as noted previously for the BSC, BEC and BIAWGN the parameter would be the channel crossover probability p, the channel erasure probability δ and the noise variance σ^2, respectively. The capacity of such a channel typically decreases with increasing γ (for the range of interest).

The notion of a threshold for a channel and decoding algorithm involves being able to conclude that if the probability of error performance converges to 0 for a given channel parameter γ, then it will converge to zero under the same conditions for all parameters $\gamma' < \gamma$. An interesting approach to this problem is as follows. Cover and Thomas discuss the notion of a *physically degraded channel* [5]. Consider a channel W with transition probabilities $p(y \mid x)$. A channel W' is said to be physically degraded with respect to W if there is a channel Q such that the transition probabilities behave as

$$p_{W'}(y' \mid x) = p_Q(y' \mid y)p_W(y \mid x).$$

In essence a degraded channel may be viewed as the catenation of the original channel with another channel that degrades the performance of the original one. The following theorem establishes the monotonicity in the performance of such channels ([20], theorem 1):

Theorem 3.5 *Let W and W' be two given symmetric memoryless channels and assume that W' is physically degraded with respect to W. For a given code and BP decoder let p be the expected fraction of incorrect messages passed at the ℓ-th iteration assuming tree-like neighborhoods with transmission over channel W and let p' be the equivalent quantity over W'. Then $p \leq p'$.*

The theorem ensures monotonicity with respect to a particular decoder if convergence for a parameter γ' implies convergence for all $\gamma \leq \gamma'$. It certifies the notion of a threshold.

An instance of such a threshold is contained in ([21], theorem 5) which asserts the error performance of a code with an edge degree distributions (λ, ρ) pair and symmetric channel noise pdf denoted $p_{vc}^{(0)}$, for a parameter r

$$r \triangleq -\ln\left(\int_{-\infty}^{+\infty} p_{vc}^{(0)}(x)e^{-x/2}dx\right)$$

then under certain conditions

(i) if $\lambda^{'}(0)\rho^{'}(1) > e^{r}$ there exists a constant ξ such that for all $\ell \in \mathbb{N},\ P_e^{(\ell)} > \xi$;
(ii) if $\lambda^{'}(0)\rho^{'}(1) < e^{r}$ there exists a constant ξ such that if, for some $\ell \in \mathbb{N},\ P_e^{(\ell)} \leq \xi$, then $P_e^{(\ell)}$ converges to 0 as ℓ tends to ∞.

Another important contribution of [20] is the notion of *concentration* of code performance. Applied to the ensemble of randomly generated ensemble of graphs $C^n(\lambda, \rho)$, it asserts that the behavior of a code chosen at random from this ensemble over the noisy channel concentrates around the average behavior of this ensemble as the code length increases. Thus if the channel parameter is below the threshold, then any small, desired error probability for such a randomly chosen graph can be achieved, with probability approaching one and exponential code length n, with a fixed number of iterations ℓ (which depends on the error probability). A converse to this statement can also be formulated. Thus the average performance of a graph in this ensemble is a good predictor of performance for any randomly chosen graph in the ensemble.

A particular instance of concentration is the following for a (d_v, d_c)-biregular Tanner graph. Let Z be the number of incorrect messages being sent out on all $d_v n$ edges from variable nodes (of a (d_v, d_c)-biregular code/graph with n variable nodes) at the ℓ-th iteration, and its expected value $E[Z]$ (over all possible such graphs and decoder inputs). Similarly let p be the expected number of incorrect messages passed along an edge from a given variable node which forms the root of a directed tree of depth 2ℓ and note that p is calculated from the derived densities given in Equation 3.27. Then it is shown ([20], theorem 2) that

$$\Pr\left(\mid Z - E[Z] \mid > n d_v \epsilon/2\right) \leq 2\exp(-\beta\epsilon^2 n) \tag{3.31}$$

where β is a function of graph parameters and iteration number ℓ. Thus finding the mean performance for the ensemble of graphs will with high probability be a good indicator of actual performance of a given graph.

Gaussian Approximation

The expressions for the pdfs of the message densities sent from variable and check nodes respectively are given in Equations 3.27, 3.29 and 3.28. These can be quite difficult to compute in practice. Fourier transforms are usually employed since the Fourier transform of a convolution is the product of Fourier transforms and the approach often leads to simpler update equations. Another possibility is to approximate the densities as normal (Gaussian) densities. Since the messages from check nodes to variable nodes involve logs of tanh functions, this is generally far from accurate for these messages. It is often approximately true for variable-to-check node messages as these involve the sums of independent random variables and if there are a large number of them, the Central Limit Theorem might make the proposition approximately true. The accuracy of the Gaussian approximation is discussed in greater detail in [1] where a technique is developed that considers outputs of variable nodes to check nodes in two steps, i.e., considering the outputs of variable nodes over a complete iteration. The approach leads to considerable improvements over the assumption of normality from both sets of nodes.

Regardless of the limitation of the Gaussian assumption noted, the approach is of interest. It has been shown that a normal pdf is symmetric iff its variance is twice its mean. Since the initial channel pdf is normal and symmetric and density evolution preserves symmetry, the assumption of normality requires only that the mean of the messages is tracked. This is an easier approach which is considered in several works and in particular [4]. Only the (d_v, d_c)-biregular case will be considered and to distinguish means from messages, means will be designated by μ and in particular the means of messages from variable and check nodes in the ℓ-th iteration are denoted $\mu_{vc}^{(\ell)}$ and $\mu_{cv}^{(\ell)}$, respectively.

From Equation 3.23 it follows that

$$\mu_{vc}^{(\ell)} = \mu_{vc}^{(0)} + (d_v - 1)\mu_{cv}^{(\ell-1)} \tag{3.32}$$

and using Equations 3.13 and 3.17 it follows that

$$\tanh\left(\frac{m_{cv}^{(\ell)}}{2}\right) = \prod_{c' \in N_{v,c}} \tanh\left(\frac{m_{vc'}^{(\ell)}}{2}\right) \tag{3.33}$$

and hence

$$E\left[\tanh\left(\frac{m_{cv}^{(\ell)}}{2}\right)\right] = E\left[\tanh\left(\frac{m_{vc}^{(\ell)}}{2}\right)\right]^{d_v-1} \tag{3.34}$$

where the expectations are with respect to the (assumed) normal densities. Since all densities are symmetric and assumed Gaussian, each must have its variance twice the mean and the expressions therefore depend only on the mean. For this last equation

$$E\left[\tanh\left(\frac{m_{cv}^{(\ell)}}{2}\right)\right] = \frac{1}{\sqrt{4\pi\mu_{cv}^{(\ell)}}}\int_{\mathbb{R}} \tanh\left(\frac{u}{2}\right)\exp\left(-\frac{\left(u-\mu_{cv}^{(\ell)}\right)^2}{4\mu_{cv}^{(\ell)}}\right)du.$$

If [4] the function

$$\phi(x) = \begin{cases} 1 - \frac{1}{\sqrt{4\pi x}}\int_{\mathbb{R}} \tanh(\frac{u}{2})\exp\left(-\frac{(u-x)^2}{4x}\right)du, & x > 0 \\ 1, & x = 0 \end{cases}$$

is defined, then the update equations for the check means are

$$\mu_{cv}^{(\ell)} = \phi^{-1}\left(1 - \left[1 - \phi(\mu_0 + (d_v - 1)\mu_{cv}^{(\ell-1)})\right]^{d_c-1}\right) \tag{3.35}$$

where μ_0 is the mean from the channel and $\mu_{cv}^{(0)} = 0$. This equation yields the updates for the means of messages from the check nodes (under the Gaussian assumption) and, along with Equation 3.32, is used to update the variable messages.

Under the Gaussian assumption, from these means, the variances can be found and the pdf of the messages at each iteration, and from this the probability of error. The thresholds below which the error probability tends to zero can thus be determined. As an example (from table II of [20]) for a $(3,6)$-biregular code (rate $1/2$) the variance achieved under Gaussian approximation is 0.8747 while the exact value from the previous analysis is 0.8809. The variance corresponding to the capacity of 0.5 (the rate of the code) is 0.979 (from Equation 3.2). From the same table, the general level of accuracy of the Gaussian approximation technique is impressive. However, in spite of these comments, it is clear in practice [1] the accuracy of the Gaussian assumption is limited, especially for messages sent from the check nodes.

EXIT Charts

The notion of an EXIT is a graphical means of tracking the process of the BP decoder as it iterates on the Tanner graph of the code. As such, it is a valuable tool for designing the code and estimating decoding thresholds. It arose in the seminal works of Stephan ten Brink and colleagues [2, 26, 27]. A variety of metrics can be used for this process, such as the mean, SNR, probability of error and mutual information between a graph message node information bit and output message node bit decision. Mutual information is often regarded as the most useful and will be discussed here.

As the decoding process progresses the improvement in the decoding performance from the input message of a node (either variable or check) and the output message, in terms of the quality of the information bit estimate is tracked. The message at the input to the node is regarded as a priori information and is designated with a subscript A. The output information is regarded as *extrinsic* and is designated with a subscript E. If the decoder is working well, one can expect the mutual information to improve with iterations and to migrate from some initial point on a plot to the point $(1,1)$ where the bit is successfully decided.

The analysis assumes a (d_v, d_c)-biregular bipartite graph, the cycle-free case and a large codeword length (number of variable nodes) and a large number of decoding iterations. It will further be assumed that the messages have a symmetric normal pdf and from the previous discussion on Gaussian approximation it is sufficient to track means since the variance of a symmetric pdf is twice the mean (see line before Equation 3.10). For a variable node with input message random variable $m_{cv}^{(\ell)}$ and output message random variable $m_{vc}^{(\ell)}$ at some iteration, the mutual information between these two random variables and the information bit associated with the variable node v are denoted $I_{A,v}$ and $I_{E,v}$, respectively. The similar quantities for a check node c are denoted $I_{A,c}$ and $I_{E,c}$, respectively.

Using Equation 3.32 it follows that

$$\sigma_v^2 = \sigma_n^2 + (d_v - 1)\sigma_A^2$$

where σ_v^2 is the variance of the message $m_{vc}^{(0)}$ leaving variable node v, σ_n^2 is the variance of the channel noise and σ_A^2 is the variance of the message $m_{cv}^{(0)}$ entering the variable node. The mutual information between the information bit associated with variable node v (say X_v) and message $m_{vc}^{(\ell)}$, which is denoted by m, is

$$\begin{aligned} I_{E,v} &= H(X) - I(X \mid M) \\ &= 1 - \sum_{x=\pm 1} \tfrac{1}{2} \int_{-\infty}^{+\infty} p(m \mid x) \log_2 \left(\frac{p(m \mid x=+1) + p(m \mid x=-1)}{p(m \mid x)} \right) dm \\ &= 1 - \int_{-\infty}^{+\infty} p(m \mid x=+1) \log_2 (1 + e^{-m}) dm \end{aligned}$$

where $p \sim N(\sigma^2/2, \sigma^2)$. It is convenient [9, 22] to designate this expression as

$$I_{E,v} = J(\sigma_n) = J\left(\sqrt{\sigma_n^2 + (d_v - 1)\sigma_A^2}\right).$$

In a similar development

$$I_{A,v} = J(\sigma_A)$$

which is the mutual information between variable node inputs $m_{cv}^{(\ell)}$ (hence a priori) and the channel bit X_v.

This function can be shown to be monotonic and has an inverse $\sigma_A = J^{-1}(I_{A,v})$ and hence

$$I_{E,v} = J\left(\sqrt{\sigma_n^2 + (d_v - 1) \mid J^{-1}(I_{A,v}) \mid^2}\right).$$

It is possible to plot the extrinsic information $I_{E,v}$ against the intrinsic information $I_{A,v}$ to monitor progress of the algorithm, i.e., monitor the improvement of the bit estimates as it goes through the iterations through variable nodes. For binary inputs, the range of both quantities is $(0,1)$ and, as the algorithm progresses, the mutual information increases horizontally or vertically if the algorithm is working correctly. In a similar fashion one can compute $I_{A,c}$ and $I_{E,c}$ through check nodes and these can also be shown on the same graph with the axes reversed. However, this relationship is a little more complicated due to the form of the messages transmitted, namely (see Equation 3.33):

$$m_{cv}^{(\ell)} = 2\tanh^{-1}\left(\prod_{v' \in N_{c,v}} \tanh\left(m_{v'c}^{(\ell)}/2\right)\right).$$

Nonetheless, accurate approximations have been developed and it is shown [9] that

$$I_{E,c} = 1 - J\left(\sqrt{(d_c - 1)(J^{-1}(1 - I_{A,c}))^2}\right).$$

Figure 3.2 shows a representative example (exaggerated somewhat for clarity) as two curves in the unit square. In order for the decoding to be successful, the chart of the variable nodes must lie above the inverse of that of the check nodes. As the channel parameter is adjusted, at the point where the curves just touch determines the threshold of the algorithm.

As the algorithm progresses (assuming successful decoding) at each step the mutual information functions increase with either a horizontal move to the right or vertical move upward, depending on the node being considered. The moves tend toward the upper right corner $(1,1)$ where certainty on the transmitted bit is achieved.

Numerous important properties of EXIT charts have been shown (see [2, 9, 26, 27]). As the noise variance on the channel increases for the BIAWGN the curves approach each other and eventually intersect. As noted, the point at which this occurs is the channel threshold as decoding then becomes impossible as the decoding trajectory cannot work its way through to the upper right-hand corner and the decoding algorithm fails. Thus the EXIT chart is

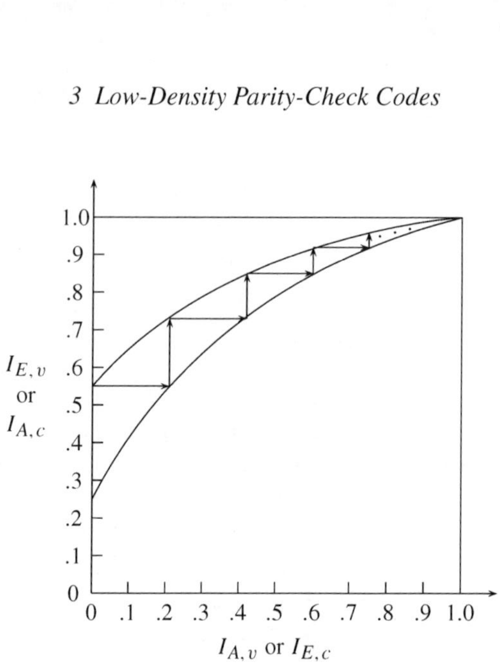

Figure 3.2 Typical shape of an EXIT chart for a biregular code

useful to determine thresholds for codes. In addition it has been used to design good codes.

While mutual information is generally regarded as the most accurate parameter to use for these charts, other parameters such as means, SNR and error probability have also been used [1, 6, 7].

Another interesting and useful aspect of EXIT charts is their area property, considered in [2] and further in [18]. Define [22] the following areas:

$$A_c = \int_0^1 I_{E,c}(I_{A,c})dI_{A,c} \quad \text{and} \quad A_v = \int_0^1 I_{E,v}(I_{A,v})dI_{A,v}$$

(on respective axes). It can be shown that, for $\bar{d}_c$ and $\bar{d}_v$, the average number of check and variable nodes,

$$A_c = 1/\bar{d}_c \quad \text{and} \quad A_v = 1 - (1 - C)/\bar{d}_v$$

for C the channel capacity. Numerous other relationships are noted in the works cited.

Comments

This chapter has been an introduction to the important class of LDPC codes. Many aspects of the message-passing variables and expressions for the probability of error have been quite technical, although the notion of estimating information bits, the estimates improving with each iteration, seems intuitive. Reading the original works is indispensable for a deeper understanding of the subject. These include the seminal works of Gallager [8] and Richardson et al. [20, 21]. The survey articles of Shokrollahi [25] and Guruswami [10] and the volumes [12, 13, 19, 22] and [9] cover a variety of other construction and performance aspects of the codes and also make for excellent reading. Other fundamental references for LDPC codes include [14, 15, 16, 17, 23, 24].

References

[1] Ardakani, M., and Kschischang, F.R. 2004. A more accurate one-dimensional analysis and design of irregular LDPC codes. *IEEE Trans. Commun.*, **52**(12), 2106–2114.

[2] Ashikhmin, A., Kramer, G., and ten Brink, S. 2004. Extrinsic information transfer functions: model and erasure channel properties. *IEEE Trans. Inform. Theory*, **50**(11), 2657–2673.

[3] Bazzi, L., Richardson, T.J., and Urbanke, R.L. 2004. Exact thresholds and optimal codes for the binary-symmetric channel and Gallager's decoding algorithm A. *IEEE Trans. Inform. Theory*, **50**(9), 2010–2021.

[4] Chung, S.-Y., Richardson, T.J., and Urbanke, R.L. 2001. Analysis of sum-product decoding of low-density parity-check codes using a Gaussian approximation. *IEEE Trans. Inform. Theory*, **47**(2), 657–670.

[5] Cover, T.M., and Thomas, J.A. 2006. *Elements of information theory*, 2nd ed. Wiley, Hoboken, NJ.

[6] Divsalar, D., Dolinar, S., and Pollara, F. 2001. Iterative turbo decoder analysis based on density evolution. *IEEE J. Select. Areas Commun.*, **19**(5), 891–907.

[7] El Gamal, H., and Hammons, A.R. 2001. Analyzing the turbo decoder using the Gaussian approximation. *IEEE Trans. Inform. Theory*, **47**(2), 671–686.

[8] Gallager, R.G. 1963. *Low-density parity-check codes.* MIT Press, Cambridge, MA.

[9] Gianluigi, L., Song, S., Lan, L, Zhang, Y., Lin, S., and Ryan, W.E. 2006. Design of LDPC codes: a survey and new results. *J. Commun. Softw. Syst.*, **2**(3), 191–211.

[10] Guruswami, V. 2006. Iterative decoding of low-density parity check codes. *Bull. Eur. Assoc. Theor. Comput. Sci. EATCS*, 53–88.

[11] Johnson, S.J. 2010. *Iterative error correction.* Cambridge University Press, Cambridge.

[12] Li, J., and Li, B. 2017. Beehive: erasure codes for fixing multiple failures in distributed storage systems. *IEEE Trans. Parallel Distrib. Syst.*, **28**(5), 1257–1270.

[13] Lin, S., and Costello, D.J. 2004. *Error control coding*, 2nd ed. Prentice Hall, Upper Saddle River, NJ.

[14] Luby, M.G., Mitzenmacher, M., Shokrollahi, M.A., Speileman, D., and Stenman, V. 1997. Practical loss-resilient codes. Pages 150–159 of: *Proceedings of the*

Twenty Ninth Annual ACM Symposium on the Theory of Computing 1997. ACM, New York.

[15] Luby, M.G., Mitzenmacher, M., Shokrollahi, M.A., and Spielman, D. 1998. Analysis of low density codes and improved designs using irregular graphs. Pages 249–258 of: *Proceedings of the Thirtieth Annual ACM Symposium on the Theory of Computing 1998*. ACM, New York.

[16] Luby, M.G., Mitzenmacher, M., and Shokrollahi, M.A. 1998. Analysis of random processes via And-Or tree evaluation. Pages 364–373 of: *Proceedings of the Ninth Annual ACM-SIAM Symposium on Discrete Algorithms (San Francisco, CA, 1998)*. ACM, New York.

[17] Luby, M.G., Mitzenmacher, M., Shokrollahi, M.A., and Spielman, D.A. 2001. Improved low-density parity-check codes using irregular graphs. *IEEE Trans. Inform. Theory*, **47**(2), 585–598.

[18] Measson, C., Montanari, A., Richardson, T.J., and Urbanke, R.L. 2009. The generalized area theorem and some of its consequences. *IEEE Trans. Inform. Theory*, **55**(11), 4793–4821.

[19] Richardson, T.J., and Urbanke, R.L. 2008. *Modern coding theory*. Cambridge University Press, Cambridge.

[20] Richardson, T.J., and Urbanke, R.L. 2001. The capacity of low-density parity-check codes under message-passing decoding. *IEEE Trans. Inform. Theory*, **47**(2), 599–618.

[21] Richardson, T.J., Shokrollahi, M.A., and Urbanke, R.L. 2001. Design of capacity-approaching irregular low-density parity-check codes. *IEEE Trans. Inform. Theory*, **47**(2), 619–637.

[22] Ryan, W.E., and Lin, S. 2009. *Channel codes*. Cambridge University Press, Cambridge.

[23] Shokrollahi, M.A. 2000. Codes and graphs. Pages 1–12 of: *In STACS 2000 (invited talk), LNCS No. 1770.*

[24] Shokrollahi, M.A. 2001. Capacity-achieving sequences. Pages 153–166 of: *Codes, systems, and graphical models (Minneapolis, MN, 1999)*. The IMA Volumes in Mathematics and Its Applications, vol. 123. Springer, New York.

[25] Shokrollahi, M.A. 2004. LDPC codes: an introduction. Pages 85–110 of: Feng, K., Niederreiter, H., and Xing, C. (eds.), *Coding, cryptography and combinatorics*. Birkhäuser, Basel.

[26] ten Brink, S. 1999. Convergence of iterative decoding. *Electron. Lett.*, **35**(10), 806–808.

[27] ten Brink, S., Kramer, G., and Ashikhmin, A. 2004. Design of low-density parity-check codes for modulation and detection. *IEEE Trans. Commun.*, **52**(4), 670–678.

4
Polar Codes

The advent of polar coding in the remarkable paper by Erdak Arıkan [3] represents a unique and important milestone in the history of coding and information theory. With almost no prior literature on such codes, it presented a theory of code construction, encoding and decoding algorithms and detailed performance analysis. Indeed, polar coding is the first known explicit construction with rigorous proofs of achieving rates within ϵ of capacity with block length, encoding and decoding complexity bounded by polynomials in $1/\epsilon$ [18]. The aim of this chapter is to introduce the essential notions involved with polar codes drawn largely from this seminal work. Besides its extraordinary technical contributions it is a beautiful exposition of information theory.

The next section outlines some of the needed information-theoretic notions required and their properties. Section 4.2 introduces the basic ideas of polar coding through properties and construction of their generating matrices. Using a series of structured simple matrix operations, a base binary-input discrete memoryless channel (BDMC), W, is transformed into a channel with $N = 2^n$ inputs and N outputs, designated W_N, referred to as the *compound channel.* The final section of the compound channel is a parallel bank of N base channels W whose outputs are the compound channel outputs. It will be noted that since the generator matrix involves 2^n information bits and a like number of "coded" bits, polar codes involve different concepts than ordinary parity-check codes.

The final Section 4.3 shows how, from the defined compound channel W_N, N binary-input subchannels, $W_N^{(i)}, i = 1, 2, \ldots, 2^n$ can be defined and it is these channels that will be shown to have been *polarized*, i.e., for large n a subset of the $N = 2^n$ subchannels will approach noiseless or perfect channels. The other channels will approach useless channels, i.e., channels that are not capable of transmitting any data reliably. Further, the number of perfect channels is such that the capacity of the base channel W can be achieved per input information bit. Thus if the binary-input base channel W has capacity $C \leq 1$ the number of perfect subchannels which each have a capacity approaching unity will be approximately NC and these channels can be used to transmit the information (without error) to achieve per input bit capacity of C. The polar code is the determination of the good channels. A decoding process, *successive cancellation decoding*, will be defined as a natural consequence of the series of transformations used in the construction of the channels, and its performance will be derived.

The coding problem will be to identify which are the good subchannels. Once identified, only the good subchannels will be used for near-perfect communication and the bad subchannels will be referred to as frozen with predetermined bits conveying no useful information. That there is a sufficient number of good channels $W_N^{(i)}$ that allows achieving overall capacity of the system per input bit is a crucial result of the work.

The basic ideas of polar codes are revolutionary and simple although the analysis requires effort. The iterative processes defined are highly structured and relatively simple and the important contribution of the work is to show how the iterative constructions and the resulting decoding strategies achieve the properties desired. Many of the proofs involve intricate information-theoretic computations. As with other chapters the emphasis will be to place the results in context rather than to trace the developments. The work is an impressive exercise in the power of information-theoretic notions.

4.1 Preliminaries and Notation

As in previous chapters, denote a row vector $\boldsymbol{a} \in \mathbb{F}^n$ over the field $\mathbb{F}$ by $\boldsymbol{a} = (a_1, a_2, \ldots, a_n), a_i \in \mathbb{F}$. A matrix $\boldsymbol{A}$ over $\mathbb{F}$ is denoted $\boldsymbol{A} = (a_{ij}), a_{ij} \in \mathbb{F}$ and the *Kronecker* or tensor product of two matrices $\boldsymbol{A}$ and $\boldsymbol{B}$ is $\boldsymbol{A} \otimes \boldsymbol{B} = (a_{ij}\boldsymbol{B})$. Simple properties of the product include

$$\begin{aligned} (\boldsymbol{A} \otimes \boldsymbol{B})(\boldsymbol{C} \otimes \boldsymbol{D}) &= (\boldsymbol{AC} \otimes \boldsymbol{BD}) \\ (\boldsymbol{A} \otimes \boldsymbol{B}) \otimes \boldsymbol{C} &= \boldsymbol{A} \otimes (\boldsymbol{B} \otimes \boldsymbol{C}) \\ (\boldsymbol{A} \otimes \boldsymbol{B})^{-1} &= \boldsymbol{A}^{-1} \otimes \boldsymbol{B}^{-1} \end{aligned} \tag{4.1}$$

assuming dimension-compatible matrices. In particular the n-fold tensor product of a matrix $\boldsymbol{A}$ is denoted $\boldsymbol{A}^{\otimes n} = \boldsymbol{A} \otimes \boldsymbol{A}^{\otimes(n-1)} = \boldsymbol{A} \otimes \boldsymbol{A} \otimes \cdots \otimes \boldsymbol{A}$ (n times) and $A^{\otimes 1} = A$.

As previously, the probability distribution of the discrete random variable X, $Pr(X = x)$ will be denoted $P_X(x)$ or as $P(x)$ when the random variable is understood. Similarly a joint discrete random variable $X \times Y$ (or XY) is denoted $Pr(X = x, Y = y) = P_{XY}(x, y)$. The conditional probability distribution is $Pr(Y = y \mid X = x) = P(y \mid x)$. A probability density function (pdf) for a continuous random variable X will be designated similarly as $p_X(x)$.

Certain notions from information theory as introduced in Chapter 1 will play a central role for polar codes and are repeated here for convenience. The entropy of a discrete ensemble $\mathcal{X} = \{x_i, P(x_i), i = 1, 2, \dots\}$ is given by

$$H(X) = -\sum_i P(x_i) \log P(x_i)$$

and similarly the joint entropy of two ensembles is given by

$$H(X, Y) = -\sum_{x,y} P(x, y) \log P(x, y).$$

The mutual information between discrete ensembles X and Y is given by

$$I(X; Y) = \sum_i \sum_j P(x_i, y_j) \log \frac{P(x_i, y_j)}{P(x_i)P(y_j)}$$

and measures the amount of information that one of the variables gives on the other. It can also be shown that (Chapter 1)

$$I(X; Y) = H(X) - H(X \mid Y).$$

A *discrete memoryless channel* (DMC), also introduced in Chapter 1, is a discrete input set $\mathcal{X}$, an output set $\mathcal{Y}$ such that for an input vector $\boldsymbol{x} = (x_1, x_2, \dots, x_n)$, $x_i \in \mathcal{X}$ the probability of receiving the DMC output vector $\boldsymbol{y} \in \mathcal{Y}^n$ is

$$P(\boldsymbol{y} \mid \boldsymbol{x}) = \prod_{i=1}^{n} W(y_i \mid x_i)$$

where $W(y_i \mid x_i)$ is a channel transition probability, the probability of output y_i given input x_i.

As has been noted, the channels BEC, BSC and general BDMC are shown in Figure 1.1 and the capacities of the BSC and BEC were noted there as (see Equation 1.15)

$$C_{BSC} = I(W_{BSC}) = 1 + p \log_2 p + (1-p) \log_2 (1-p), \quad C_{BEC} = I(W_{BEC}) = 1 - \delta.$$

This first relation is often written

$$I(W_{BSC}) = 1 - H_2(p) \quad \text{where} \quad H_2(p) = -p \log_2 p - (1-p) \log_2 (1-p)$$

is the binary entropy function of the two-element ensemble $\{p, 1-p\}$.

The notion of the BIAWGN channel, shown in Figure 1.2, has capacity

$$C_{BIAWGN} = -\int_y p(y) \log_2 p(y) dy - \frac{1}{2} \log_2(2\pi e \sigma^2)$$

where the pdf $p(y)$ is given in Equation 1.16. The general shape of these capacity functions is shown in Figure 1.3.

The notion of *computational cutoff rate* plays a role in the formulation of polar codes. This concept arose in the use of convolutional codes with sequential decoding, a particular backtracking decoding algorithm for such codes. The computational cutoff rate then is the rate at which the expected decoding complexity becomes unbounded. Polar codes arose out of considerations of the properties of this function in the research of sequential decoding [16, 35]. For a general DMC it will be defined for the channel W with input distribution $\{P(x_i)\}$, as [27]:

$$R_0(X;Y) = -\log \sum_j \left[\sum_i P(x_i) \sqrt{W(y_j \mid x_i)} \right]^2 .$$

For a BSC with crossover probability p the computation gives

$$R_0(X,Y) = 1 - \log_2 \left(1 + 2\sqrt{p(1-p)} \right).$$

The interest in these quantities for this chapter is the bounds they provide [27]:

$$I(X;Y) \geq R_0(X;Y).$$

Another parameter of interest for (binary input) BDMCs, apart from the channel capacity $I(W)$, is the *Bhattacharyya parameter* expressed as

$$Z(W) \stackrel{\Delta}{=} \sum_{y \in \mathcal{Y}} \sqrt{W(y \mid 0) W(y \mid 1)}$$

and this is often taken as an upper bound on the probability of deciding in error on a BDMC under maximum-likelihood detection. A simple argument is given to show this. For channel inputs $\mathcal{X} = \{0,1\}$ let $\mathcal{Y}_0$ be defined as

$$\mathcal{Y}_0 = \{ y \in \mathcal{Y} \mid W(y \mid 0) \leq W(y \mid 1) \}$$

Let $P_{e,0}$ be the probability of error under maximum-likelihood detection, given 0 was sent. Then

$$P_{e,0} = \sum_{y \in \mathcal{Y}_0} W(y \mid 0).$$

For $y \in \mathcal{Y}_0$, $\sqrt{W(y \mid 1)/W(y \mid 0)} \geq 1$ hence this expression can be multiplied by this amount to get the bound

$$P_{e,0} \leq \sum_{y \in \mathcal{Y}_0} \sqrt{W(y \mid 0)W(y \mid 1)}$$

and summing over the entire $\mathcal{Y}$ rather than just $\mathcal{Y}_0$ gives

$$P_{e,0} \leq \sum_{y \in \mathcal{Y}} \sqrt{W(y \mid 0)W(y \mid 1)}.$$

Finally,

$$P_e = \frac{1}{2}P_{e,0} + \frac{1}{2}P_{e,1} \leq \sum_{y \in \mathcal{Y}} \sqrt{W(y \mid 0)W(y \mid 1)} = Z(W).$$

A more refined argument [37] shows $P_e \leq \frac{1}{2}Z(W)$.

Thus $Z(W)$ can be viewed as an upper bound on the probability of error under ML decoding when the channel is used to transmit a single bit. It serves as a reliability measure for the channel since for large $Z(W)$ the channel is unreliable. Conversely, $I(W)$ is the channel symmetric capacity or rate and the closer $I(W)$ is to unity, the more reliable the channel. Intuitively one would expect $I(W) \approx 1$ when the channel is very good and hence $Z(W) \approx 0$ since the probability of error would be low. Conversely $I(W) \approx 0$ when $Z(W) \approx 1$.

It is clear that

$$Z(W_{BEC}) = \delta \quad \text{and} \quad Z(W_{BSC}) = 2\sqrt{p(1-p)}$$

(where δ is the erasure probability for the BEC and p the crossover probability for the BSC) and hence when $\delta \sim 1$, ($Z(W_{BEC}) \sim 1$) the BEC outputs mainly erasures and is useless while for $p \sim 1/2$ the BSC has $Z(W_{BSC}) \sim 1$ and produces almost random $\{0, 1\}$ outputs, hence is also useless. As shown in [3], many of the expressions of interest can be evaluated exactly for the BEC while only bounds are available for the BSC.

The following relationships between these two functions are of interest:

Proposition 4.1 ([3, 5]) *For any BDMC W*

$$\begin{aligned} &(i)\ I(W) \geq \log \tfrac{2}{1+Z(W)} \\ &(ii)\ I(W) \leq \sqrt{1 - Z(W)^2} \\ &(iii)\ I(W) + Z(W) \geq 1 \quad \textit{(equality for BEC)} \\ &(iv)\ I(W)^2 + Z(W)^2 \leq 1. \end{aligned} \tag{4.2}$$

Furthermore for a BDMC one can define a variational distance for the channel W as

$$d(W) = \frac{1}{2} \sum_{y \in \mathcal{Y}} \left| W(y \mid 0) - W(y \mid 1) \right|$$

and it can be shown that $I(W) \leq d(W)$ and $d(W) \leq \sqrt{1 - Z(W)^2}$.

The remainder of the chapter will describe the generator matrix that defines the polar codes, a decoding algorithm for them and a probability of error expression for their performance. It presents a limited overview of the original reference [3] which is essential reading for a deeper understanding of the key concepts.

4.2 Polar Code Construction

As noted, three types of channels will be defined, the (given) base channel W and two other channels defined by recursion. The base channel is a BDMC with binary input from $\mathcal{X} = \{0, 1\}$ and assumed finite number of channel outputs, $\mathcal{Y}$. The compound channel W_N, where $N = 2^n$ for some positive integer n, will have N binary inputs and N outputs, each from $\mathcal{Y}$. The construction of this compound channel is of a recursive and highly structured nature. The final stage of this compound channel will involve N copies of the base channel W in parallel whose N outputs are the compound channel outputs. From the compound channel W_N, the same number of subchannels $W_N^{(i)}, i = 1, 2, \ldots, N$ will be defined. These will be the central players of the subject. It is these subchannels that will be polarized and form the heart of polar coding. The subchannel $W_N^{(i)}$ has a single binary input u_i and $N + (i - 1)$ outputs consisting of the N compound channel outputs and the $(i - 1)$ previous input bits $u_1, \ldots, u_{i-1}$, assumed to have been provided. A decision on the i-th bit u_i will be made sequentially, based on these outputs noting that the channel outputs are functions of all input bits.

The notion of the subchannels is intimately connected to the successive cancellation decoding algorithm described in Section 4.3. While their definition using the constructed compound channel W_N is perhaps not obvious, it is the fact that their polarization can be proved in a rigorous manner that makes them of such great interest. The final segment of the compound channel W_N will involve N copies of the base channel W in parallel.

Several sets of input and output variables are defined during the development. The inputs to the compound channel will be designated $u_i, i = 1, 2, \ldots, N,\ u_i \in \{0, 1\}, N = 2^n$, or $\boldsymbol{u}_1^N = (u_1, u_2, \ldots, u_N)$, and are the information bits. Compound channel outputs will be $y_i, i = 1,2,\ldots,N$, $y_i \in \mathcal{Y}$ or $\boldsymbol{y}_1^N$. The inputs to the base channels at the final stage of the compound channel will be $x_i, = 1,2,\ldots,N$, $x_i \in \mathcal{X}$, the codewords generated from the compound channel input and the generator matrix $\boldsymbol{G}_N$. Thus $\boldsymbol{x}_1^N = \boldsymbol{u}_1^N \boldsymbol{G}_N$ for the generator matrix $\boldsymbol{G}_N$ to be described. As noted this is not the usual form of a generator matrix as in algebraic coding theory since it is a square $N \times N$ matrix that is of full rank over $\mathbb{F}_2$. A recursive figure of the situation, as shown in Figure 4.2, will be developed. Numerous sets of other variables will be introduced as the development proceeds as needed.

The remainder of this section will describe the recursive construction of the binary $N \times N = 2^n \times 2^n$ generator matrix $\boldsymbol{G}_N$ and its relevant properties via sets of other matrices performing specific actions. The relationships developed will be applied to the polar coding problem in the next section.

Denote the matrix

$$\boldsymbol{F} = \begin{bmatrix} 1 & 0 \\ 1 & 1 \end{bmatrix} \tag{4.3}$$

which will be referred to as the *kernel* matrix. Other kernels have been considered and these will be commented on later. Tensor products of this matrix with itself will figure prominently in the construction. Notice that $\boldsymbol{F}$ over $\mathbb{F}_2$ satisfies the relation

$$\boldsymbol{F}^2 = \boldsymbol{I}_2 \quad \text{or} \quad \boldsymbol{F}^{-1} = \boldsymbol{F} \quad \text{over } \mathbb{F}_2.$$

Define recursively for $N = 2^n$

$$\boldsymbol{F}_N = \boldsymbol{F}^{\otimes n} = \boldsymbol{F} \otimes \boldsymbol{F}_{N/2} = \boldsymbol{F} \otimes \boldsymbol{F}^{\otimes (n-1)}, \ \boldsymbol{F}^{\otimes 1} = \boldsymbol{F}$$

and $\boldsymbol{F}^{\otimes n}$ is a $N \times N$ binary matrix. It follows that

$$\left(\boldsymbol{F}^{\otimes n}\right)^{-1} = \boldsymbol{F}^{\otimes n}, \quad \text{over } \mathbb{F}_2.$$

Define the permutation transformation $\boldsymbol{R}_N$, an $N \times N$ permutation matrix as follows. Operating on any input row vector $\boldsymbol{a}_1^N = (a_1, a_2, a_3, \ldots, a_N)$ this permutation produces the output $\boldsymbol{b}_1^N$ as follows:

$$\boldsymbol{a}_1^N \cdot \boldsymbol{R}_N = \boldsymbol{b}_1^N = (a_1, a_3, \ldots, a_{N-1}, a_2, a_4, \ldots, a_N),$$

i.e., it rearranges the components of the input vector to have all the odd subscript coordinates sequentially first followed by the even subscript coordinates. The notation $\boldsymbol{a}_1^N$ is used rather than simply $\boldsymbol{a}$ when discussion is on the coordinates.

It will be convenient to define the $N \times N$ matrix $\boldsymbol{S}_N$ using these two matrices as:

$$\boldsymbol{S}_N = (\boldsymbol{I}_{N/2} \otimes \boldsymbol{F}) \boldsymbol{R}_N. \tag{4.4}$$

By direct computation it is verified that

$$\boldsymbol{S}_N = (\boldsymbol{I}_{N/2} \otimes \boldsymbol{F})\boldsymbol{R}_N = \boldsymbol{R}_N(\boldsymbol{F} \otimes \boldsymbol{I}_{N/2}) \tag{4.5}$$

or in terms of matrices

$$\boldsymbol{S}_N = \begin{bmatrix} \boldsymbol{F} & \boldsymbol{0} & \cdots & \boldsymbol{0} \\ \boldsymbol{0} & \boldsymbol{F} & \cdots & \boldsymbol{0} \\ \vdots & \vdots & \vdots & \vdots \\ \boldsymbol{0} & \boldsymbol{0} & \cdots & \boldsymbol{F} \end{bmatrix} \begin{bmatrix} & & \\ & \boldsymbol{R}_N & \\ & & \end{bmatrix} = \begin{bmatrix} & & \\ & \boldsymbol{R}_N & \\ & & \end{bmatrix} \begin{bmatrix} \boldsymbol{I}_{N/2} & \boldsymbol{0} \\ \boldsymbol{I}_{N/2} & \boldsymbol{I}_{N/2} \end{bmatrix}.$$

Equivalently, for any input row vector $\boldsymbol{a}_1^N$, by direct computation

$$\begin{aligned} \boldsymbol{a}_1^N \boldsymbol{S}_N &= \boldsymbol{a}_1^N (\boldsymbol{I}_{N/2} \otimes \boldsymbol{F})\boldsymbol{R}_N = \boldsymbol{a}_1^N \boldsymbol{R}_N (\boldsymbol{F} \otimes \boldsymbol{I}_{N/2}) \\ &= (a_1 \oplus a_2, a_3 \oplus a_4, \ldots, a_{N-1} \oplus a_N, a_2, a_4, \ldots, a_N). \end{aligned}$$

This structure of $\boldsymbol{S}_8$ is shown in Figure 4.1(b) and $\boldsymbol{S}_N$ in (c).

The matrices $\boldsymbol{F}, \boldsymbol{F}_N, \boldsymbol{R}_N$ and $\boldsymbol{S}_N$ are used recursively to derive binary-coded symbols via the generator matrix $\boldsymbol{G}_N$. The data binary information input sequence $\boldsymbol{u}_1^N \in \{0,1\}^N$ is encoded to yield a binary-coded sequence $\boldsymbol{x}_1^N \in \{0,1\}^N$ via an $N \times N$ generator matrix $\boldsymbol{G}_N$, i.e.,

$$\boldsymbol{x}_1^N = \boldsymbol{u}_1^N \boldsymbol{G}_N, \quad \boldsymbol{x}_1^N \in \{0,1\}^N.$$

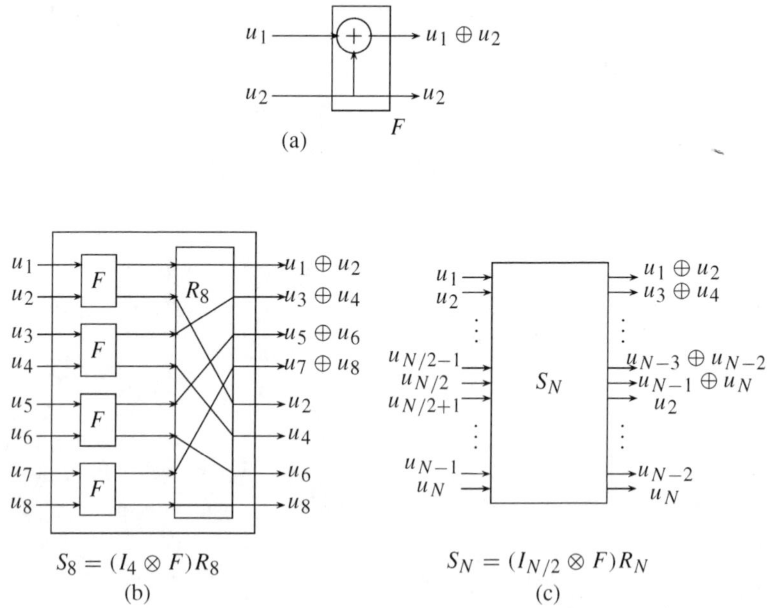

Figure 4.1 (a) F; (b) construction of S_8; (c) S_N

Each code bit is then transmitted through the base channel BDMC W to yield the compound channel outputs $\boldsymbol{y}_1^N \in \mathcal{Y}^N$. Thus the channel output symbols over the alphabet $\mathcal{Y}$ are given probabilistically by

$$W_N(\boldsymbol{y}_1^N \mid \boldsymbol{u}_1^N) = W^N(\boldsymbol{y}_1^N \mid \boldsymbol{x}_1^N = \boldsymbol{u}_1^N \boldsymbol{G}_N) = \prod_{i=1}^{N} W(y_i \mid x_i), \tag{4.6}$$

the relation that governs the overall input, coding and channel output of the system. The designation of W^N is simply W_N with the variables $\boldsymbol{x}$ rather than $\boldsymbol{u}$ as the relationship is one-to-one. Based on the received word $\boldsymbol{y}_1^N$ at the output of the compound channel W_N, decisions as to the system information inputs will be made.

The above transformations and permutations are used to recursively construct the generator matrix, using the commutativity noted above in Equation 4.5, given by ([3], section VIIA):

$$\begin{aligned} \boldsymbol{G}_N &= (\boldsymbol{I}_{N/2} \otimes \boldsymbol{F})\boldsymbol{R}_N(\boldsymbol{I}_2 \otimes \boldsymbol{G}_{N/2}), \quad N \geq 2,\ \boldsymbol{G}_1 = I_1 \\ &= \boldsymbol{S}_N(\boldsymbol{I}_2 \otimes \boldsymbol{G}_{N/2}) \\ &= \boldsymbol{R}_N(\boldsymbol{F} \otimes \boldsymbol{I}_{N/2})(\boldsymbol{I}_2 \otimes \boldsymbol{G}_{N/2}) \\ &= \boldsymbol{R}_N(\boldsymbol{F} \otimes \boldsymbol{G}_{N/2}). \end{aligned} \tag{4.7}$$

The matrix version of this recursion would be:

$$\begin{bmatrix} \boldsymbol{G}_N \end{bmatrix} = \begin{bmatrix} \boldsymbol{R}_N \end{bmatrix} \begin{bmatrix} \boldsymbol{G}_{N/2} & \boldsymbol{0} \\ \boldsymbol{G}_{N/2} & \boldsymbol{G}_{N/2} \end{bmatrix} = \begin{bmatrix} \boldsymbol{S}_N \end{bmatrix} \begin{bmatrix} \boldsymbol{G}_{N/2} & \boldsymbol{0} \\ \boldsymbol{0} & \boldsymbol{G}_{N/2} \end{bmatrix}.$$

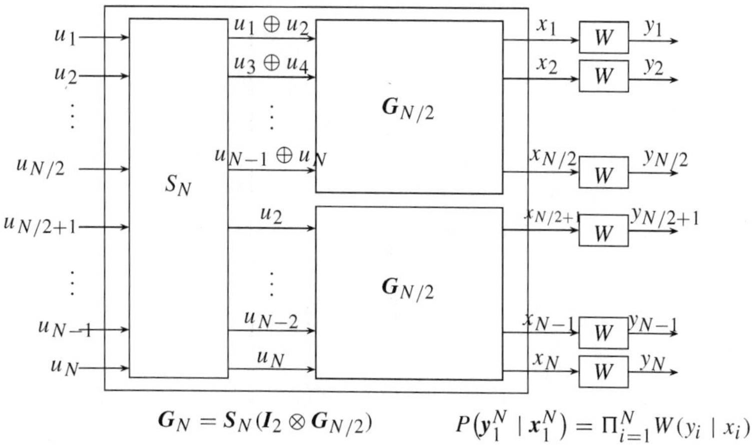

Figure 4.2 Recursive generator matrix representation of $W_N : \boldsymbol{x}_1^N = \boldsymbol{u}_1^N \boldsymbol{G}_N$

Numerous recursion relationships can be developed between many of the matrices introduced. A few selective ones are considered. Thus $\boldsymbol{G}_{N/2} = \boldsymbol{R}_{N/2}(\boldsymbol{F} \otimes \boldsymbol{G}_{N/4})$ and using Equation 4.1:

$$\begin{aligned}\boldsymbol{G}_N &= \boldsymbol{R}_N(\boldsymbol{F} \otimes \boldsymbol{R}_{N/2}(\boldsymbol{F} \otimes \boldsymbol{G}_{N/4}))\\ &= \boldsymbol{R}_N(\boldsymbol{I}_2\boldsymbol{F} \otimes \boldsymbol{R}_{N/2}(\boldsymbol{F} \otimes \boldsymbol{G}_{N/4}))\end{aligned}$$

and using Equation 4.1 (with $\boldsymbol{A} = \boldsymbol{I}_2$, $\boldsymbol{B} = \boldsymbol{F}$, $\boldsymbol{C} = \boldsymbol{R}_{N/2}$ and $\boldsymbol{D} = (\boldsymbol{F} \otimes \boldsymbol{G}_{N/4})$), this equation can be written as

$$\boldsymbol{G}_N = \boldsymbol{R}_N(\boldsymbol{I}_2 \otimes \boldsymbol{R}_{N/2})(\boldsymbol{F}^{\otimes 2} \otimes \boldsymbol{G}_{N/4}).$$

This is a convenient form to repeat the iterations to give:

$$\boldsymbol{G}_N = \boldsymbol{R}_N(\boldsymbol{I}_2 \otimes \boldsymbol{R}_{N/2})(\boldsymbol{I}_4 \otimes \boldsymbol{R}_{N/4}) \cdots (\boldsymbol{I}_{N/2} \otimes \boldsymbol{R}_2)\boldsymbol{F}^{\otimes n} = \boldsymbol{B}_N \boldsymbol{F}^{\otimes n} = \boldsymbol{B}_N \boldsymbol{F}_N, \tag{4.8}$$

where $\boldsymbol{F}_N = \boldsymbol{F}^{\otimes n}$ and

$$\boldsymbol{B}_N = \boldsymbol{R}_N(\boldsymbol{I}_2 \otimes \boldsymbol{R}_{N/2})(\boldsymbol{I}_4 \otimes \boldsymbol{R}_{N/4}) \cdots (\boldsymbol{I}_{N/2} \otimes \boldsymbol{R}_2), \quad (\boldsymbol{I}_{N/2} \otimes \boldsymbol{R}_2) = \boldsymbol{I}_N. \tag{4.9}$$

In matrix form this equation can be expressed as

$$\boldsymbol{B}_N = \boldsymbol{R}_N \begin{bmatrix} \boldsymbol{R}_{N/2} & \boldsymbol{0} \\ \boldsymbol{0} & \boldsymbol{R}_{N/2} \end{bmatrix} \begin{bmatrix} \boldsymbol{I}_2 \otimes \boldsymbol{R}_{N/4} & \boldsymbol{0} \\ \boldsymbol{0} & \boldsymbol{I}_2 \otimes \boldsymbol{R}_{N/4} \end{bmatrix} \cdots \begin{bmatrix} \boldsymbol{I}_{N/4} \otimes \boldsymbol{R}_2 & \boldsymbol{0} \\ \boldsymbol{0} & \boldsymbol{I}_{N/4} \otimes \boldsymbol{R}_2 \end{bmatrix}$$

and from the block diagonal form of this expression it follows from the "top half" of the equation that

$$\boldsymbol{B}_N = \boldsymbol{R}_N(\boldsymbol{I}_2 \otimes \boldsymbol{B}_{N/2}) \tag{4.10}$$

where the last equation is seen by writing the full matrix for $\boldsymbol{B}_N$ and observing that

$$\boldsymbol{B}_{N/2} = \boldsymbol{R}_{N/2}(\boldsymbol{I}_2 \otimes \boldsymbol{R}_{N/4}) \cdots (\boldsymbol{I}_{N/4} \otimes \boldsymbol{R}_2).$$

It is clear by definition that each term in this expression for $\boldsymbol{B}_N$ of the form $(\boldsymbol{I}_{2^i} \otimes \boldsymbol{R}_{N/2^i})$ is a permutation matrix and since $\boldsymbol{B}_N$ is a product of permutation matrices, it is a permutation matrix.

To gather the recursions Equations 4.7 and 4.10 and the trivial one for $\boldsymbol{F}_N$ we have

$$\boldsymbol{G}_N = \boldsymbol{R}_N(\boldsymbol{F} \otimes \boldsymbol{G}_{N/2}),\ \boldsymbol{B}_N = \boldsymbol{R}_N(\boldsymbol{I}_2 \otimes \boldsymbol{B}_{N/2}) \text{ and } \boldsymbol{F}_N = \boldsymbol{F} \otimes \boldsymbol{F}_{N/2}, \quad N = 2^n. \tag{4.11}$$

Example 4.2 The construction of $\boldsymbol{G}_8$ illustrates the form of the matrices. Thus

$$\boldsymbol{G}_8 = \boldsymbol{S}_8(\boldsymbol{I}_2 \otimes \boldsymbol{G}_4) = \boldsymbol{B}_8 \boldsymbol{F}_8 \quad \text{and} \quad \boldsymbol{B}_8 = \boldsymbol{R}_8(\boldsymbol{I}_2 \otimes \boldsymbol{B}_4).$$

Thus

$$
\begin{aligned}
\boldsymbol{G}_4 &= \boldsymbol{B}_4 \cdot \boldsymbol{F}_4 \\
&= \begin{bmatrix} 1 & 0 & 0 & 0 \\ 0 & 0 & 1 & 0 \\ 0 & 1 & 0 & 0 \\ 0 & 0 & 0 & 1 \end{bmatrix} \cdot \begin{bmatrix} 1 & 0 & 0 & 0 \\ 1 & 1 & 0 & 0 \\ 1 & 0 & 1 & 0 \\ 1 & 1 & 1 & 1 \end{bmatrix} = \begin{bmatrix} 1 & 0 & 0 & 0 \\ 1 & 0 & 1 & 0 \\ 1 & 1 & 0 & 0 \\ 1 & 1 & 1 & 1 \end{bmatrix}
\end{aligned}
$$

and

$$
\begin{aligned}
\boldsymbol{G}_8 &= \boldsymbol{S}_8 \cdot (\boldsymbol{I}_2 \otimes \boldsymbol{G}_4) \\
&= \begin{bmatrix} 1&0&0&0&0&0&0&0 \\ 1&0&0&0&1&0&0&0 \\ 0&1&0&0&0&0&0&0 \\ 0&1&0&0&0&1&0&0 \\ 0&0&1&0&0&0&0&0 \\ 0&0&1&0&0&0&1&0 \\ 0&0&0&1&0&0&0&0 \\ 0&0&0&1&0&0&0&1 \end{bmatrix} \cdot \begin{bmatrix} 1&0&0&0&0&0&0&0 \\ 1&0&1&0&0&0&0&0 \\ 1&1&0&0&0&0&0&0 \\ 1&1&1&1&0&0&0&0 \\ 0&0&0&0&1&0&0&0 \\ 0&0&0&0&1&0&1&0 \\ 0&0&0&0&1&1&0&0 \\ 0&0&0&0&1&1&1&1 \end{bmatrix} \\
&= \begin{bmatrix} 1&0&0&0&0&0&0&0 \\ 1&0&0&0&1&0&0&0 \\ 1&0&1&0&0&0&0&0 \\ 1&0&1&0&1&0&1&0 \\ 1&1&0&0&0&0&0&0 \\ 1&1&0&0&1&1&0&0 \\ 1&1&1&1&0&0&0&0 \\ 1&1&1&1&1&1&1&1 \end{bmatrix}
\end{aligned}
$$

Alternatively

$$
\begin{aligned}
\boldsymbol{G}_8 &= \boldsymbol{R}_8(\boldsymbol{F} \otimes \boldsymbol{G}_4) \\
&= \begin{bmatrix} 1&0&0&0&0&0&0&0 \\ 0&0&0&0&1&0&0&0 \\ 0&1&0&0&0&0&0&0 \\ 0&0&0&0&0&1&0&0 \\ 0&0&1&0&0&0&0&0 \\ 0&0&0&0&0&0&1&0 \\ 0&0&0&1&0&0&0&0 \\ 0&0&0&0&0&0&0&1 \end{bmatrix} \cdot \begin{bmatrix} 1&0&0&0&0&0&0&0 \\ 1&0&1&0&0&0&0&0 \\ 1&1&0&0&0&0&0&0 \\ 1&1&1&1&0&0&0&0 \\ 1&0&0&0&1&0&0&0 \\ 1&0&1&0&1&0&1&0 \\ 1&1&0&0&1&1&0&0 \\ 1&1&1&1&1&1&1&1 \end{bmatrix} \\
&= \begin{bmatrix} 1&0&0&0&0&0&0&0 \\ 1&0&0&0&1&0&0&0 \\ 1&0&1&0&0&0&0&0 \\ 1&0&1&0&1&0&1&0 \\ 1&1&0&0&0&0&0&0 \\ 1&1&0&0&1&1&0&0 \\ 1&1&1&1&0&0&0&0 \\ 1&1&1&1&1&1&1&1 \end{bmatrix}
\end{aligned}
$$

Additionally

$$
\begin{aligned}
\boldsymbol{B}_8 &= \boldsymbol{R}_8(\boldsymbol{I}_2 \otimes \boldsymbol{B}_4) \\
&= \begin{bmatrix} 1&0&0&0&0&0&0&0 \\ 0&0&0&0&1&0&0&0 \\ 0&1&0&0&0&0&0&0 \\ 0&0&0&0&0&1&0&0 \\ 0&0&1&0&0&0&0&0 \\ 0&0&0&0&0&0&1&0 \\ 0&0&0&1&0&0&0&0 \\ 0&0&0&0&0&0&0&0 \end{bmatrix} \cdot \begin{bmatrix} 1&0&0&0&0&0&0&0 \\ 0&0&1&0&0&0&0&0 \\ 0&1&0&0&0&0&0&0 \\ 0&0&0&1&0&0&0&0 \\ 0&0&0&0&1&0&0&0 \\ 0&0&0&0&0&0&1&0 \\ 0&0&0&0&0&1&0&0 \\ 0&0&0&0&0&0&0&1 \end{bmatrix} \\
&= \begin{bmatrix} 1&0&0&0&0&0&0&0 \\ 0&0&0&0&1&0&0&0 \\ 0&0&1&0&0&0&0&0 \\ 0&0&0&0&0&0&1&0 \\ 0&1&0&0&0&0&0&0 \\ 0&0&0&0&0&1&0&0 \\ 0&0&0&1&0&0&0&0 \\ 0&0&0&0&0&0&0&1 \end{bmatrix}
\end{aligned}
$$

More can be said of the structure of $\boldsymbol{B}_N$. Consider a vector $\boldsymbol{u}$ of dimension N and $N \times N$ matrix $\boldsymbol{M}$ with coordinate positions and rows/columns labeled with binary n-tuples. Thus, rather than index coordinate positions with integers from 1 to 2^n they can be indexed by binary n-tuples, the binary expansions of $0, 1, 2, \ldots, 2^n - 1$. The two methods are equivalent. The bit reversal of the vector $\boldsymbol{u}$ is $\boldsymbol{v}$ if $\boldsymbol{u} = (u_{00\cdots00}, u_{00\cdots01}, \ldots, u_{11\cdots10}, u_{11\cdots11})$ and $\boldsymbol{v} = (u_{00\cdots00}, u_{10\cdots00}, \ldots, u_{01\cdots11}, u_{11\cdots11})$.

The matrix $\boldsymbol{M}$ is said to be a *bit reversal matrix* if $\boldsymbol{v} = \boldsymbol{u}\boldsymbol{M}$ and $\boldsymbol{v}$ is the bit reversal vector of $\boldsymbol{u}$. Thus $v_{(i_1, i_2, \ldots, i_n)} = u_{(i_n, i_{n-1}, \ldots, u_1)}$, i.e., the binary expansion of the vector subscripts are reversed.

It is claimed that $\boldsymbol{B}_N$ is a bit reversal matrix. A recursive proof of this follows from the recursion Equation 4.9, i.e., if $\boldsymbol{B}_{N/2}$ is assumed a bit reversal matrix, then using Equation 4.9 it is shown that $\boldsymbol{B}_N$ is. The following small example will be a useful reference for the comments to follow.

Example 4.3 Consider the following array of $N = 2^4$ subscripts in the first column. Using Equation 4.9, the matrix $\boldsymbol{R}_{16}$ operates on this column to produce the second column, placing the odd row 4-tuples sequentially first (all those ending with a 0) followed sequentially by the even row 4-tuples (those ending in 1). The second column is operated on by $\boldsymbol{I}_2 \otimes \boldsymbol{R}_8$ – equivalently $\boldsymbol{R}_8$ operates on the first eight rows and the bottom eight rows independently to produce the third column. Similarly $\boldsymbol{I}_4 \otimes \boldsymbol{R}_4$ operates on the third column to produce the last column which is the bit reversal of the first column. The overall effect is to operate on the first column by $\boldsymbol{B}_{16}$ to produce the last column.

Original subscript order			Final (reverse) order
	$\boldsymbol{R}_{16}$	$\boldsymbol{I}_2 \otimes \boldsymbol{R}_8$	$\boldsymbol{I}_4 \otimes \boldsymbol{R}_4$
0000	0000	0000	0000
0001	0010	0100	1000
0010	0100	1000	0100
0011	0110	1100	1100
0100	1000	0010	0010
0101	1010	0110	1010
0110	1100	1010	0110
0111	1110	1110	1110
1000	0001	0001	0001
1001	0011	0101	1001
1010	0101	1001	0101
1011	0111	1101	1101
1100	1001	0011	0011
1101	1011	0111	1011
1110	1101	1011	0111
1111	1111	1111	1111

(4.12)

Denote a subscript as a binary n-tuple $\boldsymbol{i} = (i_1, i_1, \ldots, i_n)$ and its bit reversal as $\boldsymbol{i}_r = (i_n, i_{n-1}, \ldots, i_1)$. That $\boldsymbol{B}_N$ is a bit reversal matrix can be shown by recursion. Assume $\boldsymbol{B}_{N/2}$ is a bit reversal matrix and consider Equation 4.10

$$\boldsymbol{B}_N = \boldsymbol{R}_N (\boldsymbol{I}_2 \otimes \boldsymbol{B}_{N/2})$$

and consider a column array of binary n-tuples representing the binary expansions of the integers $0, 1, 2, \ldots, 2^n - 1$. The first bit of the elements of the top half of the array is 0 and the first bit of elements of the bottom half of the array is a 1. The impact of $\boldsymbol{R}_N$ on the top half of the array is to place this first 0 bit last in each of the top $2^{(n-1)}$ n-tuples and to place the first 1 bit of the bottom $2^{(n-1)}$ array elements last. By assumption $\boldsymbol{B}_{N/2}$ is a bit reversal matrix – consider its operation on the top half of the array. Since it achieves the bit reversal by reordering the rows of the array, the impact on the top half of the array is to reorder the rows so that the first $(n-1)$ bits are reversed. Thus the impact of $\boldsymbol{R}_N$ and $\boldsymbol{B}_{N/2}$ on each half of the array is bit reversal.

Since $\boldsymbol{B}_n$ is a permutation matrix and a bit reversal matrix it is a symmetric matrix (with ones on the main diagonal corresponding to subscript positions where $\boldsymbol{i} = \boldsymbol{i}_r$), and if $\boldsymbol{B}_N = (b_{i,j})$, then $b_{i,j} = 1$ iff $\boldsymbol{j} = \boldsymbol{i}_r$.

Thus

$$\boldsymbol{B}_N^T = \boldsymbol{B}_N \quad \text{and} \quad \boldsymbol{B}_N^{-1} = \boldsymbol{B}_N \quad \text{and} \quad \boldsymbol{B}_N^2 = \boldsymbol{I}_N \quad \text{(over both } \mathbb{F}_2 \text{ and } \mathbb{R}).$$

It was previously observed that $\boldsymbol{F}_N = \boldsymbol{F}^{\otimes n}$ is also a matrix of order 2, i.e., $\left(\boldsymbol{F}_N\right)^2 = \boldsymbol{I}_N$ over $\mathbb{F}_2$ although not symmetric.

Further properties of the matrices $\boldsymbol{F}_N, \boldsymbol{G}_N$ and $\boldsymbol{B}_N$ are noted.

Denote the matrix $\boldsymbol{F} = (f_{i_1,j_1}), i_1, j_1 \in \{0,1\}$ shown in Figure 4.1. The Boolean expression for the elements of the matrix is easily seen to be

$$f_{i_1,j_1} = 1 \oplus j_1 \oplus i_1 j_1$$

which specifies a zero element for the top right matrix entry and one for the other three elements. For the matrix $\boldsymbol{F}^{\otimes 2} = (f_{i_1 i_2, j_1 j_2})$ the corresponding Boolean expression is

$$f_{i_1 i_2, j_1 j_2} = (1 \oplus j_1 \oplus i_1 j_1)(1 \oplus j_2 \oplus i_2 j_2).$$

In this case the second subscript (i_2 and j_2) specifies the 2×2 subblock, the upper right subblock being all zero and the other three $\boldsymbol{F}$. The first subscript (i_1 and j_1) specifies the element within the 2×2 subblocks.

The extrapolation to the general case for $\boldsymbol{F}^{\otimes n} = (\boldsymbol{f}_{i_1 i_2 \dots i_n, j_1 j_2 \dots j_n})$ is immediate ([3], equations 72 and 73):

$$\begin{aligned} \boldsymbol{f}_{i_1 \dots i_n, j_1 \dots j_n} &= \prod_{k=1}^{n} \boldsymbol{f}_{i_k, j_k} \\ &= \prod_{i=1}^{n} (1 \oplus j_k \oplus i_k j_k), \end{aligned}$$

i.e., reversing the subscripts leaves the expression unchanged. Notice that from the above expression

$$\boldsymbol{f}_{i,j} = \boldsymbol{f}_{i_r, j_r}. \tag{4.13}$$

A similar expression for the generator matrix $\boldsymbol{G}_N$ is desired. Recall from Equation 4.8 that $\boldsymbol{G}_N = \boldsymbol{B}_N \boldsymbol{F}^{\otimes n} = (g_{i,j})$ and that $\boldsymbol{B}_N = (b_{i,j})$ is a symmetric permutation matrix with a 1 in row $\boldsymbol{i}$ in column $\boldsymbol{i}_r$ and $b_{i,i_r} = b_{i_r,i}$. It follows from these properties that

$$g_{i,j} = f_{i_r,j}.$$

It is interesting to note that the matrices $\boldsymbol{B}_N$ and $\boldsymbol{G}_N$ commute – consider

$$\boldsymbol{F}^{\otimes n} \boldsymbol{B}_N = (g'_{i,j}) \quad \text{and note that} \quad g'_{i,j} = f_{i,j_r}$$

since the $\boldsymbol{j}$-th column of $\boldsymbol{B}_N$ has a 1 in row $\boldsymbol{j}_r$. But from the property of Equation 4.13 it follows that

$$g'_{i,j} = f_{i,j_r} = f_{i_r,j} = g_{i,j}$$

and $\boldsymbol{B}_N$ and $\boldsymbol{F}^{\otimes n} = \boldsymbol{F}_N$ commute as do $\boldsymbol{G}_N$ and $\boldsymbol{F}_N$. It follows from this and a previous equation that

$$g_{i,j} = f_{i_r,j} = \prod_{k=1}^{n}(1 \oplus j_k \oplus i_{n-k} j_k).$$

Thus

$$\boldsymbol{G}_N = \boldsymbol{F}^{\otimes n}\boldsymbol{B}_N = \boldsymbol{B}_N\boldsymbol{F}^{\otimes n} \text{ and } \boldsymbol{G}_N^2 = \boldsymbol{B}_N\boldsymbol{F}_N\boldsymbol{B}_N\boldsymbol{F}_N = \boldsymbol{F}_N^2\boldsymbol{B}_N^2 = \boldsymbol{I}_N \text{ over } \mathbb{F}_2.$$

4.3 Subchannel Polarization and Successive Cancellation Decoding

The inputs to the final parallel base W channels in the above figures are codewords in a linear code with generator matrix $\boldsymbol{G}_N = \boldsymbol{B}_N\boldsymbol{F}_N = \boldsymbol{B}_N\boldsymbol{F}^{\otimes n}$, i.e., $\boldsymbol{x}_1^N = \boldsymbol{u}_1^N\boldsymbol{G}_N$. It follows that the transition probabilities for the overall compound channel W_N as given in Equation 4.6 are given by:

$$W_N\left(\boldsymbol{y}_1^N \mid \boldsymbol{u}_1^N\right) = W^N\left(\boldsymbol{y}_1^N \mid \boldsymbol{x}_1^N = \boldsymbol{u}_1^N\boldsymbol{G}_N\right) = \prod_{i=1}^{N} W(y_i \mid x_i), \quad \boldsymbol{y}_1^N \in \mathcal{Y}^N,\ \boldsymbol{u}_1^N \in \mathcal{X}^N$$

and the probabilistic behavior of the channel is entirely determined by the transformation process resulting in the code generator matrix $\boldsymbol{G}_N$ and the base channel W.

It remains to devise a decoding algorithm process for these codes and to determine the performance of the code and its decoding in terms of the probability of error, topics addressed in this section. The iterative and recursive structure of the generator matrix $\boldsymbol{G}_N$ leads naturally to a notion of successive cancellation decoding which will be outlined here. This will be applied, not to the compound channel W_N but to certain subchannels $W_N^{(i)}, i = 1,2,\ldots,N$, which will be defined using W_N. It is these subchannels to which successive cancellation decoding will be applied and which will be shown to have remarkable properties. The error performance will be evaluated for these subchannels.

From the compound channel W_N the BMS subchannels (also referred to as split channels) $W_N^{(i)}, i = 1,2,\ldots,N$ are defined in the following manner.

Consider a channel with single binary input u_i and outputs $\boldsymbol{y}_1^N$ and all previous inputs $\boldsymbol{u}_1^{i-1}$ which are assumed available. Thus the subchannels are defined by:

$$W_N^{(i)}\left(\boldsymbol{y}_1^N, \boldsymbol{u}_1^{i-1} \mid u_i\right) : \mathcal{X} \longrightarrow \mathcal{Y}^N \times \mathcal{X}^{i-1}$$
$$u_i \longmapsto \left(\boldsymbol{y}_1^n, \boldsymbol{u}_1^{i-1}\right).$$

It is somewhat unconventional to define a channel with an input of u_i and output consisting of all previous channel inputs as well as all channel outputs. In practice these outputs $\boldsymbol{u}_i^{i-1}$ are not available and estimates of the previous subchannel inputs are used. It is convenient for the analysis to assume a genie has provided the correct output of the previous $(i-1)$ channels at the output of the BMS channel $W_N^{(i)}$ and the subchannel will be used to derive an estimate $\hat{u}_i$ of the input u_i given knowledge of all previous inputs and all channel outputs as i goes from 1 to N. Such a process is termed *successive cancellation decoding*.

To obtain the transition probabilities for the subchannel $W_N^{(i)}$ consider the following. Let $P(\boldsymbol{y}_1^N, \boldsymbol{u}_1^N)$ denote the probability that $\boldsymbol{u}_1^N$ is chosen at the input for transmission (independently and equally likely), and $\boldsymbol{y}_1^N$ is received at the compound channel output. Then $P(\boldsymbol{u}_1^N) = 1/2^N$ and

$$P\left(\boldsymbol{y}_1^N, \boldsymbol{u}_1^N\right) = \frac{1}{2^N} W_N\left(\boldsymbol{y}_1^N \mid \boldsymbol{u}_1^N\right)$$

and

$$P\left(\boldsymbol{y}_1^N, \boldsymbol{u}_1^i\right) = \sum_{\boldsymbol{u}_{i+1}^N \in \mathcal{X}^{N-i}} \frac{1}{2^N} W_N\left(\boldsymbol{y}_1^N \mid \boldsymbol{u}_1^N\right)$$

and the transition probabilities for the subchannel, $W_N^{(i)}$ are

$$\begin{aligned} W_N^{(i)}\left(\boldsymbol{y}_1^N, \boldsymbol{u}_1^{i-1} \mid u_i\right) &= P(\boldsymbol{y}_1^N, \boldsymbol{u}_1^i)/P(u_i), \quad P(u_i) = 1/2 \\ &= \sum_{\boldsymbol{u}_{i+1}^N \in \mathcal{X}^{N-i}} \frac{1}{2^{N-1}} W_N\left(\boldsymbol{y}_1^N \mid \boldsymbol{u}_1^N\right) \end{aligned} \tag{4.14}$$

where

$$W_N\left(\boldsymbol{y}_1^n \mid \boldsymbol{u}_1^N\right) = W^N\left(\boldsymbol{y}_1^n \mid \boldsymbol{x}_1^N = \boldsymbol{u}_1^N \boldsymbol{G}_N\right) = \prod_{i=1}^N W(y_i \mid x_i). \tag{4.15}$$

Thus the probabilistic behavior of the subchannels $W_N^{(i)}\left(\boldsymbol{y}_1^N, \boldsymbol{u}_1^{i-1} \mid u_i\right)$ is entirely determined by the probabilistic behavior of the base channel W and the code specified by the matrix $\boldsymbol{G}_N$. All properties of these N subchannels such as the mutual information (symmetric capacity) $I\left(W_N^{(i)}\right)$ between the input and outputs and their Bhattacharyya parameters can be computed. It turns out that the Bhattacharyya parameter is often easier to work with:

$$Z\left(W_N^{(i)}\right) = \sum_{y_1^N \in \mathcal{Y}^N} \sum_{u_1^{i-1} \in \mathcal{X}^{i-1}} \sqrt{W_N^{(i)}\left(y_1^N, u_1^{i-1} \mid u_i = 1\right) W_N^{(i)}\left(y_1^N, u_1^{i-1} \mid u_i = 0\right)},$$
$$i = 1, 2, \ldots, N. \tag{4.16}$$

for the transition probabilities of Equation 4.15.

The information-theoretic behavior of the subchannels will be discussed later in the section where it will be noted that as $N \to \infty$ these subchannels tend to polarize in the sense that asymptotically a subchannel $W_N^{(i)}$ will tend to be either a perfect channel for which $I(W_N^{(i)}) \sim 1$ (or $\mathbb{Z}(W_N^{(i)}) \sim 0$) or a useless channel for which $I(W_N^{(i)}) \sim 0$ (or $Z(W_N^{(i)}) \sim 1$). Furthermore the fraction of the subchannels that asymptotically approach perfect tends to $I(W)$ which will imply the performance of the overall code approaches the compound channel capacity.

The computations are necessarily intricate and here we focus on their implications while preserving the thread of the argument, but omitting the details of the computations.

The central problem of polar coding will be to identify the good channels and this is a nontrivial problem. These channels will be used and the remainder of the channels, bad channels, will convey fixed or frozen bits and hence convey no useful information. More comments on this key issue will be considered after discussion of decoding and the resulting probability of error. For the moment it will be assumed that a subset $\mathcal{A}$ of the N subchannels have been chosen as the polar code, the set of good channels, the remaining useless channels $\mathcal{A}^c$ conveying only fixed information known to the receiver. Furthermore it will be convenient to identify the good channels $\mathcal{A}$ with the corresponding rows of the generator matrix $\boldsymbol{G}_N$. say $\boldsymbol{G}_{\mathcal{A}}$, and that the resulting code will be a binary $(N, K)_2$ code, $|\mathcal{A}| = K$, the actual number of useful inputs.

The likelihood ratio for the subchannels will be used to successively decode them and proceeds as follows: let

$$L_N^{(i)}\left(y_1^N, u_1^{i-1}\right) = \frac{W_N^{(i)}\left(y_1^n, u_1^{i-1} \mid u_i = 0\right)}{W_N^{(i)}\left(y_1^n, u_1^{i-1} \mid u_i = 1\right)}. \tag{4.17}$$

Using this LR the successive cancellation decoding rule is

$$\hat{u}_i \triangleq \begin{cases} u_i \text{ known}, & \text{if } \; i \in \mathcal{A}^c \\ h_i\left(y_1^N, \hat{u}_1^{i-1}\right), & i \in \mathcal{A}, \end{cases} \tag{4.18}$$

where

$$h_i\left(\boldsymbol{y}_1^N, \hat{\boldsymbol{u}}_1^{i-1}\right) \triangleq \begin{cases} 0, & \text{if } L_N^{(i)}\left(\boldsymbol{y}_1^N, \boldsymbol{u}_1^{i-1}\right) = \dfrac{W_N^{(i)}\left(\boldsymbol{y}_1^n, \hat{\boldsymbol{u}}_1^{i-1} \mid 0\right)}{W_N^{(i)}\left(\boldsymbol{y}_1^n, \hat{\boldsymbol{u}}_1^{i-1} \mid 1\right)} \geq 1 \\ 1, & \text{otherwise.} \end{cases}$$

Figure 4.3 shows the decoder for the successive cancellation decoder for the subchannel $W_N^{(i)}$. The overall successive cancellation decoder is shown in Figure 4.4. Previous input estimates are shown for this decoding – for analysis, as noted, the actual inputs u_1^{i-1} will be assumed known.

Before returning to a consideration of the important properties of the Bhattacharyya parameter, an expression for the probability of error of the successive cancellation decoder is obtained. This will further illuminate the relevance of this parameter.

Such a receiver (successive cancellation decoder) will make a block error if $\hat{\boldsymbol{u}}_{\mathcal{A}} \neq \boldsymbol{u}_{\mathcal{A}}$ since the bits of $\boldsymbol{u}_{\mathcal{A}}^c$ are assumed known. This is a one-pass algorithm.

Consider a code of length $N = 2^n$ and dimension K with the set of good channels $\mathcal{A}$. The probability of block error for such a code will be given by

$$P_e(N, K, \mathcal{A}, \boldsymbol{u}_{\mathcal{A}^c}) = \sum_{\boldsymbol{u}_{\mathcal{A}} \in \mathcal{X}^K} \frac{1}{2^K} \sum_{\substack{\boldsymbol{y}_1^N \in \mathcal{Y}^N \\ \hat{\boldsymbol{u}}_1^N \neq \boldsymbol{u}_1^N}} W_N\left(\boldsymbol{y}_1^N \mid \boldsymbol{u}_1^N\right).$$

The expression indicates the probability of error being dependent on the frozen (known) bits of the code coordinate positions of $\mathcal{A}^c$. In fact it can be shown [3] that for a symmetric channel the error performance is independent of the

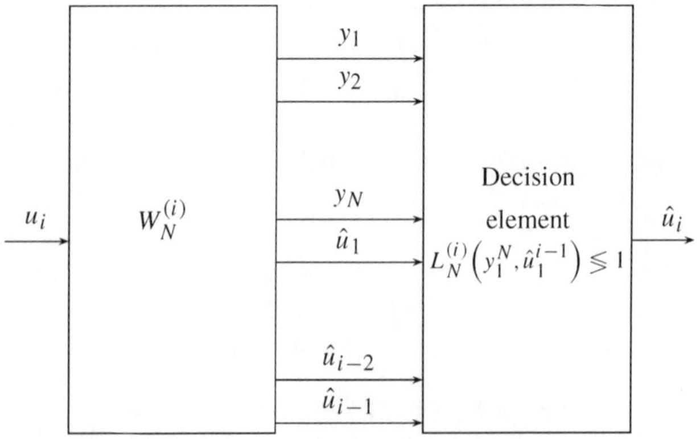

Figure 4.3 Channel splitting and decoding

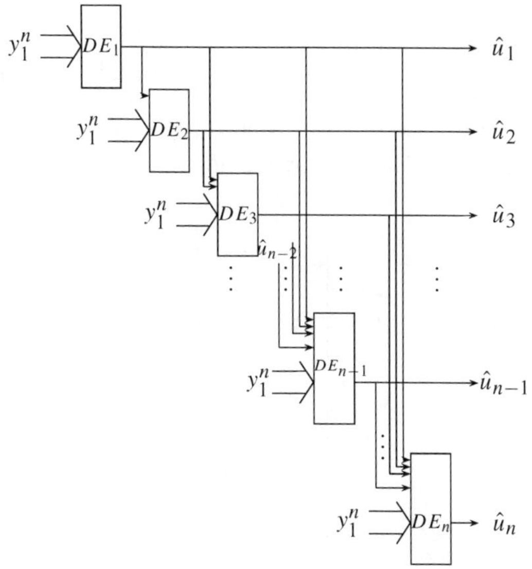

Figure 4.4 Successive cancellation decoder

chosen frozen bits and our interest is in such symmetric channels. So the frozen bits chosen will be assumed fixed, say all ones.

Interest is to derive an expression for this probability of error. Consider the block error analysis for a polar code defined by the set of positions $\mathcal{A}$. The joint probability of the chosen input $\boldsymbol{u}_1^N$ and the resulting channel output $\boldsymbol{y}_1^N$ on the space $\mathcal{X}^N \times \mathcal{Y}^N$ is given by

$$P\left(\boldsymbol{u}_1^N, \boldsymbol{y}_1^N\right) = W_N\left(\boldsymbol{y}_1^N \mid \boldsymbol{u}_1^N\right)\frac{1}{2^N}$$

and consider the decision on the i-th output bit as

$$\hat{u}_i = \begin{cases} u_i, & i \in \mathcal{A}^c \\ h_i\left(\boldsymbol{y}_1^N, \hat{\boldsymbol{u}}_1^{i-1}\right), & i \in \mathcal{A}. \end{cases}$$

since the frozen bits are known at the receiver. Likewise the decoder will make a block error if any of the bits in the set $\mathcal{A}$ are not correct – thus define the error event

$$\mathcal{E} \triangleq \left\{\left(\boldsymbol{u}_1^N, \boldsymbol{y}_1^N\right) \in \mathcal{X}^N \times \mathcal{Y}^N \mid \hat{\boldsymbol{u}}_{\mathcal{A}}\left(\boldsymbol{u}_1^N, \boldsymbol{y}_1^N\right) \neq \boldsymbol{u}_{\mathcal{A}}\right\}.$$

Define the event $\mathcal{E}_i$ as the event an error occurs at the i-th decoded bit (not necessarily the *first* decoding error in the block) – then $u_i \neq h_i(\boldsymbol{y}_1^N, \boldsymbol{u}_1^{i-1})$ or equivalently, in terms of the LR,

$$\mathcal{E}_i \triangleq \left\{ \left(\boldsymbol{u}_1^N, \boldsymbol{y}_1^N\right) \in \mathcal{X}^N \times \mathcal{Y}^N \mid W_N^{(i-1)}\left(\boldsymbol{y}_1^N, \boldsymbol{u}_1^{i-1} \mid u_i\right) \leq W_N^{(i-1)}\left(\boldsymbol{y}_1^N, \boldsymbol{u}_1^{i-1} \mid u_i \oplus 1\right) \right\}$$

since the LR implies the decision will be made in favor of $u_i \oplus 1$ and hence in error. The event there is at least one error in the received block is, by the union bound,

$$\mathcal{E} \subset \bigcup_{i \in \mathcal{A}} \mathcal{E}_i \quad \text{and} \quad P(\mathcal{E}) \leq \sum_{i \in \mathcal{A}} P(\mathcal{E}_i).$$

Let Π be a set indicator function in the sense that for a set $\mathcal{E}_i \subset \mathcal{X} \times \mathcal{Y}$

$$\Pi(\mathcal{E}_i) = \begin{cases} 1, & \text{if } \left(u_1^N, y_1^N\right) \in \mathcal{E}_i \\ 0, & \text{else.} \end{cases}$$

Then

$$\begin{aligned}
P(\mathcal{E}_i) &= \sum_{u_1^N \in \mathcal{X}^N,\, y_1^N \in \mathcal{Y}^N} \frac{1}{2^N} W_N\left(y_1^N \mid u_1^N\right) \Pi(\mathcal{E}_i) \\
&\leq \sum_{u_1^N \in \mathcal{X}^N,\, y_1^N \in \mathcal{Y}^N} \frac{1}{2^N} W_N\left(y_1^N \mid u_1^N\right) \sqrt{\frac{W_N^{(i)}\left(y_1^N, u_1^{i-1} \mid u_i \oplus 1\right)}{W_N^{(i)}\left(y_1^N, u_1^{i-1} \mid u_i\right)}} \\
&\leq \sum_{u_1^{i-1} \in \mathcal{X}^{i-1},\, y_1^N \in \mathcal{Y}^N} \left\{ \sum_{\boldsymbol{u}_{i+1}^N \in \mathcal{X}^{N-i}} \frac{1}{2^N} W_N\left(\boldsymbol{y}_1^N \mid \boldsymbol{u}_1^N\right) \right\} \sqrt{\frac{W_N^{(i)}\left(y_1^N, u_1^{i-1} \mid u_i \oplus 1\right)}{W_N^{(i)}\left(y_1^N, u_1^{i-1} \mid u_i\right)}}
\end{aligned}$$

since the argument of the square root is by definition greater than unity on $\mathcal{E}_i$ and since the expression under the square root does not involve the variables u_{i+1}^N. To show this last expression is $Z\left(W_N^{(i)}\right)$ proceed as follows. By Equation 4.14

$$\begin{aligned}
P(\mathcal{E}_i) &\leq \sum_{u_1^{i-1} \in \mathcal{X}^{i-1},\, y_1^N \in \mathcal{Y}^N} W_N^{(i)}\left(y_1^N, u_1^{i-1} \mid u_i\right) \cdot \sqrt{\frac{W_N^{(i)}\left(y_1^N, u_1^{i-1} \mid u_i \oplus 1\right)}{W_N^{(i)}\left(y_1^N, u_1^{i-1} \mid u_i\right)}} \\
&\leq \sum_{u_1^{i-1} \in \mathcal{X}^{i-1},\, y_1^N \in \mathcal{Y}^N} \sqrt{W_N^{(i)}\left(y_1^N, u_1^{i-1} \mid u_i \oplus 1\right) W_N^{(i)}\left(y_1^N, u_1^{i-1} \mid u_i\right)} \\
&= Z\left(W_N^{(i)}\right).
\end{aligned}$$

Thus

$$P(\mathcal{E}) \leq \sum_{i \in \mathcal{A}} Z\left(W_N^{(i)}\right).$$

The equation implies that to minimize the probability of error one should choose the $|\mathcal{A}|$ most reliable channels for the set $\mathcal{A}$, an obvious choice which is reflected in the definition below.

Proposition 2 of [3] follows from these observations:

Proposition 4.4 *For any BDMC W and any choice of parameters ($N = 2^n, K = |\mathcal{A}|$) the probability of block error is bounded by*

$$P_e(N,K,\mathcal{A}) \leq \sum_{i \in \mathcal{A}} Z\left(W_N^{(i)}\right). \tag{4.19}$$

Assuming the set of useful inputs $\mathcal{A}$ is known, the code with generator matrix G_N can be broken down into a "useful" part and a "useless" part as

$$\boldsymbol{x}_1^N = \boldsymbol{u}_{\mathcal{A}} \boldsymbol{G}_N(\mathcal{A}), \boldsymbol{u}_{\mathcal{A}^c} \boldsymbol{G}_N(\mathcal{A}^c)$$

where the arguments of $\boldsymbol{G}_N$ indicate corresponding submatrices of $\boldsymbol{G}_N$ and the comma indicates catenation. Thus the encoding function is from $\boldsymbol{u}_{\mathcal{A}}$ to $\boldsymbol{x}_1^N$ with $|\mathcal{A}^c|$ frozen (predetermined) variables among the N coordinate positions of $\boldsymbol{x}_1^N$. These restricted codes are termed *coset* codes for obvious reasons. In the case where the frozen bits are all set to 0 the polar code is a linear code. For $|\mathcal{A}| = K$ the overall code is referred to as an $(N, K, \mathcal{A})$ code.

The proposition suggests the following definition of polar codes:

Definition 4.5 Given a BDMC W, a $\boldsymbol{G}_N$ coset code with parameters $(N, K, \mathcal{A}, \boldsymbol{u}_{\mathcal{A}^c})$ will be called a *polar code* for W if the information set $\mathcal{A}$ is chosen as a K element subset of $\{1, 2, \ldots, N\}$ such that

$$Z\left(W_N^{(i)}\right) \leq Z\left(W_N^{(j)}\right)$$

for all $i \in \mathcal{A}$ and $j \in \mathcal{A}^c$.

Thus the $K = |\mathcal{A}|$ information positions (subchannels) are chosen corresponding to the K subchannels with the smallest values of $Z(W_N^{(i)})$.

The above material suggests several problems. It is interesting to note a polar code is chosen via the mutual information or Bhattacharyya parameter of the subchannels. It is a determination of which of the $N = 2^n$ channels are the $K = NR$ most reliable subchannels (those with lowest Bhattacharyya parameter or highest mutual information) $R = K/N$ the rate of the code, which, for N large enough, will be close to perfect, a reflection of the polarization of the subchannels, while the other channels are virtually useless. The computation of the Bhattacharyya parameter for each of the subchannels is computationally expensive (of order $O(N \log N)$). Other techniques to determine the good channels have been devised and a brief comment on these is noted below.

The above discussion has shown the crucial role of the subchannels in the formulation of polar codes. To complete the picture it is necessary to establish that the number of good subchannels is sufficient to achieve capacity on the channel. The theorem below, which is a composite of three key theorems of ([3], theorems 1, 2 and 3), stated without proof, establishes this. This theorem encapsulates the key ideas of the entire chapter: Let W be a BDMC with symmetric capacity $I(W) > 0$:

Theorem 4.6 *(i) The channels $W_N^{(i)}$ polarize in that for any fixed $\epsilon \in [0,1]$ as N goes to infinity through powers of 2 the fraction of indices $i \in [1,N]$ for which $I(W_N^{(i)}) \in (1-\epsilon, 1]$ goes to $I(W)$ and the fraction for which $I(W_N^{(i)}) \in [0,\epsilon)$ goes to $1 - I(W)$.*

(ii) For a fixed $R < I(W)$ there exists a sequence of sets $\mathcal{A}_N \subset [1,N]$ such that $|\mathcal{A}_N| \geq NR$ and $Z(W_N^{(i)}) \leq O(N^{-5/4})$ for all $i \in \mathcal{A}$.

(iii) The block error probability for polar coding under successive cancellation decoding satisfies

$$P_e(N,R) = O(N^{-1/4}).$$

It is in this sense that polar codes achieve capacity on the base channel W which has capacity C, $|C| \leq 1$. To summarize, the base channel is expanded to the compound channel W_N, $N = 2^n$ which will have capacity NC and which is comprised of approximately $NI(W)$ good subchannels, each of which is approximately perfect.

All three of these statements (theorems) are remarkable. The third statement follows from Proposition 4.4 ([3], proposition 2) and from item (ii) of the theorem in that the number of terms in the expression for the probability of error is $O(N)$, each term being $O(N^{-5/4})$. The implication of the theorem is that as N increases $I(W_N^{(i)})$ tends either to 0 or 1, and the fraction of subchannels that tend to unity is $I(W)$. Thus capacity of the compound channel W_N tends to $NI(W)$ where $I(W)$ is the (symmetric) capacity of the base channel W. The overall coding/decoding system achieves capacity. It will be noted later the error performance can be improved to $O(2^{-2^{n\beta}})$.

The highly iterative construction of the polar coder is useful in deriving recursive expressions for the transition probabilities of the subchannels $W_N^{(i)}$ and the likelihood ratios as the channel order increases through powers of two. It is also useful to give an indication of how information theory can be used to show polarization of these channels. The subchannel $W_N^{(i)}$ has an input u_i and outputs $\boldsymbol{y}_1^n, \boldsymbol{u}_1^{i-1}$ which the subchannel uses to form the estimate $\hat{u}_i$. The transition probabilities for this channel were derived in Equation 4.14 as

$$W_N^{(i)}\left(\boldsymbol{y}_1^N, \boldsymbol{u}_1^{i-1} \mid u_i\right) = \sum_{\boldsymbol{u}_{i+1}^N \in \mathcal{X}^{N-i}} \frac{1}{2^{N-1}} W_N\left(\boldsymbol{y}_1^N \mid \boldsymbol{u}_1^N\right) \tag{4.20}$$

and thus expressions for $I(W_N^{(i)})$ and $Z(W_N^{(i)})$ can be computed.

A few comments on the computation of the quantities $Z(W_N^{(i)})$ and $I(W_N^{(i)})$ and of the transition probabilities $W_N^{(i)}(\mathbf{y}_1^N, \mathbf{u}_1^{i-1} \mid u_i)$ are in order. For these quantities the highly recursive/iterative structure of the code generator matrix can be exploited effectively. The briefest of overviews of this computational issue is given.

The doubling construction, as noted in Figure 4.3, leads to the following relationships which are most useful in reducing the complexity of the required computations ([3], proposition 3):

$$W_{2N}^{(2i-1)}\left(\mathbf{y}_1^{2N}, u_1^{2i-2} \mid \mathbf{u}_{2i-1}\right) = \sum_{u_{2i}} \tfrac{1}{2} W_N^{(i)}\left(\mathbf{y}_1^N, \mathbf{u}_{1,o}^{2i-2} \oplus \mathbf{u}_{1,e}^{2i-2} \mid u_{2i-1} \oplus u_{2i}\right) \cdot W_N^{(i)}\left(\mathbf{y}_{N+1}^{2N}, u_{1,e}^{2i-2} \mid u_{2i}\right) \tag{4.21}$$

and

$$W_{2N}^{(2i)}\left(\mathbf{y}_1^{2N}, \mathbf{u}_1^{2i-1} \mid u_{2i}\right) = \tfrac{1}{2} W_N^{(i)}\left(\mathbf{y}_1^N, \mathbf{u}_{1,o}^{2i-2} \oplus \mathbf{u}_{1,e}^{2i-2} \mid u_{2i-1} \oplus u_{2i}\right) \cdot W_N^{(i)}\left(\mathbf{y}_{N+1}^{2N}, \mathbf{u}_{1,e}^{2i-2} \mid u_{2i}\right), \tag{4.22}$$

thus leveraging the i-th computation to give the $2i$-th one.The subscripts "e" and "o" indicate even and odd subscripts are to be taken. Furthermore it can be shown that

$$\frac{1}{2^{2N-1}} W_{2N}\left(y_1^{2N} \mid u_1^{2N}\right) = \frac{1}{2} \cdot \frac{1}{2^{N-1}} \cdot W_N\left(y_1^N \mid u_{1,o}^{2N} \oplus u_{1,e}^{2N}\right) \cdot \frac{1}{2^{N-1}} W_N\left(y_{N+1}^{2N} \mid u_{1,e}^{2N}\right).$$

Equations 4.21 and 4.22 do not directly lead to the previous statements on polarization of the subchannels but give an indication of how polarization occurs as N increases. Numerous very interesting information-theoretic relations are established in [3] that are useful in computing this doubling transition probability and other relationships but are not considered here. Again, the BEC is a special case of these where exact relationships rather than just bounds can be established.

As noted, the central problem of polar coding is the identification of the good subchannels – those with $I(W_N^{(i)}) \sim 1$ (or equivalently those with $Z(W_N^{(i)}) \sim 0$). Since the original paper [3] considerable progress has been made on this and related aspects of polar coding. Limited comments on these problems are noted below. Of course conceptually one could consider computing the Bhattacharyya parameters for all N subchannels and choosing those with values close to 0 but this is prohibitively computationally expensive.

There is a connection of polar codes to Reed–Muller (RM) codes as follows. For an RM code of length $N = 2^n$, a parity-check matrix for an (N, K) RM code by choosing an $(N-K) \times N$ submatrix of $F^{\otimes n}$ of all rows of Hamming weight above some minimum weight. It is not hard to determine the minimum distance of such a code and its dimension following from the fact that the matrix $F^{\otimes n}$ is of full rank over a field of characteristic 2. However, it is shown ([3], section X) that this leads to the inclusion of some bad subchannels and is not a good scheme.

The next subsections comment on certain aspects of polar coding.

The Special Case of the BEC

It is clear that the BEC is a special case of the above relationships, where equality is often achieved while inequality exists for more general channels. Before returning to general BMSs the situation is commented on further. In particular from the above, we have for a BEC

$$Z\left(W_{2N}^{(2i-1)}\right) = 2Z\left(W_N^{(i)}\right) - Z\left(W_N^{(i)}\right)^2 \quad \text{and} \quad Z\left(W_{2N}^{(2i)}\right) = 2Z\left(W_N^{(i)}\right)^2.$$

A probability of erasure of δ will be assumed. These recurrences define a sequence of polynomials that, given the polarized nature of the channel noted in the above theorem, should have interesting properties. Such properties are pursued in the interesting study [31, 32]. Indeed it is shown there how one might use the properties of the polynomials to determine the good channels. The approach is not pursued here.

Construction of Polar Codes

The problem of constructing polar codes is considered further. The original work of Arıkan [3] suggested a technique based on the computation of Bhattacharrya parameters for the subchannels leading to the probability of error. The work of Mori and Tanaka [27, 28] suggested interpreting the polar code as an LDPC code with Tanner graph and determine the probability of error via pdfs of LLRs and using density evolution as described in Chapter 3. Those channels with small probability of error are identified as the good channels. The probability of error calculations use the recursive forms of likelihood ratios noted above for the polar code case and drawing on results of Chapter 3 for the other computations, including density evolution and Gaussian approximation.

Polar codes defined with this density evolution technique are referred to as density evolution polar codes. It is anticipated that since they are defined using the probability of error rather than the Bhattacharyya parameters, one can expect better performance and experimental evidence bears this out [29].

As with LDPC codes one can use the method of Gaussian approximation [11]. This and other aspects of designing polar codes using density evolution and/or Gaussian approximations are considered in the works [13, 24, 27, 28, 29, 44, 45]. The situation parallels the techniques for the LDPC codes in Chapter 3.

Tal and Vardy [41] also address the problem of the efficient construction of polar codes. They use an approximation technique that brackets the original bit channel, a degrading and upgrading quantization method, a concept hinted at previously in connection with LDPC codes. The degrading quantization transforms the original subchannel into one with a smaller output alphabet (hence simpler calculations) while the original subchannel is a degraded version of the upgraded one. The result is an interesting technique that yields polar codes in linear time ([41], theorem 14), with the upgraded and degraded channel versions being very close in terms of performance. The notions of such channels were also discussed in [12] and [37].

The result of the algorithms ([41], theorem 1) is to show that for a BMS W of capacity $I(W)$ and for fixed $\epsilon > 0$ and $\beta < 1/2$ there is an even integer μ_0 such that for all even integers $\mu \geq \mu_0$ and all sufficiently long code lengths $N = 2^n$ there is a code construction algorithm with running time $O(N \cdot \mu^2 \log \mu)$ that produces a polar code of rate $R \geq I(\mathcal{W}) - \epsilon$ such that

$$P_{\text{block}} \leq 2^{-N^{\beta}}$$

under successive cancellation decoding, a result also achieved in [7] and noted below. The complexity of the algorithm is linear in code length. Note the improvement in this probability of error from the original $O(N^{-1/4}) = O(2^{-n/4})$ to $O(2^{-2^{n\beta}})$. Further comments on the probability of error for polar codes are given below.

Kernels and the Rate of Polarization

An important problem for the construction of polar codes is the speed at which the channels polarize or the speed with which $Z(W_N^{(i)})$ approaches 0 or 1 as $n \longrightarrow \infty, N = 2^n$, since this is directly connected to the probability of error via Proposition 4.4 and the feasibility of implementation of such codes. From Theorem 4.6, the rate of polarization achieved in the original work led to a probability of error that was $O(N^{-1/4}) = O(2^{-n/4})$. As noted above, this was improved with the following result ([7], theorem 1):

Theorem 4.7 *Let $\mathcal{W}$ be a BMS with $I(\mathcal{W}) > 0$. Let $R < I(\mathcal{W})$ and $\beta < 1/2$ be fixed. Then for $N = 2^n$ the best achievable block error probability for polar coding under successive cancellation decoding at block length N and rate R satisfies*

$$P_e(N,R) = o\left(2^{-N^{\beta}}\right).$$

A question arises as to whether successive cancellation decoding can be improved upon. The following implies not.

Theorem 4.8 ([22], theorem 5) *Let $R > 0, \beta > 1/2, N = 2^n, n \geq n(\beta, R, \mathcal{W})$ and $\mathcal{W}$ be any BMS. The probability of block error under maximum a posteriori (MAP) decoding satisfies*

$$P_e(N,R) > 2^{-N^{\beta}}.$$

The implication of this is that successive cancellation decoding performance is comparable to that of maximum a posteriori (MAP) decoding.

The above bounds on the error probability are valid for any code rate. One might naturally expect to achieve a better error probability at lower code rates. A step in this direction is the result of Tanaka and Mori [43] who showed for the same parameters as above that

$$P_e(N,R) = o\left(2^{-2^{(n+t\sqrt{n})/2}}\right),$$

where $t < Q^{-1}(R/I(\mathcal{W}))$ and $Q(x) = \int_x^{\infty} e^{-u^2/2}du/\sqrt{2\pi}$. As with the code independent result one might wonder if a MAP decoder, rather than an SC decoder, might have a better performance. This is answered negatively in [20] which gives a lower bound of the same form for MAP decoding.

The original work of Arıkan [3] noted it would be natural to consider kernels for polar codes larger than the 2×2 considered (i.e., the $\boldsymbol{F}$ matrix). It is somewhat remarkable that such good results were obtained in the original paper for such a small kernel. The problem of $\ell \times \ell$ (binary) kernels was considered in [23]. The work gives necessary and sufficient conditions for a kernel to be polarizing, and in particular that any $\ell \times \ell$ matrix, none of whose column permutations gives an upper triangular matrix, will polarize a symmetric channel (the only kind under consideration).

It is also shown there is no polarizing kernel of size less than 16 with an exponent which exceeds $1/2$. Using BCH codes and shortened versions of them, a polar code with $\ell = 16$ is defined with exponent greater than $1/2$ is achieved (with exponent 0.51828). It is also shown the exponent converges to unity as $\ell \to \infty$ for larger values of ℓ.

It is natural to also formulate polar codes for nonbinary channels and numerous authors have considered these including [27, 30, 33, 39, 40]. The thesis of Mori [27] addresses the extension to nonbinary codes and their rate of polarization. It is shown that an asymptotic polarization rate approaching unity as block size increases is achievable using Reed–Solomon matrices in the construction processes which, it is argued, are a natural generalization of the kernel used for the binary case. It is also shown that the polarization constant achieved with an RS kernel of size 4 is larger than the largest exponent achievable with a binary kernel of size 16. Interestingly it is also shown that using Hermitian codes (from algebraic geometry) leads to even better exponents.

Another interesting aspect of polar codes is to determine how fast their rate approaches capacity as a function of their block length, for a given probability of block error, an aspect often referred to as *scaling* [15, 17, 18, 19, 21, 26, 36, 46]. Only the work of [18] is briefly mentioned. It shows that for a fixed probability of error on a BMS channel W, polar codes of rates with a gap to capacity ($I(W)$) of ϵ with construction complexity and decoding complexity that are polynomial in $1/\epsilon$ are possible. Specifically the following theorem is shown:

Theorem 4.9 ([18], theorem 1) *There is an absolute constant μ (the scaling constant) such that the following holds. Let W be a binary-input memoryless output-symmetric channel with capacity $I(W)$. Then there exists $a_W < \infty$ such that for all $\epsilon > 0$ and all powers of two $N \geq a_W(1/\epsilon)^{\mu}$, there is a deterministic poly(N/ϵ) time construction of a binary linear code of block length N and rate at least $I(W) - \epsilon$ and a deterministic Npoly$(\log N)$ time decoding algorithm for the code with block error probability at most $2^{-N^{.49}}$ for communication over W.*

Other work ([17], proposition 3) shows that for a q-ary channel W for transmission at rate R with error probability at most P_e then if $q = 3$ a code length of

$$N = \frac{\beta}{(I(W) - R)^{5.571}}$$

is sufficient where β is a constant that depends on P_e. A similar result for $q = 5$ and scaling exponent 6.177 is given.

Comments

The original article on polar codes by their inventor, [3], is essential reading. It is a beautiful exposition on information theory and coding and should be read for a deeper understanding of the subject. This chapter is a limited introduction to the central ideas of that work.

The performance of polar codes has generated a very large literature and it is difficult to convey the depth and elegance of the many aspects investigated in a short essay. Initially it was felt that while the theoretical results on polar codes were unquestioned, their performance at practical block lengths might not compete with other coding systems such as LDPC. This concern appears to have diminished considerably and they are under consideration for standards for 5G networks [8]. They will continue to attract intense interest.

An aspect of polar codes is that in many if not most instances, code design is not universal in that a code designed for a channel with one parameter may not be suitable for a channel with a different parameter. A technique to overcome this is considered in [38].

While successive cancellation decoding has a natural relationship to the construction of polar codes, other decoding techniques have been considered with a view to achieving superior performance or a better complexity/performance trade-off. As an example, Ahmed et al. [1] consider the use of belief propagation decoding for such a purpose.

The 2×2 matrix $\boldsymbol{F}$ used in the construction of binary polar codes has a natural connection to the $(u, u+v)$ construction used in the construction of RM codes and a few limited comments on this connection were given. Also, some of the inequalities established on the channel reliability and performance or symmetric capacity functions, $Z(W)$ and $I(W)$, as in Equation 4.2, as noted, achieve equality for the BEC. The large literature on RM codes and the equalities in these equations have led to numerous interesting contributions which include [2, 4, 25, 31].

The use of list decoding with polar codes has also been considered. A stronger notion of polarization than used here has been of interest (e.g., [9, 10]). Other interesting contributions to the problem include [6, 14, 22, 34, 35, 40, 41, 42, 44].

References

[1] Ahmed, E.A., Ebada, M., Cammerer, S., and ten Brink, S. 2018. Belief propagation decoding of polar codes on permuted factor graphs. *CoRR*, abs/1801.04299.

[2] Arıkan, E. 2008. A performance comparison of polar codes and Reed-Muller codes. *IEEE Commun. Lett.*, **12**(6), 447–449.

[3] Arıkan, E. 2009. Channel polarization: a method for constructing capacity-achieving codes for symmetric binary-input memoryless channels. *IEEE Trans. Inform. Theory*, **55**(7), 3051–3073.

[4] Arıkan, E. 2010. A survey of Reed-Muller codes from polar coding perspective. Pages 1–5 of: *2010 IEEE Information Theory Workshop on Information Theory (ITW 2010, Cairo)*, Cairo, Egypt.

[5] Arıkan, E. 2016. Polar codes. In: *2016 JTG / IEEE Information Theory Society Summer School, Department of Electrical Communication Engineering, Indian Institute of Science, Bangalore, India, 27 June–1 July 2016.*

[6] Arıkan, E. 2016. On the origin of polar coding. *IEEE J. Select. Areas Commun.*, **34**(2), 209–223.

[7] Arıkan, E., and Telatar, E. 2008. On the rate of channel polarization. *CoRR*, abs/0807.3806.

[8] Bioglio, V., Condo, C., and Land, I. 2018. Design of polar codes in 5G new radio. *CoRR*, abs/1804.04389.

[9] Blasiok, J., Guruswami, V., Nakkiran, P., Rudra, A., and Sudan, M. 2018. General strong polarization. Pages 485–492 of: *Proceedings of the 50th Annual ACM SIGACT Symposium on Theory of Computing*. STOC '18. ACM, New York.

[10] Blasiok, J., Guruswami, V., and Sudan, M. 2018. Polar Codes with exponentially small error at finite block length. *APPROX-RANDOM*, arXiv:1810.04298.

[11] Chung, S.-Y., Richardson, T.J., and Urbanke, R.L. 2001. Analysis of sum-product decoding of low-density parity-check codes using a Gaussian approximation. *IEEE Trans. Inform. Theory*, **47**(2), 657–670.

[12] Cover, T.M., and Thomas, J.A. 2006. *Elements of information theory*, 2nd ed. Wiley, Hoboken, NJ.

[13] Dai, J., Niu, K., Si, Z., Dong, C., and Lin, J. 2017. Does Gaussian approximation work well for the long-length polar code construction? *IEEE Access*, **5**, 7950–7963.

[14] Eslami, A., and Pishro-Nik, H. 2011. A practical approach to polar codes. *CoRR*, abs/1107.5355.

[15] Fazeli, A., and Vardy, A. 2014 (September). On the scaling exponent of binary polarization kernels. Pages 797–804 of: *2014 52nd Annual Allerton Conference on Communication, Control, and Computing (Allerton)*, Monticello, IL.

[16] Gallager, R.G. 1968. *Information theory and reliable communication*. John Wiley & Sons, New York.

[17] Goldin, D., and Burshtein, D. 2018. On the finite length scaling of q-ary polar codes. *IEEE Trans. Inform. Theory*, **64**(11), 7153–7170.

[18] Guruswami, V., and Velingker, A. 2015. An entropy sumset inequality and polynomially fast convergence to Shannon capacity over all alphabets. Pages 42–57 of: *Proceedings of the 30th Conference on Computational Complexity*. CCC '15. Schloss Dagstuhl–Leibniz-Zentrum fuer Informatik, Germany.

[19] Guruswami, V., and Wang, C. 2013. Linear-algebraic list decoding for variants of Reed-Solomon codes. *IEEE Trans. Inform. Theory*, **59**(6), 3257–3268.

[20] Hassani, S.H., and Urbanke, R.L. 2010 (June). On the scaling of polar codes: I. The behavior of polarized channels. Pages 874–878 of: *2010 IEEE International Symposium on Information Theory*, Austin, TX.

[21] Hassani, S.H., Alishahi, K., and Urbanke, R.L. 2014. Finite-length scaling for polar codes. *IEEE Trans. Inform. Theory*, **60**(10), 5875–5898.

[22] Hussami, N., Korada, S.B., and Urbanke, R.L. 2009. Polar codes for channel and source coding. *CoRR*, abs/0901.2370.

[23] Korada, S.B., Sasoglu, E., and Urbanke, R.L. 2009. Polar codes: characterization of exponent, bounds, and constructions. *CoRR*, abs/0901.0536.

[24] Li, H., and Yuan, Y. 2013 (April). A practical construction method for polar codes in AWGN channels. Pages 223–226 of: *IEEE 2013 Tencon – Spring*.

[25] Mondelli, M., Hassani, S.H., and Urbanke, R.L. 2014. From polar to Reed-Muller codes: a technique to improve the finite-length performance. *IEEE Trans. Commun.*, **62**(9), 3084–3091.

[26] Mondelli, M., Hassani, S.H., and Urbanke, R.L. 2016. Unified scaling of polar codes: error exponent, scaling exponent, moderate deviations, and error floors. *IEEE Trans. Inform. Theory*, **62**(12), 6698–6712.

[27] Mori, R. 2010. Properties and construction of polar codes. *CoRR*, abs/1002.3521, 1–24.

[28] Mori, R., and Tanaka, T. 2009. Performance and construction of polar codes on symmetric binary-input memoryless channels. *CoRR*, abs/0901.2207.

[29] Mori, R., and Tanaka, T. 2009. Performance of polar codes with the construction using density evolution. *IEEE Commun. Lett.*, **13**(7), 519–521.

[30] Mori, R., and Tanaka, T. 2014. Source and channel polarization over finite fields and Reed-Solomon matrices. *IEEE Trans. Inform. Theory*, **60**(5), 2720–2736.

[31] Ordentlich, E., and Roth, R.M. 2017 (June). On the pointwise threshold behavior of the binary erasure polarization subchannels. Pages 859–863 of: *2017 IEEE International Symposium on Information Theory (ISIT)*.

[32] Ordentlich, E., and Roth, R.M. 2019. On the pointwise threshold behavior of the binary erasure polarization subchannels. *IEEE Trans. Inform. Theory*, **65**(10), 6044–6055.

[33] Park, W., and Barg, A. 2013. Polar codes for q-ary channels, $q = 2^r$. *IEEE Trans. Inform. Theory*, **59**(2), 955–969.

[34] Pedarsani, R., Hassani, S.H., Tal, I., and Telatar, E. 2012. On the construction of polar codes. *CoRR*, abs/1209.4444.

[35] Pfister, H. 2014. A brief introduction to polar codes. Tutorial paper, Duke University. http://pfister.ee.duke.edu/courses/ecen655/polar.pdf.

[36] Pfister, H.D., and Urbanke, R.L. 2016 (July). Near-optimal finite-length scaling for polar codes over large alphabets. Pages 215–219 of: *2016 IEEE International Symposium on Information Theory (ISIT)*.

[37] Richardson, T., and Urbanke, R. 2008. *Modern coding theory*. Cambridge University Press, Cambridge.

[38] Şaşoğlu, E., and Wang, L. 2016. Universal polarization. *IEEE Trans. Inform. Theory*, **62**(6), 2937–2946.

[39] Şaşoğlu, E., Telatar, E., and Arikan, E. 2009. Polarization for arbitrary discrete memoryless channels. *CoRR*, abs/0908.0302.

[40] Tal, I. 2017. On the construction of polar codes for channels with moderate input alphabet sizes. *IEEE Trans. Inform. Theory*, **63**(3), 1501–1509.

[41] Tal, I., and Vardy, A. 2013. How to construct polar codes. *IEEE Trans. Inform. Theory*, **59**(10), 6562–6582.

[42] Tal, I., and Vardy, A. 2015. List decoding of polar codes. *IEEE Trans. Inform. Theory*, **61**(5), 2213–2226.

[43] Tanaka, T., and Mori, R. 2010. Refined rate of channel polarization. *CoRR*, abs/1001.2067.

[44] Trifonov, P. 2012. Efficient design and decoding of polar codes. *IEEE Trans. Commun.*, **60**(11), 3221–3227.

[45] Wu, D., Li, Y., and Sun, Y. 2014. Construction and block error rate analysis of polar codes over AWGN channel based on Gaussian approximation. *IEEE Commun. Lett.*, **18**(7), 1099–1102.
[46] Yao, H., Fazeli, A., and Vardy, A. 2019. Explicit polar codes with small scaling exponent. *CoRR*, abs/1901.08186.

5
Network Codes

The first notion of nodes in a network performing computation on packets before forwarding in the network to increase network throughput seems to have risen explicitly in the work [9] in the context of satellite repeaters.[1] The notion of a network code was formalized in the influential paper [1]. The idea is simply to show how, by allowing nodes in a communications network to process packets as they arrive at servers, rather than just store and forward them, the throughput of the network can be improved significantly and, indeed, be shown to achieve an upper bound. Such codes have clear possibilities for existing networks and the theoretical and practical literature on them is now large. This work will largely focus on the fundamental papers [1, 34, 36] which set the stage for many of the advances in the area.

The next section introduces the terminology of the problem to be used and discusses the celebrated max-flow min-cut theorem for networks. This theorem has been proved by several authors and in many books on graph theory (e.g., [13, 17, 18, 49]). The simple proof of [17] is discussed here. The section also considers the important example of the so-called *butterfly network* contained in the seminal paper [1], a simple but convincing example of the value of network coding.

A more detailed consideration of the work [1] is given in Section 5.2. The work is information-theoretic in nature and introduces the fundamental aspects of network coding. It establishes the existence of network codes that achieve the max-flow min-cut bound for certain types of networks. This was a foundational paper for virtually all subsequent work. The work [34] shows how the network coding problem can be formulated in an algebraic manner which is mathematically convenient for both the discussion and the proofs. It is an important contribution to the area as it allowed a greater insight into the

[1] The author is grateful to Raymond Yeung for this reference.

network coding problem. The section also includes a discussion of the paper [36] on linear network codes which establishes the important result that it is sufficient to consider linear coding (which essentially involves a linear sum of packets of some fixed length entering a node, over some finite field) to obtain a maximum flow for the multicast case (although it is noted in [30] that network coding does not increase the achievable rate when the one source node multicasts to all other nodes in the network). The max-flow min-cut theorem does not hold for the case of multiple sources and multiple sinks in the network.

Section 5.3 considers a few examples of actual construction techniques for network codes and their performance on networks. Given the enormous theoretical advantages proved for network codes, it is not surprising there is a considerable literature on the problems associated with the implementation of network coding and their performance on peer-to-peer networks, such as overlay networks on the Internet.

This chapter considers only network codes on networks that are delay-free and acyclic (i.e., contain no cycles). Thus as packets are launched into the network, they appear sequentially at each node and the final sink without delay. Under certain conditions, virtually all of the results, however, can also be shown to hold for networks with delays and cycles, with further analytical techniques.

5.1 Network Flows

A network is viewed as a directed graph (digraph) G without loops on a set of nodes V and edges E such that each edge is assigned a nonnegative *capacity* $c(e) > 0, e \in E$. For the sake of simplicity, only integral capacities will be considered here. It will be assumed the flows are in bits per unit time. Often it is assumed each edge of the network has a unit capacity and allows for multiple edges between nodes to account for various integer capacities. More recent literature (e.g., [59]) refers to edges as channels and the terms are regarded as interchangeable.

In a *unicast network* there are two distinguished nodes, a source node $s \in V$ and a sink or *terminal* node $t \in V$, while the *multicast network* has a single source and multiple sinks. The case of multiple sources and sinks is also of interest although not discussed further except for noting that [17] for some purposes the case of multiple sources and multiple sinks can be handled by adding a new source which has an outgoing edge to each of the original sources and a new sink which has an incoming edge from each of the original sinks, thus converting the system back to a single source, single sink case. However, for many questions the multiple sink case is more complicated.

In the unicast network, single source/single sink, it is of interest to determine the maximum flow achievable through the network from source to sink, so that the capacity constraint on any edge on the flow is not violated, referred to as a *feasible* flow.

If the directed edge $e \in E$ is $(u, v) \in E$ (directed from u to v), the node u is referred to as the tail of the edge e and denoted e_- and v is the head of the edge and denoted e_+. The *source* vertex is s and *sink* or terminal vertex is t. A *flow* in a network (integer flows on edges are considered here) is defined as a function f from the set of edges E to the positive integers $\mathbb{Z}_{\geq 0}$ satisfying the properties:

(i) $0 \leq f(e) \leq c(e)$ for all $e \in E$,
(ii) $\sum_{e_-=v} f(e) = \sum_{e_+=v} f(e)$ for $v \in V \backslash \{s, t\}$.

The second item above asserts the conservation of flow for all vertices except the source and sink, i.e., the flow out of node v is equal to the flow into node v. (Some authors allow flow into a source and out of a sink.) The term $c(e)$ is the capacity of the edge e, the maximum flow allowed on that edge at any instant of time.

The *value* of a flow is defined as

$$\mathrm{val}(f) = \sum_{e_-=s} f(e) = \sum_{e_+=t} f(e) \tag{5.1}$$

which is simply the flow out of the source node s which is also the flow into the sink node t.

The problem of interest is to determine the maximum value that a flow in a given directed network may achieve in such a manner that the flow on any edge does not exceed the capacity $c(e)$ of that edge.

Define an (S, T) *cut* of the network, for $S \subset V$, $T = V \backslash S = \bar{S}$ as the set of edges with one end in S and the other in T, $s \in S, t \in T$, with the property that any path from a source s to sink t traverses an edge in (S, T). For the flow assigned to edge $e \in E$ with capacity $c(e)$ and flow $f(e)$, define the value and capacity of a cut as

$$\mathrm{val}(S,T) = \sum_{e \in (S,T)} f(e) \quad \text{and} \quad \mathrm{cap}(S,T) = \sum_{e \in (S,T)} c(e).$$

The celebrated max-flow min-cut theorem can be found in many texts on graph theory, algorithms and combinatorics (e.g., [8, 11, 13, 31, 48, 52]). The proof of [17] is instructive.

Theorem 5.1 (Max-Flow Min-Cut) ([17]) *The maximum value of a flow in a network for any source s and sink t is equal to the minimum capacity of any cut of the network and this maximum flow is achievable.*

Proof: To show that the flow of a minimum valued cut-set cannot be exceeded, consider the cut $C = (S, T)$ to be such a minimum cut and denote a given flow in the network as f. Then $\text{val}(f) \leq \text{cap}(C)$ since capacity of the cut would only be achieved if all directed branches from S to T were carrying capacity and those from T to S were carrying 0. It is clear that the flow across the cut from left to right is the flow into the sink node t – since all nodes in T except the terminal t have a net flow of zero. Thus $\text{val}(f) \leq \text{cap}(C)$.

To show that if C is a minimum cut-set (min-cut), then a flow exists that achieves $\text{cap}(C)$. The notion of a *reduced network* is introduced. Consider the following process to construct a reduced network from a given network. If there is an edge in the network which is not in some minimum cut-set, reduce its capacity until it is in some minimum cut-set or its flow value is 0 (equivalent to eliminating it from the network). Continue to look for such edges until there are no edges not in some min-cut. The reduced network then has the properties:

(i) the graph of the reduced network is that of the original except possibly some edges are missing;
(ii) each edge of the reduced network has capacity $\leq$ that edge of the original network;
(iii) each edge of the reduced network is in at least one min-cut of the reduced network.

It is clear that a valid flow in the reduced network is a valid flow in the original. Thus proving the theorem for reduced networks is sufficient to prove it for the original network. The proof is by induction on the number of nodes in the network. It is clear that any network (of any number of nodes) with path lengths from s to t of length at most 2 will satisfy the theorem.

Suppose the theorem is true for any reduced network with n or fewer nodes. We show it true for reduced networks with $n + 1$ nodes. Suppose the reduced network with $n+1$ nodes has a path of length at least 3. There exists an internal edge (end vertices neither s nor t) on this path and this edge is in some min-cut C (since the network is reduced). On each edge of the cut-set introduce a middle node such that each resulting edge has the same capacity as the original edge. Merge these new nodes into one node (say v^*) – the effect is to create two networks, say N_1 and N_2 in series with the node v^* serving as the sink for the first network and source for the second network. Each of these networks has fewer than $n+1$ nodes and every path from s to t goes through v^*. Each of these networks contains a min-cut corresponding to C and neither can contain a lower valued min-cut as this process can only eliminate cut-sets – not create new ones. Since by induction the theorem is true for each of the networks it is true for the original network and hence for all networks with $n+1$ nodes. By separating the node v^* into the original nodes and eliminating the introduced nodes on the cut-set edges, any flow in the two networks corresponds to a flow in the original network and the theorem is proven. ■

The more conventional approach to proving this theorem, which mirrors somewhat the notion of a reduced network in the above proof, uses the notion of *augmented flows*. To give a flavor of this approach only integer flows and capacities (the extension to noninteger flows is not difficult) are considered and for an arbitrary set of vertices S of the network G, $S \subset V, s \in S, t \in \bar{S} = V \backslash S$, define [8]:

$$\begin{aligned} S^* &= \{e \in E : e^- \in S, e^+ \in \bar{S}\}, \\ S_* &= \{e \in E : e^- \in \bar{S}, e^+ \in S\}. \end{aligned}$$

From these definitions and the above discussion it is easy to see that

$$\sum_{e \in S^*} f(e) - \sum_{e \in S_*} f(e) = \mathrm{val}(f) \leq \mathrm{cap}(C)$$

where C is the capacity of the cut $(S, \bar{S})$. Let f be an (integer) flow. Then either there is a cut C with $\mathrm{cap}(C) = \mathrm{val}(f)$ or there is a flow $f^{'}$ with $\mathrm{val}(f^{'}) = \mathrm{val}(f) + 1$.

For the integer flow f on the digraph G with source and sink nodes s and t, construct the set S as follows:

> Set $S = \{s\}$ and while there is an edge $e = (v, w)$ with either
> (i) $f(e) < c(e), v \in S, w \notin S$ or
> (ii) $f(e) > 0, v \notin S, w \in S$ add to S the vertex either v or w not already in S.

The set S so created can be used to either increase the flow or show that the flow is already at capacity. If $t \in S$, then there is a path from s to t, say $v_0 = s, v_1, \ldots, v_k = t$ such that

(i) (v_i, v_{i+1}) is a (forward) edge e such that $f(e) < c(e)$ or
(ii) (v_{i+1}, v_i) is a (backward) edge e such that $f(e) > 0$.

Thus the path can contain both "forward" and "backward" edges (going with or against the edge direction). On the forward edges increase the flow by 1 (since the flows and capacities are integers and by assumption $f(e) < c(e)$ this can be done). On backward edges decrease the flow by 1 (other edges are not affected). It is clear that all nodes not in the path are on edges that are not affected. It is easily shown that these operations preserve conservation of flow at nodes that are neither s nor t and the flow is increased by 1, i.e., $\text{val}(f^{'}) = \text{val}(f) + 1$.

Similarly it is shown that if $t \notin S$, then $C = (S^*, S_*)$ is a cut and the original flow achieves capacity, i.e., $\text{val}(f) = \text{cap}(C)$, which finishes the proof. Thus a maximum flow is achieved if and only if there is no path which can be augmented.

The above theorem shows that a maximal flow of a single source, single-sink network (a *unicast network*) always exists and the flow is given by the flow across a min-cut. While integer flows have been assumed, it is easily generalized to real flows.

Consider now a *multicast* situation where there is a single source node s and multiple sink nodes, say $T_L = \{t_1, t_2, \ldots, t_L\}$ (referred to as a *multicast network*). It is clear that if in such a network the maximum flow possible, say ω, so that *every* sink node receives all information from the source at a rate of at least ω, then it must be true that

$$\omega \leq \min_{t_\ell \in T_L} \text{max-flow}(s, t_\ell).$$

Until now it has been assumed that in the network the nodes are only capable of either storing or forwarding messages they receive (replicating them if necessary for each forwarding edge from the node). In this case the above upper bound on received rate is not generally possible to achieve. In [1] it is shown that if one allows the nodes to perform *network coding*, then the bound is achievable. Furthermore in [36] it is shown that *linear coding* is sufficient to achieve capacity (max-flow) in multicast networks. By this it is meant that packets arriving at a node are linearly combined, treating the packets as a set of symbols in a finite field and allowing multiplication of packets by elements of the field and summing packets over that finite field. Generally the point is to allow each node to perform operations on the information it receives on each of its incoming edges and transmit functions of the received information on its outgoing edges. The situation is discussed further in the next section. It will be shown how allowing such operations can increase information flow. The messages transmitted on the edges are often assumed to be of the same length and referred to as packets or symbols, depending on the discussion.

The example below, the famous *butterfly network*,[2] contained in the original network coding paper [1], illustrates the point convincingly. (It is included in numerous papers on the subject.)

[2] It is remarked in [60] that this term was first suggested by Michelle Effros of CalTech.

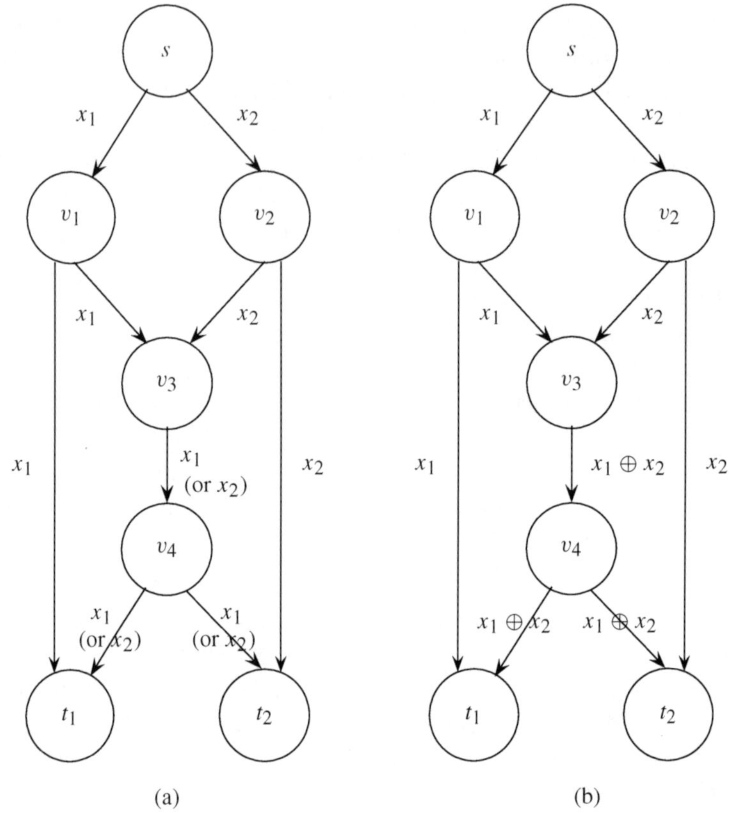

Figure 5.1 Butterfly network: (a) no node processing and (b) with node processing

Example 5.2 In the butterfly network of Figure 5.1 a single source, two-sink network, the object is to transmit two bits to each of the sinks in a delay-less manner. In Figure 5.1(a) it is assumed that each node is capable of only replicating and forwarding the received data on an outgoing edge and each edge has unit capacity. The cut-set bound for each sink is 2. Node v_3 has to make a decision as to which bit, either x_1 or x_2 to forward to v_4. The result is that one sink node receives both bits while the other receives only one. In Figure 5.1(b) nodes are allowed to process the received information before forwarding. If node v_3 computes $x_1 \oplus x_2$, it is clear both sink nodes t_1 and t_2 can recover both bits by computing, respectively, $x_1 \oplus (x_1 \oplus x_2) = x_2$ and $x_2 \oplus (x_1 \oplus x_2) = x_1$, in the same number of transmissions.

The example is both simple and elegant and a convincing demonstration that allowing processing in the node can significantly improve the transmission performance of networks.

As noted, only the case of multicasting is considered in this work where there is a single source node and multiple sink nodes and the objective is to send the same information to all sinks. In this case it is known [30, 58] that a rate that is the minimum of all the max-flow min-cut rates from the source to the individual sinks is achievable with *linear coding multicast* to all sinks in this multicast situation. Although Jaggi et al. [30] note that "a classical result of Edmonds [14] shows that network coding does not increase the achievable rate in the special case where one node is the source and all other nodes in the network are sinks" ($V = \{s\} \cup T$ where T is the set of sinks). In other words, if a certain flow is achievable between a single source and several sink nodes individually, then with linear coding this (minimum) rate is achievable simultaneously.

The next section gives an overview of the information theoretic basis of network coding of the paper [1]. The algebraic approach to representing this network coding paradigm contained in [34] is also discussed as well as the approach of [46, 61]. Section 5.3 will consider representative explicit network coding schemes and some of the problems faced in implementing network codes on real networks such as the Internet.

5.2 Network Coding

The description of a network code for the multicast situation with a single source node s and L sink nodes $T = \{t_1, t_2, \ldots, t_L\}$ is necessarily involved since it must encompass a description of the coding process at each intermediate node from s to t_ℓ for each such path. The aim is to describe a process that gives a prescription of how to code at each node that allows a proof that the capacity (max-flow) can be achieved and that will allow each sink node to retrieve all information transmitted by the source at the maximum rate. The situation in real networks will involve adjustments to the ideas used here in that the networks are seldom delayless and synchronous. In practice servers/nodes might well combine packets that are not synchronous and operate with time tolerances that are not considered in this work.

First a formal model will be introduced as originally conceived ([1] and [58]) and based on this model a more familiar linear system-theoretic model will be discussed. While this model involves a certain amount of notation and terminology, it captures the essence of the network coding problem. Section 5.3 considers a few actual constructions of network codes found in the literature as examples of the requirements. It is to be kept in mind that the goal of any network coding system is to ensure that each terminal node is able to decode the received packets into the set of transmitted packets at the maximum rate. This generally means that a sufficient supply of combined packets is received by each terminal node that allows the node to solve for the transmitted packets, meaning that as the network nodes combine packets (usually linearly), a sufficient supply of received combined packets forms a matrix equation of full rank that will allow for solution for the unknown information packets. Naturally, identification of all the combining processes of the nodes traversed is stored in the headers of transmitted packets.

It is assumed all edges (communication links between nodes) are error-free and the directed network is acyclic and delayless. Thus the packets are transmitted sequentially along a path in that a node can transmit a packet only after it has received it, but the process takes no delay. The introduction of directed cycles in such a network might lead to instabilities in the processes.

Consider first a network G [58] with a single source node $\{s\}$ that generates a certain amount of information that must be transmitted to each of the terminal nodes T_L. Let $\boldsymbol{R} = (R_{i,j})$ denote the capacity of the directed graph edge $(i,j) \in E$. Clearly the max-flow bound from the source to sink t_ℓ is governed by the max-flow min-cut bound of the previous section. The bound

$$\omega \leq \min_{t_\ell \in T_L} \text{max-flow}(s, t_\ell) \tag{5.2}$$

is referred to as the max-flow bound of the multicast network. It is the largest rate possible to each terminal node, governed by the smallest (maximum) rate to any of the terminal nodes. That such a rate is achievable, in the sense there will exist a network code that can achieve such a network rate, is one of the important results of network coding.

For a formal and somewhat abstract model of network coding [1, 24, 58], consider a source of rate τ bits/use and view a message source X which is sampled to give $\boldsymbol{x}$, viewed as a sampling of the set $\chi = \left\{1, 2, \ldots, r = \lceil 2^{n\tau} \rceil\right\}$ messages, sent as codewords of length n over a symbol alphabet, usually a finite field of some order . The coding strategy will be such that each node will choose coefficients to linearly combine arriving packages for forwarding to the next node so that all r messages can be decoded from the received packets at all L sinks at a rate approaching capacity (max-flow) of the network. An $(n, (\eta_{ij} : (i,j) \in E), \tau)$-code is defined, where E is the set of directed edges in the network, n is the length of the block code and for $(i,j) \in E$ and τ is the information rate of the source. A network code is a set of server/node encoding functions. It will be referred to as *feasible* if it is capable of transmitting the messages from the source to the sinks within the edge capacities. Functions are defined to account for the construction of messages formed at each node, based on the messages received at that node. Define a *transaction* to be the sending of a message from one node to an adjacent node and T_{ij} the set of transactions sent from node i to node j, $(i, j) \in E$ an edge and let V denote the set of network nodes. It is assumed there are a total of K transactions in the entire coding session from the source to all sinks and the edge on which the k-th transaction is sent is specified by two functions u and v:

$$u\colon \{1,2,\dots K\} \longrightarrow V$$
$$v\colon \{1,2,\dots K\} \longrightarrow V$$

such that the k-th transaction takes place on the edge $(u(k), v(k)) \in E$ and

$$T_{ij} = \{1 \leq k \leq K : (u(k), v(k)) = (i, j)\},$$

the set of transactions from node i to node j. Define sets of indices

$$A_k = \{1,2,\dots, |A_k|\}, \quad 1 \leq k \leq K,$$

where $\mid A_k \mid$ indicates the number of indices possible. Similarly η_{ij} is the number of possible index tuples that can be sent from node i to node j (hence on edge $(i, j) \in E$) and the average bit rate of the code on this edge is given by $n^{-1} \log \eta_{ij}$. The total number of index tuples that can be sent on edge (i, j) during the coding session is η_{ij}. In the k-th transaction, the node $u(k)$ encodes the information it receives on its incoming edges according to the function f_k and sends an index in A_k to $v(k)$, on an edge which connects the two nodes. If $u(k) = s$, then

$$f_k\colon \chi \longrightarrow A_k,$$

where, as noted, χ is the set of possible messages, $|\chi| = r$. If $u(k) \neq s$ define the set of nodes that transmit messages to $u(k)$ as Q_k where

$$Q_k = \{1 \leq k' < k : v(k') = u(k)\},$$

i.e., the set of nodes that send messages to $u(k) \in V$. If this set is nonempty, then

$$f_k\colon \prod_{k' \in Q_k} A_{k'} \longrightarrow A_k,$$

which determines the messages sent to node $v(k)$ from $u(k)$. As noted, T_{ij} is the set of transactions sent from node i to node j (which determines the edge

$(u(k), v(k)))$ for some k via the maps u and v, η_{ij} is the number of all possible index tuples sent from i to j.

Finally, for each $\ell \in \{1,2,\ldots,L\}$ the set W_ℓ is the set of vertices sending data to a sink node t_ℓ, i.e.,

$$W_\ell = \{1 \leq k \leq K : v(k)\} = t_\ell, \quad \ell = 1,2,\ldots,L$$

and the function g_ℓ maps the information received on all incoming branches to t_ℓ and retrieves the information transmitted at the source in χ, i.e.,

$$g_\ell : \prod_{k' \in W_\ell} A_{k'} \longrightarrow \chi .$$

This definition of a network code, although somewhat abstract and complex, captures all the necessary ingredients of specifying at each node the action necessary for the transmission of information to a sink. Importantly, it allows an information-theoretic proof of the fundamental theorem of network coding.

The following definition establishes the notion of an achievable edge rate [1, 58]:

Definition 5.3 For a network G with rate constraints $\boldsymbol{R} = \{R_{ij}, (i,j) \in E\}$, an information rate ω is *asymptotically achievable* if, for any $\epsilon > 0$ there exists for sufficiently large n an $(n, (\eta_{ij} : (i,j) \in E), \tau)$-code on G such that

$$n^{-1} \log_2 \eta_{ij} \leq R_{ij} + \epsilon$$

for all $(i,j) \in E$ where $n^{-1} \log_2 \eta_{ij}$ is the average bit rate of the code on channel (i,j) and

$$\tau \geq \omega - \epsilon .$$

An asymptotically achievable information rate will simply be referred to as an achievable information rate.

The *fundamental theorem of network coding*, stated without proof, is ([1], theorem 1, [58], theorems 11.3 and 11.4):

Theorem 5.4 *For a multicast network G with rate constraints $\boldsymbol{R}$, ω is an achievable information rate if and only if*

$$\omega \leq \min_\ell \textit{max-flow}(s, t_\ell).$$

where the minimum is over all sink nodes $t_\ell \in T_L$.

The proofs of such a statement usually treat the delayless *acyclic* (no cycles in the digraph) case and cyclic case separately as further care must be taken in treating sequences of nodes, in particular cycles, among other considerations.

While only the single source multicast case has been considered it is generally possible to extend the analysis to cover the multiple source case as well.

An algebraic formulation of the network coding problem is given in [34] and discussed further in [20, 24, 25, 28], that makes more concrete the above approach. The following discussion describes a linear network coding scheme following the approach of [24].

It is assumed the source node is s, with source messages designated $X = (X_1, X_2, \dots, X_r)$, and each of these independent messages must be communicated from the single source to the set of sink nodes $T_L = \{t_1, t_2, \dots, t_L\}$. To discuss the notion of a linear operation of the nodes it is convenient to assume the bitstreams of the sources and edge processes are divided into symbols or packets, say of m bits each, allowing their representation as elements of the finite field $\mathbb{F}_q$ where $q = 2^m$, perhaps even a single field element per message. It is also assumed the network is acyclic and delayless so all transmissions occur instantaneously and simultaneously although sequentially in that a node transmits information only after it has received information on its input edges.

For a nonterminal node $v \in V \backslash \{s, T_L\}$ denote by $N_i(v)$ the set of input edges to node v and by $N_o(v)$ the set of output edges. It is assumed there are no self-loops on nodes and that $N_i(s) = N_o(t_\ell) = \phi$, i.e., there are no input edges to the source node and no output edges from any sink node. It is desired to determine conditions to network code the messages of $\chi = \{X_1, X_2, \dots, X_r\}$ for transmission so that each of the L terminal nodes receives all the messages. Denote the signal on edge ℓ of the network to be Y_ℓ where $\ell \in N_o(v)$ and

$$Y_\ell = \begin{cases} \sum_{i=1}^{r} a_{i,\ell} X_i \,, & \text{if } \ell \in N_o(s) \\ \sum_{k \in N_i(v)} f_{k,\ell} Y_k \,, & \ell \in N_o(v),\ v \in V \backslash \{s, T_L\}. \end{cases}$$

As discussed, the coefficients are in a suitable finite field $\mathbb{F}_{2^m}$ for some symbol size m. The coefficients used by each node as it transmits messages to the next node are stored in some manner in the message header and modified by each node as it forms the messages it transmits on its out edges. A successful coding system will ensure the coding is such that it enables the terminal node to recover all of the messages sent, i.e., the coefficients chosen by the nodes as the messages traverse the network are such that the messages received at each terminal $t_\ell \in T_L$ node allow solution for all r of the original messages. Thus at the sink node $t_\ell \in T_L$, the encoding process, with node coefficients used in each linear combining stored in packet headers, has been such that each source message input X_i, $i = 1, 2, \dots, r$, $\boldsymbol{X} = (X_1, \dots, X_r)$, can be recovered from the signals on the edges entering each terminal node t_ℓ, $N_i(t_\ell)$, by finding appropriate coefficients b_{t_ℓ} such that

$$Z_{t_\ell,i} = \sum_{j \in N_i(t_\ell)} b_{t_\ell,(i,j)} Y_j, \quad i = 1,2,\ldots,r,\ t_\ell \in T_L \tag{5.3}$$

and that $Z_{t_\ell,i} = X_i$, $i = 1,2,\ldots,r$, i.e., it is the i-th message reproduced at the terminal node t_ℓ. Equivalently the matrix of such coefficients $\boldsymbol{B}_{t_\ell} = b_{t_\ell,(i,j)}$ an $r \times E$ matrix is of full rank for each terminal node t_ℓ where $\mathcal{E}$ is the set of edges and $|\mathcal{E}| = E$. The matrix $\boldsymbol{B}_{t_\ell}$ is nonzero only for edges (columns of $\boldsymbol{B}_{t_\ell}$) into the terminal node t_ℓ. It is the matrix of coefficients that restores the messages at terminal node t_ℓ from the input messages.

Let the $r \times E$ matrix involved in describing the messages on edges emanating from the source node be denoted $\boldsymbol{A} = (a_{i,j})$ and the $E \times E$ matrix describing messages on edges out of internal nodes be $\boldsymbol{F} = (f_{i,j})$, i.e., $f_{i,j}$ is the coefficient used to multiply the message on edge i into a node to form part (a component) of the message output on edge j. Let $\boldsymbol{X} = (X_1, X_2, \ldots, X_r)$ be the vector of messages.

The *transfer function* from s to t_ℓ is required which will describe the transfer from the source messages to the sink outputs, i.e., it tracks the paths of the packets and the specific linear operations performed on the packets at each node as the packets traverse the network. The matrix product $\boldsymbol{X} \cdot \boldsymbol{A}$, $\boldsymbol{A}$ an $r \times E$ matrix then gives the signals on the edges of $N_0(s)$, the output edges of the source.

The $E \times E$ matrix $\boldsymbol{F} = (f_{i,j})$ gives the transfer of signals on edges from one instant to the next. Label the rows and columns of the matrix $\boldsymbol{F}$ by edges $e_1, e_2, \ldots, e_{|E|}$. The matrix will have the coefficient f_{ij} at row e_i and column e_j if and only if $e_{i,+} = e_{j,-}$ and the message into node $e_{i,+}$ on edge e_i is multiplied by f_{ij} before combining with other messages and transmitted out on edge e_j. It will have zeros in matrix positions not corresponding to edges. If the signals on the edges are given by $\boldsymbol{Y} = (Y_{e_1}, Y_{e_2}, \ldots, Y_{e_E})$ at one stage, at the next stage they are given by $\boldsymbol{Y}' = \boldsymbol{Y} \cdot \boldsymbol{F}$. Notice that

$$(\boldsymbol{I} - \boldsymbol{F})^{-1} = \boldsymbol{I} + \boldsymbol{F} + \boldsymbol{F}^2 + \cdots .$$

For an acyclic network the matrix links can be indexed and ordered in a sequential manner such that if e_1 and e_2 are two edges such that $e_{1,+} = e_{2,-}$, then e_1 has a lower index than e_2 and it follows that the matrix $\boldsymbol{F}$ is upper triangular (and in particular, zeroes on the main diagonal – no loops). Hence $\boldsymbol{F}$ is upper triangular and hence nilpotent (i.e., $\boldsymbol{F}^m = \boldsymbol{O}$ for some positive integer m) and the matrix $(\boldsymbol{I} - \boldsymbol{F})^{-1}$ is nonsingular. The matrices $\{\boldsymbol{A}, \boldsymbol{F}, \boldsymbol{B}_{t_1}, \ldots, \boldsymbol{B}_{t_L}\}$ form a *linear network code* as it is clear that the sum of any two codewords is a codeword [20, 25, 29, 34]. The transfer function from the inputs $\boldsymbol{X} =$

$(X_1, X_2, \ldots, X_r)$ to the output processes $\boldsymbol{Z}_{t_\ell} = (Z_{t_\ell,1}, \ldots, Z_{t_\ell,r})$ is given by the *transfer matrix* $\boldsymbol{M}_{t_\ell} = \boldsymbol{A}(\boldsymbol{I}-\boldsymbol{F})^{-1}\boldsymbol{B}_{t_\ell}^T$ in that $\boldsymbol{Z}_{t_\ell} = \boldsymbol{X} \cdot \boldsymbol{A}(\boldsymbol{I}-\boldsymbol{F})^{-1}\boldsymbol{B}_{t_\ell}^T$.

Thus [29] the set $(\boldsymbol{A}, \boldsymbol{F}, \boldsymbol{B}_{t_1}, \ldots, \boldsymbol{B}_{t_L})$ is a network code and a feasible solution to the multicast coding problem. The ℓ-th column of the matrix $\boldsymbol{A}(\boldsymbol{I}-\boldsymbol{F})^{-1}$ specifies the map from input signals to the signal on edge ℓ. If the set of edges into sink t_ℓ are denoted τ_ℓ and the corresponding columns of the matrix by $\boldsymbol{A}_{\tau_\ell}$, the network has a feasible solution iff this matrix has full rank for all sinks in T_L.

To examine this nonsingularity define the *extended transfer matrix* [20] (Edmonds matrix in [25]) as

$$\boldsymbol{N}_{t_\ell} = \begin{bmatrix} \boldsymbol{A} & \boldsymbol{O} \\ \boldsymbol{I}-\boldsymbol{F} & \boldsymbol{B}_{t_\ell}^T \end{bmatrix} \tag{5.4}$$

a form that shows certain properties of the transfer matrix more clearly. A simple proof of the following lemma is given:

Lemma 5.5 ([20], lemma 2.2, [25]) $\det \boldsymbol{M}_{t_\ell} = \pm \det \boldsymbol{N}_{t_\ell}$.

Proof: From the matrix equation

$$\begin{bmatrix} \boldsymbol{A} & \boldsymbol{O} \\ \boldsymbol{I}-\boldsymbol{F} & \boldsymbol{B}_{t_\ell}^T \end{bmatrix} \cdot \begin{bmatrix} (\boldsymbol{I}-\boldsymbol{F})^{-1}\boldsymbol{B}_{t_\ell}^T & (\boldsymbol{I}-\boldsymbol{F})^{-1} \\ -\boldsymbol{I} & \boldsymbol{O} \end{bmatrix} = \begin{bmatrix} \boldsymbol{M}_{t_\ell} & \boldsymbol{A}(\boldsymbol{I}-\boldsymbol{F})^{-1} \\ \boldsymbol{O} & \boldsymbol{I} \end{bmatrix}$$

and taking determinants of both sides gives

$$\pm \det \boldsymbol{N}_{t_\ell} \cdot \det(-\boldsymbol{I}) \cdot \det(\boldsymbol{I}-\boldsymbol{F})^{-1} = \det \boldsymbol{M}_{t_\ell} \cdot \det \boldsymbol{I}.$$

■

By examining the structure of the determinant of the above matrices and the polynomials in the matrix indeterminates, one can show it is always possible to choose matrix elements so this is true, given a large enough finite field for the coefficients:

Lemma 5.6 ([29], theorem 1) *For a feasible multicast network problem with independent or linearly correlated sources and L sinks, in both the acyclic delay-free case and general case with delays, there exists a solution*

$$(\boldsymbol{A}, \boldsymbol{F}, \boldsymbol{B}_{t_1}, \ldots, \boldsymbol{B}_{t_L})$$

over the finite field if $q > L$.

The proof of this lemma is rather involved and beyond our purpose (see [26], theorem 4, [25], theorem 1, [29], lemma 2), involving the products of gains along the edge-disjoint paths and the determinant of subsets of columns of the matrix **A** as well as the degree of the polynomial of certain variables in the determinants. Thus if the entries of the matrices are treated as variables and the determinant is expanded into the variables, a multivariable polynomial results. If the original entries are substituted for the variables, the polynomial evaluation should be nonzero or a nonsingular matrix.

The above linear network coding is interesting in that it illuminates the requirements of network codes from a linear system point of view, yet falls short of actually constructing network codes. The following section considers a few examples of such constructions.

Further aspects of properties of network codes, such as generic linear network codes, linear dispersion codes, linear broadcast and multicast codes, are considered in [36, 59, 61].

While only the delay-free and acyclic cases have been considered here, the extension of the results to networks with cycles and delays is generally possible by introducing a delay variable and rational functions.

5.3 Construction and Performance of Network Codes

To recap the previous two sections, assume an acyclic multicast delayless directed network is assumed with one source node s and L sink nodes T_L and the same source information is required at each sink. More general situations with multiple sources and sinks with different sources transmitting different information and sinks requiring different information from the different sources are possible but the analysis of such general situations becomes even more complex and the special case of multicast considered here gives an indication of the techniques used and insight of network performance.

The capacity of a multicast network was given by Equation 5.2

$$\omega = \min_{t_\ell \in T_L} \text{max-flow}(s, t_\ell),$$

the minimum of the capacities of the L unicast paths. For a given source-sink pair, if all edges are assumed to have unit capacity and parallel edges are allowed, then the maximum number of pairwise edge-disjoint paths from source to sink is equal to the capacity of the min-cut, a result referred to as Menger's theorem. The Ford–Fulkerson algorithm can be used to find such paths [18]. The assumption of unit capacity edges is often easier to handle but the generalization to integer capacities is straightforward.

The previous section noted that a rate approaching ω is achievable asymptotically by using coding at the nodes [1]. In [36] it was shown that this rate can in fact be achieved with linear coding, assuming coding over some finite field. Such a result is not valid in general networks with multiple sources and sinks.

The question as to how such practical and specific network codes can be constructed is commented on below with three particular constructions from the literature.

Deterministic Constructions of Network Codes

The construction of linear network codes is necessarily algorithmic since an arbitrary but given network is assumed and the coding scheme is a function of the network topology. Specification of the operations required at each node of the network is required. The works of [20, 30, 42] assume unit capacity edges and find edge-disjoint paths from s to t_ℓ for each single source-sink pair, $t_\ell \in T_L$. Techniques to overcome the restriction to unit capacity edges are given there but not considered here. The informal discussion of the proof for the basic construction of such codes is outlined.

The basic algorithm of [30], referred to as the *linear information flow* (LIF) algorithm or the Jaggi algorithm, is a centralized algorithm in that the topology of the network is known and the code is designed beforehand. It assumes a network of unit capacity edges with multiple edges allowed, which is able to model networks with integer flows.

Assume the network has a max-flow of r symbols per channel session (rather than the previous ω to emphasize integer value). Note this implies at least r edges in $N_o(s)$ and in $N_i(t_\ell)$ for each $t_\ell \in T_L$. The objective of the algorithm is to have a nonsingular matrix transmitted with the information packets so the terminal node t_ℓ is able to retrieve the r information packets from the r received packets. Recall for this discussion it is assumed each edge of the graph has unit capacity and parallel edges are allowed.

The case of only one terminal node t_ℓ is considered – the extension to multiple terminal nodes is obvious. The algorithm proceeds in two stages. In the first stage a set of r edge-disjoint paths from source s to t_ℓ is found. The Ford–Fulkerson algorithm can be used for such a purpose. That such a set of paths exist follows from Menger's theorem, previously mentioned, which proves for unit capacity edges the maximum number of edge-disjoint paths between any two nodes is the capacity of the min-cut between those nodes, in our case assumed to be r.

The following is a slightly modified version of the LIF in [30].

Consider the r edge-disjoint paths $\mathcal{P}_1, \ldots, \mathcal{P}_r$ from s to t_ℓ. Messages (or symbols or packets) $X_1, X_2, \ldots, X_r$ are to be transmitted to t_ℓ. Denote the message sent on edge e by Y_e where

$$Y_e = \sum_{i=1}^{r} m_{e,i} X_i.$$

(If not all packets are represented by the incoming packets to the node, zero coefficients occur.) The edges on the r paths are examined sequentially both across the paths and along the paths in such a manner that if edges e_j and e_k are on a particular path $\mathcal{P}_i$ and e_j precedes e_k in travel from s to t_ℓ on $\mathcal{P}_i$, then e_j is considered before e_k, i.e., as each path is considered in the algorithm, the "first" nonvisited edge along the path is considered.

Two properties of the transmission algorithm are tracked as each path is visited and updated as each edge is considered: an $r \times r$ invertible matrix $\boldsymbol{M}_\ell$ (over the symbol field, often $\mathbb{F}_{2^s}$ – for some suitable s) and a set of edges C_ℓ.

Consider visiting each path sequentially in some fixed order. As a path $\mathcal{P}_i$ is visited the first edge along the path from source s to terminal node t_ℓ not yet visited is considered, say edge $e = (u,v) \in \mathcal{P}_i$. The set C_ℓ keeps track of the edges visited, one from each path. As each new edge $e = (u,v) \in \mathcal{P}_i$ is considered, the previous edge on $\mathcal{P}_i$ is dropped from C_ℓ and the new edge added.

The $r \times r$ matrix $\boldsymbol{M}_\ell$ keeps track of the coefficients used to form the message on each edge in the set C_ℓ, the i-th column containing the coefficients used for forming the message on the last edge considered on $\mathcal{P}_i$. As the edge $e \in \mathcal{P}_i$ is visited the corresponding column of $\boldsymbol{M}_\ell$ (representing the previous edge on $\mathcal{P}_i$ visited) is dropped and replaced with a new column representing the coefficients used for the message on the new edge e in such a way that the matrix $\boldsymbol{M}_\ell$ is nonsingular. The next path (in some sequence) is then visited, considering the first edge not yet visited on that path. As the algorithm evolves eventually a stage is reached where it reaches the terminal node t_ℓ and the fact the matrix $\boldsymbol{M}_\ell$, which represents the coefficients used to form the messages on the r input edges to t_ℓ, is nonsingular means the r original messages $X_1, \ldots, X_r$ can be retrieved. In the case the r paths are not of equal length, the algorithm is easily modified.

The properties of this LIF algorithm are described in ([30], theorem 3). The algorithm has complexity $O(|E| \cdot |T_L| \cdot (h + |T_L|))$ and any field of size $q \geq 2 \; | T_L |$ can be used for the encoding. A node needs time $O\big(\min(|T_L|, \; |N_i(e_-|)\big)$ to compute the symbol to be sent on edge e.

While this simple argument pertains to unicast acyclic networks with unit capacity edges, it is shown [30] that it can be extended to integer capacity edges and multicast networks and even to algorithms that deal with edge failures. Network codes that can deal with edge failures are often referred to as *robust* network codes.

Numerous other interesting results are shown in [30] including the simple demonstration (theorem 2) that there are acyclic networks with unit capacity edges where multicasting with coding allows a factor $\Omega(\log |V|)$ larger rate than multicasting without coding, V the set of nodes of the network.

Random Linear Codes

The previous approach to constructing an effective network code required a degree of centralization and knowledge of the network topology since this must be known in the construction. This can be a drawback in situations where the entire network structure may be difficult to determine or is evolving. An approach that removes this constraint is to choose the message coefficients from $\mathbb{F}_q$ at random. At each sink then, for the process to be successful, the random matrix of coefficients must be invertible, i.e., have full rank. The approach was first considered in [25, 26, 27] and further developed in [29]. Only the single-source unicast delay-free case is discussed here although the work [29] considers general networks with cycles and delays.

Suppose, as above, there are r information sources, which for convenience are considered packets, at the source node s. At each node along the path to the sink, a number of information packets are received. The node forms a linear (over $\mathbb{F}_q$) combination of the packets to transmit on each out edge, each element of $\mathbb{F}_q$ being chosen uniformly at random. The sink node t must receive a sufficient number of linearly independent coded packets in order to determine the r information packets, i.e., each node and in particular the sink node receives at least r (usually more) linear sums of the r information packets from which it is to solve for the information packets.

Expressions for the probability of success in retrieving the information packets are readily obtained. Suppose the sink node obtains $r + m, m > 0$ packets, each packet containing a random linear sum of the information packets, $X_1, X_2, \ldots, X_r$, each coefficient chosen uniformly at random from $\mathbb{F}_2$. The probability the $r \times (r + m)$ coefficient matrix over $\mathbb{F}_2$ is of full rank (r) has been derived (e.g., Appendix A, Equation A.7) to be

$$\prod_{\ell=m+1}^{r+m} \left(1 - \frac{1}{2^\ell}\right).$$

The $m = 0$ case corresponds to exactly r packets being received. For large r and $m > 1$ the probabilities can be shown to converge to unity rapidly with m. This approach is pursued in [47]. Thus, typically very few extra packets m are required to ensure nonsingularity of the matrix.

Another approach to determining the performance of random linear coding is to consider again the (ordinary) transfer matrices of Lemma 5.5 and whether or not they are of full rank and hence allow solution, where some or all of the matrix entries are chosen at random. A more detailed look at the expression for the determinants of these matrices (as, e.g., given in lemma 2 of [29]) leads to the following:

Theorem 5.7 ([24], theorem 2.6, [29], theorem 2) *Consider a multicast connection problem on an arbitrary network with independent or linearly correlated sources and a network code in which some or all of the network code coefficients* $\{\mathbf{a}, \mathbf{f}\}$ *are chosen uniformly at random from the finite field* $\mathbb{F}_q$ *where* $q > L$ *and the remaining code coefficients, if any, are fixed. If there exists a solution to the network connection problem with the same values for the fixed code coefficients, then the probability the random network code is valid for the problem is at least* $(1 - L/q)^{\eta}$ *where* η *is the number of edges with the associated random coefficients.*

As before, this theorem follows from considering the expansion of the transfer matrix determinants viewing the entries as variables and considering the number of variable values for which the determinant is nonzero. The expansion of the determinant is viewed as a multivariate polynomial in the indeterminates and the question of nonsingularities becomes one of polynomial zeros. The actual assigned values can be considered as random values and hence the probability mentioned in the previous theorem.

It is argued in [35] that random network coding (RNC) (usually random linear network coding – RLNC) is likely to be most effective in peer-to-peer (P2P) networks where overlay networks are formed among participating users and where centralized coordination for transmission and coding strategies is not needed. Further, such peers are more likely to be able to afford the computing power necessary to implement finite field arithmetic and matrix inversion. That work considers progress on the use of RLNC in both large file download and streaming applications. It also reports on the use of gossiping techniques, where each user transmits to selected neighbors to reach all required participants after a number of rounds. An advantage of RLNC, similar to rateless coding techniques, is that a receiver needs only to collect the required number of *any* coded packets before applying Gaussian elimination to decode, rather than wait for *specific* packets it may be missing. A disadvantage is the complexity of inverting a matrix over the finite field in order to obtain a solution. Interesting references on RLNC and packet networks more generally include [12, 37, 53].

Batched Sparse (BATS) Codes

BATS codes are a special class of interesting network codes designed for the efficient transmission of packets through a network. They combine aspects of fountain coding and linear network coding to achieve certain performance characteristics that make them attractive for implementation. A brief overview of their construction is described. The notation of [54, 55, 56] is followed.

Suppose a file is divided into K packets each represented by T symbols over the alphabet $\mathbb{F}_q$ and define the $T \times K$ matrix whose columns are packets (column vectors):

$$\boldsymbol{B} = [\boldsymbol{b}_1, \boldsymbol{b}_2, \ldots, \boldsymbol{b}_K].$$

A subset of the columns (packets) of $\boldsymbol{B}$, $\boldsymbol{B}_i$ is formed as follows. Much as for fountain codes (see Chapter 2) a discrete distribution $\Psi = \{\Psi_i\}_{i=0}^{K}$ is assumed. The distribution will be optimized later. The distribution is sampled to give integer d_i with probability Ψ_{d_i}. From $\boldsymbol{B}$, d_i packets (columns) are chosen uniformly at random from the K packets and used as columns to form the $T \times d_i$ matrix $\boldsymbol{B}_i$. For a fixed integer M a *batch* $\boldsymbol{X}_i$ of M packets is formed by the matrix operation

$$\boldsymbol{X}_i = \boldsymbol{B}_i \boldsymbol{G}_i$$

where $\boldsymbol{G}_i$ is a $d_i \times M$ matrix over $\mathbb{F}_q$. This matrix $\boldsymbol{G}_i$ can be either predesigned or randomly generated as needed. Only the random case is considered here and the elements of $\boldsymbol{G}_i$ are chosen uniformly at random from $\mathbb{F}_q$. Thus the batch $\boldsymbol{X}_i$ consists of the M packets formed by random linear combinations of the (original uncoded) d_i packets of $\boldsymbol{B}_i$. The notion of batches seems related to the notion of "generations" used by [10] and "chunks" used by [39] in other contexts. Suppose n (not fixed) batches $\{\boldsymbol{X}_1, \boldsymbol{X}_2, \ldots, \boldsymbol{X}_n\}$ are formed in this manner. The situation is represented in Figure 5.2(a) as a Tanner graph with the K variable nodes of the original packets and the check nodes as the n (coded) batches, each batch containing M random linear combinations of a randomly selected subset of certain packets of a certain number d_i (for $\boldsymbol{B}_i$) determined by sampling the distribution Ψ.

The packet header contains information as to the batch number and original packets. The source transmits the packets in a batch randomly into the network. An intermediate node receives the packets of a batch over incoming links and applies (random) linear network coding of packets into new packets for outgoing links. It is important to note that in this operation only packets from the same batch are linearly combined.

The specific types of operations allowed at the intermediate nodes might depend on the type of network and the application. However, the transformations on the packets as they traverse the network do not depend on the specifics of the network operations. Let $\boldsymbol{Y}_i$ be the set of packets received by the destination node, all from the same batch. As in a previous section, the combined effect of the linear combining operations used at the various intermediate nodes of the network can be realized by a transfer matrix $\boldsymbol{H}_i$. Thus $\boldsymbol{Y}_i$ can be expressed by

$$Y_i = B_i G_i H_i$$

where the matrix $\boldsymbol{H}_i$ is the transfer matrix that reflects the operations performed by the intermediate network nodes. Such channels are referred to as *linear operator channels* (LOC) [57]. It is a matrix with M rows and a number of columns that depends on the number of distinct packets received – say L. Thus $\boldsymbol{Y}_i$ is a $T \times L$ matrix over $\mathbb{F}_q$. The matrix $\boldsymbol{G}_i\boldsymbol{H}_i$ is known to the receiver by contents of the packet headers. The equation is to be solved for the d_i data packets that were used in the formation of $\boldsymbol{B}_i$. This will be possible if and only if the matrix $\boldsymbol{G}_i\boldsymbol{H}_i$ is of full rank d_i.

The decoding operation is depicted in Figure 5.2 (only the edges for check node $\boldsymbol{G}_i\boldsymbol{H}_i$ are shown for clarity) and is as follows. If rank of the $d_i \times n$ matrix $\boldsymbol{G}_i\boldsymbol{H}_i$ is full (i.e., d_i and assuming all batches are represented), then the above equation will have a unique solution for the involved d_i original packets obtained by Gaussian elimination. The solution for the d_i packets $\boldsymbol{b}_{i_j}, j = 1, 2, \ldots, d_i$ can be found and substituted for the unknown variable nodes on the left-hand side of the graph. These now-known packet values can then be used by each neighbor check node to reduce the size of the matrix equations (by removing the corresponding columns) – thus reducing the complexity of further matrix inversions. The process is repeated until all variable nodes are resolved.

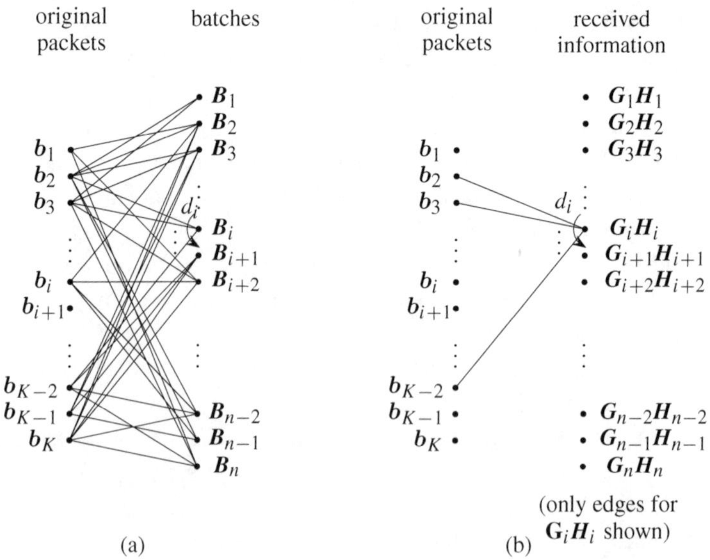

Figure 5.2 BATS code representations: (a) encoding and (b) decoding

Further details on the construction, analysis and performance analysis of these BATS codes may be found in the papers [54, 55] and especially the detailed monograph [56]. Conditions are developed on how to choose the distribution to ensure decoding success with high probability.

Index Codes

The notion of index coding arose first in the context of source coding with side information (e.g., Birk and Kot [5, 6]). The form of the problem discussed in [3, 4] is briefly outlined here to give an indication of the results available. In general the situation is that a server has n clients, $R_1, R_2, \ldots, R_n$, and a set of information blocks $\chi = \{X_1, X_2, \ldots, X_L\}$. Each client already contains a subset of the blocks and is able via a low-rate back channel to request other blocks from the server. The contents of each client's cache are known to the server and the problem is for the server to transmit (broadcast) a sufficient minimum amount of information so that each client receives the blocks requested. There is no client-to-client communication and all communication is error-free.

It is clear ([6], lemma 1) that one can limit interest to the case where only one block is requested by each client in that a client requesting r blocks could be "split" into r subclients. The following model is of interest [4, 5, 6]. The data held by the server is regarded as an n-bit string $\boldsymbol{x} \in \{0,1\}^n \sim \mathbb{F}_2{}^n$ and there are n receivers. Receiver $R_i, i = 1, 2, \ldots, n$ is interested in receiving x_i and possesses a subset of x (although not x_i). The side information of the clients is conveniently represented by a directed graph $\mathcal{G}$ (referred to as the *side information graph*) on n nodes V with no self-loops (node i does not have x_i as side information) and no parallel edges. The graph has a directed edge (i, j) if receiver R_i has bit x_j. The case of more information bits than clients is easily taken care of.

For a subset $S \subseteq [n]$ let $x[S]$ be the subset of x corresponding to the coordinates of S. The side information contained by client R_i is denoted

$$N_i \stackrel{\Delta}{=} \{j \in V \mid (i, j)) \in E\}, \quad i \notin N_i,$$

where E is the set of directed edges of the graph $\mathcal{G}$. Note that one could also have used a directed bipartite graph (as in [32]) with the set of clients as left vertices and information bits as right vertices with directed edges indicating side information and information requests of clients.

Definition 5.8 ([4]) A deterministic *index code* C for a string $\boldsymbol{x} \sim \mathbb{F}_2^n$ with side information graph $\mathcal{G}$ is a set of codewords over $\mathbb{F}_2$ of total length ℓ together with

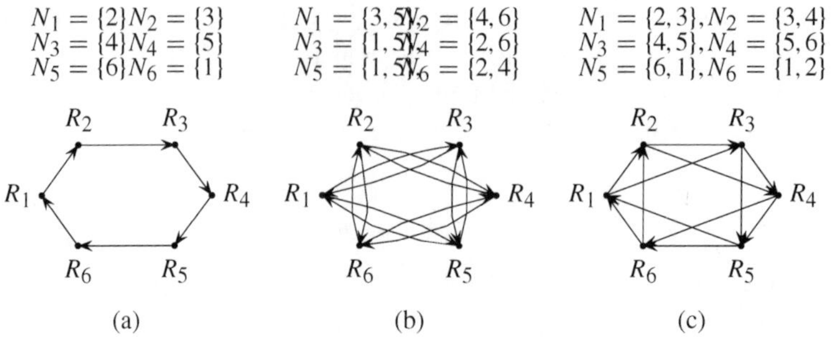

Figure 5.3 Example side information graphs

(i) an encoding function E that maps $\{0,1\}^n$ to a set of codewords;
(ii) a set of decoding functions $D_1, D_2, \ldots, D_n$ such that $D_i(E(x), x[N_i]) = x_i,\ i = 1, 2, \ldots, n.$

Example 5.9 Figure 5.3 shows some small examples of nodes containing files and their associated side information graphs to illustrate the concept. The corresponding side information matrices are discussed in Example 5.13.

The following definition will prove important to the subject.

Definition 5.10 Let $\mathcal{G}$ be a directed graph on n vertices without self-loops or parallel (directed) edges. The $\{0,1\}$ matrix $\boldsymbol{A} = (a_{ij})$ is said to fit $\mathcal{G}$ if for all i and j (1) $a_{ii} = 1$ and (2) $a_{ij} = 0$ whenever (i,j) is *not* an edge of $\mathcal{G}$. The matrix $\boldsymbol{A}$ is the *side information matrix* of the side information graph $\mathcal{G}$.

Notice that for such a matrix $\boldsymbol{A} - \boldsymbol{I}_n$ for $\boldsymbol{I}_n$ the $n \times n$ identity matrix is an adjacency matrix for the edge subgraph of $\mathcal{G}$. Denote by rk_2 the rank of a $\{0,1\}$ matrix $\boldsymbol{A}$ over $\mathbb{F}_2$.

Definition 5.11 $\text{minrk}_2(\mathcal{G}) \triangleq \min \left\{ \text{rk}_2(\boldsymbol{A}) \mid \boldsymbol{A} \text{ fits } \mathcal{G} \right\}.$

This quantity $\text{minrk}_2(\mathcal{G})$ is shown to be related to the clique number of $\mathcal{G}$ (the largest complete subgraph in $\mathcal{G}$), the Shannon capacity of $\mathcal{G}$ and the chromatic number of $\mathcal{G}$ (smallest number of colors needed so that no two neighboring vertices have the same color). An *index code* then is a set of codewords (bits in our case) the server can transmit so that each client $R_i, i = 1, 2, \ldots, n$ is able to retrieve x_i from the transmitted codewords given their side information. The code is called *linear* if the encoding functions of the server are linear over $\mathbb{F}_2$.

The important property of $\text{minrk}_2(\mathcal{G})$ for our interest is the following:

Theorem 5.12 ([3, 4]) *For any side information graph $\mathcal{G}$ there exists a linear index code for $\mathcal{G}$ whose length is $\text{minrk}_2(\mathcal{G})$. The bound is optimal for all linear index codes for $\mathcal{G}$.*

Example 5.13 The side information matrices of the examples in Example 5.9 are as follows:

$$\boldsymbol{A}_{(a)} = \begin{bmatrix} 1&1&0&0&0&0 \\ 0&1&1&0&0&0 \\ 0&0&1&1&0&0 \\ 0&0&0&1&1&0 \\ 0&0&0&0&1&1 \\ 1&0&0&0&0&1 \end{bmatrix}, \quad \boldsymbol{A}_{(b)} = \begin{bmatrix} 1&0&1&0&1&0 \\ 0&1&0&1&0&1 \\ 1&0&1&0&1&0 \\ 0&1&0&1&0&1 \\ 1&0&1&0&1&0 \\ 0&1&0&1&0&1 \end{bmatrix}$$

$$\boldsymbol{A}_{(c)} = \begin{bmatrix} 1&1&1&0&0&0 \\ 0&1&1&1&0&0 \\ 0&0&1&1&1&0 \\ 0&0&0&1&1&1 \\ 1&0&0&0&1&1 \\ 1&1&0&0&0&1 \end{bmatrix}$$

A code for each of the cases is simple to find. The matrix $\boldsymbol{A}_{(a)}$ is of rank 5 and a coding scheme that achieves the bound of Theorem 5.12 is $x_1 \oplus x_2, x_2 \oplus x_3, x_3 \oplus x_4, x_4 \oplus x_5$ and $x_5 \oplus x_6$ (each of the five codewords is one bit). R_1 receives x_1 by XORing its side information with $x_1 \oplus x_2$ retrieves x_2 and similarly for $R_2, \ldots, R_5$. R_6 receives x_6 by XORing all the codewords together to give $x_1 \oplus x_6$ and XORing this with its side information x_1.

Another "extreme" contrived case (not shown in the figure) might be where each receiver has as side information all bits except its own, i.e., $R_i, i = 1, 2, \ldots, n$ has side information $N_i = V \backslash \{i\}$. In this case the side information matrix would be the all-ones matrix of rank 1 and a code would be the single bit $x_1 \oplus \cdots \oplus x_n$.

The matrix $\boldsymbol{A}_{(b)}$ is of rank 2 – the graph is the union of two disjoint graphs, each with three vertices. The two codewords are $x_1 \oplus x_3 \oplus x_5$ and $x_2 \oplus x_4 \oplus x_6$.

The matrix $A_{(c)}$ can be shown to have rank 4 and four codewords are $y_1 = x_1 \oplus x_2 \oplus x_3,\ y_2 = x_2 \oplus x_3 \oplus x_4,\ y_3 = x_3 \oplus x_4 \oplus x_5, y_4 = x_1 \oplus x_5 \oplus x_6$.

The notion of random codes for index codes is also considered [2]. It is clear there are close connections between index coding and the more general network coding and a few of these are explored in [15, 16]. The papers [7, 32] consider index coding with error correction.

The monograph [2] is an excellent and comprehensive view of this subject that considers determining capacity regions for such codes and code constructions.

Network Codes on Real Networks

Given the theoretical results on network coding that imply very significant performance gains on network performance, it is not a surprise there has been considerable effort to realize these gains on real networks. This includes the pervasive Internet where such gains would have a major impact on performance. This would be at both the IP layer and the application layer where overlay networks are of particular interest. Other packet networks such as ATM networks, wireless ad hoc networks and many others are potential customers for the benefits of network coding.

However, there are major obstacles standing in the way of these performance gains. Typically in real networks, packets experience random delays on edges whose capacities may be unknown and may be bidirectional. Hence it may not be realistic to compute the capacity of the network with any degree of confidence. Typically there will be many cycles, contradicting the assumption of an acyclic network, although the precise network topology may be unknown. Edges and nodes may fail requiring continuous adjustment of the transmission strategy. Transmitting a large file on a network coded system would generally require breaking the file into manageable segments which introduces the problem of latency at the nodes as the nodes wait for a sufficient number of packets to combine for their next linear combining for transmission.

The work of Chou et al. [10] introduces two ideas to address some of these problems. First they consider appending a *global encoding vector* which, in essence, gives the coefficients of the information packets transmitted from a given node on a given outgoing edge. Thus each packet that arrives at a node contains a global encoding vector. As each node forms a linear encoding of the packets arriving, it is also able to translate those coefficients into the coefficients that would be needed to form the outgoing packet in terms of the information packets themselves. Thus if a sink node receives a sufficient number of linearly independent packets, it is able to compute from the coefficients in the arriving packets, via Gaussian elimination, the information packets themselves. This solution has the desirable properties that it does not depend on network topology and is unaffected by link and node failures. In addition the work [10] suggests dividing an information source into *generations* (and packets within a generation), a generation number appearing in a packet header. Only packets within a generation are then combined which limits the size of the Gaussian elimination needed at the receiver.

Numerous other issues concerning the advantages and practicality of network coding are considered in the extensive literature, including the references [10, 19, 21, 22, 23, 35, 38, 39, 40, 41, 44, 45, 50, 51].

Comments

The subject of network coding embraces a wide variety of techniques and disciplines and this chapter is limited in its scope. The seminal papers of the subject include [1] the paper that first outlined the possible gains to be had through network coding [36], the paper that showed that linear coding was sufficient to achieve these gains (under certain circumstances) and [28, 29] the papers that described the important notion of RLNC. The monographs [24] and [59] are very informative, as is the article [60]. The paper [34] described the algebraic approach to network coding that is widely followed. The monographs [2] on index coding and [56] on BATS codes are detailed and authoritative expositions on these topics.

The important technique of random subspace coding for networks, developed in [33, 43], is discussed in Chapter 12 due to its connections with rank-metric codes.

References

[1] Ahlswede, R., Cai, N., Li, S. R., and Yeung, R.W. 2000. Network information flow. *IEEE Trans. Inform. Theory*, **46**(4), 1204–1216.

[2] Arbabjolfaei, F., and Kim, Y.-H. 2018. *Foundations and trends in communications and information theory*. NOW Publisher, Boston, MA.

[3] Bar-Yossef, Z., Birk, Y., Jayram, T.S., and Kol, T. 2006. Index coding with side information. Pages 197–206 of: *2006 47th Annual IEEE Symposium on Foundations of Computer Science (FOCS'06)*.

[4] Bar-Yossef, Z., Birk, Y., Jayram, T.S., and Kol, T. 2011. Index coding with side information. *IEEE Trans. Inform. Theory*, **57**(3), 1479–1494.

[5] Birk, Y., and Kol, T. 1998. Informed-source coding-on-demand (ISCOD) over broadcast channels. Pages 1257–1264 of: *Proceedings. IEEE INFOCOM '98, the Conference on Computer Communications. Seventeenth Annual Joint Conference of the IEEE Computer and Communications Societies. Gateway to the 21st Century (Cat. No. 98, San Francisco, CA)*, vol. 3.

[6] Birk, Y., and Kol, T. 2006. Coding on demand by an informed source (ISCOD) for efficient broadcast of different supplemental data to caching clients. *IEEE Trans. Inform. Theory*, **52**(6), 2825–2830.

[7] Byrne, E., and Calderini, M. 2017. Error correction for index coding with coded side information. *IEEE Trans. Inform. Theory*, **63**(6), 3712–3728.

[8] Cameron, P.J. 1994. *Combinatorics: topics, techniques, algorithms*. Cambridge University Press, Cambridge.

[9] Celebiler, M., and Stette, G. 1978. On increasing the down-link capacity of a regenerative satellite repeater in point-to-point communications. *Proc. IEEE*, **66**(1), 98–100.

[10] Chou, P.A., Wu, Y., and Jain, K. 2003 (October). Practical network coding. In: *Allerton Conference on Communication, Control, and Computing*. Invited paper.

[11] Cormen, T.H., Leiserson, C.E., Rivest, R.L., and Stein, C. 2009. *Introduction to algorithms*, 3rd ed. MIT Press, Cambridge, MA.

[12] Dana, A.F., Gowaikar, R., Palanki, R., Hassibi, B., and Effros, M. 2006. Capacity of wireless erasure networks. *IEEE Trans. Inform. Theory*, **52**(3), 789–804.

[13] Diestel, R. 2018. *Graph theory*, 5th ed. Graduate Texts in Mathematics, vol. 173. Springer, Berlin.

[14] Edmonds, J. 1965. Minimum partition of a matroid into independent subsets. *J. Res. Nat. Bur. Standards Sect. B*, **69B**, 67–72.

[15] Effros, M., El Rouayheb, S., and Langberg, M. 2015. An equivalence between network coding and index coding. *IEEE Trans. Inform. Theory*, **61**(5), 2478–2487.

[16] El Rouayheb, S., Sprintson, A., and Georghiades, C. 2010. On the index coding problem and its relation to network coding and matroid theory. *IEEE Trans. Inform. Theory*, **56**(7), 3187–3195.

[17] Elias, P., Feinstein, A., and Shannon, C. 1956. A note on the maximum flow through a network. *IRE Trans. Inform. Theory*, **2**(4), 117–119.

[18] Ford, Jr., L.R., and Fulkerson, D.R. 1956. Maximal flow through a network. *Canad. J. Math.*, **8**, 399–404.

[19] Gkantsidis, C., Miller, J., and Rodriguez, P. 2006 (February). Anatomy of a P2P content distribution system with network coding. In: *IPTPS'06*.

[20] Harvey, N.J.A., Karger, D.R., and Murota, K. 2005. Deterministic network coding by matrix completion. Pages 489–498 of: *Proceedings of the Sixteenth Annual ACM-SIAM Symposium on Discrete Algorithms*. SODA '05, Philadelphia, PA.

[21] Heide, J., Pedersen, M.V., Fitzek, F.H.P., and Larsen, T. 2008. Cautious view on network coding From theory to practice. *J. Commun. Net.*, **10**(4), 403–411.

[22] Heide, J., Pedersen, M.V., Fitzek, F.H.P., and Medard, M. 2014 (May). A perpetual code for network coding. Pages 1–6 of: *2014 IEEE 79th Vehicular Technology Conference (VTC Spring)*.

[23] Heide, J., Pedersen, M.V., F.H.P., and Médard, M. 2015. Perpetual codes for network coding. *CoRR*, abs/1509.04492.

[24] Ho, T., and Lun, D.S. 2008. *Network coding*. Cambridge University Press, Cambridge.

[25] Ho, T., Karger, D.R., Medard, M., and Koetter, R. 2003 (June). Network coding from a network flow perspective. Page 441 of: *IEEE International Symposium on Information Theory, 2003. Proceedings*.

[26] Ho, T., Koetter, R., Medard, M., Karger, D.R., and Effros, M. 2003 (June). The benefits of coding over routing in a randomized setting. Page 442 of: *IEEE International Symposium on Information Theory, 2003. Proceedings*.

[27] Ho, T., Médard, M., Shi, J., Effros, M., and Karger, D.R. 2003. On randomized network coding. In: *2010 48th Annual Allerton Conference on Communication, Control, and Computing (Allerton)*.

[28] Ho, T., Médard, M., Koetter, R.L., Karger, D.R., Effros, M., Shi, J., and Leong, B. 2004. Toward a random operation of networks. *IEEE Trans. Inform. Theory*, 2004, 1–8.

[29] Ho, T., Médard, M., Koetter, R., Karger, D.R., Effros, M., Shi, J., and Leong, B. 2006. A random linear network coding approach to multicast. *IEEE Trans. Inform. Theory*, **52**(10), 4413–4430.

[30] Jaggi, S., Sanders, P., Chou, P.A., Effros, M., Egner, S., Jain, K., and Tolhuizen, L.M.G.M. 2005. Polynomial time algorithms for multicast network code construction. *IEEE Trans. Inform. Theory*, **51**(6), 1973–1982.

[31] Jungnickel, D. 2013. *Graphs, networks and algorithms*, 4th ed. Algorithms and Computation in Mathematics, vol. 5. Springer, Heidelberg.

[32] Kim, J.W., and No, J.-S. 2017. Index coding with erroneous side information. *IEEE Trans. Inform. Theory*, **63**(12), 7687–7697.

[33] Koetter, R., and Kschischang, F.R. 2008. Coding for errors and erasures in random network coding. *IEEE Trans. Inform. Theory*, **54**(8), 3579–3591.

[34] Koetter, R., and Médard, M. 2003. An algebraic approach to network coding. *IEEE/ACM Trans. Netw.*, **11**(5), 782–795.

[35] Li, B., and Niu, D. 2011. Random network coding in peer-to-peer networks: from theory to practice. *Proc. IEEE*, **99**(3), 513–523.

[36] Li, S.-R., Yeung, R.W., and Cai, N. 2003. Linear network coding. *IEEE Trans. Inform. Theory*, **49**(2), 371–381.

[37] Lun, D.S., Médard, M., Koetter, R., and Effros, M. 2008. On coding for reliable communication over packet networks. *Phys. Commun.*, **1**(1), 3 – 20.

[38] Maymounkov, P., and Mazières, D. 2003. Rateless codes and big downloads. Pages 247–255 of: Kaashoek, M.F., and Stoica, I. (eds.), *Peer-to-peer systems II*. Springer, Berlin, Heidelberg.

[39] Maymounkov, P., and Harvey, N.J.A. 2006. Methods for efficient network coding. In: *In Allerton Conference in Communication, Control and Computing*.

[40] Médard, M., and Sprintson, A. 2011. *Network coding: fundamentals and applications*. Academic Press, New York/London.

[41] Médard, M., Fitzek, F., Montpetit, M.-J., and Rosenberg, C. 2014. Network coding mythbusting: why it is not about butterflies anymore. *Commun. Mag., IEEE*, **52**(07), 177–183.

[42] Sanders, P., Egner, S., and Tolhuizen, L. 2003. Polynomial time algorithms for network information flow. Pages 286–294 of: *Proceedings of the Fifteenth Annual ACM Symposium on Parallel Algorithms and Architectures*. SPAA '03. ACM, New York.

[43] Silva, D., Kschischang, F.R., and Koetter, R. 2008. A rank-metric approach to error control in random network coding. *IEEE Trans. Inform. Theory*, **54**(9), 3951–3967.

[44] Silva, D., Zeng, W., and Kschischang, F.R. 2009 (June). Sparse network coding with overlapping classes. Pages 74–79 of: *2009 Workshop on Network Coding, Theory, and Applications*.

[45] Sundararajan, J.K., Shah, D., Medard, M., Jakubczak, S., Mitzenmacher, M., and Barros, J. 2011. Network coding meets TCP: theory and implementation. *Proc. IEEE*, **99**(3), 490–512.

[46] Tan, M., Yeung, R.W., Ho, S., and Cai, N. 2011. A unified framework for linear network coding. *IEEE Trans. Inform. Theory*, **57**(1), 416–423.

[47] Trullols-Cruces, O., Barcelo-Ordinas, J.M., and Fiore, M. 2011. Exact decoding probability under random linear network coding. *IEEE Commun. Lett.*, **15**(1), 67–69.

[48] van Leeuwen, J. (ed.). 1990. *Handbook of theoretical computer science. Vol. A.* Elsevier Science Publishers, B.V., Amsterdam; MIT Press, Cambridge, MA.

[49] van Lint, J.H., and Wilson, R.M. 1992. *A course in combinatorics*. Cambridge University Press, Cambridge.

[50] Wang, M., and Li, B. 2006 (June). How practical is network coding? Pages 274–278 of: *2006 14th IEEE International Workshop on Quality of Service*.

[51] Wang, M., and Li, B. 2007. R2: random push with random network coding in live peer-to-peer streaming. *IEEE J. Sel. Areas Commun.*, **25**(9), 1655–1666.

[52] West, D.B. 1996. *Introduction to graph theory*. Prentice Hall, Upper Saddle River, NJ.

[53] Wu, Y. 2006. A trellis connectivity analysis of random linear network coding with buffering. Pages 768–772 of: *2006 IEEE International Symposium on Information Theory*.

[54] Yang, S., and Yeung, R.W. 2011. Coding for a network coded fountain. Pages 2647–2651 of: *2011 IEEE International Symposium on Information Theory Proceedings*.

[55] Yang, S., and Yeung, R.W. 2014. Batched sparse codes. *IEEE Trans. Inform. Theory*, **60**(9), 5322–5346.

[56] Yang, S., and Yeung, R.W. 2017. BATS codes: theory and practice. *Synth. Lect. Commun. Netw.*, **10**(09), 1–226.

[57] Yang, S., Ho, S.-W., Meng, J., and Yang, E.-H. 2010. Linear operator channels over finite fields. *CoRR*, abs/1002.2293.

[58] Yeung, R.W. 2002. *A first course in information theory*. Information Technology: Transmission, Processing and Storage. Kluwer Academic/Plenum Publishers, New York.

[59] Yeung, R.W. 2008. *Information theory and network coding*, 1st ed. Springer, Boston, MA.

[60] Yeung, R.W. 2011. Network coding: a historical perspective. *Proc. IEEE*, **99**(3), 366–371.

[61] Yeung, R.W., Li, S.-Y. Robert, Cai, N., and Zhang, Z. 2006. Network coding theory. *Found. Trends Commun. Inform. Theory*, **2**(4), 241–329.

6

Coding for Distributed Storage

Over the past two decades a considerable effort has been made in developing coding techniques in connection with storing and retrieving information in a network environment, clearly an area of practical importance where significant impact of research may be realized. This chapter and the next four chapters are closely linked. This chapter considers the coding of information for storing on networked servers in such a manner that if a server fails a new one can be regenerated by downloading from other servers to restore the contents of the failed server. The aim will be to code the information stored on the servers (also referred to as nodes) in such a way that the amount of information stored on the servers and the total amount that needs to be downloaded in order to restore the failed server are, in a sense to be specified, minimized. Such coded systems can be shown to have substantial gains over simple replication of the server contents. The next section discusses the theoretical gains that might be realized from such a system and the subsequent section considers explicit coding systems to realize these gains.

The next chapter introduces codes that are *locally repairable* in the sense that an erasure in a codeword coordinate position can be restored (or interpolated) by contacting a limited number of other coordinate positions. Here the term repairable refers to a coordinate position being erased as opposed to being in error, which is considered in the subsequent chapter on *locally decodable codes* (LDCs). Chapter 9 considers *private information retrieval* (PIR) codes. For these systems the information is coded and stored on servers in such a way that users can query the database and not have the servers obtain any information (in an information-theoretic sense) on which information is being sought. It will be shown there is an intimate connection with LDCs and PIR codes and this is explored in that chapter.

Finally, in Chapter 10, ways of encoding information on the servers in such a way that users retrieving information from them, no matter which information is being sought, no server is required to download more than some fixed number of symbols. Such a code is referred to as a *batch code*. It is a load-sharing type of coding ensuring that no server is overloaded with download requests, regardless of the information being requested.

As noted, these four topics are interrelated and all have been vigorously researched over the past two decades. The literature on them is now large and, in keeping with the purpose of this volume, these four chapters are merely intended to introduce the topics by providing some of the basic concepts and a few examples with pointers to the more general results available. Each chapter discusses some of their interrelationships.

For the present chapter the performance limits and techniques laid out in the seminal paper [5] have generated great interest in the research community addressing such problems and generated an unprecedented amount of research. An overview of this important work is given in the next section.

The number of coding techniques developed to achieve the performance limits given in Section 6.1 is very large. The use of Reed–Solomon codes for this problem is natural and recent work has shown that modifications of them can be very effective.

6.1 Performance Limits on Coding for Distributed Storage

It is difficult to overestimate the importance of the problem of storing and retrieving information from a network of servers to the global society. Most governments and large corporations maintain significant data centers and the cost of creating, powering, updating and maintaining them to manage this data is incalculable, not to mention the telecommunications infrastructure needed to use them. Further, the servers and memory devices used in these systems fail with some predictability and the failed units must be restored or replaced. The simple solution of replicating data on backup servers is not efficient and the role of coding has become central to such systems. In a sense it is remarkable that the application of coding-theoretic techniques to this problem did not arise sooner.

As noted, the paper [5] gives fundamental limits for the basic storage and repair models assumed here and is the starting point of most of the work in this area during the past decade. The purpose of the chapter is to discuss the approach of this paper and some of the subsequent work on the construction of codes for the models of interest. There will be two criteria of interest for such systems. The first is that of *minimum storage* of the coded system, the minimum of the total amount of storage for the file among all servers for the reliable operation of repairing a failed server, and the second one is *minimum download bandwidth* or *repair bandwidth*, the total amount of information that must be downloaded from adjacent servers to restore a failed server. Much of the notation in that influential paper is adapted with a view to making a full reading of that work (recommended) more amenable. There are numerous variations of this basic model considered in the literature.

The following basic model will be convenient for our purpose. Consider a file $\mathcal{M}$ of $|\mathcal{M}| = m$ symbols. The term symbol will be defined in various contexts and is left vague for the moment as is the term segment, meant to be some fixed number of symbols. It will at times be taken to be a single or multiple elements of a fixed finite field. These m file symbols are divided into k segments, each of m/k symbols. One possibility then is that these segments are encoded with a code to give n coded segments and each such coded segment is stored on one of n storage nodes or servers (terms used interchangeably), one segment per node. For this chapter it will often be the case that a segment size will be the symbol (element) size of a finite field (hence field size) and thus may be of a very large order. A server then may hold either a single or multiple finite field elements. A single finite field element may also be regarded as several subfield elements. The work [5] uses bits rather than symbols. When a node fails, it will be assumed the coding is such that the entire file $\mathcal{M}$ can be regenerated by downloading the contents of any k of the surviving servers. It is established that while this model is good in terms of the total amount of storage required, it may not be good in terms of the number of symbols that must be downloaded in order to repair or regenerate a failed server, the repair bandwidth, in the sense that k segments must be downloaded to repair just one segment (in the failed server). Indeed it is shown that if each of n servers is able to store slightly more than m/k symbols, this repair bandwidth can be significantly decreased.

As above, assume the original file has $|\mathcal{M}| = m$ symbols, and a symbol is considered an element of a finite field $\mathbb{F}_q$ for some prime power order q. The file is divided into k segments of $m/k = \alpha$ finite field symbols each. These segments are encoded into n segments by an (n,k) code (possibly over an extension field of order the size of segments) and the n segments are stored on n servers, i.e., each of the n servers is able to store α finite field symbols, one segment. Much of the research assumes an $(n,k,d_{\min})_q$ MDS code for which $d_{\min} = n - k + 1$. For this chapter the parameter d will indicate the number of helper nodes approached to regenerate the contents of a failed node, rather than code minimum distance which will be specified as $d_{\min}$. Thus, if a node fails, a replacement node can contact any d surviving nodes and download $\beta \leq \alpha$ symbols from each of them. The *repair bandwidth* is the total number of symbols downloaded to this replacement node, $\gamma = d\beta$. Such a model will be described as an (n,k,d,α,γ)-tuple. A code will be referred to as *feasible* if there exists a coding system with storage α symbols in each node and repair bandwidth γ symbols, each of the d helper nodes downloading β symbols for the regeneration process. The literature contains numerous other models with differing meaning of the parameters. The following captures the intent of the model of interest here.

For future reference it will be convenient to record these parameters:

$\mathcal{M}$:	original file of m $\mathbb{F}_q$ elements (= symbols)
m :	$\lvert\mathcal{M}\rvert$, size of original file in $\mathbb{F}_q$ symbols
k :	information in any k servers is sufficient to restore original file
n :	number of active servers in the system at any time
d :	number of surviving servers contacted by new node to regenerate
α :	the number of symbols stored in each server = m/k
β :	the number of symbols downloaded from each of the d helper nodes contacted for the regeneration process
γ :	the bandwidth of the regeneration process = total number of symbols downloaded = $d\beta$

In the literature of coding for distributed storage, different repair strategies [5, 6] are considered. In *exact repair* the contents of the failed server are exactly reproduced. In *functional repair* the contents of the restored server may differ from the original as long as the overall MDS property of the information is maintained, i.e., the ability of using the contents of any k of the n servers to reconstruct the original file. In functional repair, e.g., the reconstructed contents of the new server may simply be a scalar multiple of the original contents. In some works *systematic exact repair* of the systematic information is exact while the parity information may be functionally repaired. Only exact repair is considered in this work.

For each set of parameters (n,k,d,α,γ) it is of interest to determine conditions and parameter sets corresponding to feasible solutions to the problem, i.e., conditions which guarantee the existence of feasible codes. Furthermore it is of interest to determine coding schemes that correspond to *minimum storage regeneration* (MSR) codes and *minimum bandwidth regeneration* (MBR) codes. The following technique is developed [5] to determine such conditions.

The notion of an *information flow graph* is introduced to tackle the problem. The idea, borrowing from techniques in network coding, is that the replacement node will be able to repair the information lost by a server failure with the parameter set posed, only if the associated information flow graph has a sufficiently large min-cut with an accompanying max-flow. Thus the server content reconstruction problem can be viewed as the flow problem of a multicast network considered in Chapter 5. A description of the problem formulation and solution is given with reference to [5] for details.

The process is described. Initially assume a source node S containing the entire file of $|\mathcal{M}| = m\ \mathbb{F}_q$ symbols. Assume the k segments, each of size m/k symbols, are encoded into n segments according to some $(n,k)_q$ code (which is not yet specified – the construction of such codes will be of interest for suitable parameters). Each of these n segments will be stored in one of the n servers – there are n servers active at any given time. Designate the n servers as $x^{(i)}, i = 1, 2, \ldots, n$ – except that each server will be viewed as a pair of nodes $x_{in}^{(i)}$ and $x_{out}^{(i)}$. For each node the edge between $x_{in}^{(i)}$ and $x_{out}^{(i)}, i = 1, 2, \ldots, n$ is assumed to have a transmission capacity of α symbols, the capacity assumed of each server. By definition of the assumed code, the contents of any k of the servers is sufficient to reconstruct the original file. However, only the contents of the failed server are to be reconstructed and it will be assumed $d \leq k$, i.e., it is sufficient to contact d helper nodes to reconstruct the contents of the failed node. The edges between the source S and the n active server input nodes $x_{in}^{(i)}$ are assumed to have infinite capacity. The edges between each node $x_{out}^{(i)}$ and the sink node, called *data collector* node DC, or terminal node T which becomes the regenerated node at the end of the process, are also assumed to have infinite capacity at this stage. None of the edges of infinite capacity will be involved in the graph cuts noted.

After initially loading the n servers, the source node becomes inactive. Assume there have been multiple node failures which have been repaired (exactly) so there remain at any time n active nodes. Assume for simplicity by renaming reconstructed nodes, at some point in the process, the currently active nodes are $x^{(1)}, x^{(2)}, \ldots, x^{(n)}$ (and each such node is actually an (input, output) pair with a single edge of capacity α between the input node and output node). When a node fails, a new data collector node DC is formed which downloads β symbols from each node of some set of d active nodes (a total of $\gamma = d\beta$ symbols downloaded) and from these $d\beta$ symbols the contents of the failed node can be reconstructed. Assume the d helper nodes involved are $x^{(i_1)}, \ldots, x^{(i_d)}$.

It is desired to determine the capacity of any min-cut of the resulting flow graph, noting that the only edges in the cut $\mathscr{C}$ are those of the form either $x_{in}^{(i)}$ to $x_{out}^{(i)}$ (of capacity α) or $x_{out}^{(i)}$ to $x_{in}^{(j)}, i \neq j$ (of capacity β), either from a helper node to helper node or non-helper node to helper node. Each active node contains α symbols. As mentioned, the capacity of the edges from the original source to an active node and from an active helper node to the terminal node (data collector DC) have infinite capacity and are not included in the min-cut computation.

The argument to determine the capacity of a min-cut is as follows [5]. Consider a min-cut of the signal flow graph representing the current server storage configuration from the original source S and the current data collector DC (i.e., reconstruction node) with the restriction that the min-cut does not include paths to/from source or terminal DC. The source S initially loads information into n servers and after that plays no role in the argument. When a server fails a new data collector node DC is formed and connects to d servers downloading β symbols from each in order to regenerate the failed server. The links between the DC and servers are assumed to have infinite capacity and are not part of any cut considered below. After completion of the process the DC becomes the new active server, replacing the failed one. The similarity of the situation for network coding in Chapter 5 is immediate although it may not be clear how this formulation solves the regeneration problem. Recall the max-flow min-cut theorem of Chapter 5:

Theorem 6.1 (Max-Flow Min-Cut Theorem) *In any flow network with source S and terminal node $T = DC$, the value of the maximum (S,T) flow is equal to the capacity of the (S,T) min-cut.*

The point of the argument is to realize the value of the max-flow through the min-cut must be at least m, the size of the file, in order for a file or server content to be regenerated. The repair bandwidth (the total number of symbols downloaded to regenerate the contents of a failed node) is $\gamma = d\beta$.

For a given system parameter set it can be shown the minimum amount of storage needed for each server, α, must satisfy

$$m \leq \sum_{i=0}^{\min\{d,k\}-1} \min\left\{\left(1-\frac{i}{d}\right)\gamma,\alpha\right\} \tag{6.1}$$

or correspondingly

$$\text{min-cut} \geq \sum_{i=0}^{\min\{d,k\}-1} \min\left\{(d-i)\beta,\alpha\right\}.$$

Recall that while the contents of any k servers can be used to reconstruct the entire file, only β symbols are downloaded from each of the d helper nodes from which the contents of the failed server are reconstructed. Also note that if $d < k$ the above equation implies that any d nodes will have a flow of m in which case, by the assumptions made, it would imply $d = k$. Thus later on it will be assumed $d \geq k$. The minimum storage problem [5] is then, for a given $d, \gamma = d\beta$, to determine the smallest server storage α, i.e.,

$$\alpha * (d,\gamma) \triangleq \min \alpha \quad \text{subject to} \sum_{i=0}^{\min\{d,k\}-1} \min\left\{\left(1-\frac{i}{d}\right)\gamma, \alpha\right\} \geq m. \quad (6.2)$$

For this reason the quantity α is often referred to as the capacity for the codes. The argument below will apply to any signal flow graph. It is not difficult to construct a signal flow graph which reflects the above equation although it seems difficult to display such a graph in a useful manner. Such a graph appears in [5] and ([7], chapter by Liu and Oggier) who also give an argument as to how Equation 6.1 or 6.2 arises.

Consider a min-cut between the source S and data collector DC, using no edges emanating from either S or DC. All edges between nodes have capacity β and those from the input to output within a node capacity α. Define the cut $\mathscr{C}$ by the disjoint vertex sets of the n active nodes, U, V such that $U \cup V$ contains all the vertices $x_{in}^{(i)}, x_{out}^{(i)}, i = 1,2,\ldots,n$. For the first helper node $\left(x_{in}^{(1)}, x_{out}^{(1)}\right)$, if $x_{in}^{(1)}$ is in U, then the edge (of capacity α) between $x_{in}^{(1)}$ and $x_{out}^{(1)}$ is in the cut $\mathscr{C}$ since the edge between $x_{out}^{(1)}$ and DC is not in the cut, by definition. If $x_{in}^{(1)} \in V$, then the d input edges of $x_{in}^{(1)}$ are in $\mathscr{C}$ and it is clear the first helper node contributes $\min\{d, \alpha)\beta$ to the sum.

For the second helper node $\left(x_{in}^{(2)}, x_{out}^{(2)}\right)$ if $x_{in}^{(2)} \in U$, then $x_{out}^{(2)}$ must be in V and the edge between the input and output parts is in the cut. If $x_{in}^{(2)} \in V$, then $x_{out}^{(2)}$ is also and at least $d-1$ edges, each of capacity β must be in the cut since there could be an edge between $x_{out}^{(2)}$ and $x_{in}^{(1)}$. Hence the contribution of the helper node to the sum is at least $\min\{(d-1)\beta, \alpha\}$. The argument continues to give the sum in Equation 6.1 [5, 7].

A failed server cannot be regenerated if the min-cut is smaller than the original file size m and that if the min-cut of the signal flow graph does satisfy the above equation, then there will exist ([5], proposition 1), for a sufficiently large field, a linear code that will allow regeneration. Also ([5], lemma 2) there will exist a signal flow graph for coding parameters (n,k,d,α,γ) if the above bound is met with equality.

It is of interest to determine all points on the (α, γ) plane that satisfy the above Equations 6.1 and 6.2. It is noted ([7], chapter by Liu and Oggier) that for given values of α, $\gamma = d\beta$ and d, the inequality of Equation 6.1 can be uniquely determined in a closed-form equation. Before giving this solution note first that if $\alpha \leq \left(1-(k-1)/d\right)\gamma$, then α is less than each term in the sum and hence Equation 6.1 yields

$$m = k\alpha$$

and hence

$$\gamma \geq \frac{\alpha d}{d-(k-1)} = \frac{m}{k} \cdot \frac{d}{d-(k-1)}.$$

It is clear that this corresponds to a minimum storage code since we have the restriction that the contents of any k nodes can be used to regenerate the entire file, each node must store at least m/k symbols. Thus a code achieving this storage and download point (α, γ) is referred to as the *minimum storage repair* (MSR) bound and designated

$$(\alpha_{MSR}, \gamma_{MSR}) = \left(\frac{m}{k}, \frac{md}{k(d-k+1)}\right) \qquad \text{MSR bound.} \tag{6.3}$$

For the minimum storage solution the minimum download is achieved with $d = n-1$, i.e., when a node fails, the minimum download is achieved by downloading from all remaining active nodes.

At the other end of the storage range suppose $\alpha \geq \gamma$ and hence the larger term for each term of the summation, and consider the solution to Equation 6.1, i.e.,

$$\begin{aligned} m &= \sum_{i=0}^{\min\{d,k\}-1} \min\left\{\left(1-\tfrac{i}{d}\right)\gamma, \alpha\right\} \\ &= \sum_{i=0}^{\min\{d,k\}-1} \left(1-\tfrac{i}{d}\right)\gamma = \gamma\left(k - \tfrac{k(k-1)}{2d}\right) = k\gamma\left(1-\tfrac{k-1}{2d}\right) \end{aligned}$$

and in this case

$$\gamma = \frac{2md}{k(2d-(k-1))}.$$

Since this derivation assumed $\alpha \geq \gamma$ the minimum storage possible for this minimum download γ is $\alpha = \gamma$ and the corresponding bound is referred to the *minimum bandwidth repair* (MBR) bound and the code with these parameters is referred to as an MBR code and designated:

$$(\alpha_{MBR}, \gamma_{MBR}) = \left(\frac{2md}{k(2d-(k-1))}, \frac{2md}{k(2d-(k-1))}\right), \qquad \text{MBR bound.} \tag{6.4}$$

Notice that since the two terms of this relation are the same, i.e., $\alpha_{MBR} = \gamma_{MBR}$ for the MBR regime, the amount of download is minimized since the contents of each server are α symbols. The expression for α (and hence γ) is minimized with $d = n-1$, i.e., as noted [5] even though the number of nodes from which information is downloaded from increases, the amount downloaded from each node is less and the product $\gamma = d\beta$ decreases.

For the case between the two extremes of $\alpha \leq (1-(k-1)/d)\gamma$ and $\alpha \geq \gamma$, discussed above, the situation is slightly more complicated but straightforward – the solution is given in the following theorem (see appendix of [5]).

Theorem 6.2 ([5], theorem 1) *For any $\alpha \geq \alpha^*(n,k,d,\gamma)$ the points (n,k,d,α,γ) are feasible and linear network codes suffice to achieve them. It is information theoretically impossible to achieve points with $\alpha < \alpha^*(n,k,d,\gamma)$. The threshold function $\alpha^*(n,k,d,\gamma)$ is the following:*

$$\alpha^*(n,k,d,\gamma) = \begin{cases} \dfrac{m}{k}, & \gamma \in [f(0), +\infty) \\ \dfrac{m - g(i)\gamma}{k-i}, & \gamma \in [f(i), f(i-1)) \end{cases} \tag{6.5}$$

where

$$f(i) = \frac{2md}{(2k-i-1)i + 2k(d-k+1)}$$
$$g(i) = \frac{(2d-2k+i+1)i}{2d}$$

where $d \leq n-1$. For d,n,k given, the minimum repair bandwidth γ is

$$\gamma_{MBR} = f(k-1) = \frac{2md}{2kd - k^2 + k}. \tag{6.6}$$

Good discussions of the signal flow graph and the related issues are contained in the references [5, 6, 7, 14].

Over the past decade the search for coding and storage schemes that approach the bounds of Equations 6.3 and 6.4 has been intense with contributions by a very large number of researchers.

The quantity $\alpha^*(n,k,d,\gamma)$ will be referred to as $\alpha^*(d,\gamma)$ when n and k are understood. It is clear that $\alpha^*(d,\gamma)$ is a decreasing function of d since accessing more nodes could result in lower storage per node.

In the case where $d = n-1$, the case where the replacement node contacts all other active nodes, the previous MSR bound yields the point

$$\left(\alpha_{\text{MSR}}^{\min}, \gamma_{\text{MSR}}^{\min}\right) = \left(\frac{m}{k}, \frac{m}{k}\left(\frac{n-1}{n-k}\right)\right). \tag{6.7}$$

This regime requires a factor of $\left(\frac{n-1}{n-k}\right)$ more download than each node stores which is a fundamental requirement of this regime, achieved by MDS codes.

Similarly for the MBR regime, Equation 6.4 for the extreme case where $d = n-1$ the minimum is given by

$$\left(\alpha_{\text{MBR}}^{\min}, \gamma_{\text{MBR}}^{\min}\right) = \left(\frac{m}{k} \cdot \frac{2(n-1)}{2n-k+1}, \frac{m}{k} \cdot \frac{2(n-1)}{2n-k-1}\right). \tag{6.8}$$

Thus, as expected, the MBR case requires slightly less download bandwidth compared to the MSR case but requires more storage for each node.

The parameter α is often referred to as the *subpacketization level* taken in the following sense. If $\alpha = 1$, then one finite field symbol is stored in each node. For many cases it is possible to construct more efficient codes by using a smaller field than the symbol field size and storing more than one subfield element in each node, giving more flexibility for coding options. More comments on this issue are discussed in the next section. The parameter β, the amount transferred by a helper node to a regenerating node, is often much smaller than α, the amount stored in each node ([12], chapter by Ramkumar et al., section 31.2.5 and also [7], chapter by Liu and Oggier, section 2.2). These references also contain interesting observations on the subtleties of the achievability of the bounds discussed above for functional repair as opposed to exact repair. In particular, constructions of exact repair MSR and MBR pairs (α, γ) can always be constructed. Other pairs, referred to as interior points, which do not correspond to MSR or MBR points, may not be possible to achieve.

The next two sections demonstrate aspects of a few of the many coding schemes that have been reported in the literature for these cases. Although the sections are necessarily limited in scope, they are indicative of the large number of results available.

Section 6.2 considers the problem of constructing codes for distributed storage that either meet various criteria or achieve the MBR or MSR cut-set bounds. The final section of the chapter considers array-type constructions of such codes. The distinction between the two types is artificial. In addition the particular constructions chosen for these sections from the vast array of possibilities are arbitrary, being mainly the ones the author was familiar with or found interesting. They tend to be from the more recent literature and by authors who have contributed significantly to the field. The selection of construction techniques chosen by another author might well be disjoint from those given here.

The next chapter considers the construction of *locally repairable codes* (LRCs). These codes are erasure-correcting codes which are able to correct an erased coordinate in a codeword by accessing a limited number of other nonerased coordinate positions. While they represent a separate research effort from the distributed coding techniques of this chapter it appears to be appropriate to study them in this context. The constructions of such codes tend to involve intricate uses of finite field and combinatorial structures.

6.2 Regenerating Codes for Distributed Storage and Subpacketization

This section describes a few of the numerous approaches to constructing codes for distributed storage. The MSR or MBR cut-set bounds which for convenience are repeated as:

$$\begin{aligned}(\alpha_{MSR}, \gamma_{MSR}) &= \left(\frac{m}{k}, \frac{md}{k(d-k+1)}\right),\\(\alpha_{MBR}, \gamma_{MBR}) &= \left(\frac{2md}{k(2d-(k-1))}, \frac{2md}{k(2d-(k-1))}\right).\end{aligned} \tag{6.9}$$

The aim of the work is to construct codes that either achieve one of the bounds or whose parameters approach the bound as well as describe the repair process. Of course, codes whose parameters lie in between these bounds can also be of interest.

Recall the model from the previous section: An information file of m finite field $\mathbb{F}_q$ symbols is encoded in some manner and α $\mathbb{F}_q$ symbols are stored on each of n servers in such a way that the information on any of the k servers is sufficient to regenerate the entire file.

In many of the coding schemes in the literature this model is specialized by using a finite field $\mathbb{F}_q$ and an extension field of order ℓ, $\mathbb{F}_{q^\ell}$. The original data file consists of $m = k$, $\mathbb{F}_{q^\ell}$ symbols which are encoded using a linear (over $\mathbb{F}_{q^\ell}$) $(n,k,d_{min})_{q^\ell}$ code and one $\mathbb{F}_{q^\ell}$ symbol is stored on each of n servers. The one $\mathbb{F}_{q^\ell}$ symbol can be viewed as ℓ $\mathbb{F}_q$ symbols and some subset of β of these can be downloaded from each of the d servers contacted during the regenerating process. For obvious reasons the parameter ℓ is referred to as the *subpacketization level*. Such a code is termed an (n,k,d,q,α,γ) code. Recall the parameter d in this work denotes the number of helper nodes contacted for repair or the download quantity in terms of $\mathbb{F}_q$ symbols. When the code minimum distance is required it will be referred to as d_{min} explicitly.

Only the case of a single node failure and exact repair is considered here. Other works consider multiple node failures and cooperative repair. It is natural, from the reconstruction requirement, to consider MDS codes, $(n,k,d_{min})_{q^\ell}$ codes for which the minimum distance is $d_{min} = n-k+1$. This might often refer to Reed–Solomon codes but variations are often considered.

As a matter of notation a few standard observations on codes are noted. Let $\mathbb{F}_q$ be a subfield of the finite field $\mathbb{F}_{q^\ell}$. Consider an $\mathscr{C}' = (n,k,d_{min})_{q^\ell}$ MDS code over $\mathbb{F}_{q^\ell}$. Such a code is referred to as a *scalar code* (linear over $\mathbb{F}_{q^\ell}$). For $\boldsymbol{c} \in \mathscr{C}', \boldsymbol{c} = (c_1, c_2, \ldots, c_n)$, $c_i \in \mathbb{F}_{q^\ell}$, one can replace each coordinate element c_i by a column ℓ-tuple over $\mathbb{F}_q$ to form the code $\mathscr{C}$ which might be viewed as a $\mathscr{C} = (n\ell, k\ell)_q$ code. Such a code can be referred to as a *vector* code over $\mathbb{F}_q$ or *array* code since a codeword can be represented as an $\ell \times n$ array of $\mathbb{F}_q$ symbols. The notions of MSR and MBR bounds also apply to such a model.

It is convenient to reproduce some coding results from Chapter 1. As in Chapter 1, a standard construction of Reed–Solomon codes over any $\mathbb{F}_q$ is as follows. Let $\boldsymbol{u} = \{u_1, u_2, \ldots, u_n\}$ be an evaluation set of $n \leq q$ distinct evaluation elements of $\mathbb{F}_q$ and $\mathbb{F}_q^{<k} \subseteq \mathbb{F}_q[x_1, \ldots, x_n]$ be the set of polynomials over $\mathbb{F}_q$ of degree less than k. Then one incarnation of a Reed–Solomon code can be taken as

$$RS_{n,k}(u,q) = \left\{ \boldsymbol{c}_f = (f(u_1), f(u_2), \ldots, f(u_n)), f \in \mathbb{F}_q^{<k} \right\}.$$

The code is MDS and the dual of such a code is also MDS. While the RS code always contains the all-ones vector (since $f(x) = 1$ is a valid polynomial) the dual code in general does not and the dual code is not generally an RS code, although it is an MDS code. Of course the construction is valid for any finite field, e.g., $\mathbb{F}_{q^\ell}$.

From Chapter 1, a generalization of this code is the *Generalized Reed–Solomon* (GRS) code denoted $GRS_{n,k}(\boldsymbol{u}, \boldsymbol{v}, q)$, where $\boldsymbol{u}$ is the evaluation set of distinct field elements as above and $\boldsymbol{v} = \{v_1, v_2, \ldots, v_n\}$, $v_i \in \mathbb{F}_q^*$ is the multiplier set of nonzero elements. The $GRS_{n,k}(\boldsymbol{u}, \boldsymbol{v}, q)$ is the (linear) set of codewords:

$$GRS_{n,k}(\boldsymbol{u}, \boldsymbol{v}, q) = \left\{ \boldsymbol{f} = (v_1 f(u_1), v_2 f(u_2), \ldots, v_n f(u_n)), \ f \in \mathbb{F}_q^{<k} \right\}.$$

An RS code is a $GRS_{n,k}(\boldsymbol{u}, \boldsymbol{v}, q)$ code with $\boldsymbol{v} = (1, 1, \ldots, 1)$.

The dual of a GRS code is also GRS. In particular, given $GRS_{n,k}(\boldsymbol{u}, \boldsymbol{v}, q)$ with evaluation set $\boldsymbol{u}$ and multiplier set $\boldsymbol{v}$ there exists a multiplier set $\boldsymbol{w} \in (\mathbb{F}_q^*)^n$ such that

$$\begin{aligned} GRS_{n,k}^{\perp}(\boldsymbol{u}, \boldsymbol{v}, q) &= GRS_{n,n-k}(\boldsymbol{u}, \boldsymbol{w}, q) \\ &= \left\{ (w_1 f(u_1), w_2 f(u_2) \ldots, w_n f(u_n), \ f \in \mathbb{F}_q^{<(n-k)} \right\}. \end{aligned}$$

Note in particular that

$$RS_{n,k}^{\perp}(\boldsymbol{u}, q) = GRS_{n,k}(\boldsymbol{u}, \boldsymbol{w}, q)$$

for some multiplier set $\boldsymbol{w}$.

It was noted in Chapter 1 that if $\mu = \{\mu_1, \ldots, \mu_n\}$ is a basis of $\mathbb{F}_{q^n}$ over $\mathbb{F}_q$ and $\{\nu_1, \ldots, \nu_n\}$ is a trace dual basis for the basis, then

$$y = \sum_{i=1}^{n} a_i \mu_i = \sum_{i=1}^{n} \mathrm{Tr}_{q^n|q}(y\nu_i)\mu_i. \tag{6.10}$$

Thus an element $y \in \mathbb{F}_{q^n}$ can be represented by the traces $\mathrm{Tr}_{q^n|q}(y\nu_j) \in \mathbb{F}_q,\ j = 1,2,\ldots,n$. This will be of importance for the code constructions to follow.

For the remainder of the section a few constructions of codes for distributed storage are considered. As noted, the number of these that can be found in the literature is very large and many important constructions are not mentioned here.

To initiate the discussion, two explicit constructions of codes that achieve either the MSR or MBR bounds are mentioned to show that there are simple schemes for both regimes ([7], chapter by Liu and Oggier).

Consider a file $\mathcal{M}$ of $|\mathcal{M}| = m = k$ $\mathbb{F}_q$ symbols. The k $\mathbb{F}_q$ symbols are MDS encoded into n $\mathbb{F}_q$ symbols and each symbol is stored on one of the n servers. To restore a failed server, the replacement node could contact k other servers and download their contents to restore the entire file (by definition).This is clearly wasteful since it involves the download of k symbols to restore a single symbol. However, note that this is an MSR solution since $d = k,\ \beta = 1,\ \gamma = d\beta = k,\ \alpha = 1$ and hence by Equation 6.9

$$(\alpha_{MSR}, \gamma_{MSR}) = \left(\frac{m}{k}, \frac{md}{k(d-k+1)}\right) = (1,k).$$

To show a simple scheme that achieves the MBR parameters consider the following [15]. Consider only a single node failure. For this case the data file $\mathcal{M}$ consists of $m = k(k+1)$ $\mathbb{F}_q$ symbols divided into $(k+1)$ data segments, $f^{(i)}, i = 0,1,\ldots,k$, each segment containing k $\mathbb{F}_q$ symbols. There are $n = k+1$ nodes and each node will contain $2k$ $\mathbb{F}_q$ symbols in the following manner. Let $g^{(i)}, i = 1,2,\ldots,k$ be k linearly independent vectors over $\mathbb{F}_q$ (k-tuples in $\mathbb{F}_q^k$). The node i stores the data segment of k $\mathbb{F}_q$ symbols $f^{(i)}$ and the k inner product $\mathbb{F}_q$ symbols $\{(f^{(i+1)}, g^{(1)}), (f^{(i+2)}, g^{(2)}), \ldots, (f^{(i+k)}, g^{(k)})\}$ (superscripts taken $\mod (k+1)$). Suppose node η^* has failed. The regenerating node contacts the k other nodes and downloads sufficient information to generate $f^{(\eta^*)}$ and the inner products $(f^{(\eta^*+j)}, g^{(j)}), j = 1,2,\ldots,k,$ superscripts taken $\mod (k+1)$. From each node it downloads 2 $\mathbb{F}_q$ symbols. From node $i \neq \eta^*$ it downloads the two symbols $(f^{(\eta^*)}, g^{(\eta^*-i)})$ and $(f^{(i)}, g^{(i+(n-\eta^*))}) = (f^{(\eta^*+j)}, g^{(j)}), j = 1,2,\ldots,k$. The segment $f^{(\eta^*)}$ is reconstructed by a simple matrix equation since the vectors $g^{(i)}$ are linearly independent and the other k symbols are as required.

The system parameters are $n = (k+1)$, $d = k$, and $m = k(k+1), \alpha = 2k$ and hence $\alpha = \gamma = 2k$. From Equation 6.9

$$\alpha_{MBR} = \frac{2md}{k(2d-k+1)} = \frac{2k^2(k+1)}{k(k+1)} = 2k \quad \text{and}$$
$$\gamma_{MBR} = \frac{2md}{k(2d-(k-1))} = \frac{2k^2(k+1)}{k(k+1)} = 2k$$

and it follows the system described achieves the MBR parameters.

The construction contained in the work of Guruswami and Wootters [8, 9, 10] uses the trace representation of field elements as in Equation 6.10 and has been influential in generating a considerable number of further constructions. The basic construction of this work is described using also the approach in [4].

Let $\mathcal{M}$ consist of $m = k$ $\mathbb{F}_{q^\ell}$ symbols which are encoded by an $RS_{n,k}(\boldsymbol{u}, q^\ell)$ code into n $\mathbb{F}_{q^\ell}$ symbols, with the n distinct nonzero evaluation set elements $\boldsymbol{u} = \{u_1, u_2, \ldots, u_n\}$. Assume $f(x) \in \mathbb{F}_{q^\ell}[x]$ is of degree less than k is used for encoding the database, i.e., the coded word is $(f(u_1), f(u_2), \ldots, f(u_n))$ where the original data is the set of k coefficients of $f(x)$. Each such symbol $f(u_i) \in \mathbb{F}_{q^\ell}$ is stored in one of n servers which for convenience are labeled with the corresponding evaluation set elements u_i. These $\mathbb{F}_{q^\ell}$ symbols are viewed as ℓ $\mathbb{F}_q$ symbols. Let $r = n - k$ denote the redundancy of the code. The dual of the code is a $GRS_{n,n-k}(\boldsymbol{u}, \boldsymbol{w}, q^\ell)$ code, for some multiplier set $\boldsymbol{w}$, whose codewords can be viewed as being generated by (and identified by) polynomials over $\mathbb{F}_{q^\ell}$ of degree less than $n-k$. Thus for an evaluation set $\boldsymbol{u} \in \mathbb{F}_{q^\ell}^n$ and for polynomials $f(x)$ and $g(x)$ of degrees less than k and $n-k$, respectively, it follows that

$$\sum_{u_i \in \boldsymbol{u}} f(u_i) w_i g(u_i) = 0.$$

It can be shown that when $n = |\mathbb{F}_{q^\ell}| = q^\ell$ then the multiplier set is $w_i = 1\ \forall i$. It will be convenient to make this assumption regardless of the size of the evaluation set since evaluation of a code component $w_i g(u_i)$ is equivalent to finding the value $g(u_i)$.

Suppose $f(x) \in \mathbb{F}_{q^\ell}[x]$ of degree at most $k-1$ corresponds to a codeword in the MDS $(n,k,d_{min})_{q^\ell}$ code $\mathscr{C}$ and $g(x) \in \mathbb{F}_{q^\ell}[x]$ of degree at most $n-k-1 = r-1$, in $\mathscr{C}^\perp$ and that (assuming a multiplier set of unity elements for convenience)

$$\sum_{i=1}^{n} f(u_i) g(u_i) = 0. \tag{6.11}$$

Assume $r = n - k \geq q^{\ell-1}$ and consider the polynomial $g(x) \in \mathbb{F}_{q^\ell}[x]$ as a codeword in the dual code and let u^* correspond to the failed server (also codeword position or evaluation point).

As noted in Equation 6.10, for a basis $\sigma = \{\sigma_1, \sigma_2, \ldots, \sigma_\ell\}$ of $\mathbb{F}_{q^\ell}$ over $\mathbb{F}_q$ and trace dual basis $\rho = \{\rho_1, \rho_2, \ldots, \rho_\ell\}$ of $\mathbb{F}_{q^\ell}$ over $\mathbb{F}_q$, any element $\alpha \in \mathbb{F}_{q^\ell}$ can be expressed as

$$\alpha = \sum_{i=1}^{\ell} \mathrm{Tr}_{q^\ell|q}(\alpha\sigma_i)\rho_i. \tag{6.12}$$

For the basis $\boldsymbol{\sigma} = \{\sigma_1, \ldots, \sigma_\ell\}$ of $\mathbb{F}_{q^\ell}$ over $\mathbb{F}_q$ and evaluation set $\boldsymbol{u} = \{u_1, u_2, \ldots, u_n\}$, $u^* \in \boldsymbol{u}$ has failed and the contents $f(u^*)$ are to be regenerated by contacting the $(n-1)$ other servers. Here $f \in \mathbb{F}_{q^\ell}[x],\ \deg(f) < k$, as noted above, is some fixed polynomial used in the storage system, corresponding to the codeword in use. The strategy will be to define a suitable set of polynomials $g \in \mathbb{F}_{q^\ell}[x],\ \deg g < n-k$, associated with codewords in the dual code, so the orthogonality condition in Equation 6.11 is satisfied.

Define the polynomials

$$\begin{aligned} g_i(x) &= \mathrm{Tr}_{q^\ell|q}(\sigma_i(x-u^*))/(x-u^*),\ i = 1, 2, \ldots, \ell, \quad u \in \boldsymbol{u} \\ &= \sum_{j=0}^{\ell-1} \left(\sigma_i(x-u^*)\right)^{q^j} / (x-u^*) \\ &= \sum_{j=0}^{\ell-1} \sigma_i^{q^j} \left(x-u^*\right)^{q^j-1} \end{aligned} \tag{6.13}$$

and

$$(x-u^*)g_i(x) = \sum_{j=0}^{\ell-1} \sigma_i^{q^j} \left(x-u^*\right)^{q^j} = \mathrm{Tr}_{q^\ell|q}\left(\sigma_i(u-u^*)\right).$$

(Note that $(x - u^*)$ divides each term of the trace function – hence the result $g_i(x)$ is a polynomial.) It is clear that each of these ℓ polynomials has the properties that

$$\deg(g_i(x)) = q^{\ell-1} - 1 \quad \text{and} \quad g_i(u^*) = \sigma_i,\ i = 1, 2, \ldots, \ell.$$

Assuming $r - 1 \geq q^{\ell-1} - 1$ these ℓ polynomials can be viewed as generators of codewords in the dual code to the $RS_{n,k}(\boldsymbol{u}, q^\ell)$ code $(GRS_{n,n-k}(\boldsymbol{u}, \boldsymbol{w}, \mathbb{F}_{q^\ell}))$ and hence for each such polynomial for $f \in \mathbb{F}_{q^\ell}^{<k}[x]$ (assuming unit multipliers)

$$\sum_{j=1}^{n} f(u_j)g_i(u_j) = 0 \quad \text{or} \quad g_i(u^*)f(u^*) = -\sum_{u \in \boldsymbol{u}\setminus\{u^*\}} g_i(u)f(u),\ i = 1, 2, \ldots, \ell.$$

Since the data file of k symbols over $\mathbb{F}_{q^\ell}$ was RS encoded with the polynomial $f(x) \in \mathbb{F}_{q^\ell}^{<k}[x]$, these ℓ polynomials $g_i(x)$ give conditions which allow regeneration of the contents of the failed server in the following manner. Taking traces of the above equation gives

$$\mathrm{Tr}_{q^\ell|q}\big(g_i(u^*)f(u^*)\big) = -\sum_{u\in \boldsymbol{u}\backslash\{u^*\}} \mathrm{Tr}_{q^\ell|q}\big(g_i(u)f(u)\big),\ i = 1,2,\ldots,\ell.$$

From the above properties of the polynomials $g_i(x)$ and the linearity property of the trace function, and substituting Equation 6.13 for $g_i(u)$, this can be expressed as

$$\mathrm{Tr}_{q^\ell|q}\big(\sigma_i f(u^*)\big) = -\sum_{u\in \boldsymbol{u}\backslash\{u^*\}} \mathrm{Tr}_{q^\ell|q}\big(\sigma_i(u-u^*)\big)\cdot \mathrm{Tr}_{q^\ell|q}\left(\frac{f(u)}{(u-u^*)}\right), \quad i = 1,2,\ldots,\ell.$$

Each server is able to compute the first trace in the summation $\mathrm{Tr}_{q^\ell|q}\big(\sigma_i(u - u^*)\big)$. Thus if the server corresponding to evaluation element $u \in \boldsymbol{u}\backslash\{u^*\}$ transmits the quantity

$$\mathrm{Tr}_{q^\ell|q}\left(\frac{f(u)}{u-u^*}\right) \in \mathbb{F}_q,$$

(for a total of $(n-1)$ $\mathbb{F}_q$ elements), the regenerating server is able to reconstruct the contents of the failed server via Equation 6.12, i.e.,

$$f(u^*) = \sum_{i=1}^{\ell} \mathrm{Tr}_{q^\ell|q}(\mu_i f(u^*))\rho_i.$$

This method of reconstruction thus requires each of $n-1$ servers to download a single $\mathbb{F}_q$ element in order to reconstruct a single $\mathbb{F}_{q^\ell}$ element (or ℓ $\mathbb{F}_q$ elements), as opposed to each server downloading a single element of $\mathbb{F}_{q^\ell}$ previously. This construction (from [4]) is a convenient description of the theorem 4 and algorithm 1 of [10], an example of the deeper results of that work.

This approach of linear repair with the use of trace functions appears to have inspired numerous other works. The optimality of the approach is discussed in [4, 8, 10]. In particular, since n is typically quite large in practice and the restriction of the argument that $r \geq q^{\ell-1}$, the level of subpacketization ℓ can be large and numerous efforts have been reported to reduce this level. In particular a large number of $(n,k)_{q^\ell}$ coding schemes with subpacketization ℓ are available for a wide variety of parameter sets in the literature.

A novel and interesting variation of the above construction is given in [18] and the elements of it are discussed, preserving as far as possible the notation. The technique involves the distribution of primes and has the feature that the number of helper nodes required to repair a node depends on the node. An overview of it is given.

Let $\pi(r)$ be the number of primes less than or equal to r, a well-studied function in number theory, and if $\pi(r) = m$ let the primes be $p_1, p_2, \ldots, p_m$. Let

$$\eta = \prod_{i=1}^{m} p_i \quad \text{and} \quad \eta_i = \eta / p_i.$$

Several fields are defined. Let $q \geq n - m$, $\mathbb{F} = \mathbb{F}_q$ and $\mathbb{F}_i = \mathbb{F}_{q^{\eta_i}}$ and $\mathbb{E} = \mathbb{F}_{q^\eta}$ so that $[\mathbb{E} : \mathbb{F}_i] = p_i$. Let $\mathbb{F}_{q^{p_i}} = \mathbb{F}(\alpha_i), \alpha_i \in \mathbb{E}$ and

$$\mathbb{E} = \mathbb{F}_{q^\eta} = \mathbb{F}_q(\alpha_1, \ldots, \alpha_m).$$

Adjoin to the set of above elements $\{\alpha_1, \alpha_2, \ldots, \alpha_m\}$ a further set of $n - m$ distinct elements of $\mathbb{E}$ and let $\boldsymbol{\alpha} = \{\alpha_1, \alpha_2, \ldots, \alpha_n\}$ be an evaluation set for an RS code $\mathscr{C}$ over $\mathbb{E}$ of dimension k and $\mathscr{C}^\perp$ its dual (GRS) code, say with multiplier set $\boldsymbol{v}$. The work shows that assuming the i-th node has failed, for $i = 1, 2, \ldots, m$ (i.e., one of the first m nodes) its contents can be optimally repaired from any $d_i = p_i + k - 1$ helper nodes.

To see this choose a subset $\mathcal{R}_i$ of d_i helper nodes, $\mathcal{R}_i \subseteq [n] \backslash \{i\}$ and note that $|d_i| \leq n - 1$. Define $h(x)$ to be the annihilator polynomial

$$h(x) = \prod_{j \in [n] \backslash \{\mathcal{R}_i \cup \{i\}\}} (x - \alpha_j)$$

and note the degree of $h(x)$ is $n - (k + p_i)$ and so $\deg(x^s h(x)) < r$ for $s = 0, 1, \ldots, p_i - 1$. Thus the codeword in $\mathscr{C}^\perp$ (multiplier set $\boldsymbol{v}$) corresponding to such a polynomial is

$$\left(v_1 \alpha_i^s h(\alpha_1), \ldots, v_n \alpha_n^s h(\alpha_n) \right) \in \mathscr{C}^\perp, \ s = 0, 1, \ldots, p_i - 1.$$

Suppose the polynomial $f(x)$, $\deg(f(x)) < k$ corresponds to the codeword $(f(\alpha_1), \ldots, f(\alpha_n)) \in \mathscr{C}$ used for the distributed data storage system. Then, by definition

$$\sum_{j=1}^{n} v_j \alpha_j^s h(\alpha_j) f(\alpha_j) = 0, \quad s = 0, 1, \ldots, p_i - 1.$$

The argument is as in [10]. Taking traces of this expression ($\mathbb{E}$ over $\mathbb{F}_i$) and rearranging terms gives

$$\mathrm{Tr}_{\mathbb{E}/\mathbb{F}_i}\left(v_i \alpha_i^s h(\alpha_i) f(\alpha_i) \right) = - \sum_{j \in \mathcal{R}_i} \alpha_j^s \mathrm{Tr}_{\mathbb{E}/\boldsymbol{F}_i}(v_j h(\alpha_j) f(\alpha_j)).$$

(Note that for $j \neq i$, $\alpha_j \in \mathbb{F}_i$.) Thus the j-th helper node downloads the single $\mathbb{F}_i$ element to the i-th node:

$$\mathrm{Tr}_{\mathbb{E}/\mathbb{F}_i}(v_j h(\alpha_j) f(\alpha_j)).$$

The elements $\{v_1 h(\alpha_1), v_i\alpha_i h((\alpha_i), \ldots, v_i\alpha_i^{p_i-1} h(\alpha_i)\}$ form a basis of $\mathbb{E}$ over $\mathbb{F}_i$ and let $\theta_1, \theta_2, \ldots, \theta_{p_i-1}$ be the trace dual basis. As before the i-th node then is able to compute

$$f(\alpha_i) = \sum_{s=0}^{p_s-1} \mathrm{Tr}_{\mathbb{E}/\mathbb{F}_i}(v_i\alpha^s h(\alpha_i) f(\alpha_i))\theta_s$$

to restore the node. It is argued [18] that this scheme attains the cut-set bound of Equation 6.9 (or Equation 6.14). Further developments of this technique with linear combining and trace functions primes are considered in [19, 23].

Another interesting construction considered here is that of Ye and Barg [22]. It is relatively simple and representative. The work of Guruswami and Wootters [8] with a similar approach is also of interest.

As in the previous example, it will be assumed the file consists of k $\mathbb{F}_{q^\ell}$ symbols MDS encoded into n $\mathbb{F}_{q^\ell}$ symbols, each stored on a separate server. When a node fails, a replacement node appears and must restore its contents by downloading information from the remaining $n-1$ (or fewer) nodes. Interest will only be in the case of exact regeneration and using all $n-1$ surviving nodes. From Equation 6.9 if download for repair is obtained from all $(n-1)$ nodes, the repair bandwidth of a replacement node must be at least

$$\frac{m(n-1)}{k(n-k)} = \frac{\ell(n-1)}{(n-k)}. \tag{6.14}$$

when each node is viewed as containing a single symbol of a field $\mathbb{F}_{q^\ell}$ or ℓ $\mathbb{F}_q$ symbols.

6.3 Array-Type Constructions of Regenerating Codes

As noted earlier, an $(n,k)_{q^\ell}$, $r = (n-k)$, *array code* $\mathscr{C}$ is constructed as follows. A codeword $C \in \mathscr{C}$ will be of the form $C = (\boldsymbol{C}_1, \boldsymbol{C}_2, \ldots, \boldsymbol{C}_n)$ where each component $\boldsymbol{C}_i$ is an ℓ column vector over $\mathbb{F}_q$. Many works in the literature refer to this as an $(n,k,\ell)_{q^\ell}$ code. Since this might be confused with the more common $(n,k,d_{\min})_{q^\ell}$ the preference here will be to state explicitly it is an $(n,k)_{q^\ell}$ code with subpacketization level ℓ. The usual RS code over q^ℓ is an example of such a code. Another is the code construction below. While quite intricate in its requirements the manner in which it achieves reconstruction is of interest.

The code is defined as:

$$\mathscr{C} = \left\{(\boldsymbol{C}_1, \boldsymbol{C}_2, \ldots, \boldsymbol{C}_n) : \sum_{i=1}^{n} \boldsymbol{A}_{s,i}\boldsymbol{C}_i = \boldsymbol{0},\ s = 1,2,\ldots,r\right\} \tag{6.15}$$

where $\{\boldsymbol{A}_{s,i}\}$ is a set of rn $\ell \times \ell$ matrices, thought of as an $r \times n$ array of $\ell \times \ell$ matrices, over $\mathbb{F}_q$ and

$$\boldsymbol{C}_i = (c_{i,0}, c_{i,1}, \ldots, c_{i,\ell-1})^t,\ c_{i,j} \in \mathbb{F}_q.$$

The several code constructions in [22] differ in their choice of these matrices.

For the construction of interest here ([22], section III) choose

$$\boldsymbol{A}_{s,i} = \boldsymbol{A}_i^{s-1}, s \in [r],\ i \in [n]$$

for diagonal matrices $\boldsymbol{A}_i^{s-1}$.

The redundancy of the storage scheme is $r = n - k$ and assume $q > rn$ and let $\ell = r^n$ and note the exponential dependence of ℓ on r. Each coordinate of the array code is a column vector of length ℓ over $\mathbb{F}_q$. Each integer a in the range $\{0, 1, 2, \ldots, \ell-1\}$ can be expressed (represented) as an r-ary n-tuple: $a = (a_n, a_{n-1}, \ldots, a_i, \ldots, a_1)$ where each a_i is an integer in the set $\{0, 1, 2, \ldots, r-1\}$, i.e.,

$$a = \sum_{i=1}^{n} a_i r^i \sim (a_n, a_{n-1}, \ldots, a_1),\ a = 0, 1, \ldots, \ell-1 = r^n-1, 0 \le a_i \le r-1,\ i \in [n].$$

Thus as a runs through the integers 0 to $r^n - 1$, the representation runs through all possible r-ary n-tuples. We will often not distinguish between the integer a and its r-ary vector representation in the sense that if a and its r-ary representation $(a_1, \ldots, a_n)$ occur in an expression, it will be assumed they correspond.

Let $\left\{\lambda_{i,j}, i = 1, 2, \ldots, n, j = 0, 1, \ldots, r-1\right\}$ be an array of rn distinct nonzero elements of $\mathbb{F}_q^*$ $(q > rn)$.

Let $\boldsymbol{e}^{(a)}, a = 0, 1, \ldots, \ell - 1$ be the standard basis of column vectors of $\mathbb{F}_{q^\ell}$ over $\mathbb{F}_q$, i.e., $\boldsymbol{e}^{(a)}$ contains a single 1 in the a-th position and all other elements are 0. Define the matrices above as:

$$\boldsymbol{A}_i = \sum_{a=0}^{\ell-1} \lambda_{i,a_i} \boldsymbol{e}^{(a)} \boldsymbol{e}^{(a)t}, \quad i = 1, 2, \ldots, n.$$

Thus

$$\boldsymbol{A}_i = \mathrm{diag}\{\lambda_{i,a_0}, \lambda_{i,a_1}, \ldots, \lambda_{i,a_{\ell-1}}\}, \quad i = 1, 2, \ldots, n.$$

These matrices are diagonal and the matrix $\boldsymbol{e}^{(a)}\boldsymbol{e}^{(a)t}$ is an all-zero matrix except for a single 1 on the diagonal in the a-th row and column. The code defined by Equation 6.15 can then be defined as: $C = (\boldsymbol{C}_1, \boldsymbol{C}_2, \ldots, \boldsymbol{C}_n) \in \mathscr{C}$, each $\boldsymbol{C}_i$ a column vector of dimension ℓ over $\mathbb{F}_q$, if

$$\begin{bmatrix} \boldsymbol{I} & \boldsymbol{I} & \cdots & \boldsymbol{I} \\ \boldsymbol{A}_1 & \boldsymbol{A}_2 & \cdots & \boldsymbol{A}_n \\ \boldsymbol{A}_1^2 & \boldsymbol{A}_2^2 & \cdots & \boldsymbol{A}_n^2 \\ \vdots & \vdots & \vdots & \vdots \\ \boldsymbol{A}_1^{r-1} & \boldsymbol{A}_2^{r-1} & \cdots & \boldsymbol{A}_n^{r-1} \end{bmatrix} \begin{bmatrix} \boldsymbol{C}_1 \\ \boldsymbol{C}_2 \\ \vdots \\ \boldsymbol{C}_n \end{bmatrix} = \begin{bmatrix} \boldsymbol{0} \\ \boldsymbol{0} \\ \vdots \\ \boldsymbol{0} \end{bmatrix}. \tag{6.16}$$

The zeros in the right-hand column vector are vectors of all zeros of dimension ℓ, and $\boldsymbol{C}_i$ is the column vector over $\mathbb{F}_q$ of dimension ℓ, $\boldsymbol{C}_i = (c_{i,0}, c_{i,1}, \ldots, c_{i,\ell-1})^t$.

This is equivalent to the system of equations:

$$\sum_{i=1}^{n} \lambda_{i,a_i}^{s} c_{i,a} = 0, \quad s = 0, 1, \ldots, r-1, \quad a = 0, 1, \ldots, \ell - 1 \tag{6.17}$$

where the a_i subscripts in the λ's in the summation correspond to the a in the $c_{i,a}$. All $\ell \times \ell$ matrices are viewed as having rows and columns indexed by r-ary n-tuples.

It is clear Equation 6.16, by a permutation of rows and columns, can be expressed as the block diagonal equation:

$$\begin{bmatrix} \boldsymbol{B}_0 & \boldsymbol{0} & \cdots & \boldsymbol{0} \\ \boldsymbol{0} & \boldsymbol{B}_1 & \cdots & \boldsymbol{0} \\ \boldsymbol{0} & \boldsymbol{0} & \cdots & \boldsymbol{0} \\ \vdots & \vdots & \vdots & \vdots \\ \boldsymbol{0} & \boldsymbol{0} & \cdots & \boldsymbol{B}_{\ell-1} \end{bmatrix} \begin{bmatrix} \boldsymbol{C}'_0 \\ \boldsymbol{C}'_1 \\ \\ \vdots \\ \boldsymbol{C}'_{\ell-1} \end{bmatrix} = \begin{bmatrix} \boldsymbol{0} \\ \boldsymbol{0} \\ \vdots \\ \boldsymbol{0} \end{bmatrix}, \tag{6.18}$$

where zeros indicate zero matrices/column vectors of appropriate dimension and the matrices B_a are $r \times n$ matrices and the a-th row of this equation is:

$$B_a \cdot C'_a = \begin{bmatrix} 1 & 1 & \cdots & 1 \\ \lambda_{1,a_1} & \lambda_{2,a_2} & \cdots & \lambda_{n,a_n} \\ \lambda_{1,a_1}^2 & \lambda_{2,a_2}^2 & \cdots & \lambda_{n,a_n}^2 \\ \vdots & \vdots & \vdots & \vdots \\ \lambda_{1,a_1}^{r-1} & \lambda_{2,a_2}^{r-1} & \cdots & \lambda_{n,a_n}^{r-1} \end{bmatrix} \cdot \begin{bmatrix} c_{1,a} \\ c_{2,a} \\ c_{3,a} \\ \vdots \\ c_{n,a} \end{bmatrix} = \begin{bmatrix} 0 \\ 0 \\ \vdots \\ 0 \end{bmatrix}, \quad a = 0, 1, \ldots, \ell - 1 \tag{6.19}$$

and

$$\boldsymbol{C}'_a = (c_{1,a}, c_{2,a}, \ldots, c_{n,a})^t, \quad a = 0, 1, \ldots, \ell - 1$$

and the equivalence between a and its r-ary expansion is noted. It follows directly from this equation that if $r = n-k$ components of the original codeword $\boldsymbol{C} \in \mathscr{C}, \boldsymbol{C} = (C_1, C_2, \ldots, C_n), C_i \in \mathbb{F}_{q^\ell}$ are erased, they can be reconstructed since the matrix in Equation 6.19 has the property that any r columns have full rank r. Thus the codes are MDS. For an equivalent view, consider the original codeword as an $\ell \times n$ array of $\mathbb{F}_q$ elements. Equation 6.19 then implies that each row of the array is a codeword in an RS code with evaluation set $\{\lambda_{1,a_1}, \lambda_{2,a_2}, \ldots, \lambda_{n,a_n}\}$ and in this sense the MDS property of the overall code $\boldsymbol{C}$ is inherited from the RS/MDS property of its rows.

It is important to note that as the r-ary n-tuples $\boldsymbol{a}$ vary there will be repetition of elements and hence in any set such as $\{\lambda_{1,a_1}, \lambda_{2,a_2}, \ldots, \lambda_{n,a_n}\}$ there will also be repetition. This is used below to effect a repair process. Also it is tacitly assumed for the matrices or vectors indexed by ℓ there is a fixed order – say lexicographic – to the n-tuples that need not be specifically noted.

To consider the repair process, suppose the i-th server fails and a replacement server is to replace the data lost by downloading certain information from the remaining $n-1$ nodes. It will be useful for the following discussion to be reminded of the structure of a codeword $C = (\boldsymbol{C}_1, \boldsymbol{C}_2, \ldots, \boldsymbol{C}_n)$ where $\boldsymbol{C}_i = (c_{i,0}, c_{i,1}, \ldots, c_{i,\ell-1})^t$, $c_{i,j} \in \mathbb{F}_q$, $i = 1, 2, \ldots, n$ and $\boldsymbol{C}_i \in \mathbb{F}_{q^\ell} \sim \mathbb{F}_q^{\ \ell}$. The set of integer subscripts $\mathcal{L} = \{0, 1, \ldots, \ell - 1\}$ is associated in some one-to-one manner with the set of $\ell = r^n$ n-tuples in the set $\mathcal{R}^n = \{a = (a_n, a_{n-1}, \ldots, a_1),\ a_i \in \mathcal{R}\}$ where $\mathcal{R} = \{0, 1, \ldots, r-1\}$. The mapping between the two sets is not specified.

It is clear the set $\mathcal{R}^n$ can be decomposed into r^{n-1} subsets $\mathcal{R}_i, i = 1, 2, \ldots, r^{n-1}$, $|\mathcal{R}_i| = r$ so that within a set $\mathcal{R}_i$ the n-tuples differ only in the i-th component for i fixed. For an n-tuple $a \in \mathcal{R}^n$ define

$$a(i,u) \stackrel{\Delta}{=} (a_n, a_{n-1}, \ldots, a_{i+1}, u, a_{i-1}, \ldots, a_1),$$

i.e., an n-tuple whose i-th component is replaced by a variable u. The variable u can assume the values $u = 0, 1, \ldots, r-1$. Clearly there are r^{n-1} such sets, corresponding to the possibilities of $(n-1)$-tuples $\{a_n, \ldots, a_{i+1}, a_{i-1}, \ldots a_1\}$.

Equation (6.17) is

$$\sum_{i=1}^{n} \lambda_{i,a_i}^{s} c_{i,a} = 0,\ s = 0, 1, \ldots, r-1,\ a = 0, 1, \ldots, \ell - 1. \tag{6.20}$$

Assume node i fails and it is required to reconstruct the code column $\boldsymbol{C}_i$ by downloading information for the $n-1$ other nodes, i.e., it is required to determine the codeword $(c_{i,0}, c_{i,1}, \ldots, c_{i,\ell-1})$ where in $c_{i,j}$ the equivalent

r-ary n-tuple is used. In the above equation choose a to be $a(i,u)$ (leaving u as variable for the moment and noting that $a(i,u)$ is a valid r-ary n-tuple):

$$\lambda^s_{i,u} c_{i,a(i,u)} + \sum_{\substack{j=1 \\ j\neq i}}^{n} \lambda^s_{j,a_j} c_{j,a(i,u)} = 0.$$

Suppose the j-th node $j \neq i$ computes the quantity

$$\mu^{(a)}_{j,i} = \sum_{u=0}^{r-1} c_{j,a(i,u)}, \quad j \in [n]\backslash\{i\} \tag{6.21}$$

which is possible since it contains the j-th coordinate of the coded database $\boldsymbol{C}_j$. Rewriting Equation 6.20 over $u = 0, 1, \ldots, r-1$ gives

$$\begin{aligned}
\sum_{u=0}^{r-1} \lambda^s_{i,u} c_{i,a(i,u)} &+ \sum_{u=0}^{r-1} \sum_{\substack{j=1 \\ j\neq i}}^{n} \lambda^s_{j,a_j} c_{j,a(i,u)} \\
&= \sum_{u=0}^{r-1} \lambda^s_{i,u} c_{i,a(i,u)} + \sum_{\substack{j=1 \\ j\neq i}}^{n} \lambda^s_{j,a_j} \sum_{u=0}^{r-1} c_{j,a(i,u)} \\
&= \sum_{u=0}^{r-1} \lambda^s_{i,u} c_{i,a(i,u)} + \sum_{\substack{j=1 \\ j\neq i}}^{n} \lambda^s_{j,a_j} \mu^{(a)}_{j,i} = 0, \; s = 0, 1, \ldots, r-1.
\end{aligned}$$

For fixed values of i and u there are r^{n-1} possibilities for $a(i,u)$ corresponding to coordinates $j \in [n]\backslash\{i\}$. In matrix form these equations may be written:

$$\begin{bmatrix} 1 & 1 & \ldots & 1 \\ \lambda_{i,0} & \lambda_{i,1} & \ldots & \lambda_{i,r-1} \\ \vdots & \vdots & \vdots & \vdots \\ \lambda^{r-1}_{i,0,} & \lambda^{r-1}_{i,1} & \ldots & \lambda^{r-1}_{i,r-1} \end{bmatrix} \begin{bmatrix} c_{i,a(i,0)} \\ c_{i,a(i,1)} \\ \vdots \\ c_{i,a(i,r-1)} \end{bmatrix} = - \begin{bmatrix} \sum_{j=1,j\neq i}^{n} \mu^{(a)}_{j,i} \\ \sum_{j=1,j\neq i}^{n} \lambda_{j,a_j} \mu^{(a)}_{j,i} \\ \vdots \\ \sum_{j=1,\neq i}^{n} \lambda^{r-1}_{j,a_j} \mu^{(a)}_{j,i} \end{bmatrix}. \tag{6.22}$$

Since the matrix has a Van der monde form and the elements are chosen to be distinct, it may be solved for the r elements of $\boldsymbol{C}_i$. Thus as each of the $(n-1)$ surviving nodes transmits a single $\mathbb{F}_q$ symbol in the coordinated manner shown, the replacement node is able to compute r coordinates of its codeword $\boldsymbol{C}_i$. Each coordinate corresponds to $\ell = r^n$ $\mathbb{F}_q$ symbols and each surviving node must transmit a total of r^{n-1} $\mathbb{F}_q$ symbols for the regenerating node to compute $\boldsymbol{C}_i$.

Thus in this scenario each node downloads a total of r^{n-1} $\mathbb{F}_q$ symbols and the total repair bandwidth is

$$\gamma = (n-1)r^{n-1} = (n-1)\ell/r \quad \mathbb{F}_q \quad \text{symbols},$$

which by Equation 6.14 gives an optimal bandwidth scheme.

A variant of the above approach is to consider the notion of repair spaces rather than matrices. First considered by [17], it was also considered in [20] and more recently by [2, 3]. Chowdhury and Vardy [2] summarize this approach which is not described here.

In the literature of code constructions a problem of particular interest has been to devise constructions that yield optimal storage requirements while keeping the packetization level low – many of the optimal or near-optimal performances in terms of download bandwidth and/or constructions have large sub-packetization levels. To address this problem, [13] recently introduced the notion of an ϵ-MSR code defined as follows:

Definition 6.3 Let $\epsilon > 0$ and $\mathscr{C}$ be an $(n, k\ell, d_{min} = n - k + 1)_q$ MDS code. The code $\mathscr{C}$ is an $(n,k,d,\ell)_q$ ϵ-MSR code if, for every $i \in [n]$ there is a repair scheme to repair the i-th code block $c^{(i)}$ with

$$\beta_{j,i} \leq (1+\epsilon) \cdot \frac{\ell}{d-k+1} \text{ symbols (over } \mathbb{F}_q)$$

for all $j \in \mathcal{R} \subseteq [n]\backslash\{i\}$ such that $\mid \mathcal{R} \mid = d$. Here $\beta_{j,i}$ denotes the number of symbols that the code block $c^{(j)}$ contributes during the repair of the code block $c^{(i)}$.

The notion of relaxing the requirement of optimal repair bandwidth to within ϵ of the optimal in order to achieve lower packetization levels for MSR codes has been pursued to advantage also in [11].

The idea of interference alignment arose in the context of wireless communications where techniques were developed to design multiuser signal systems such that signals from unwanted users occupy a subspace leaving the remainder of the space for the intended users. This idea has been adopted in the distributed coding domain to develop effective coding strategies. Here the interference refers to the download of symbols not used by the repair node (interference) and the techniques are developed to mitigate their effect. Numerous works develop these techniques including [16, 21] and [1].

Comments

The past two decades has seen considerable progress in the areas of code constructions and protocols for the distributed storage and retrieval of information. The economic importance of the problem is clear and only a few of the many techniques developed are noted here. The gains that can be realized with these ideas are very significant.

References

[1] Cadambe, V.R., Jafar, S.A., Maleki, H., Ramchandran, K., and Suh, C. 2013. Asymptotic interference alignment for optimal repair of MDS codes in distributed storage. *IEEE Trans. Inform. Theory*, **59**(5), 2974–2987.

[2] Chowdhury, A., and Vardy, A. 2018. New constructions of MDS codes with asymptotically optimal repair. Pages 1944–1948 of: *2018 IEEE International Symposium on Information Theory (ISIT).*

[3] Chowdhury, A., and Vardy, A. 2021. Improved schemes for asymptotically optimal repair of MDS codes. *IEEE Trans. Inform. Theory*, **67**(8), 5051–5068.

[4] Dau, H., and Milenkovic, O. 2017. Optimal repair schemes for some families of full-length Reed–Solomon codes. *CoRR*, abs/1701.04120.

[5] Dimakis, A.G., Godfrey, P.B., Wu, Y., Wainwright, M.J., and Ramchandran, K. 2010. Network coding for distributed storage systems. *IEEE Trans. Inform. Theory*, **56**(9), 4539–4551.

[6] Dimakis, A.G., Ramchandran, K., Wu, Y., and Suh, C. 2011. A survey on network codes for distributed storage. *Proc. IEEE*, **99**(3), 476–489.

[7] Greferath, M., Pavčević, M.O., Silberstein, N., and Vázquez-Castro, M.A. (eds.). 2018. *Network coding and subspace designs*. Signals and Communication Technology. Springer, Cham.

[8] Guruswami, V., and Wootters, M. 2016. Repairing Reed-Solomon codes. Page 216–226 of: *Proceedings of the Forty-Eighth Annual ACM Symposium on Theory of Computing*. STOC '16. Association for Computing Machinery, New York.

[9] Guruswami, V., and Wootters, M. 2016. Repairing Reed–Solomon codes. *CoRR*, abs/1509.04764v2.

[10] Guruswami, V., and Wootters, M. 2017. Repairing Reed–Solomon codes. *IEEE Trans. Inform. Theory*, **63**(9), 5684–5698.

[11] Guruswami, V., Lokam, S.V., and Jayaraman, S.V.M. 2020. MSR codes: contacting fewer code blocks for exact repair. *IEEE Trans. Inform. Theory*, **66**(11), 6749–6761.

[12] Huffman, W.C., Kim, J.L., and Solé, P. Eds. 2021. *A concise encyclopedia of coding theory*. CRC Press, Boca Raton, FL.

[13] Rawat, A.S., Tamo, I., Guruswami, V., and Efremenko, K. 2018. MDS code constructions with small sub-packetization and near-optimal repair bandwidth. *IEEE Trans. Inform. Theory*, **64**(10), 6506–6525.

[14] Sasidharan, B., Senthoor, K., and Kumar, P.V. 2014. An improved outer bound on the storage-repair-bandwidth tradeoff of exact-repair regenerating codes. Pages 2430–2434 of: *2014 IEEE International Symposium on Information Theory.*

[15] Shum, K.W., and Hu, Y. 2011. Exact minimum-repair-bandwidth cooperative regenerating codes for distributed storage systems. Pages 1442–1446 of: *2011 IEEE International Symposium on Information Theory Proceedings.*

[16] Suh, C., and Ramchandran, K. 2011. Exact-repair MDS code construction using interference alignment. *IEEE Trans. Inform. Theory*, **57**(3), 1425–1442.

[17] Tamo, I., Wang, Z., and Bruck, J. 2014. Access versus bandwidth in codes for storage. *IEEE Trans. Inform. Theory*, **60**(4), 2028–2037.

[18] Tamo, I., Ye, M., and Barg, A. 2017. Optimal repair of Reed–Solomon codes: achieving the cut-set bound. Pages 216–227 of: *58th Annual IEEE Symposium on Foundations of Computer Science – FOCS '17*. IEEE Computer Society, Los Alamitos, CA.

[19] Tamo, I., Ye, M., and Barg, A. 2019. The repair problem for Reed-Solomon codes: optimal repair of single and multiple erasures with almost optimal node size. *IEEE Trans. Inform. Theory*, **65**(5), 2673–2695.

[20] Wang, Z., Tamo, I., and Bruck, J. 2016. Explicit minimum storage regenerating codes. *IEEE Trans. Inform. Theory*, **62**(8), 4466–4480.

[21] Wu, Y., and Dimakis, A.G. 2009. Reducing repair traffic for erasure coding-based storage via interference alignment. Pages 2276–2280 of: *2009 IEEE International Symposium on Information Theory*.

[22] Ye, M., and Barg, A. 2017. Explicit constructions of high-rate MDS array codes with optimal repair bandwidth. *IEEE Trans. Inform. Theory*, **63**(4), 2001–2014.

[23] Ye, M., and Barg, A. 2017. Repairing Reed–Solomon codes: universally achieving the cut-set bound for any number of erasures. *CoRR*, abs/1710.07216.

7
Locally Repairable Codes

The term *repairable* here refers to erasure correction in that the coordinate position where the erasure occurs is known and contains a special symbol indicating erasure. This is in contrast to locally decodable codes considered in the next chapter where a symbol (coordinate position in a codeword) of interest may be in error which can be corrected by computation with a few other coordinate symbols.

This topic is related to the material of the previous and following two chapters. It is perhaps closest to the material of the previous chapter where the repair of a failed server is of interest, a notion which is almost identical to the repair of an erased codeword position. The focus of this chapter is on minimizing the number of codeword positions (or servers) that must be contacted to retrieve information from which the repair is effected. Otherwise the two themes have many similarities. We retain the distinction in this chapter, based largely on the literature being quoted, although the techniques developed here are applicable to the previous chapter as well.

The codes of these four topics are all related to the storage and retrieval of information with certain properties and have been vigorously researched over the past two decades. The literature on them is now large and, in keeping with the purpose of this volume, these four chapters are merely intended to introduce the topics by providing some of the basic concepts and a few examples with pointers to the more general results available. Interrelationships between the four topics are commented upon.

7.1 Locally Repairable Codes

This section is concerned with the reconstruction of codewords that have erasures in them with an emphasis on code constructions that require access to as few other codeword coordinate positions as possible. The restrictions such conditions place on the code construction are of interest.

The notion of a *locally repairable code* (LRC) arose in the works [13, 25]. To be consistent with the notation of the previous chapter the minimum distance of a code is denoted explicitly as $d_{\min}$ (rather than d, which was the number of helper nodes in the previous chapter). Also the parameter r in this section refers to the number of symbols needed to reconstruct an erased symbol as opposed to $n-k$ in the previous section. The changes in notation reflect that of the literature in the respective areas.

The definition of a locally repairable code from [25].

Definition 7.1 An $(n,r,d_{\min},m,\alpha)$ *locally repairable code* (LRC) is a code that takes a file $\mathcal{M}$ of size m bits and creates k symbols, each of size $\alpha = |\mathcal{M}|\ /k = m/k$ bits and encodes it into n coded symbols, such that each of the coded symbols can be reconstructed by accessing at most r other symbols. It is often assumed the size of a symbol is that of a finite field q. Moreover the minimum distance of the code is $d_{\min}$ implying a file of size m symbols can be reconstructed by accessing any of the $(n - d_{\min} + 1)$ coded symbols.

For the last sentence of this definition, note that for linear $(n,k,d_{\min})_q$ codes, a property of a $k \times n$ generator matrix of the code with minimum distance $d_{\min}$ is that every $k \times (n - d_{\min} + 1)$ submatrix is of full rank since if this were not the case there would exist a nonzero linear combination of rows with zeros in $(n - d_{\min} + 1)$ coordinates giving a nonzero codeword of weight at most $d_{\min} - 1$ and a contradiction. Thus from any $(n - d_{\min} + 1)$ coordinates, the remaining $(d_{\min} - 1)$ coordinates can be reconstructed.

Codes that satisfy the above definition are referred to as codes with locality r. The work [13] refers to such codes as local $(r,d_{\min})$ codes with the view that such codes have the property that a given coordinate erasure can be recovered by access to at most r other coordinate positions. This notion will be generalized to the notion of an (r,δ) code, which allows recovery even in the presence of other codeword position failures in a manner to be discussed shortly.

It is natural that the additional constraint for an LRC that each codeword position can be reconstructed by contacting at most r other positions imposes limits on the code parameters. The first general study of codes with locality is [13] although the work [18] contained similar ideas in their definition of Pyramid codes discussed later in the section. For example [13], suppose $\mathscr{C}$ is an $(n,k,d_{\min})_q$ maximum distance separable (MDS) code and hence the number of parity relationships in the code is $n-k = d_{\min} - 1$. Take one of these parity equations (leaving $(d-2)$ others and replace it with $\left\lceil \frac{k}{r} \right\rceil$ parity checks, each of size at most r, i.e., each parity sums r other information coordinate positions). This would yield a code with

$$n - k = \left\lceil \frac{k}{r} \right\rceil + d_{\min} - 2,$$

a modification of the usual MDS condition, indicating that repair locality incurs a cost in code parameters. It turns out this relationship for linear codes extends to more general codes than MDS as the following theorem shows:

Theorem 7.2 ([13], theorem 5) *For any $(n,k,d_{\min})_q$ linear code with information locality r*

$$n - k \geq \left\lceil \frac{k}{r} \right\rceil + d_{\min} - 2. \tag{7.1}$$

Note that for $k = r$ the bound reduces to the Singleton bound. Codes that achieve this bound with equality will be referred to as *distance optimal LRCs*, noting that the introduction of (r,δ) LRCs and *maximally recoverable* (MR) LRCs later in the chapter will result in modified bounds leading to ambiguity in the term optimal.

The following is of interest:

Theorem 7.3 ([13], theorem 17) *Let n,k,r and $d \geq 2$ be positive integers. Let $q > kn^k$ be a prime power. Suppose $(r+1) \mid n$ and*

$$n - k = \left\lceil \frac{k}{r} \right\rceil + d_{\min} - 2.$$

Then there exists an $(n,k,d_{\min})_q$ LRC code where all code symbols have locality r.

Other works considering bounds for locally repairable codes include [5, 6]. More generally for a file of size m bits that maps it into $k = m/\alpha$ symbols of size α bits with a code of minimum distance $d_{\min}$ and locality r satisfies ([25], theorem 1)

$$d_{\min} \leq n - \left\lceil \frac{m}{\alpha} \right\rceil - \left\lceil \frac{m}{r\alpha} \right\rceil + 2 \tag{7.2}$$

which reduces to the previous bound for $m/\alpha = k$.

As variations of LRC codes are considered in the remainder of the section, the above bound will be changed to accommodate the variations considered.

Two constructions for locally repairable codes are considered here. While not representative of the wide variety of constructions available they give a brief indication of some of the approaches taken to the problem.

The first construction, due to [25], is simpler than many yet effective and intuitive. The construction uses MDS codes, achieves an MDS distance and suffers only an additive factor of $1/r$ reduction in the code rate. It will be convenient to use the code described to regenerate all the elements of a failed server while the definition of an LRC code only requires the regeneration of a single coordinate position. Thus we return briefly to the problem of the previous chapter.

Denote by $y_j^{(i)} \in \mathbb{F}_q$ the j-th component of the n-tuple $\boldsymbol{y}^{(i)}$ over $\mathbb{F}_q$. For simplicity the approach of [25] is slightly adapted for the first construction.

Let a file consist of $M = rk$ symbols of the finite field $\mathbb{F}_q$, $\boldsymbol{x} = [\boldsymbol{x}^{(1)}, \ldots, \boldsymbol{x}^{(r)}]$ and $\boldsymbol{x}^{(i)} \in \mathbb{F}_q^k$ row vectors. These k-tuples over $\mathbb{F}_q$ are $(n,k,d_{\min})_q$ MDS encoded ($q > n$) as

$$\boldsymbol{y}^{(i)} = \boldsymbol{x}^{(i)}G, \quad \boldsymbol{x}^{(i)} \in \mathbb{F}_q^k, \quad \boldsymbol{y}^{(i)} \in \mathbb{F}_q^n, \ i = 1,2,\ldots,r,$$

where G is a $k \times n$ generator matrix of the code. Generate the overall parity check

$$\boldsymbol{s} = \oplus_{i=1}^{r} \boldsymbol{y}^{(i)} \in \mathbb{F}_q^n.$$

Among the codewords and parity-check word, there is a total of $(r+1)n$ $\mathbb{F}_q$ symbols. Each of the n servers will store $\alpha = (r+1)$ symbols chosen from distinct codewords and from the parity-check word. Assume $(r+1) \mid n$ and $\eta = n/(r+1)$. The n servers will be divided into η groups of servers and a failure of a server within a group will be restored entirely by information from other servers within the group. To facilitate node restoration the symbols are arranged in a cyclic fashion, within a group of nodes. A simple example (a particular case of an example in ([25], section 5) will easily expand to illustrate the general procedure.

Example 7.4 (Refer to Figure 7.1) Consider the case of $k = 4 \times 12$ data symbols over a field of characteristic 2 (the more general case is easily considered). The data are arranged into $r = 4$ words of 12 $\mathbb{F}_q$ symbols each. The 4 words $\boldsymbol{x}^{(i)}, i = 1,2,3,4$ are MDS encoded using a $(15,12,4)_q, q > 15$ into the 4 words $\boldsymbol{y}^{(i)}, i = 1,2,3,4$. The coordinates of the 4 codewords and the one parity-check word are stored on 15 servers, each server containing $(r+1) = 5$ $\mathbb{F}_q$ symbols. The servers are divided into $\eta = 15/5 = 3$ groups each group containing five servers. The situation is depicted in Figure 7.1. As noted

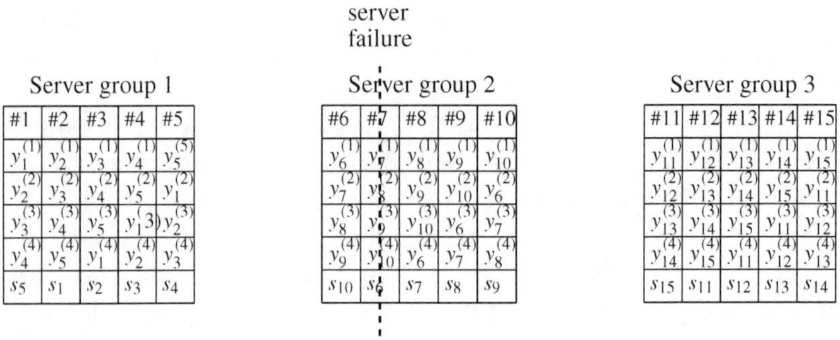
server failure
Server group 1
#1 #2 #3 #4 #5
$y_1^{(1)}$ $y_2^{(1)}$ $y_3^{(1)}$ $y_4^{(1)}$ $y_5^{(5)}$
$y_2^{(2)}$ $y_3^{(2)}$ $y_4^{(2)}$ $y_5^{(2)}$ $y_1^{(2)}$
$y_3^{(3)}$ $y_4^{(3)}$ $y_5^{(3)}$ $y_1^{(3)}$ $y_2^{(3)}$
$y_4^{(4)}$ $y_5^{(4)}$ $y_1^{(4)}$ $y_2^{(4)}$ $y_3^{(4)}$
s_5 s_1 s_2 s_3 s_4
Server group 2
#6 #7 #8 #9 #10
$y_6^{(1)}$ $y_7^{(1)}$ $y_8^{(1)}$ $y_9^{(1)}$ $y_{10}^{(1)}$
$y_7^{(2)}$ $y_8^{(2)}$ $y_9^{(2)}$ $y_{10}^{(2)}$ $y_6^{(2)}$
$y_8^{(3)}$ $y_9^{(3)}$ $y_{10}^{(3)}$ $y_6^{(3)}$ $y_7^{(3)}$
$y_9^{(4)}$ $y_{10}^{(4)}$ $y_6^{(4)}$ $y_7^{(4)}$ $y_8^{(4)}$
s_{10} s_6 s_7 s_8 s_9
Server group 3
#11 #12 #13 #14 #15
$y_{11}^{(1)}$ $y_{12}^{(1)}$ $y_{13}^{(1)}$ $y_{14}^{(1)}$ $y_{15}^{(1)}$
$y_{12}^{(2)}$ $y_{13}^{(2)}$ $y_{14}^{(2)}$ $y_{15}^{(2)}$ $y_{11}^{(2)}$
$y_{13}^{(3)}$ $y_{14}^{(3)}$ $y_{15}^{(3)}$ $y_{11}^{(3)}$ $y_{12}^{(3)}$
$y_{14}^{(4)}$ $y_{15}^{(4)}$ $y_{11}^{(4)}$ $y_{12}^{(4)}$ $y_{13}^{(4)}$
s_{15} s_{11} s_{12} s_{13} s_{14}

Figure 7.1 Example 7.4 of LRC code, $k = 12, n = 15, r = 4, \eta = 3$, failure of server 7

$$\boldsymbol{y}^{(i)} = \boldsymbol{x}^{(i)} \cdot G \in \mathbb{F}_q^{15}, \quad i = 1,2,3,4, \quad \boldsymbol{s} = \oplus_{i=1}^{4} \boldsymbol{y}^{(i)} \in \mathbb{F}_q^{15}.$$

Suppose server 7 fails. The original contents of the server are to be reconstructed using only symbols from the other servers within group 2, reconstructing each symbol by using no more than $r = 4$ other symbols. The reconstruction of the required symbols is as follows:

$$\begin{aligned}
y_7^{(1)} &= s_7 \oplus y_7^{(2)} \oplus y_7^{(3)} \oplus y_7^{(4)} \\
y_8^{(2)} &= s_8 \oplus y_8^{(1)} \oplus y_8^{(3)} \oplus y_8^{(4)} \\
y_9^{(3)} &= s_9 \oplus y_9^{(1)} \oplus y_9^{(2)} \oplus y_9^{(4)} \\
y_{10}^{(3)} &= s_{10} \oplus y_{10}^{(1)} \oplus y_{10}^{(2)} \oplus y_{10}^{(3)}
\end{aligned}$$

and these equations satisfy the requirements.

As noted above, the coding scheme has been described as a scheme to regenerate the entire contents of a failed server, containing many $\mathbb{F}_q$ symbols. However, it is clear the code satisfies the definition of an LRC code, regenerating a single coordinate position.

The scheme in the above example is easily generalized to more parameters. In that case each node would contain r $\mathbb{F}_q$ code symbols from each of the r codewords (same set of subscripts) plus the same set of subscripts of the overall parity-check word. Also the circular pattern of indices for each row/column in the same group is easily derived from the first group and the above example. The fact the code has locality r follows directly from the construction.

Since the codewords $\mathbf{y}^{(i)}$ are from an MDS generator matrix, these codewords have minimum distance $d_{\min} = n - k + 1$. The code rate is

$$R = \frac{rk}{(r+1)n} = \frac{k}{n} - \frac{k}{(r+1)n}$$

showing a slight reduction of the overall code rate in order to accommodate local repairability.

Also, since the total number of overall data is $M = rk$ $\mathbb{F}_q$ symbols and $(r+1)$ $\mathbb{F}_q$ symbols are stored on each server, it is straightforward to verify that for the case $(r+1) \nmid k$ the bound of Equation 7.1 is satisfied in that [25]

$$n - \left\lceil \frac{M}{\alpha} \right\rceil - \left\lceil \frac{M}{r\alpha} \right\rceil + 2 = n - \left\lceil \frac{rk}{r+1} \right\rceil - \left\lceil \frac{k}{r+1} \right\rceil + 2 = n - k + 1$$

(noting that $\lceil rk/(r+1) \rceil + \lceil k/(r+1) \rceil = k + \lceil -k/(r+1) \rceil + \lceil k/(r+1) \rceil = k + 1$).

For the second LRC code construction, recall the bound 7.1 and that codes that meet this bound are referred to as *distance-optimal LRC codes*. The codes of [31] achieve this bound through an interesting and novel construction. Recall first that a polynomial $f(x)$ that interpolates through the set of n points $(x_i, y_i),\ i = 1,2,\ldots,n$, i.e., $y_i = f(x_i)$ is of degree at most $n-1$ can be constructed of the form

$$f(x) = \sum_{i=1}^{n} y_i \prod_{\substack{j=1 \\ j \neq i}}^{n} \frac{(x - x_j)}{(x_i - x_j)}.$$

The polynomial can be of degree less than $n-1$ depending on the x, y values.

A code of length n over $\mathbb{F}_q$, $q \geq n$ is constructed. It will be an evaluation code on a set $U \subset \mathbb{F}_q$, $|U| = n$ of distinct elements of $\mathbb{F}_q$. It is assumed that $r \mid k$ and that $(r+1) \mid n$ (restrictions that further constructions in [31] remove). The set U is divided into equal-sized subsets U_i, $i = 1,2,\ldots,U_\ell$, $U = \cup_{i=1}^{\ell} U_i$ such that the subsets are each of size $(r+1)$. The reason for the $(r+1)$ subsets is that, in constructing codes of locality r, if a server (codeword position) within a subset fails, the code construction will be such that contacting the r other nonfailed servers (codeword positions) within the subset will be sufficient to allow server regeneration. The construction assumes the existence of a polynomial $g(x) \in \mathbb{F}_q[x]$ of degree $r+1$ which is constant on the subsets. Such a polynomial, referred to as a *good polynomial*, will be considered shortly. The construction will lead to optimal (n,k,r) LRC codes where r is the locality of the code, i.e., any missing symbol can be reconstructed with knowledge of r other code symbols.

The codeword corresponding to an information sequence $\boldsymbol{a} \in \mathbb{F}_q^k$ (k $\mathbb{F}_q$ symbols) is constructed as follows. Assume that $r \mid k$ and write the information vector $\boldsymbol{a}$ as the $r \times (k/r)$ rectangular array

$$a_{ij},\ i = 0, 1, \ldots, r-1, \quad j = 0, 1, \ldots, \frac{k}{r} - 1.$$

For $g(x)$ a good polynomial, constant on each of the sets U_i, construct the two polynomials:

$$f_a(x) = \sum_{i=0}^{r-1} f_i(x)x^i, \quad f_i(x) = \sum_{j=0}^{\frac{k}{r}-1} a_{ij}g(x)^j,\ i = 0, 1, \ldots, r-1. \tag{7.3}$$

Notice immediately that since $g(x)$ is constant on each set U_j, $j = 1, 2, \ldots, \ell$, so is the polynomial $f_i(x)$.

The code $\mathscr{C}$ then is defined by evaluating polynomial $f_{\boldsymbol{a}}$ on the evaluation set U, one codeword for each of the possible q^k polynomials $f_{\boldsymbol{a}}$:

$$\mathscr{C} = \left\{ \boldsymbol{c}_{f_{\boldsymbol{a}}} = (f_{\boldsymbol{a}}(u_1), f_{\boldsymbol{a}}(u_2), \ldots, f_{\boldsymbol{a}}(u_n)),\ \boldsymbol{a} = (a_1, \ldots, a_k) \in \mathbb{F}_q{}^k \right\}. \tag{7.4}$$

The degrees of these polynomials $f_{\boldsymbol{a}}$ are at most $(k/r - 1)(r + 1) + r - 1 = k + k/r - 2$, and hence the minimum distance of the code is n minus this quantity, i.e., two such polynomials corresponding to different information k-tuples cannot interpolate to the same codeword and there are q^k codewords. Since the code is linear the minimum distance is at least as large as n less the maximum degree of the polynomials (the maximum number of zeros), i.e.,

$$d_{\mathscr{C}} \geq n - \left(k + \frac{k}{r} - 2 \right)$$

which achieves the minimum distance bound for codes of locality r of Equation 7.2.

To see that the codes have locality r suppose the symbol (server) $u^* \in U_j$ is erased. We show how such a symbol can be regenerated by use of the remaining r symbols in U_j, i.e., it is desired to find $f_{\boldsymbol{a}}(u^*)$ where u^* is known and the remaining values of $f_a(u)$, $u \in U_j \backslash \{u^*\}$ are given. Consider the two polynomials

$$\eta_1(x) = \sum_{u \in U_j \backslash \{u^*\}} c_u \prod_{v \in U_j \backslash \{u^*, u\}} \frac{x - v}{u - v}, \quad c_u = f_{\boldsymbol{a}}(u). \tag{7.5}$$

and

$$\eta_2(x) = \sum_{i=0}^{r-1} f_i(u^*)x^i, u^* \in U_j \tag{7.6}$$

where $f_i(x)$ is defined in Equation 7.3 (and u^* is the erased position in U_j). Notice that the polynomial η_1 can be computed by the regeneration process since all quantities involved are known, while polynomial η_2 requires knowledge of all polynomials f_i and hence the information sequence $\boldsymbol{a} \in \mathbb{F}_q^k$ which is unknown to the regeneration process. Both of these polynomials have degree at most $r - 1$. It is claimed that the polynomials are equal since by definition

$$\eta_1(\gamma) = c_\gamma = f_{\boldsymbol{a}}(\gamma), \quad \gamma \in U_j \backslash \{u^*\},$$

and similarly by Equation 7.3

$$\eta_2(\gamma) = \sum_{i=0}^{r-1} f_i(u^*)\gamma^i = c_\gamma, \quad \gamma \in U_j$$

and the two polynomials of degree at most $r-1$ assume the same values on the set $U_j \backslash \{u^*\}$ of r values. Hence they are equal. Thus the value $c_{u^*} = \eta_2(u^*)$, the erased value sought, can be evaluated by polynomial $\eta_1(u^*)$ $(= \eta_2(u^*))$ which can be formed from known quantities.

A key aspect of the above construction is the need to form a polynomial $g(x)$ that is constant on the subsets of the evaluation set U_i where $U = \cup_{i=1}^{\ell} U_i$. A simple suggestion in [31] is to consider cosets of a subgroup of $\mathbb{F}_q$, either additive or multiplicative. For example, if η is a primitive element of $\mathbb{F}_q$ and $s \mid q-1, t = (q-1)/s$, then the subgroup $G = \langle \eta^s \rangle$ is cyclic of order t and the (multiplicative) cosets of $\mathbb{F}_q{}^*$ are $\eta^i G, i = 0, 1, \ldots, s-1$ and the polynomial

$$g(x) = \prod_{\psi \in G} (x - \psi) = x^t - 1$$

is easily shown to be constant on the cosets. The following extended example illustrates these constructions:

Example 7.5 Let $q = 2^4$ and base the arithmetic on the primitive polynomial $f(x) = x^4 + x + 1$ over $\mathbb{F}_2$ with the table shown allowing easy arithmetic in the field where α is a primitive element (zero of $f(x)$). Define the subsets of interest to be the cosets U_i of the cyclic subgroup of $\mathbb{F}_q{}^*$ of order 5 generated by α^3, i.e., $G = \langle \alpha^3 \rangle = \{1, \alpha^3, \alpha^6, \alpha^9, \alpha^{12}\}$, i.e.

$$\begin{aligned} g(x) &= x^5 + 1 \\ U_1 &= \{1, \alpha^3, \alpha^6, \alpha^9, \alpha^{12}\}, \ g(U_1) = 0 \\ U_2 &= \{\alpha, \alpha^4, \alpha^7, \alpha^{10}, \alpha^{13}\}, \ g(U_2) = \alpha^{10} \\ U_3 &= \{\alpha^2, \alpha^5, \alpha^8, \alpha^{11}, \alpha^{14}\}, \ g(3) = \alpha^5 \end{aligned}$$

$$\begin{array}{ll} 1+\alpha = \alpha^4 & 1+\alpha^8 = \alpha^2 \\ 1+\alpha^2 = \alpha^8 & 1+\alpha^9 = \alpha^7 \\ 1+\alpha^3 = \alpha^{14} & 1+\alpha^{10} = \alpha^5 \\ 1+\alpha^4 = \alpha & 1+\alpha^{11} = \alpha^{12} \\ 1+\alpha^5 = \alpha^{10} & 1+\alpha^{12} = \alpha^{11} \\ 1+\alpha^6 = \alpha^{13} & 1+\alpha^{13} = \alpha^6 \\ 1+\alpha^7 = \alpha^9 & 1+\alpha^{14} = \alpha^3 \end{array}$$

Consider the $(15,8)_q$ code with information symbol array (chosen arbitrarily), with the associated polynomials:

$$A = (a_{ij}) = \begin{bmatrix} \alpha^4 & \alpha \\ \alpha^{13} & \alpha^7 \\ \alpha^2 & \alpha^{11} \\ 0 & 1 \end{bmatrix} \qquad \begin{array}{ll} & \text{associated polynomials:} \\ f_0(x) & = \alpha^4 + \alpha g(x) = \alpha x^5 + 1 \\ f_1(x) & = \alpha^{13} + \alpha^7 g(x) = \alpha^7 x^5 + \alpha^5 \\ f_2(x) & = \alpha^2 + \alpha^{11} g(x) = \alpha^{11} x^5 + \alpha^9 \\ f_3(x) & = 0 + x^5 + 1. \end{array}$$

These computations lead to the codeword evaluation polynomial and codeword:

$$\begin{aligned} f_a(x) &= x^8 + \alpha^{11}x^7 + \alpha^7 x^6 + \alpha x^5 + x^3 + \alpha^9 x^2 + \alpha^5 x + 1 \\ c_{f_a} &= (f_a(1), f_a(\alpha), f_a(\alpha^2), f_a(\alpha^3), f_a(\alpha^4), f_a(\alpha^5), f_a(\alpha^6), f_a(\alpha^7), \\ &\qquad f_a(\alpha^8), f_a(\alpha^9), f_a(\alpha^{10}), f_a(\alpha^{11}), f_a(\alpha^{12}), f_a(\alpha^{13}), f_a(\alpha^{14})) \\ &= (\alpha^9, \alpha^{10}, \alpha^4, \alpha^2, 0, \alpha^2, \alpha^{14}, \alpha^{10}, 0, \alpha^{11}, \alpha^8, \alpha^9, \alpha^9, \alpha^3, \alpha). \end{aligned}$$

Suppose the codeword position of $u^* = \alpha^{10}$ in the subset U_2 is erased and it is desired to regenerate the codeword value in this position (α^8) using Equation 7.5 which refers to the codeword values at the other coordinate positions of subset U_2, i.e., the coordinate positions of $\{\alpha, \alpha^4, \alpha^7, \alpha^{13}\}$ – the codeword values at these positions are respectively $\{\alpha^{10}, 0, \alpha^{10}, \alpha^3\}$. Equation 7.5 yields:

$$\begin{aligned} \eta_1(x) &= \sum_{\beta \in U_2 \backslash \{\alpha^*\}} c_\beta \prod_{\beta' \in U_2 \backslash \{\alpha^*, \beta\}} \frac{(x - \beta')}{(\beta - \beta')} \\ &= \alpha^{10} \cdot \left(\prod_{\beta' \in \{\alpha^4, \alpha^7, \alpha^{13}\}} \frac{(x + \beta')}{(\beta - \beta')} \right) + 0 \cdot \left(\prod_{\beta' \in \{\alpha^1, \alpha^7, \alpha^{13}\}} \frac{(x + \beta')}{(\beta - \beta')} \right) \\ &\quad + \alpha^{10} \cdot \left(\prod_{\beta' \in \{\alpha^1, \alpha^4, \alpha^{13}\}} \frac{(x + \beta')}{(\beta - \beta')} \right) + \alpha^3 \cdot \left(\prod_{\beta' \in \{\alpha^1, \alpha^4, \alpha^7\}} \frac{(x + \beta')}{(\beta - \beta')} \right) \\ \eta_1(\alpha^{10}) &= \alpha^{10} \cdot \frac{\alpha^2 \cdot \alpha^6 \cdot \alpha^9}{1 \cdot \alpha^{14} \cdot \alpha^{12}} + 0 + \alpha^{10} \cdot \frac{\alpha^8 \cdot \alpha^2 \cdot \alpha^9}{\alpha^{14} \cdot \alpha^3 \cdot \alpha^5} + \alpha^3 \cdot \frac{\alpha^8 \cdot \alpha^2 \cdot \alpha^6}{\alpha^{12} \cdot \alpha^{11} \cdot \alpha^5} \\ &= \alpha + 0 + \alpha^7 + \alpha^6 = \alpha^8, \end{aligned}$$

as required. By construction the code is a $(15,8)_{2^4}$ LRC code with $r = 4$.

The work [31] develops many other interesting aspects of this construction. We limit ourselves to the following discussion for which it is not necessary for the m partition sets U_i to be of equal size. Denote the partition set $U = \{U_1, U_2, \ldots, U_m\}$ and let $h(x) = \prod_{u \in U}(x - u)$, the annihilator polynomial of U. Denote by $\mathbb{F}_U[x]$ the set of polynomials that are constant on the sets of U taken modulo $h(x)$, i.e.,

$$\mathbb{F}_U[x] = \{f \in \mathbb{F}_q[x] : f \text{ is constant on the sets } U_i, i = 1, 2, \ldots, m, \deg f < |U|\}.$$

The set is closed under addition and multiplication modulo $h(x)$ and can be shown to be a commutative algebra with identity and has the following properties ([31], proposition 3.5):

(i) A nonconstant $f \in \mathbb{F}_U[x]$ satisfies $\max_i |U_i| \leq \deg(f) < |U|$.
(ii) The m polynomials

$$f_i(x) = \sum_{u \in U_i} \prod_{v \in U \backslash u} \frac{x - v}{u - v}, \; i = 1, 2, \ldots, m$$

form a basis of $\mathbb{F}_U[x]$, $\dim(\mathbb{F}_U[x]) = m$, and indeed

$$f_i(U_j) = \delta_{ij}$$

where δ_{ij} is the Kronecker delta function. It follows immediately that:
(iii) if $g(x)$ takes on the value u_i on set U_i, $i = 1, 2, \ldots, m$, $\deg(g) < |U|$, then $g(x)$ can be written

$$g(x) = \sum_{i=1}^{m} u_i f_i(x)$$

and the polynomials $1, g(x), g^2(x), \ldots, g^{m-1}(x)$ form a basis of $\mathbb{F}_U[x]$.
(iv) There exist m integers $0 = d_0 < d_1 < \cdots < d_{m-1} < |U|$ such that the degree of each polynomial in $\mathbb{F}_U[x]$ is d_i for some i and $d_i = \deg f_i$.

In the case the subsets U_i are of equal size $(r+1)$, then $d_i = i(r+1)$, $i = 0, 1, \ldots, m-1$.

More properties of this interesting algebra $\mathbb{F}_U[x]$ are established in [31] and used for further constructions of LRCs.

Given the nature of the coding problem it is not surprising that combinatorial configurations have been used to construct interesting classes of locally repairable codes with their use. These include resolvable designs [29], partial geometries [24], Hadamard designs [26] and matroid theory [26, 33] among many other configurations.

The construction techniques and the properties of the resulting codes available in the literature vary a great deal and the examples discussed here give a brief look into a few of the techniques used.

For the remainder of the section some classes of LRC codes are considered as well as classes of codes specifically designed for the recovery of certain types of storage systems. It is clear that the notion of optimality can only be discussed in reference to the type of constraints of these systems and it differs for each class.

(r, δ) locally repairable codes

The notion of an (r, δ) code is introduced:

Definition 7.6 ([28]) The i-th code symbol c_i, $1 \leq i \leq n$ of an $(n,k,d)_q$ linear code $\mathscr{C}$, is said to have locality (r, δ) if there exists a punctured subcode of $\mathscr{C}$ with support containing position i whose length is at most $r + \delta - 1$ and whose minimum distance is at least δ. Equivalently there exists a subset $\mathcal{S}_i \subseteq [n] = \{1, 2, \ldots, n\}$ such that

(i) $i \in \mathcal{S}_i$ and $|\mathcal{S}_i| \leq r + \delta - 1$;
(ii) the minimum distance of the code $\mathscr{C}_{\mathcal{S}_i}$ obtained by deleting code symbols $c_j, j \in [n] \backslash \mathcal{S}_i$ is at least δ.

In terms of parity-check matrices, if the linear code with parity-check matrix $\boldsymbol{H}$ has all coordinate positions in $[n] \backslash \mathcal{S}_i$ removed, the remaining $\eta_i \times |\mathcal{S}_i|$ matrix $\boldsymbol{H}_i$, for some $\eta_i \leq n - k$, has the property that any $\delta - 1$ columns are linearly independent.

If all information positions i of the code $\mathscr{C}$ have locality (r, δ), it is referred to as having information locality and designated an $(r, \delta)_i$ code. If all positions (information and parity) have (r, δ) locality, it is referred to as an $(r, \delta)_a$ code. The point of the definition is that if a code has information locality, then a systematic coordinate position can be repaired by using r other positions even in the presence of $\delta - 2$ coordinate erasures. The previous (r, d) codes correspond to $(r, 2)_i$ codes (i.e., $\delta = 2$).

With the extra constraint imposed by such codes it is shown [28] that the minimum distance of an $(r, \delta)_i$ code must be upper bounded by

$$d_{\min} \leq n - k + 1 - \left(\left\lceil \frac{k}{r} \right\rceil - 1 \right) (\delta - 1). \tag{7.7}$$

Under certain conditions the bound also applies to $(r, \delta)_a$ codes and codes meeting this bound can be constructed. In particular it is shown:

Theorem 7.7 ([28], theorem 9) *Let $q > kn^k$ and $(r + \delta - 1) \mid n$. Then there exists an $(r, \delta)_a$ $(n, k, d_{\min})_q$ linear code with $\delta - 1 < d$ satisfying the bound Equation 7.7.*

Some nonexistence results are also available ([30], theorems 1 and 11):

Theorem 7.8 *If $(r + \delta - 1) \nmid n$ and $r \mid k$, then there exists no $(r, \delta)_a$ linear code which achieves the bound Equation 7.7.*

Further aspects of these codes are discussed in [20]. Cyclic versions of the codes are considered in [10] and codes with differing localities, an aspect of interest in practice, in [9].

7.2 Maximally Recoverable Codes

The idea of maximally recoverable (MR) LRC codes arose in the work [12] as a stronger notion than LRC. The notion in that work is quite general. It defines the notion of an erasure configuration Ω and the set of all recoverable erasures $E(\Omega)$ and a code as being MR if it is able to recover any erasure $e \in E(\Omega)$, i.e., it is able to recover all erasure patterns it is information theoretically capable of correcting.

It will be helpful to consider the discussion of a somewhat narrower class of MR LRCs definition in ([14], definition 5):

Definition 7.9 Let $\mathscr{C}$ be a linear systematic (n,k) code. $\mathscr{C}$ is a (k,r,h) *local code* if the following conditions are satisfied:

(i) There are k data symbols and h heavy parity symbols (which may depend on all data symbols).
(ii) Assume $r \mid (k+h)$ and $n = k+h+(k+h)/r$.
(iii) The $(k+h)$ symbols are grouped into $\ell = (k+h)/r$ groups of size r and an overall parity check is added to each group (hence the length of the code is n).
(iv) The local code is said to be MR if for any set $E \subseteq [n]$ where E is obtained by choosing one coordinate from each of the groups, puncturing $\mathscr{C}$ in the positions of E yields an MDS $(k+h,k)$ code.

Notice the definition implies such a code has locality r for all code symbols and that after correction of a single erasure in each of the groups, from the MDS property of the punctured code, it is able to correct any h remaining erasures. It is clear that such a code is by definition capable of correcting any pattern $E \subset [n]$ of erasures where E is formed by picking one coordinate from each of the $(k+h)/r$ groups and any h additional positions. These codes are also called partial MDS or PMDS codes in [3]. These are discussed further later in the section as PMDS codes as there is a considerable literature under that name. The distinction is preserved for reference to the literature as a matter of convenience. They are, however, MR codes.

An example of the construction of a local MR LRC is as follows. The work [14] refers to *multisets*, where set elements may be repeated although attention here is restricted to sets (no repetition of elements). Let $\mathbb{F}_q$ be a finite field of characteristic 2 and let $S = \{a_1,a_2,\ldots,a_n\}$, $a_i \in \mathbb{F}_q$ and denote the $h \times n$ parity-check matrix $\boldsymbol{A}(S,h) = (a_{gj})$ by

$$a_{gj} = a_j^{2^{g-1}}, \ g = 1,2,\ldots,h, \ j = 1,2,\ldots,n.$$

Definition 7.10 The set $S \subseteq \mathbb{F}_q$ is said to be t-wise independent over a field $\mathbb{F}' \subseteq \mathbb{F}_q$ if every subset of S of size at most t is linearly independent over $\mathbb{F}'$.

Denote the code $\mathscr{C}(S,h)$ containing $\boldsymbol{x} = (x_1, x_2, \ldots, x_n) \in \mathbb{F}_q{}^n$ if

$$\sum_{j=1}^{n} a_j^{2^{g-1}} x_j = \boldsymbol{A}(S,h) \cdot \boldsymbol{x}^t = 0, \; g = 1,2,\ldots,h.$$

The following lemma ([14], lemma 10) is immediate:

Lemma 7.11 *The code $\mathscr{C}(S,h)$ has distance $h+1$ if and only if the set S is h-wise independent over $\mathbb{F}_2 \subset \mathbb{F}_q$.*

This follows since every $h \times h$ submatrix of $\boldsymbol{A}(S,h)$ is nonsingular if and only if any h elements of S are linearly independent over $\mathbb{F}_q$ and by lemma 3.51 of [22] it is sufficient they are linearly independent over $\mathbb{F}_2$.

Let

$$n = k + h + (k+h)/r = (r+1)(k+h)/r \quad \text{and let} \quad \ell = \frac{n}{r+1} = \frac{k+h}{r}.$$

Let the i-th group contain the $r+1$ symbols $\{x_{i,1}, x_{i,2}, \ldots, x_{i,r+1}\}$, $i = 1,2,\ldots,\ell$, a relabeling of the components of $\boldsymbol{x}$. In each of the ℓ groups, the variables satisfy a parity check $\sum_{s=1}^{r+1} x_{i,s} = 0$, i.e., one symbol is the overall parity check of the other r symbols in the group. It follows that all code coordinates have locality r. To facilitate discussion of these groups write the set S as

$$S = \{a_{i,s}\} \text{ for } i \in [\ell], \; s \in [r+1].$$

With this notation the parity checks may be written as:

$$\sum_{i=1}^{\ell} \sum_{s=1}^{r+1} a_{i,s}^{2^{g-1}} x_{i,s} = 0 \quad \text{for} \quad g = 1,2,\ldots,h$$

$$\sum_{s=1}^{r+1} x_{i,s} = 0 \quad \text{for} \quad i = 1,2,\ldots,\ell,$$

and these equations define the code $C(S,r,h)$.

Let $\boldsymbol{e} = (e_1, e_2, \ldots, e_\ell) \in [r+1]^\ell$ be an indicator vector indicating a single position within each of the ℓ code groups and let $\mathscr{C}^{-e}(S,r,h)$ be the code obtained by puncturing $\mathscr{C}(S,r,h)$ in positions $\{i, e_i\}, i = 1,2,\ldots,\ell$, one position in each group. By the Definition 7.9(iv) the code $\mathscr{C}(S,r,h)$ is maximally recoverable if and only if the code $\mathscr{C}^{-e}(S,r,h)$ is MDS for every $e \in [r+1]^\ell$. A criterion to achieve this is the following ([14], proposition 11):

Proposition 7.12 *The code $\mathscr{C}(S,r,h)$ is maximally recoverable if and only if for every $e \in [r+1]^\ell$ the set*

$$T(S,e) = \{a_{i,s} + a_{i,e_i}, i \in [\ell], s \in [r+1]\backslash e_i\}$$

is h-wise independent.

Thus the construction of maximally recoverable $\mathscr{C}(S,r,h)$ codes is reduced to the construction of h-wise independent sets. Several constructions of such sets are given in [14]. The following theorem results from using binary BCH codes to derive such sets:

Theorem 7.13 ([14], theorem 13) *Let k,r,h be positive integers such that $r \mid (k+h)$. Let m be the smallest integer such that*

$$n = k + h + \frac{(k+h)}{r} \leq 2^m - 1.$$

Then there exists an MR local (k,r,h) code over the characteristic 2 field $\mathbb{F}_{2^{hm}}$.

As with previous types of codes, combinatorial configurations have been used to construct interesting classes of LRCs and MR LRCs (e.g., [26, 33]). The topic of MR codes has a large and interesting literature and the above is but a brief look at the variety of such constructions. As noted, the PMDS codes to be discussed are also MR codes.

7.3 Other Types of Locally Repairable Codes

There have been interesting classes of codes proposed for certain types of storage devices to address specific requirements. For example, the code construction might depend on the fact that the storage device has a redundant array of independent disks (RAID) structure. A view of this architecture [27] is as an array of n disks, referred to as a stripe, each disk containing a certain number of sectors or symbols, each symbol comprised of a number of bits. The various RAID architectures vary in the number of disks and sectors devoted to parities.

A few of the types of coding related to these structures are discussed in this section since they do not fall easily into the previous sections. In particular the notions of partial MDS (PMDS) and sector-disk (SD) codes arose in response to these architectures [2, 3, 4, 27]. A few comments on such codes are given below.

Partial MDS and Sector-Disk Codes

It has been noted that PMDS codes are in the class of MR LRC codes discussed previously. The reason the previous material and the material here was not combined is to access the literature under the PMDS name. These PMDS codes have a particular structure and, since they are capable of correcting all erasures which they are information theoretically capable of, they are MR codes. Certain aspects of these codes are more relevant to the RAID storage architecture noted and to the SD codes discussed below. The definition of the codes used here (equivalent to the definition of MR codes considered earlier in the chapter) is as follows [11]:

Definition 7.14 An $m \times n$ array code $\mathscr{C}$ over $\mathbb{F}_q$ is called an (r,s) PMDS code if for positive integers r,s:

(i) $\mathscr{C}$ is a linear $(mn, m(n-r)-s)_q$ code.
(ii) The set of rows of the code $\mathscr{C}$ (a codeword as an $m \times n$ array) is an $(n, n-r, r+1)_q$ MDS code.
(iii) For any m nonnegative integers $s_1, s_2, \ldots, s_m$, $s_1 + s_2 + \cdots + s_m = s$ the array code can correct up to $s_i + r$ erasures in row $i, i = 1, 2, \ldots, m$. (More precisely, an (r,s) PMDS code is capable of correcting r erasures in each of the m rows and any other s erasures in the array.)

Sector-disk (SD) codes arise from considering types of failure of storage systems. Array codes often consider a sector failure of a disk as an entire disk failure. The SD codes arise from differentiating these types of failures. They are a special case of PMDS codes.

Definition 7.15 An $m \times n$ array code $\mathscr{C}$ over $\mathbb{F}_q$ is called an (r,s) SD code if for positive integers r,s:

(i) $\mathscr{C}$ is a linear $(mn, m(n-r)-s)_q$ code.
(ii) Each row of the code $\mathscr{C}$ is an $(n, n-r, r+1)_q$ MDS code.
(iii) $\mathscr{C}$ can correct up to $mr + s$ erasures provided that among the erasures mr of them are in m columns (i.e., r of the columns are completely erased, the other s erasures are unrestricted).

Thus an SD code is a restricted type of PMDS code where certain of the erasures are constrained to all be in a column.

An example of an $(r, 1)$ PMDS code is given [11]:

Example 7.16 Consider information symbols $\{\sigma_1, \sigma_2, \ldots, \sigma_{mk-1}\}$ over $\mathbb{F}_q$ and form the $m \times k$ array

$$S = \begin{bmatrix} \sigma_1 & \sigma_2 & \cdots & \sigma_{k-1} & \sigma_k \\ \sigma_{k+1} & \sigma_{k+2} & \cdots & \sigma_{2k-1} & \sigma_{2k} \\ \vdots & \vdots & \vdots & \vdots & \vdots \\ \sigma_{(m-1)k+1} & \sigma_{(m-1)k+2} & \cdots & \sigma_{mk-1} & \sigma_{mk} \end{bmatrix}$$

where the $\sigma_{mk} = \delta$ element is the only noninformation symbol in the array given by

$$\begin{aligned} \delta = \sigma_{mk} &= -(\sigma_k + \sigma_{2k} + \cdots + \sigma_{(m-1)k}) \quad \text{or} \\ \sigma_{ik} &= -(\sigma_k + \cdots + \sigma_{(i-1)k} + \sigma_{(i+1)k} + \cdots \sigma_{(m-1)k} + \delta), \end{aligned} \tag{7.8}$$

i.e., δ is chosen to make the sum of the last column 0. Let $\alpha_1, \ldots, \alpha_n$ be distinct nonzero elements of $\mathbb{F}_q$ (hence $q > n$) and encode the above information symbol array into the code array:

$$X = S \cdot \begin{bmatrix} 1 & 1 & \cdots & 1 \\ \alpha_1 & \alpha_2 & \cdots & \alpha_n \\ \vdots & \vdots & \vdots & \vdots \\ \alpha_1^{k-1} & \alpha_2^{k-1} & \cdots & \alpha_n^{k-1} \end{bmatrix} = \begin{bmatrix} f_1(\alpha_1) & f_1(\alpha_2) & \cdots & f_1(\alpha_n) \\ f_2(\alpha_1) & f_2(\alpha_2) & \cdots & f_2(\alpha_n) \\ \vdots & \vdots & \vdots & \vdots \\ f_m(\alpha_1) & f_m(\alpha_2) & \cdots & f_m(\alpha_n) \end{bmatrix}$$

where $f_i(x) = \sum_{j=0}^{k-1} \sigma_{ik+j+1} x^j$, $i = 1,2,\ldots,m$.

Suppose the matrix $Y = (y_{ij}), i = 1,2,\ldots,m,\ j = 1,2,\ldots,n$ is the array X with at most r erasures introduced into $m-1$ of the rows and at most $r+1$ erasures introduced into the other row. Suppose the rows with at most r erasures are the first $m-1$ and the last row contains $r+1$ erasures. It can be verified that these assumptions can be made without loss of generality, i.e., the argument is easily changed to accommodate alternate assumptions. Since the rows are defined by evaluating a polynomial of degree at most $k-1$ at the points α_i, they may be viewed as a codeword in an RS code with parameters $(n,k,n-k+1)_q$ and $r = n-k$. Thus the first $m-1$ rows of the array can be decoded to give the corresponding information values. The value y_{mk} can be decoded as minus the sum of the (correctly) decoded elements in the last column in the top $(m-1)$ rows. This value is σ_{mk}. The value $\sigma_{mk}\alpha_n^{k-1}$ from each of the nonerased entries of the last row of Y now correspond to evaluations of polynomials of degree at most $k-2$ and hence to codewords of an RS coded $(n,k-1,n-k+2)_q$ and thus capable of correcting $(r+1)$ erasures. Thus the code is an $(r,1)$ PMDS code.

The above is a simple example of a PMDS code. The following construction is general and somewhat intricate but of significant interest, yielding (r,s) PMDS codes for arbitrary positive integers r,s. The construction is based on rank-metric codes (as in Chapter 12). The notation of [8] is preserved except the roles of r and m are reversed here from that work.

Let $\mathfrak{g}_1, \mathfrak{g}_2, \ldots, \mathfrak{g}_N$ be N linearly independent elements of $\mathbb{F}_{q^M}$ over $\mathbb{F}_q$ and encode information symbols $u_0, u_1, \ldots, u_{K-1} \in \mathbb{F}_{q^M}$ into

$$\boldsymbol{x} = (x_1, x_2, \ldots, x_N) = (f(\mathfrak{g}_1), f(\mathfrak{g}_2), \ldots, f(\mathfrak{g}_N)) \in \mathbb{F}_{q^M}^N,$$
$$\text{where} \quad f(x) = \sum_{j=0}^{K-1} u_i x^{q^i} \in \mathbb{F}_{q^M}[x].$$

Note that $\boldsymbol{x}$ is a codeword in a rank-metric Gabidulin code introduced in Chapter 12. The code is clearly linear and the parameters chosen will give an $(N = K + s, K, D = s + 1)_{q^M}$ rank-metric MDS (MRD) code. Thus the rank of the difference of any two distinct codewords over $\mathbb{F}_{q^M}$ is at least $s + 1$. These codes will be used to produce an (r, s) PMDS code.

Assume the code length N can be written as $N = K + s = m(n - r) = mk, r = n - k$, and the codeword $\boldsymbol{x}$ is written as an $m \times k$ array/matrix $\boldsymbol{X}$ (over $\mathbb{F}_{q^M}$). For this it will be convenient to identify

$$x_{ik+j} \to x_{i,j} \quad \text{and} \quad \mathfrak{g}_{ik+j} \to \mathfrak{g}_{i,j}, \; i = 0, 1, \ldots, m-1, \; j = 1, 2, \ldots, k.$$

Define the matrices

$$\boldsymbol{X} = \begin{bmatrix} x_{0,1} & x_2 & \cdots & x_{0,k} \\ x_{1,1} & x_{1,2} & \cdots & x_{1,k} \\ \vdots & \vdots & \vdots & \vdots \\ x_{m-1,1} & x_{m-1,2} & \cdots & x_{m-1,k} \end{bmatrix} \quad \text{and} \quad \mathfrak{G} = \begin{bmatrix} \mathfrak{g}_{0,1} & \mathfrak{g}_{0,2} & \cdots & \mathfrak{g}_{0,k} \\ \mathfrak{g}_{1,1} & \mathfrak{g}_{1,2} & \cdots & \mathfrak{g}_{1,k} \\ \vdots & \vdots & \vdots & \vdots \\ \mathfrak{g}_{m-1,1} & \mathfrak{g}_{m-1,2} & \cdots & \mathfrak{g}_{m-1,k} \end{bmatrix}$$

Denote by $V_j, j = 0, 1, \ldots, m-1$ the vector subspaces of the vector space of $\mathbb{F}_{q^M}$ over $\mathbb{F}_q$, each of dimension k, generated by the elements of $\boldsymbol{x}_j$, i.e., the subspace of linear sum of elements over $\mathbb{F}_q$ of $x_{jk+1}, \ldots, x_{(j+1)k} = x_{j,1}, \ldots, x_{j,k}$ which are linearly independent over $\mathbb{F}_q$. By assumption any two distinct subspaces V_i and $V_j, j \neq i$ intersect only in the zero element.

These $m \times k$ matrices will both be extended to $m \times n$ matrices. Denote the j-th row of $\boldsymbol{X}$ as $\boldsymbol{x}_j, j = 0, 1, 2, \ldots, m-1$, $\boldsymbol{x}_j = (x_{j,1}, \ldots, x_{j,k}) \in \mathbb{F}_{q^M}^k, j = 0, 1, \ldots, m-1$. Consider an $(n, k, n-k+1)_q$ MDS code over $\mathbb{F}_q$ with $k \times n$ generator matrix $\boldsymbol{G}$ assumed to be in systematic form (the rows of $\boldsymbol{X}$ are over $\mathbb{F}_{q^M}$). Any such generator matrix $\boldsymbol{G}$ has the property that any k columns are linearly independent over $\mathbb{F}_q$ since otherwise it would be possible to generate a nonzero codeword of weight $n - k$, a contradiction.

Let

$$\boldsymbol{Y} = \boldsymbol{X}\boldsymbol{G}, \quad \boldsymbol{G} \in \mathbb{F}_q^{k \times n}.$$

Since the generator matrix is assumed systematic, the result is

$$\boldsymbol{Y} = \boldsymbol{XG} = \begin{bmatrix} x_{0,1} & x_{0,2} & \cdots & x_{0,k} & y_{0,k+1} & \cdots & y_{0,n} \\ x_{1,1} & x_{1,2} & \cdots & x_{1,k} & y_{1,k+1} & \cdots & y_{1,n} \\ \vdots & \vdots & \vdots & \vdots & \vdots & \vdots & \vdots \\ x_{(m-1),1} & x_{(m-1),2} & \cdots & x_{m(-1),k} & y_{m-1,k+1} & \cdots & y_{m-1,n} \end{bmatrix}$$

where

$$y_{i,j} = \sum_{\ell=1}^{k} x_{i,\ell} g_{\ell,j} = \sum_{\ell=1}^{k} f(\mathfrak{g}_{i,\ell}) g_{\ell,k+j} = f\Big(\sum_{\ell=1}^{m-1} \mathfrak{g}_{i,\ell} g_{\ell,k+j}\Big)$$
$$i = 0,1,\ldots,m-1, j = k+1,\ldots,n.$$

From this equation define new evaluation points as

$$\tilde{\mathfrak{g}}_{i,j} = \sum_{\ell=1}^{m-1} \mathfrak{g}_{i,\ell} g_{\ell,k+j}\,, i = 0,1,\ldots,m-1, j = k+1,\ldots,n.$$

The matrix $\mathfrak{G}$ above can be extended to

$$\mathfrak{G}_e = \begin{bmatrix} \mathfrak{g}_{0,1} & \mathfrak{g}_{0,2} & \cdots & \mathfrak{g}_{0,k} & \tilde{\mathfrak{g}}_{0,k+1} & \cdots & \tilde{\mathfrak{g}}_{0,n} \\ \mathfrak{g}_{1,1} & \mathfrak{g}_{1,2} & \cdots & \mathfrak{g}_{1,k} & \tilde{\mathfrak{g}}_{1,k+1} & \cdots & \tilde{\mathfrak{g}}_{1,n} \\ \vdots & \vdots & \vdots & \vdots & \vdots & \vdots & \vdots \\ \mathfrak{g}_{m-1,1} & \mathfrak{g}_{m-1,2} & \cdots & \mathfrak{g}_{m-1,k} & \tilde{\mathfrak{g}}_{m-1,k+1} & \cdots & \tilde{\mathfrak{g}}_{m-1,n} \end{bmatrix}.$$

Thus, any set of elements of the matrix $\mathfrak{G}_e$ with at most k elements from each row will result in a set of linearly independent elements of $\mathbb{F}_{q^M}$ over $\mathbb{F}_q$.

For the polynomial $f(x) = \sum_{j=0}^{K-1} u_i x^{q^i} \in \mathbb{F}_{q^M}[x]$ formed from the information symbols $u_i \in \mathbb{F}_{q^M}, i = 0,1,\ldots,K-1$, form the corresponding $m \times n$ codeword as

$$\boldsymbol{c}_f = \begin{bmatrix} f(\mathfrak{g}_{0,1}) & f(\mathfrak{g}_{0,2}) & \cdots & f(\mathfrak{g}_{0,k}) & f(\tilde{\mathfrak{g}}_{0,k+1}) & \cdots & f(\tilde{\mathfrak{g}}_{0,n}) \\ f(\mathfrak{g}_{1,1}) & f(\mathfrak{g}_{1,2}) & \cdots & f(\mathfrak{g}_{1,k}) & f(\tilde{\mathfrak{g}}_{1,k+1}) & \cdots & \tilde{\mathfrak{g}}_{1,n} \\ \vdots & \vdots & \vdots & \vdots & \vdots & \vdots & \vdots \\ f(\mathfrak{g}_{m-1,1}) & f(\mathfrak{g}_{m-1,2}) & \cdots & f(\mathfrak{g}_{m-1,k}) & f(\tilde{\mathfrak{g}}_{m-1,k+1}) & \cdots & f(\tilde{\mathfrak{g}}_{m-1,n}) \end{bmatrix}.$$

Suppose there are $r = n - k$ erasures in each of the m rows and a further s erasures anywhere in the array. Suppose specifically there are s_i erasures in the i-th row and $\sum_{i=1}^{m} s_i = s$. Under these conditions let ℓ_i be the number of unerased positions in the i-th row and note that by construction $\ell_i \leq k$. It follows that the number of linearly independent evaluations (since at most k evaluations from each row are used) of the polynomial f that can be formed is

$$\sum_{i=1}^{m} \min\{\ell_i, k\} = \sum_{i=1}^{m} \ell_i = mk - \sum_{i=1}^{m} s_i = mk - s = K.$$

The linear independence of the evaluations ensures that in setting up a matrix equation between the K linearized polynomial coefficients u_i and the evaluations, the matrix will have full rank and hence is solvable for a unique solution. The conclusion then is the code is an (r,s) PMDS code which can be formed for any positive integers r,s.

More recent work [23] uses rank-metric codes in an interesting and deeper technique to construct MR LRCs. Further aspects of both SD and PMDS codes can be found in [1, 2, 3, 4, 7, 11, 15, 16, 17, 21].

Pyramid Codes

The class of pyramid codes is introduced in [19] as particularly effective erasure-correcting codes for commercial application. The classes of *basic pyramid codes* (BPC) and *generalized pyramid codes* (GPC) are introduced there but only BPCs are considered here. The BPCs are modified MDS codes and while not MDS they have several important properties making them attractive for disk storage arrays.

Denote a codeword $\boldsymbol{c} = (x_1, x_2, \ldots, x_k, p_1, \ldots, p_{n-k}) \in \mathbb{F}_q^n$ in an $(n,k)_q$ MDS code, $\mathscr{C}$, $r = n-k$, for some finite field $\mathbb{F}_q$ for a set of k information symbols $\boldsymbol{x} \in \mathbb{F}_q^k$ and parity symbols $\boldsymbol{p} \in \mathbb{F}_q^r$ and $\boldsymbol{c} = (\boldsymbol{x},\boldsymbol{p})$ (catenation). Separate the k data symbols into a disjoint union of subgroups $S_i, i = 1,2,\ldots,L$ of not necessarily the same size, where $| S_i |= k_i$, $\sum_{i=1}^{L} k_i = k$ and denote by $\boldsymbol{x}_i$ the set of information symbols corresponding to S_i, $i = 1,2,\ldots,L$. Without loss of generality let $\boldsymbol{x} = (\boldsymbol{x}_1, \boldsymbol{x}_2, \ldots, \boldsymbol{x}_L)$.

From the codewords of this code a new set of codewords is constructed as follows. For any parity subvector $\boldsymbol{p} \in \mathbb{F}_q^r, r = n-k$, let $\boldsymbol{p}' \in \mathbb{F}_q^{r_0}$, denote the r_0-tuple of the first r_0 elements of $\boldsymbol{p}$, $r_0 < r$. For the construction any fixed r_0 parity coordinates could be chosen but wlog the set of first r_0 coordinates is convenient. Further, let $\boldsymbol{p}''$ denote the last $r_1 = r - r_0$ elements of $\boldsymbol{p}$ (thus $\boldsymbol{p} = (\boldsymbol{p}',\boldsymbol{p}'')$). For a subgroup S_i of information symbols $\boldsymbol{x}_i, i = 1,2,\ldots,L$, compute a parity vector $\boldsymbol{p}_i \in \mathbb{F}_q^r$ by using the parity-check matrix of the full code $\mathscr{C}$ by assuming all other subgroup information symbols $\boldsymbol{x}_j, j \neq i$ are all-zero k_j-tuples (over $\mathbb{F}_q$). Notice that the set of all $k_i + r$-tuples $(\boldsymbol{x}_i,\boldsymbol{p}_i)$ corresponding to the q^{k_i} codewords for all possible q^{k_1}-tuples over $\mathbb{F}_q$ for $\boldsymbol{x}_i$ form a $(k_i + r, k_i)_q$ MDS code since it is a subcode of an MDS code. If this code is punctured in its last r_1 parity positions of $\boldsymbol{p}_i$ to give $\boldsymbol{p}_i'$, the result is an MDS $(k_i + r_0, k_i)_q$ code.

The BPC code is then the set of q^k codewords of the form

$$\mathscr{C}' = \left\{ \boldsymbol{c}' = \left(\boldsymbol{x}_1, \boldsymbol{p}_1', \boldsymbol{x}_2, \boldsymbol{p}_2', \ldots, \boldsymbol{x}_L, \boldsymbol{p}_L', \boldsymbol{p}'' \right) \right\}.$$

The code is clearly a $(k + r_0L + r_1, k)_q$ BPC code although not an MDS code.
The property of interest for this code construction is

Theorem 7.17 ([19], theorem 1) *The $(k+r_0L+r_1,k)_q$ BPC code constructed above can recover from $r = n - k = r_0 + r_1$ erasures.*

Proof: Assume r erasures have occurred and let e be the number of erasures among the r_0L positions of the subgroup parities $\boldsymbol{p}'_1,\ldots,\boldsymbol{p}'_L$, $e \leq r$. Consider two cases: $e \geq r_0$ and $e < r_0$.

If $e \geq r_0$ puncture the code in all r_0L group parity positions. The code remaining is by definition a $(k + r_1, k)_q$ MDS code which can correct r_1 erasures. The actual number of erasures among the $k + r_1$ places of the punctured code is $r - e \leq r - r_0 = r_1$ which the punctured code is able to correct.

In the case $e < r_0$ first note that the subgroup parities $\boldsymbol{p}'_i$ satisfy the relationship

$$\bigoplus_{i=1}^{L} \boldsymbol{p}'_i = \boldsymbol{p}',$$

i.e., the sum of the first r_0 positions of each of the subgroup parities add to the global parity of the original codeword. Then e of the erasures in the subgroup parity symbols can affect at most $e < r_0$ of the first r_0 positions of the global parity $\boldsymbol{p}$. At least $r'_0 > r_0 - e$ of the first r_0 positions of $\boldsymbol{p}$ can be computed correctly by the above equation. These parity positions together with the k information symbols and the r_1 last global parity symbols of $\boldsymbol{p}$ form a $(k + r'_0 + r_1, k)$ MDS code which can correct $r'_0 + r_1$ erasures. Since $r'_0 + r_1 \geq r_0 - e + r_1 = r - e$ which is the actual number of erasures remaining which the code can correct. ■

For the construction and decoding of GPCs the reader is referred to [19]. It is noted that both BPCs and GPCs require slightly more storage space than the equivalent MDS codes, but have significantly better access efficiency, the reconstruction read cost or the average number of symbols that must be read to access a missing symbol. The update complexity, the number of symbols that need to be accessed to update an information symbol, is the same as for MDS codes.

Zigzag Codes

Zigzag codes are introduced in [32] as codes that achieve a lower bound on the fraction of information that must be accessed in the use of MDS codes to repair a given number of erasures, referred to as the *rebuilding ratio*. The *optimal update* property is also of interest there, the minimum number of rewrites when an information symbol is changed. Here the term *zigzag* refers to the formation of parity symbols from an information array by using a single element from each row and column to form the parities – rather than simple row or column parities as is common.

Recall that the cut-set MBR bound for contacting $n-1$ other nodes to repair a single failed node using (n,k) MDS codes is

$$\frac{m(n-1)}{k(n-k)}.$$

Thus to repair a single node requires contacting a fraction of $1/(n-k) = 1/r$ of the total data stored in all nodes other than the failed node.

Additionally it is of interest to determine the number of stored symbols that are required to change when an information symbol is changed or updated. The optimal value for this is $r+1$ achieved by the use of MDS codes.

Much of the work of [32] is concerned with constructing $(k+2,k)_q$ MDS codes over some alphabet $\mathbb{F}$ to be specified. A brief description of the technique is noted.

For integers p,k consider a $p \times k$ array of information symbols over $\mathbb{F}$, say $\{a_{i,j}, i = 0,1,\ldots,p-1, j = 0,1,\ldots,k-1\}$. Two parity columns are added to the array and the overall $(k+2,k)_q$ code is MDS in the sense that the failure of any two columns can be restored. The two added parity columns are constructed in the following manner. The first parity column, designated the row parity column $\boldsymbol{R} = \{r_0, r_1, \ldots, r_{p-1}\}^t$ is simply a linear sum of the information elements in the same row, i.e., $r_i = \sum_{j=0}^{k-1} \alpha_{i,j} a_{i,j}$ for some set of coefficients $\alpha_{i,j} \in \mathbb{F}$.

For the second parity column, denoted the zigzag parity column

$$\boldsymbol{Z} = \{z_0, z_1, \ldots, z_{p-1}\}^t,$$

a set of permutations is first developed. These permutations are then used to determine the data symbols to be used in forming the zigzag parities – specifically for zigzag parity element z_j the data symbols used will be $\{a_{j_0,0}, a_{j_1,1}, \ \ldots, a_{j_{k-1},p-1}\}$ where the subscripts j_i are chosen using the permutations to ensure that each ensemble has one element from each row of the data array. The element z_j is then given by $z_j = \sum_{i=0}^{k-1} \beta_{j_i,i} a_{j_i,i}$ for some set of constants $\beta_{j_i,i}$.

The coefficients in these parity sums are chosen to ensure the resulting code is MDS so that any two column erasures can be repaired. None of the parity coefficients can be 0 and it seems sufficient to choose the row parity coefficients as unity, 1. It is shown in [32] that a field size of 3 (e.g., $\mathbb{F}_3$) is sufficient to achieve the properties required for the $(k+2, k)$ MDS code and renders the arithmetic involved simple. Generalizations of the construction for more parity checks are also considered in that work. It is further noted that a field size of 4 is sufficient for the construction of a $(k+3, 3)$ MDS code. Notice that such codes are array codes and hence the alphabet when calling it an MDS code is as a column over a small finite field and the usual restriction of lengths of MDS codes less than the alphabet size has to be adjusted.

The general construction gives an (n, k) MDS code (over a certain size field) and all the constructions have an optimal rebuild ratio of $1/r$ under certain conditions.

Comments

The past two decades have seen considerable progress in the areas of code constructions and repairs and protocols for the distributed storage and retrieval of information. The economic importance of the problem and the gains that can be realized with these techniques are significant. Only a few of the many techniques developed are noted here.

References

[1] Blaum, M. 2013. Construction of PMDS and SD codes extending RAID 5. *CoRR*, abs/1305.0032.

[2] Blaum, M., and Plank, J.S. 2013. Construction of two SD codes. *CoRR*, abs/1305.1221.

[3] Blaum, M., Hafner, J.L., and Hetzler, S. 2013. Partial-MDS codes and their application to RAID type of architectures. *IEEE Trans. Inform. Theory*, **59**(7), 4510–4519.

[4] Blaum, M., Plank, J.S., Schwartz, M., and Yaakobi, E. 2016. Construction of partial MDS and sector-disk codes with two global parity symbols. *IEEE Trans. Inform. Theory*, **62**(5), 2673–2681.

[5] Cadambe, V., and Mazumdar, A. 2013. An upper bound on the size of locally recoverable codes. Pages 1–5 of: *2013 International Symposium on Network Coding (NetCod)*.

[6] Cadambe, V.R., and Mazumdar, A. 2015. Bounds on the size of locally recoverable codes. *IEEE Trans. Inform. Theory*, **61**(11), 5787–5794.

[7] Cai, H., and Schwartz, M. 2021. On optimal locally repairable codes and generalized sector-disk codes. *IEEE Trans. Inform. Theory*, **67**(2), 686–704.

[8] Calis, G., and Koyluoglu, O.O. 2017. A general construction for PMDS codes. *IEEE Commun. Letters*, **21**(3), 452–455.

[9] Chen, B., Xia, S.-T., and Hao, J. 2017. Locally repairable codes with multiple (r_i, δ_i)-localities. Pages 2038–2042 of: *2017 IEEE International Symposium on Information Theory (ISIT)*.

[10] Chen, B., Xia, S.-T., Hao, J., and Fu, F.-W. 2018. Constructions of optimal cyclic (r, δ) locally repairable codes. *IEEE Trans. Inform. Theory*, **64**(4), 2499–2511.

[11] Chen, J., Shum, K.W., Yu, Q., and Sung, Chi W. 2015. Sector-disk codes and partial MDS codes with up to three global parities. Pages 1876–1880 of: *2015 IEEE International Symposium on Information Theory (ISIT)*.

[12] Chen, M., Huang, C., and Li, J. 2007. On the maximally recoverable property for multi-protection group codes. Pages 486–490 of: *2007 IEEE International Symposium on Information Theory*.

[13] Gopalan, P., Huang, C., Simitci, H., and Yekhanin, S. 2012. On the locality of codeword symbols. *IEEE Trans. Inform. Theory*, **58**(11), 6925–6934.

[14] Gopalan, P., Huang, C., Jenkins, R., and Yekhanin, S. 2014. Explicit maximally recoverable codes with locality. *IEEE Trans. Inform. Theory*, **60**(9), 5245–5256.

[15] Holzbauer, L., Puchinger, S., Yaakobi, E., and Wachter-Zeh, A. 2020. Partial MDS codes with local regeneration. *CoRR*, abs/2001.04711.

[16] Horlemann-Trautmann, A.-L., and Alessandro, N. 2017. A complete classification of partial-MDS (maximally recoverable) codes with one global parity. *CoRR*, abs/1707.00847.

[17] Hu, G., and Yekhanin, S. 2016. New constructions of SD and MR codes over small finite fields. Pages 1591–1595 of: *2016 IEEE International Symposium on Information Theory (ISIT)*.

[18] Huang, C., Chen, M., and Li, J. 2007. Pyramid codes: flexible schemes to trade space for access efficiency in reliable data storage systems. Pages 79–86 of: *Sixth IEEE International Symposium on Network Computing and Applications (NCA 2007)*.

[19] Huang, C., Chen, M., and Li, J. 2013. Pyramid codes: flexible schemes to trade space for access efficiency in reliable data storage systems. *ACM Trans. Storage*, **9**(1).

[20] Kamath, G.M., Prakash, N., Lalitha, V., and Kumar, P.V. 2014. Codes with local regeneration and erasure correction. *IEEE Trans. Inform. Theory*, **60**(8), 4637–4660.

[21] Li, J., and Li, B. 2017. Beehive: erasure codes for fixing multiple failures in distributed storage systems. *IEEE Trans. Parallel Distribut. Syst.*, **28**(5), 1257–1270.

[22] Lidl, R., and Niederreiter, H. 1997. *Finite fields*, 2nd ed. Encyclopedia of Mathematics and Its Applications, vol. 20. Cambridge University Press, Cambridge.

[23] Martínez-Peñas, U., and Kschischang, F.R. 2019. Universal and dynamic locally repairable codes with maximal recoverability via sum-rank codes. *IEEE Trans. Inform. Theory*, **65**(12), 7790–7805.

[24] Pamies-Juarez, L., Hollmann, H.D.L., and Oggier, F. 2013. Locally repairable codes with multiple repair alternatives. *CoRR*, abs/1302.5518.

[25] Papailiopoulos, D.S., and Dimakis, A.G. 2014. Locally repairable codes. *IEEE Trans. Inform. Theory*, **60**(10), 5843–5855.

[26] Papailiopoulos, D.S., Dimakis, A.G., and Cadambe, V.R. 2013. Repair optimal erasure codes through Hadamard designs. *IEEE Trans. Inform. Theory*, **59**(5), 3021–3037.

[27] Plank, J.S., Blaum, M., and Hafner, J.L. 2013. SD codes: erasure codes designed for how storage systems really fail. Pages 95–104 of: *11th USENIX Conference on File and Storage Technologies (FAST 13)*. USENIX Association, San Jose, CA.

[28] Prakash, N., Kamath, G.M., Lalitha, V., and Kumar, P.V. 2012. Optimal linear codes with a local-error-correction property. Pages 2776–2780 of: *2012 IEEE International Symposium on Information Theory Proceedings*.

[29] Ravagnani, A. 2016. Rank-metric codes and their duality theory. *Des. Codes Cryptogr.*, **80**(1), 197–216.

[30] Song, W., Dau, S.H., Yuen, C., and Li, T.J. 2014. Optimal locally repairable linear codes. *IEEE J. Select. Areas Commun.*, **32**(5), 1019–1036.

[31] Tamo, I., and Barg, A. 2014. A family of optimal locally recoverable codes. *IEEE Trans. Inform. Theory*, **60**(8), 4661–4676.

[32] Tamo, I., Wang, Z., and Bruck, J. 2013. Zigzag codes: MDS array codes with optimal rebuilding. *IEEE Trans. Inform. Theory*, **59**(3), 1597–1616.

[33] Tamo, I., Papailiopoulos, D.S., and Dimakis, A.G. 2016. Optimal locally repairable codes and connections to matroid theory. *IEEE Trans. Inform. Theory*, **62**(12), 6661–6671.

8
Locally Decodable Codes

Chapter 6 considered codes designed to be efficient in repairing a failed node or server in a distributed storage system by either minimizing the amount of information that must be downloaded from adjacent nodes to allow exact reconstruction of the missing data, the minimum repair bandwidth situation, or minimizing the amount of information that must be stored, the minimum storage situation. The notion of a locally repairable code was considered in the previous chapter as well as the notion of maximum recoverability. These are related to erasures of information. As seen in the previous chapter, repairing erasures is often related to solving matrix equations.

This chapter introduces the notion of *locally decodable codes* (LDCs) which allow the decoding of a few bits of a codeword that may be corrupted (with errors) by accessing a few coordinate positions of the codeword without decoding the entire codeword. The corrupted positions are thus unknown and any given position may contain an incorrect symbol. Thus with an LDC, if a database is encoded using very long codewords for efficiency, the retrieval of a single information symbol would be possible without the more extensive procedure of decoding the entire codeword. If the symbol of interest is in error the decoding technique of LDCs may allow its correction under certain conditions. The price to be paid for such a property is that the codes constructed are longer and less efficient. Such systems have a natural association with *private information retrieval* (PIR) schemes (considered in Chapter 9) which allow the retrieval of information from an array of servers without revealing to the servers which bits are of interest. The relationships between LDCs and PIRs will be discussed later in the chapter.

Efficient and practical schemes for local decoding are considered in this chapter. The next section discusses the basic properties of LDCs as well as their relation to a variety of allied concepts such as locally correctable codes (LCCs) and locally testable codes (LTCs). Section 8.2 describes several coding/decoding schemes for the local decoding setting – a rather extensive section. The final Section 8.3 considers the interesting and seminal work of Yekhanin on the construction of 3-query LDCs and related work, results which have had a significant impact on both LDCs and PIR schemes.

8.1 Locally Decodable Codes

Efficient codes in terms of error-correcting capability can have long block lengths. Thus it might be advantageous if a small number of information symbols of interest could be computed from a few codeword symbols extracted from the possibly corrupted codeword without having to decode the entire word.

In this work, it will be assumed that n information symbols over an alphabet Σ are encoded into codewords of length m over an alphabet Γ:

$$\begin{aligned} C\colon \Sigma^n &\longrightarrow \Gamma^m \\ x &\longmapsto \boldsymbol{c}(x). \end{aligned}$$

Although at odds with usual coding notation, this notation has become standard in the literature and will be used here.

Informally, a (q,δ,ϵ) LDC is one that makes at most q queries to positions of the codeword and computes the value of the i-th information symbol from a corrupted version of the codeword with the probability ϵ of the retrieved symbol being correct, provided that the fraction of errors in the word does not exceed δ. Thus the overworked symbol δ, used for various purposes in other chapters, including fraction of erasures allowed, represents the largest fractional number of errors to be allowed in a codeword for this chapter. Similarly, it is common in the LDC literature to use q for the number of queries and thus q is serving double duty as the finite field size when necessary. It will be clear from the context which it refers to.

Often the alphabets will be finite fields. As before, denote for any positive integer n, the set $[n] = \{1,2,\ldots,n\}$, and the Hamming distance on a vector space over $\mathbb{F}_q$ by $d(\boldsymbol{x},\boldsymbol{y})$ or, for normalized distance $\Delta(\boldsymbol{x},\boldsymbol{y}) = d(\boldsymbol{x},\boldsymbol{y})/m$ for a code of length m. It is emphasized the symbol δ in this chapter is reserved for the fraction of errors in a codeword.

Definition 8.1 ([42]) A code $C\colon \Sigma^n \longrightarrow \Gamma^m$ is said to be a (q,δ,ϵ) LDC if there exists a randomized decoding algorithm $\mathcal{A}$ which produces an estimate of an information symbol of the codeword such that

(i) For all information sequences $\boldsymbol{x} = (x_1, x_2, \ldots, x_n) \in \mathbb{F}_q^n$, with corresponding codeword $\boldsymbol{c}(\boldsymbol{x}) \in \mathbb{F}_q^m$, and all vectors $\boldsymbol{y} \in \mathbb{F}_q^m$ such that $d(\boldsymbol{c}(\boldsymbol{x}), \boldsymbol{y}) \leq \delta m$

$$\Pr[\mathcal{A}^{y}(i) = x_i] \geq 1 - \epsilon,$$

where the probability is taken over the random coin tosses of the algorithm $\mathcal{A}$.

(ii) $\mathcal{A}$ makes at most q queries to the word $\boldsymbol{y}$.

Definition 8.2 The (q, δ, ϵ) LDC code C is called *smooth* if the queries to locations of the database (codeword) $\boldsymbol{x}$ are probabilistically uniform.

It is noted there is also the notion of a smooth (q, c, ϵ) LDC/decoding algorithm [17] where at most q queries are made and the probability any given coordinate position of the codeword is queried is at most c/m and the probability a position can be recovered is $\geq 1/\left|\Sigma\right| + \epsilon$.

A closely related concept is the following.

Definition 8.3 ([21, 42]) A code $C \colon \Sigma^n \longrightarrow \Gamma^m$ is said to be a (q, δ, ϵ) LCC if there exists a randomized decoding algorithm $\mathcal{A}$ which produces an estimate of a coded symbol such that

(i) For all codewords $\boldsymbol{c} = (c_1, c_2, \ldots, c_m) \in C \subset \Sigma^m$ and all vectors $\boldsymbol{y} \in \Gamma^m$ such that $d(\boldsymbol{c}(x), \boldsymbol{y}) \leq \delta m$

$$\Pr[\mathcal{A}^{y}(i) = c_i] \geq 1 - \epsilon,$$

where the probability is taken over the random coin tosses of the algorithm $\mathcal{A}$.

(ii) $\mathcal{A}$ makes at most q queries to the word $\boldsymbol{y}$.

Note the difference in the two definitions is the estimation of an information symbol (LDC) versus a codeword symbol (LCC). There are slight differences in the definitions of these concepts by various authors. The following is noted ([21], claim 2.5) that a (q, δ, ϵ) LCC is a (q, δ, ϵ) LDC and that a code being an LDC does not imply it is an LCC.

The two results below follow easily ([17], theorem 2.7, [19], theorem 1):

Theorem 8.4 *Let C be a (q, δ, ϵ) LCC. Then C is also a $(q, \delta q, \epsilon)$ LCC code.*

Theorem 8.5 ([21]) *If $C \subseteq \mathbb{F}^m$ is a (q, δ, ϵ) LCC, then C is a (q, δ, ϵ) LDC.*

The notion of an LTC is similar in that in such a code it is of interest to query q coordinates of a word to determine probabilistically if it is in a given code. More formally [25]:

Definition 8.6 A code C is called an LTC with q queries if there is a randomized algorithm $\mathcal{T}$ which, given oracle access to a word $\mathbf{y} \in \Gamma^m$, makes at most q queries to $\mathbf{y}$ and has the following behavior:

(i) If $y \in C$, then $Pr(\mathcal{T}\text{accepts}) = 1$.
(ii) $Pr(\mathcal{T}\text{rejects}) \geq \frac{1}{4}\Delta(\mathbf{y}, C)$ (normalized distance).

The notion of local testability is intimately connected with probabilistically checkable proof (PCP) systems, which is a proof that allows the correctness of a claim with some probability by looking at a constant number of symbols of the proof. This is an important connection for certain aspects of computer science and there is a large literature on the connection.

In general the notion of an LTC is quite different from that of an LDC [21] although certain codes (such as the Hadamard code described in the next section) are in both categories. That said, it appears that sparse codes (polynomial number of codewords in code length) and random linear codes [21] are both LDC and LTC. The class of codes investigated there are codes of large length m and minimum distance $m/2 - \Theta(\sqrt{m})$. This includes a class referred to as dual-BCH (m,t) where the codes for $t = 1$ correspond to the Hadamard codes described in the next section. An interesting result ([20], theorem 1) is that a linear code C with distance at least $m/2 - \sqrt{tm}$ is LTC using $O(t/\epsilon)$ queries. In [20] such codes are referred to as *almost orthogonal* binary codes.

A notion related to LDCs is that of *locally updatable* codes (which are also LDC) is introduced in [4]. The notion asks to construct codes for which changing a few information bits in a codeword results in changing only a few bits of the corresponding codeword. In a sense the two notions of local updatability and local decodability are in opposition since one often requires a large distance for codes – hence changing a single bit will change many codeword bits – yet such codes are feasible.

An interesting related problem is the *polynomial identity testing* (PIT) problem where one is given a circuit computing a multivariable polynomial over a field and one is to determine if the polynomial is identically zero. Its relation to the LDC problem is considered in [6].

The first notion of an LDC appears in [1] and [34] and was first formally defined in [19]. As noted, much of the work on LDCs arises from work on PIR schemes.

An important aspect of LDCs is the trade-off between the local correctability properties of the code, the query length and the code length. For a given number of queries and fraction of errors allowed, it is important to minimize the length of a code that achieves the given parameters. For the remainder of the section bounds on the lengths achievable for LDCs with a given set of parameters are discussed.

From [19] it is shown there is an upper bound on the length of the data sequence n for which a $(1, \delta, \epsilon)$ LDC code can exist that is independent of the codeword length. An argument is given showing that from this observation, no such $(1, \delta, \epsilon)$ LDC codes can exist.

In general the length of the codes needed for a (q, δ, ϵ) LDC is of interest. Lower bounds on the length m of codewords for given data length n have been derived with particular interest in 2 and 3 query codes including such works as [19, 24, 36] and [29] where it is shown that for C a binary linear $(2, \delta, \epsilon)$ LDC we must have[1]

$$m \geq 2^{\Omega\left(\frac{\delta n}{1-2\epsilon}\right)}.$$

Also [12] for linear binary codes (i.e., $C\colon \{0,1\}^n \longrightarrow \{0,1\}^m$) if C is a linear binary $(2, \delta, \epsilon)$ LDC and $n \geq 8/\epsilon\delta,$ then

$$m \geq 2^{(\epsilon\delta n/4)}.$$

This is improved in ([6], theorem 1.2) to

$$m \geq 2^{(\epsilon\delta n)/4-1}.$$

a bound which is also expressed as $m = 2^{O(n)}$.

The 3-query case has attracted considerable attention. The best lower bound for linear codes for this case, for many years, was

$$m = \Omega\big(n^2/\log\log n\big).$$

However, somewhat dramatically it was shown in [41] that 3-query LDCs exist, assuming there are an infinite number of Mersenne primes, and that for a Mersenne prime of the form $p = 2^t - 1$, the code length satisfies

$$m = \exp\left(n^{1/t}\right)$$

and since the largest known such prime has $t = 32{,}582{,}657 > 10^7$ there exists a 3-query LDC for every n of length

$$m = \exp\left(n^{10^{-7}}\right)$$

and more generally of length

$$m = \exp\left(n^{1/\log\log n}\right),$$

[1] Recall that $f(n) = \Omega(g(n))$ if, for $k > 0, \exists n_0 \; st \; \forall n > n_0, \; |f(n)| \geq k(g(n)$.

a very significant improvement over the above bound. This 3-query case is discussed in Section 8.3.

It is interesting to recall that a polynomial $f(x) \in \mathbb{F}_q[x]$ of degree $d \leq q/2$ can be recovered from its values at $2d$ points even if a small fraction of its values are corrupted (in error). It is shown [33] that a k-sparse polynomial of degree $d \leq q/2$ can be recovered from its values at $O(k)$ randomly chosen points even if a small fraction of them are corrupted. The result can be shown from the discussion in Chapter 13.

Section 8.2 considers a few of the techniques to encode and decode LDCs. In keeping with the aim of this volume, this is a rather extensive section, hopefully of interest to readers.

The final Section 8.3 is on the class of 3-query codes of Yekhanin [41] which represents a breakthrough in the construction of such codes in terms of the short lengths they achieve compared to prior constructions. The work of Efremenko [9] and Gopalan [13] and in particular Raghavendra [32] on the problem is also noted.

Many of the results of the chapter assume linear codes but this will be stated explicitly for the results.

8.2 Coding for Local Decodability

The interest in constructing LDCs is to have the code rate high, the number of queries low and the probability of correct decoding high. Of course the parameters are conflicting and the trade-off is of interest.

Several classes of interesting LDCs have been investigated in the literature, including Hadamard codes, generalized Reed–Muller (GRM) codes, multiplicity codes and matching vector codes. The construction and local decoding of them is briefly discussed. The reference [44] is a concise reference for such topics.

Hadamard Code

Hadamard matrices arise out of the following ([3], Hadamard's theorem 16.6.1): If $A = (a_{ij})$ over $\mathbb{R}$ is such that $|a_{ij}| \leq 1$ for all i, j, then $|\det(\boldsymbol{A})| \leq n^{n/2}$ and equality is achieved iff $a_{ij} = \pm 1$ for all i, j and $\boldsymbol{A}\boldsymbol{A}^t = nI_n$. Such a ± 1 matrix is called a *Hadamard matrix*.

Suppose A is an $n \times n$ Hadamard matrix. Any column can be multiplied by ± 1 without changing it being a Hadamard matrix and hence without loss of generality it can be assumed the first row of the matrix contains all $+1$'s.

The signs in any given column in rows 2 and 3 are either $\{+,+\},\{+,-\},\{-,+\}$ or $\{-,-\}$. If a,b,c,d denote the number of columns containing these combinations, by the orthogonality assumption between pairs of the first three rows, it can be shown that $a = b = c = d = n/4$ and hence $4 \mid n$ for the order of any Hadamard matrix.

It is long conjectured that Hadamard matrices exist for all orders divisible by 4. Denote by $\boldsymbol{H}_n$ a $2^n \times 2^n$ Hadamard matrix. The following construction can be used to give Hadamard matrices of order a power of 2:

$$\boldsymbol{H}_1 = \begin{bmatrix} 1 & 1 \\ 1 & -1 \end{bmatrix}$$

$$\boldsymbol{H}_n = \boldsymbol{H}_2 \otimes \boldsymbol{H}_{n-1},$$

and construct recursively the $2^n \times 2^n$ Hadamard matrices, where $\otimes$ denotes tensor product. This construction yields a sparse sequence of such matrices. Numerous other constructions, however, give other matrices whose orders are divisible by 4. The smallest order for which a Hadamard matrix of order n has not been shown to exist currently appears to be $n = 668$.

Another construction of a Hadamard matrix $\boldsymbol{H}'_n$ of order 2^n that will be convenient for the discussion on Hadamard codes is as follows. Label the rows and columns of $\boldsymbol{H}'_n$ with all 2^n binary (0,1) n-tuples and let

$$\boldsymbol{H}'_n = (h'_{\boldsymbol{u},\boldsymbol{v}}),\ \boldsymbol{u},\boldsymbol{v} \in \mathbb{F}_2^n, \quad h'_{\boldsymbol{u},\boldsymbol{v}} = (-1)^{(\boldsymbol{u},\boldsymbol{v})}$$

and let $\boldsymbol{h}'_{\boldsymbol{u}}$ be the row corresponding to the label $\boldsymbol{u}$. Clearly each row and column contains an equal number of $+1$s and -1s and any two distinct columns are orthogonal as are any two distinct columns and $\boldsymbol{H}'_n$ is a Hadamard matrix of order 2^n. A related matrix is

$$\boldsymbol{H}_n = (h_{\boldsymbol{u},\boldsymbol{v}}), \quad h_{\boldsymbol{u},\boldsymbol{v}} = (\boldsymbol{u},\boldsymbol{v}) \in \{0,1\},\ \boldsymbol{u},\boldsymbol{v} \in \mathbb{F}_2^n \quad \text{(inner product)}.$$

Clearly each row and column of this matrix has the same number of zeros as ones and any two distinct rows or columns have a Hamming distance of 2^{n-1}.

The two matrices are related by

$$\begin{array}{rcl} \boldsymbol{H}'_n & \longrightarrow & \boldsymbol{H}_n \\ (-1)^{(\boldsymbol{u},\boldsymbol{v})} & \mapsto & (\boldsymbol{u},\boldsymbol{v}) \\ -1 & \mapsto & 1 \\ +1 & \mapsto & 0. \end{array}$$

Encode a binary $\{0,1\}$ information n-tuple $\boldsymbol{u} \in \mathbb{F}_2^n$ to be the 2^n-tuple $\boldsymbol{h}_{\boldsymbol{u}}$, the row of $\boldsymbol{H}_n$ corresponding to $\boldsymbol{u}$.

Let C be the set of codewords which are the rows of $\boldsymbol{H}_n$. Suppose the binary 2^n-tuple codeword $\boldsymbol{h}_m$ of the information n-tuple $\boldsymbol{m} \in \mathbb{F}_2^n$ is transmitted and the word $\boldsymbol{r} \in \mathbb{F}_2^{2^n}$ is received with at most $\delta 2^n$ errors. To show the code is an LDC suppose the bit in the coordinate position $\boldsymbol{w} \in \mathbb{F}_2^n$ of $\boldsymbol{h}_{\boldsymbol{m}}$, i.e., $h_{\boldsymbol{m},\boldsymbol{w}}$ is sought. Choose at random $\boldsymbol{v} \in \mathbb{F}_2^n$ and query positions $\boldsymbol{w}$ and $\boldsymbol{v} \oplus \boldsymbol{w}$ of the received word $\boldsymbol{r}$, i.e., $r_{\boldsymbol{v}}$ and $r_{\boldsymbol{v}\oplus\boldsymbol{w}}$. These two received word positions are uncorrupted with probability $(1-\delta)^2 \approx 2\delta$. In this case

$$r_{\boldsymbol{v}} \oplus r_{\boldsymbol{v}\oplus\boldsymbol{w}} = h_{\boldsymbol{m},\boldsymbol{v}} \oplus h_{\boldsymbol{m},\boldsymbol{v}\oplus\boldsymbol{w}} = (\boldsymbol{m},\boldsymbol{v}) \oplus (\boldsymbol{m},\boldsymbol{v}\oplus\boldsymbol{w}) = (\boldsymbol{m},\boldsymbol{w}) = h_{\boldsymbol{m},\boldsymbol{w}}.$$

By choosing $\boldsymbol{m}$ appropriately one can retrieve information bits and it is seen the Hadamard code is a $(2,\delta,1-2\delta)$ LDC.

The Hadamard code is also a 3-query LTC [21]. To test whether the received word $\boldsymbol{r} \in \mathbb{F}_2^{2^n}$ is a codeword, query the word in coordinate positions $\boldsymbol{u},\boldsymbol{v} \in \mathbb{F}_2^n$ and $\boldsymbol{u} \oplus \boldsymbol{v}$ and accept it as a codeword if

$$r_{\boldsymbol{u}} \oplus r_{\boldsymbol{v}} = r_{\boldsymbol{u}\oplus\boldsymbol{v}}.$$

The analysis of this test is more complicated as various errors can result in acceptance and is not considered here. While the simplicity of the decoding procedure is attractive, the low rate of such a code is a severe limitation.

Generalized Reed–Muller Codes

It is claimed the class of GRM codes $GRM_q(d,m)$ (d the degree of polynomial used, not minimum distance) can be used as a locally correctable $(d+1,\delta,(d+1)\delta)$ code, i.e., as a code for which a coordinate position in position $\boldsymbol{\alpha} \in \mathbb{F}_q^m$ can be decoded by making at most $(d+1)$ queries to other codeword positions with a probability of being correct of at least $1-(d+1)\delta$. Recall from Chapter 1, such codes have length q^m, the coordinate positions labeled with $\boldsymbol{a} \in \mathbb{F}_q^m$ and using m-variate monomials of degree at most d to generate rows of the generator matrix yields an $(n,k,d_{min})_q$ codes where

$$n = q^m,\ k = \binom{m+d}{d} \quad \text{and} \quad d_{min} = q^{m-1}(1-d/q),\ d < q-1.$$

To see how local correctability is achieved with such codes, consider a codeword $\boldsymbol{c}_f = (f(\boldsymbol{u}_1), f(\boldsymbol{u}_2), \ldots, f(\boldsymbol{u}_{q^m}))$ (for some ordering of the points $\boldsymbol{u}_1,\boldsymbol{u}_2,\ldots,\boldsymbol{u}_{q^m}$ of F_q^m) for some m-variable polynomial f of total degree $\leq d$. It is desired to retrieve the coordinate position represented by the element $\boldsymbol{u}$, i.e., the value $f(\boldsymbol{u}), \boldsymbol{u} \in \mathbb{F}_q^m$ of the transmitted codeword by querying a number of other positions of the received word $\boldsymbol{y} \in \mathbb{F}_q^{q^m}$.

Consider the bivariate case ($m = 2$). Since the code is coordinated by a two-dimensional vector space choose another point $\boldsymbol{v} \in \mathbb{F}_q^2$ and consider those $q-1$ coordinate positions that fall on the line $\{\boldsymbol{u} + \lambda\boldsymbol{v}, \lambda \in \mathbb{F}_q^*\}$. Choose a subset S of these $(q-1)$ points, $|S| = d+1$, i.e., $(d+1)$ distinct points $\boldsymbol{u} + \lambda_i \boldsymbol{v}, i = 1,2,\ldots,d+1$ on the line and let $z_i = f(\boldsymbol{u} + \lambda_i \boldsymbol{v})$. Determine the univariate polynomial $h(x)$ of degree d over $\mathbb{F}_q$ that interpolates these $(d+1)$ points in that $h(\lambda_i) = z_i$, $i = 1,2,\ldots,d+1$. Thus h is a univariate polynomial of degree at most d whose values are known at $d+1$ elements and hence can be evaluated. It follows that $h(0) = f(\boldsymbol{u})$. This value $h(0)$ will be the correct value if the values $z_i, i = 1,2,\ldots,d+1$ are uncorrupted, i.e., are the correct values, and this happens with probability $(1-\delta)^{d+1} \approx 1-(d+1)\delta$. It easily follows that the GRM code $GRM_q(d,m)$ is a $((d+1),\delta,(d+1)\delta)$ LCC. (Recall that δ is the maximum fraction of received code symbols that are corrupted.) The decoding of the multiplicity codes of the next subsection will involve a similar process.

That GRM codes are LTC codes is established in a similar manner (see, e.g., [25]).

The construction of the *multiplicity codes* to be considered next, as with certain of the codes treated in Chapter 13, depends on the theory of multiplicities of zeros of multivariable functions over a finite field. For the properties needed the notion of the *Hasse derivatives* is useful and a brief account of these is given in Appendix B.

Multiplicity Codes

Multiplicity codes, proposed by Kopparty et al. [26, 27], can be viewed as an interesting generalization of the $GRM_q(d,m)$ codes. While somewhat technical to describe, they have excellent properties and local decoding algorithms reminiscent of those for GRM codes. The original papers of Kopparty et al. [26, 27] develop their basic properties and local decoding algorithms. Chapter 3 of [42] contains proofs of some of the background material.

The bivariate case will give an easy introduction to these codes [26]. Results for the more general m-variate case will then be stated. The code will be indexed by elements of $\mathbb{F}_q^2$ – hence this bivariate code is of length q^2 – and let $f(x_1,x_2)$ be a bivariate polynomial of total degree at most $d = 2q(1-\delta)$ for some fixed δ (error fraction). The codeword corresponding to $f \in \mathbb{F}_q[x]$ will be

$$\boldsymbol{c}_f(\boldsymbol{a}) = \left(f(\boldsymbol{a}), \frac{\partial f(\boldsymbol{a})}{\partial x}, \frac{\partial f(\boldsymbol{a})}{\partial y} \right), \ \boldsymbol{a} \in \mathbb{F}_q^2, \text{ and } \boldsymbol{c}_f = (\boldsymbol{c}_f(\boldsymbol{a}_1), \ldots, \boldsymbol{c}_f(\boldsymbol{a}_{q^m-1}))$$

and each coordinate position of a codeword is a 3-tuple of elements of $\mathbb{F}_q$. (As noted in Appendix B the first partial ordinary and Hasse derivatives are identical.) Denote the codeword coordinate corresponding to $\boldsymbol{a} \in \mathbb{F}_q^2$ as $c_f(\boldsymbol{a})$ for f of degree at most d. The code symbols are now triples over $\mathbb{F}_q$ corresponding to the function f and its two partial derivatives, and the code length is q^2, corresponding to the elements of $\mathbb{F}_q^2$. The symbol alphabet for this code designated $\Sigma = \mathbb{F}_q^3$.

To compute the Hamming distance of the code, consider a codeword $\boldsymbol{c}_f$, $f \in \mathbb{F}_q{}^{\leq d}[x_1,x_2]$ of degree at most d. What is the minimum weight such a codeword can have? For the codeword $\boldsymbol{c}_f$ to have a zero at position $\boldsymbol{a} \in \mathbb{F}_q^2$ the polynomial f must have a zero of multiplicity 2 (in order to have the polynomial and its two partial derivatives to be zero) and hence the codeword symbol over Σ to be 0. The maximum number of zeros of multiplicity 2 that f can have is governed by Equation B.2 of Appendix B, namely

$$\frac{d}{2q}.$$

Thus the minimum fractional weight of a codeword in this code (hence the minimum fractional distance of the code) is

$$1 - \frac{d}{2q},$$

recalling again that the d here is the total degree of the bivariate polynomials used in the code construction. In the literature the fractional distance quantity is often designated δ. For this chapter δ is reserved for the assumed maximum fraction of corrupted codeword positions of a received word in error under local decoding (and also not to be confused with the erasure probability used in previous chapters for the BEC). This gives the relationship between the minimum fractional distance of the code and the degree of polynomials used. The total number of monomials of degree at most d is $\binom{d+2}{2}$ (Equation 1.11) and the space of such monomials is clearly a vector space of this dimension over $\mathbb{F}_q$, i.e., the number of codewords corresponds to the number of $\mathbb{F}_q$ coefficients of such monomials which is $q^{\binom{d+2}{2}}$. Thus this code has the parameters

$$\left(q^2, \binom{d+2}{2}, d_{\min} \geq 1 - d/2q\right)_{\Sigma}, \quad \text{over } \Sigma = \mathbb{F}_q^3.$$

Since each coordinate position of the code has a symbol over Σ, it can also be viewed as code of dimension $\binom{d+2}{2}/3$ over $\mathbb{F}_q$. Note that these code parameters can be compared to those of the GRM code for $m = 2$, namely $\left(q^2, \binom{d+2}{2}, 1 - d/q\right)_q$.

The local decoding of the multiplicity codes will be considered below. The code can be viewed as a $3 \times q^2$ array over $\mathbb{F}_q$ and has rate

$$\frac{k}{n} = \frac{\binom{d+2}{d}}{3q^2} \approx \frac{1}{6}\left(\frac{d}{q}\right)^2 = 2(1-\Delta)^2/3 \quad (\text{as } d/q = 2(1-\Delta)), \tag{8.1}$$

a rate that approaches 2/3 (as opposed to the maximum rate of 1/2 for the bivariate GRM case considered earlier) where Δ is the fractional minimum distance. Thus the process of requiring codeword polynomials to have zeros of higher multiplicities has led to codes with improved minimum distance and rate at the expense of larger field sizes.

Before considering the decoding techniques for these codes and the fact they are locally decodable, the general construction of multiplicity codes is considered.

As before, let $\boldsymbol{i} = (i_1, i_2, \ldots, i_m)$ be a set of nonnegative integers and its weight be $\mathrm{wt}(\boldsymbol{i}) = \sum_{j=1}^m i_j$. As before, denote by $\boldsymbol{x}$ the set $x_1, x_2, \ldots, x_m$ and $\mathbb{F}_q[x_1, x_2, \ldots, x_m]$ by $\mathbb{F}_q[\boldsymbol{x}]$. For $\boldsymbol{i} = \{i_1, i_2, \ldots, i_m\}$ let $\boldsymbol{x}^{\boldsymbol{i}}$ denote the monomial $\prod_{j=1}^m x_j^{i_j}$ and its total degree by $\mathrm{wt}(\boldsymbol{i})$. For the general case, rather than each codeword symbol being just a polynomial evaluation along with first derivatives as in the bivariate case considered above, higher-order Hasse derivatives are used (see Appendix B). Specifically we have:

Definition 8.7 Let $f(\boldsymbol{x}) \in \mathbb{F}_q[\boldsymbol{x}]$ be an m-variable polynomial over $\mathbb{F}_q$ of degree at most d. Then the s-evaluation of f at a point $\boldsymbol{a} \in \mathbb{F}_q^m$ is the column vector of all Hasse derivatives of order less than s. There are $\binom{m+s-1}{m}$ such derivatives (i.e., positive integer m-tuples $\boldsymbol{i}$ with $\mathrm{wt}(\boldsymbol{i}) < s$) (from Equation 1.11 with $d = (s-1)$) and such a vector will be denoted $f^{[<s]}$. Denote the alphabet for the set of such vectors by $\Sigma = \mathbb{F}_q^{\binom{m+s-1}{m}}$ viewed as a column vector of length $\binom{m+s-1}{m}$ over $\mathbb{F}_q$, the vector of all Hasse derivatives of f of order less than s.

Note that $f^{[<s]}$ denotes a vector of length $\binom{m+s-1}{m}$ over $\mathbb{F}_q$ while $\mathbb{F}_q^{<k}[x]$ denotes polynomials over $\mathbb{F}_q$ of degree less than k.

The parameters of the multiplicity codes are determined by the positive integers m, s and d and the base field $\mathbb{F}_q$ (with d the maximum total degree of polynomials – not code distance). The codes will be of length q^m, coordinate positions indexed by the elements of $\mathbb{F}_q^m$. The information symbols corresponding to a codeword or m-variable polynomial $f \in \mathbb{F}_q[\boldsymbol{x}]$ will serve as the coefficients of the monomials of the polynomial, each such polynomial generating a codeword by evaluation, and for a polynomial of degree at most d there are $\binom{m+d}{m}$ such coefficients/monomials (Equation 1.11). Clearly the set of all such polynomials is a vector space of this dimension over $\mathbb{F}_q$. Furthermore the codewords generated by the set of all such monomials are linearly independent and could be used as rows for the generating matrix for the code.

For a polynomial $f \in \mathbb{F}_q[\boldsymbol{x}]$ of degree at most d the corresponding codeword will be denoted

$$\begin{aligned} \boldsymbol{c}_f &= \left\{ (f^{[<s]}(\boldsymbol{a}), \boldsymbol{a} \in \mathbb{F}_q^m \right\} \\ &= \left\{ (f^{[<s]}(\boldsymbol{a}_1), f^{[<s]}(\boldsymbol{a}_2), \ldots, f^{[<s]}(\boldsymbol{a}_{q^m})), \ \boldsymbol{a}_1, \boldsymbol{a}_2, \ldots, \boldsymbol{a}_{q^m} \in \mathbb{F}_q^m \right\}, \end{aligned}$$

(for some ordering of the elements of $\mathbb{F}_q^m$). The information symbols are taken as the coefficients of the monomials in forming a codeword polynomial. In determining the code rate it is convenient to use the $\mathbb{F}_q$ symbols. There are $\binom{m+d}{m}$ $\mathbb{F}_q$ coefficients available for the monomials and the number of $\mathbb{F}_q$ symbols in a codeword, viewing the codeword as a $\binom{m+s-1}{m} \times q^m$ array of such symbols, gives the code rate as

$$\frac{\binom{m+d}{m}}{\binom{m+s-1}{m} q^m} .$$

The codewords are viewed as having coordinate symbols that are over $\Sigma = q^{\binom{m+s-1}{m}}$, or equivalently as $\binom{m+s-1}{m}$-tuples over $\mathbb{F}_q$, corresponding to the Hasse derivatives of all orders less than s. In computing the code minimum distance as a code over Σ, meaning coordinate positions of two distinct codewords are the same only if all Hasse derivatives have the same value. The normalized minimum distance of the code is obtained as for the previous case, namely using the result of Equation B.1 of Appendix B, the maximum fraction of zeros in a codeword is at most d/sq and hence the normalized minimum weight of a nonzero codeword (codeword polynomial), the normalized distance of the code is at least $1 - d/sq$.

The properties of the code are summarized in the following:

Lemma 8.8 ([26], lemma 9, [42], lemma 3.7) *Let $C_{m,d,s,q}$ be the multiplicity code of m-variable polynomials of degree at most d and order s-evaluations over $\mathbb{F}_q$. Then the code over has:*

$$\textit{length} = q^m, \quad \textit{rate} = \frac{\binom{m+d}{m}}{\binom{m+s-1}{m} q^m}, \quad \textit{normalized distance} = 1 - \frac{d}{sq}.$$

The code rate can be lower bounded as follows:

$$\begin{aligned}\frac{\binom{m+d}{m}}{\binom{m+s-1}{m}q^m} &= \frac{\prod_{j=0}^{m-1}(d+m-j)}{\prod_{j=1}^{m}(s+m-j)q} \\ &= \frac{\prod_{j=0}^{m-1}d(1+(m-j)/d)}{\prod_{j=1}^{m}(qs(1+(m-j)/s)} \\ &\geq \left(\frac{1}{1+m/s}\right)^m \left(\frac{d}{qs}\right)^m \\ &\geq \left(1-\tfrac{m^2}{s}\right)\left(\frac{d}{qs}\right)^m .\end{aligned}$$

The approximation used in the last line of the development includes

$$\left(\frac{1}{1+x}\right)^m \geq (1-x)^m \geq (1-mx),\ 0 \leq x \leq 1.$$

Consider now the local decoding of the bivariate multiplicity codes discussed earlier. The code has length q^2, dimension $\binom{d+2}{d}$ and minimum distance $(1-d/2q)$, consisting of all bivariate polynomials of degree at most d and derivatives less than $s=2$. Denote partial derivatives (Hasse derivative is the usual derivative for $s=2$) of $f(x_1,x_2)$ wrt to $x_i, i=1,2$ by $f_{x_i} = \partial f/\partial x_i$. As noted earlier the coordinate position corresponding to $\boldsymbol{a} \in \mathbb{F}_q^2$ in the codeword corresponding to the bivariate polynomial $f(x_1,x_2) \in \mathbb{F}_q[x_1,x_2]$ is

$$c_f(\boldsymbol{a}) = \left(f(a), f_{x_1}(\boldsymbol{a}), f_{x_2}(\boldsymbol{a})\right)^t = \left(f(a), \frac{\partial f(\boldsymbol{a})}{\partial x_1}, \frac{\partial f(\boldsymbol{a})}{\partial x_2}\right)^t = \left(f^{[<2]}(\boldsymbol{a})\right)^t.$$

Consider the line $\mathcal{L} = \{\boldsymbol{a}+\lambda\boldsymbol{b}, \lambda \in \mathbb{F}_q, \boldsymbol{a},\boldsymbol{b} \in \mathbb{F}_q^2\}$ and define the univariate polynomial

$$h(\lambda) = f(\boldsymbol{a}+\lambda\boldsymbol{b}),\ \lambda \in \mathbb{F}_q.$$

An estimate of the codeword coordinate position $\boldsymbol{a} \in \mathbb{F}_q^2$ is desired: $c_f(\boldsymbol{a}) = (f(\boldsymbol{a}), f_{x_1}(\boldsymbol{a}), f_{x_2}(\boldsymbol{a}))^t$ is desired. Notice that ([45])

$$\begin{aligned} h(\lambda) &= f(\boldsymbol{a}+\lambda\boldsymbol{b}), \boldsymbol{a},\boldsymbol{b} \in \mathbb{F}_q^2 \\ \frac{\partial h(\lambda)}{\partial \lambda}(\lambda) &= b_1 f_{x_1}(\boldsymbol{a}+\lambda\boldsymbol{b}) + b_2 f_{x_2}(\boldsymbol{a}+\lambda\boldsymbol{b}) \end{aligned}$$

where $h(\lambda)$ is of degree at most $d = 2q(1-\Delta)$. If we knew the function h, then

$$h(0) = f(\boldsymbol{a}) \quad \text{and} \quad \frac{\partial h(\lambda)}{\partial \lambda}(0) = b_1 f_{x_1}(\boldsymbol{a}) + b_2 f_{x_2}(\boldsymbol{a}).$$

However, the partial derivatives of the function appear as a sum and their individual values are required. To overcome this problem a second "direction vector" $\boldsymbol{b}'$ is chosen and the above process is repeated to give

$$h(0) = f(\boldsymbol{a}) \quad \text{and} \quad \frac{\partial h(\lambda)}{\partial \lambda}(0) = b'_1 f_{x_1}(\boldsymbol{a}) + b'_2 f_{x_2}(\boldsymbol{a})$$

and if the two vectors $\boldsymbol{b}, \boldsymbol{b}' \in \mathbb{F}_q^2$ are chosen not to be colinear with $\boldsymbol{a}$, the equations can be solved to give

$$c_f(\boldsymbol{a}) = (f(\boldsymbol{a}), f_{x_1}(\boldsymbol{a}), f_{x_2}(\boldsymbol{a}))^t.$$

Thus if at most $2q$ queries are made and these $2q$ queries are to uncorrupted places of the codeword, which occurs with probability at least $(1-\delta)^{2q}$, the above procedure yields a $(2q, \delta, 1-(1-\delta)^{2q})$ LDC code. It is noted [26, 27, 45] that these bivariate multiplicity codes for polynomial degree $2q(1-\delta)$ (see Equation 8.1) have rates approximately

$$\frac{2}{3}(1-\Delta)^2,$$

or a maximum rate of 2/3 compared to the GRM codes which had rates only up to 1/2.

The decoding of the general multiplicity code with the parameters of Lemma 8.8 is a generalization of the bivariate case and ordinary partial derivatives ($s = 2$) case to higher-order Hasse derivatives. The procedure is technical but straightforward and requires obtaining values of a sufficient number of direction vectors that solutions for the individual Hasse derivatives (rather than linear sums of them) can be obtained via Gaussian elimination of a matrix equation. To get rates approaching 1 requires increasing both the number of variables and the number of Hasse derivatives considered. The query complexity compared to the higher-dimension space is improved. Local decoding of such codes, while a straightforward generalization of the bivariate case, is quite technical. Good discussions of the technique appear in several places, including [27, 42, 45]. The following results are representative (notation adapted to that used here):

Lemma 8.9 ([27], theorem 3.6) *Let $C_{m,d,s,q}$ be a multiplicity code of order s evaluations of degree d polynomials in m variables over $\mathbb{F}_q$. Let $\Delta = 1 - d/sq$ be the lower bound for the code's normalized minimum distance. Suppose*

$$q > \max\left\{10m, \frac{d+6s}{s}, 12(s+1)\right\}$$

then C is locally correctable from a fraction $\Delta/10$ errors with $(O(s)^m \cdot q)$ queries.

The multiplicity codes will also be used to construct codes for PIR retrieval (see Section 9.2) as well as batch codes (see Section 10.1).

Matching Vector Codes

Matching vector codes [44, 45] share common elements with the GRM code decoding procedure discussed above. For positive integer m, $m \mid (q-1)$ let $C_m = \langle g \rangle \subset \mathbb{F}_q$ be a cyclic multiplicative subgroup of order m of the finite field $\mathbb{F}_q{}^*$ generated by $g = \alpha^{(q-1)/m}$ for $\alpha \in \mathbb{F}_q$ a primitive element. The exponents of g are in $\mathbb{Z}_m$, the integers modulo m, and the following definition describes certain subsets of such exponent vectors:

Definition 8.10 Let $S \subset \mathbb{Z}_m \backslash \{0\}$, $\mid S \mid = s$ and consider two families of subsets of vectors of $\mathbb{Z}_m^n$, $\mathcal{U} = \{\boldsymbol{u}^{(1)}, \ldots, \boldsymbol{u}^{(k)}\}$, $\mathcal{V} = \{\boldsymbol{v}^{(1)}, \ldots, \boldsymbol{v}^{(k)}\}$. The sets $\mathcal{U}$, $\mathcal{V}$ are said to form an *S-matching family* if the following conditions are satisfied:

(i) $(\boldsymbol{u}^{(i)}, \boldsymbol{v}^{(i)}) = 0$, $i = 1, 2, \ldots, k$, the inner product in $\mathbb{Z}_m$
(ii) For $i \neq j$, $i, j \in [k]$, $(\boldsymbol{u}^{(i)}, \boldsymbol{v}^{(j)}) \in S \subseteq \mathbb{Z}_m$.

The set S is to be specified. The definitive works for the construction of such matching sets are [14, 15].

The coordinates of the vectors will only appear in exponents of $g \in \mathbb{F}_q$, an element of order m, $g^m = 1$, for $m \mid (q-1)$.

For a vector $\boldsymbol{w} \in \mathbb{Z}_m^n$, $\boldsymbol{w} = (w_1, w_2, \ldots, w_n)$ define the exponential

$$\boldsymbol{g}^{\boldsymbol{w}} \triangleq \left(g^{w_1}, g^{w_2}, \ldots, g^{w_n}\right) \quad \text{(exponents modulo } m)$$

and the corresponding set on n-tuples over $\mathbb{Z}_m$:

$$\boldsymbol{M}_{\boldsymbol{w},\boldsymbol{v}} = \left\{\boldsymbol{g}^{\boldsymbol{w}+\lambda\boldsymbol{v}} \mid \lambda \in \mathbb{Z}_m\right\}.$$

Notice that $\boldsymbol{g}^{w}$ and the set $\boldsymbol{M}_{\boldsymbol{w},\boldsymbol{v}}$ are in $C_m^n = \langle g \rangle^n \subset \mathbb{F}_q^n$ (i.e., they are n-tuples of $\mathbb{F}_q$ elements). Similarly define the monomial in the variables $z = (z_1, \ldots, z_n)$

$$m^{\boldsymbol{u}}(z) = \prod_{i=1}^{n} z_i^{u_i} \in \mathbb{F}_q[z] = \mathbb{F}_q[z_1, \ldots, z_n]$$

and notice that

$$m^{\boldsymbol{u}}(\boldsymbol{g}^{\boldsymbol{w}+\lambda\boldsymbol{v}}) = \prod_{i=1}^{n} g^{u_i(w_i+\lambda v_i)} = g^{(\boldsymbol{u},\boldsymbol{w})} \cdot (g^{\lambda})^{(\boldsymbol{u},\boldsymbol{v})}.$$

Consider the following univariate polynomial

$$m_{\boldsymbol{u},\boldsymbol{v},\boldsymbol{w}}(y) = g^{(\boldsymbol{u},\boldsymbol{w})} y^{(\boldsymbol{u},\boldsymbol{v})} \in C_m[y] \subset \mathbb{F}_q[y],\ y = g^{\lambda}.$$

To encode the information vector $\boldsymbol{x} \in \mathbb{F}_q^k$ evaluate the polynomial

$$F(z) = F(z_1, \ldots, z_n) = \sum_{j=1}^{k} x_i \cdot m^{\boldsymbol{u}^{(j)}}(z) = \sum_{j=1}^{k} x_i \prod_{\ell=1}^{k} z_\ell^{u_\ell^{(j)}} \tag{8.2}$$

at all m^n points of C_m^n and for the sequences $\boldsymbol{u}^{(j)}$ in the S matching family. Notice the role of only one set $\mathcal{U}$ of the set of matching vectors here. Notice also that while the cyclic group is multiplicatively closed it is not necessarily additively closed and so the evaluations of $F(z)$ are in $\mathbb{F}_q$.

To show the decoding process, assume the information symbol $x_i \in \mathbb{F}_q$ is required from the codeword corresponding to $F(z)$. A random $\boldsymbol{w} \in \mathbb{Z}_m^n$ is chosen and the valuations of $F(z)$ are retrieved at the $s+1$ locations

$$\left\{ \boldsymbol{g}^{\boldsymbol{w}+\lambda \boldsymbol{v}^{(i)}} \right\},\ \lambda \in \{0, 1, 2, \ldots, s\} \subseteq \mathbb{Z}_m$$

(the inner products of the matching sets are in S, $|S| = s$) and suppose

$$F\left(\boldsymbol{g}^{\boldsymbol{w}+\lambda \boldsymbol{v}^{(i)}}\right) = c_\lambda \in \mathbb{F}_q.$$

Form the unique polynomial $h(y)$ of degree s such that $h(g^{\lambda}) = c_\lambda$. It follows that

$$h(y) = \sum_{j=1}^{k} x_j g^{(\boldsymbol{u}^{(j)},\boldsymbol{w})} y^{(\boldsymbol{u}^{(j)},\boldsymbol{v}^{(i)})},\ \boldsymbol{u}^{(i)} = \left\{ u_1^{(i)}, u_2^{(i)}, \ldots, u_n^{(i)} \right\},\ u_j^{(i)} \in \mathbb{Z}_m.$$

In the summation when $j = i$ the exponent of y is $(\boldsymbol{u}^{(i)}, \boldsymbol{v}^{(i)}) = 0$ by construction of the matching sets and hence corresponds to the constant term in the polynomial, i.e.,

$$h(0) = x_i g^{(\boldsymbol{u}^{(i)},\boldsymbol{w})} \quad \text{or} \quad x_i = h(0) \big/ g^{(\boldsymbol{u}^{(i)},\boldsymbol{w})}.$$

If the $(s+1)$ places of the code are error-free, the correct symbol is retrieved by this process. If the error rate of the code symbols is uniformly δ, the probability all places of the code queried are error-free is $(1-\delta)^{s+1}$ which is at least $1-(s+1)\delta$ and the probability the retrieved symbol is in error is at most $(s+1)\delta$. Hence the matching vector codes are $(s+1, \delta, (s+1)\delta)$ LDCs.

These codes are studied further in [5, 7, 8, 10, 23, 32, 33, 40].

Lifted Codes

The interesting notion of a lifted code was introduced in [2] and studied further in [16] and is briefly noted here. From the definition of a Reed–Solomon code (Chapter 1) as

$$\mathscr{C}' = RS(d,q) = \left\{ \boldsymbol{c}_f = (f(a_1), f(a_2), \ldots, f(a_n)), f \in \mathbb{F}_q^{\leq d}[x], \deg(f) \leq d \right\}$$

the evaluations of all polynomials over $\mathbb{F}_q$ of degree at most d at n distinct points of $a_i \in \mathbb{F}_q$. Consider also the GRM code

$$\mathscr{C} = GRM(d,q,m) = \Big\{ (f(\boldsymbol{a}_1), f(\boldsymbol{a}_2), \ldots, f(\boldsymbol{a}_{q^m})), \\ f \in \mathbb{F}_q^{\leq d}[x_1, x_2, \ldots, x_{q^m}],\ \deg f \leq d, \boldsymbol{a}_j \in \mathbb{F}_q^m \Big\},$$

where the $\boldsymbol{a}_i$ are the q^m distinct elements of the vector space of m-tuples over $\mathbb{F}_q$. The affine transformation

$$\boldsymbol{y} = \boldsymbol{A}\boldsymbol{x} + \boldsymbol{b}, \quad \boldsymbol{A} \text{ an } m \times m \text{ nonsingular matrix over } \mathbb{F}_q,\ \boldsymbol{b} \in \mathbb{F}_q^m$$

is viewed as a transformation (a permutation) on the coordinate positions of the code $\mathscr{C}$. If $\boldsymbol{c}_f(\boldsymbol{a})$ is a codeword in $\mathscr{C}$, then $\boldsymbol{c}_f(\boldsymbol{A}\boldsymbol{a} + \boldsymbol{b})$ (the coordinate positions are permuted) is also a codeword of $\mathscr{C}$, the code is referred to as being affine-invariant.

In the vector space of m-tuples over $\mathbb{F}_q$ consider the set

$$\mathcal{L} = \left\{ \boldsymbol{a}t + \boldsymbol{b}, \boldsymbol{a}, \boldsymbol{b} \in \mathbb{F}_q^m, t \text{ a variable with values in } \mathbb{F}_q \right\}.$$

If one restricts the code $\mathscr{C}$ to these q points (to obtain a code of length q by deleting all other coordinates) one obtains a Reed–Solomon code for every such affine transformation. The minimum distance of such a "projected" code is determined by the maximum degree of the polynomial interpolating the points on the line.

It is noted [28] that a limitation of GRM codes is that they have low rate. If one could determine a code $\mathscr{C}_\ell$ by restricting the code to use only m-variate evaluation polynomials (rather than m-variate polynomials of degree at most d) which are low-degree univariate polynomials when restricted to every line of the affine geometry, one might obtain a more efficient code. This is in fact the case: [16, 28]: if $\mathscr{C}_\ell$ is defined as the code resulting by using all polynomials whose restrictions to lines of the geometry are low-degree univariate polynomials, one obtains a code that is much larger than the corresponding GRM code. Such a code $\mathscr{C}_\ell$ is referred to as a lifted Reed–Solomon code.

Such lifted codes find application to codes with disjoint repair groups [28, 30] and in batch codes [18, 31].

8.3 Yekhanin's 3-Query LDCs

In a groundbreaking contribution [40, 41] Yekhanin constructed a class of 3-query LDCs. Previous 3-query LDCs were known with exponential length (in data length) and a lower bound that was approximately the square of the length. The work of Yekhanin lowered these code lengths considerably. The original work uses notions of algebraically nice and combinatorically nice sets which require some effort to describe. It also required the existence of an infinite number of Mersenne primes (long conjectured to be true) to yield 3-query codes of length $\exp(n^{O(1/\log\log n)})$ for data length n. Efremenko [9, 10] modified the construction to give an unconditional proof of 3-query codes of length $\exp\left(\exp\left(O\left(\sqrt{\log n \log\log n}\right)\right)\right)$. Gopalan [13] showed Efremenko's work can be seen from the viewpoint of Reed–Muller codes. The work of Kedlaya and Yekhanin [22, 23] further expands on [41].

Raghavendra [32] captures the essence of the Yekhanin work in a simple and elegant manner (although still requiring Mersenne primes) and this section follows that approach. Interestingly, these works all use the notions of matching vector sets discussed earlier for the matching vector codes with a specific definition/restriction of the set S. To recall the definition, two families of vectors (m-tuples over $\mathbb{F}_p$ for some prime p) $U = \{\boldsymbol{u}^{(1)}, \boldsymbol{u}^{(2)}, \ldots, \boldsymbol{u}^{(n)}\}$, $V = \{\boldsymbol{v}^{(1)}, \boldsymbol{v}^{(2)}, \ldots, \boldsymbol{v}^{(n)}\}$, each vector of length m, $\boldsymbol{u}^{(i)}, \boldsymbol{v}^{(j)} \in \mathbb{F}_p^m$, are said to be matching if:

(i) For all $i \in [n]$, $(\boldsymbol{u}^{(i)}, \boldsymbol{v}^{(i)}) = 0$.
(ii) For all $i, j \in [n]$ such that $i \neq j$, $(\boldsymbol{u}^{(i)}, \boldsymbol{v}^{(j)}) = 2^{r_{ij}} \mod p$ for some integer r_{ij}, i.e., the set S contains powers of two.

The construction uses two finite related fields $\mathbb{F}_p$ and $\mathbb{F}_{2^t}$ and it will be assumed $2^t - 1 = p$ is a prime. Primes of the form $2^t - 1$ are referred to as *Mersenne primes* and as noted, a long-standing conjecture is that there are an infinite number of such primes. The largest currently known such prime appears to be $2^{82,589,933} - 1$ (some 24,862,048 digits – although the search for such primes is an active area and a distributive computational effort known as the Great Internet Mersenne Prime Search (GIMPS) is ongoing). It is noted that if $2^t - 1$ is a (Mersenne) prime, then t is a prime and $t \mid (2^t - 2)$ (e.g., $2^5 - 1 = 31$ is a Mersenne prime and $5 \mid (2^5 - 2)$).

The structure of the two finite fields and the matching vector sets are described. Assume that $p = (2^t - 1)$ is a Mersenne prime and consider the two finite fields $\mathbb{F}_p$ and $\mathbb{F}_{2^t}$. The nonzero elements of $\mathbb{F}_{2^t}^*$ form a cyclic group $\langle g \rangle$ of order $p = (2^t - 1)$ assumed to be generated by the primitive element $g \in \mathbb{F}_{2^t}^*$, $g^p = 1$.

In $\mathbb{F}_p^*$ the integer 2 generates the group $G = \langle 2 \rangle = \{1, 2, 2^2, \ldots, 2^{t-1}\}$ (multiplicative) of order t since by assumption $2^t = 1$ in $\mathbb{F}_p$. Notice that since it will be assumed that $p = (2^t - 1)$ is prime and $\mathbb{F}_p^*$ is a cyclic group, if $\ell = \frac{p-1}{t}$, then the ℓ-th power of any element in $\mathbb{F}_p^*$ is of order t and hence in G, i.e., a power of 2.

It is desired to define matching sets of vectors U and V such that $(\boldsymbol{u}^{(i)}, \boldsymbol{v}^{(i)}) = 0$ and for $i \neq j$, $((\boldsymbol{u}^{(i)}, \boldsymbol{v}^{(j)}) = 2^r$ for some integer r (which may depend on the particular vectors). The reason for this is that for the proof to follow it will be useful to invoke the property of fields $\mathbb{F}_{2^t}$ of characteristic 2 that

$$a^{2^r} + b^{2^r} = (a+b)^{2^r}.$$

To define the matching vector sets, let M be an integer larger than p. The sets will each contain $n = \binom{M}{p-1}$ vectors, the vectors will each be of length $m = M^{(p-1)/t} = M^\ell, \ell = (p-1)/t$ and will be defined over $\mathbb{F}_p$.

It is noted [32] that $n = \binom{M}{p-1} \geq \left(\frac{M}{p}\right)^{p-1}$ and since $m = M^{(p-1)/t}$ then $m = O(n^{1/t})$ and if we choose $M = 2^p$ the code length is $m = n^{O(1/\log\log n)}$, a significant shortening over previous constructions.

Let $\mathbf{1}_M \in \mathbb{F}_p^M$ be the M-tuple of all ones. Consider the set of $n = \binom{M}{p-1}$ binary (0, 1) incidence vectors of length M, of all subsets of the set $[M]$ of size $p-1$, denoted by $\{\boldsymbol{\mu}^{'(i)}, i = 1, 2, \ldots, n\}$ and define a second set by $\boldsymbol{\nu}^{'(i)} = \mathbf{1}_M - \boldsymbol{\mu}^{'(i)}$ (over $\mathbb{F}_p$) – the binary complements. Consider the ordinary tensor product of vectors (of length M^2):

$$\begin{aligned}\mu^{'(i)} \otimes \mu^{'(i)} &= \Big(\mu_1^{'(i)}\Big(\mu_1^{'(i)}, \mu_2^{'(i)}, \ldots, \mu_M^{'(i)}\Big), \mu_2^{'(i)}\Big(\mu_1^{'(i)}, \mu_2^{'(i)}, \ldots, \mu_M^{'(i)}\Big), \ldots \\ &\qquad \ldots, \mu_M^{'(i)}\Big(\mu_1^{'(i)}, \mu_2^{'(i)}, \ldots, \mu_M^{'(i)}\Big)\Big) \\ &\triangleq (\boldsymbol{\mu}^{'(i)})^{\otimes 2}.\end{aligned}$$

The extension to an ℓ-fold tensor – $(\boldsymbol{\mu}^{'(i)})^{\otimes \ell}$ is clear.

Define the sets

$$\begin{aligned} U &= \Big\{\boldsymbol{u}^{(i)} = \Big(\boldsymbol{\mu}^{'(i)}\Big)^{\otimes \ell}, \ i = 1, 2, \ldots, n\Big\} \text{ and} \\ V &= \Big\{\boldsymbol{v}^{(i)} = \Big(\boldsymbol{\nu}^{'(i)}\Big)^{\otimes \ell}, \ i = 1, 2, \ldots, n\Big\},\end{aligned}$$

the ℓ-fold tensor product of the vectors $\boldsymbol{\mu}^{'(i)}$ and $\boldsymbol{\nu}^{'(i)}$, respectively. That the sets U and V are matching follows from the fact that

$$\Big(\boldsymbol{u}^{(i)}, \boldsymbol{v}^{(j)}\Big) = \Big(\Big(\boldsymbol{\mu}^{'(i)}\Big)^{\otimes \ell}, \Big(\boldsymbol{\nu}^{'(j)}\Big)^{\otimes \ell}\Big) = \Big(\boldsymbol{\mu}^{'(i)}, \boldsymbol{\nu}^{'(j)}\Big)^{\ell}$$

and that $(\boldsymbol{\mu}^{(i)}, \boldsymbol{\nu}^{(i)}) = 0$ and that the offset values $(\boldsymbol{\mu}^{(i)}, \boldsymbol{\nu}^{(i)})$, $i \neq j$ are ℓ-th powers of an element of $\mathbb{F}_p^*$ and hence, as an ℓ-th power, is in G, i.e., a power of 2 – which is the reason for the particular parameters chosen. The sets U, V are matching sets of vectors of length $m = M^\ell$ and size $n = |U| = |V| = \binom{M}{p-1}$ over $\mathbb{F}_p$.

To consider the encoding and decoding algorithms for the LDC codes to be defined and their properties, recall $p = (2^t - 1)$ is a prime and g a primitive element of the finite field of characteristic 2, $\mathbb{F}_{2^t}^*$, and note that $g^{2^t-1} = g^p = 1$. Define the integer γ such that

$$1 + g + g^\gamma = 0, \; \gamma < 2^t - 1. \tag{8.3}$$

Let U, V be matching sets (over $\mathbb{F}_p$) derived above and define the functions

$$\begin{aligned} f_i : \mathbb{F}_p^m &\longrightarrow \mathbb{F}_{2^t}^*, \quad i = 1, 2, \ldots, n \\ \boldsymbol{x} &\longmapsto g^{(\boldsymbol{u}^{(i)}, \boldsymbol{x})} \end{aligned}$$

and note the function is a homomorphism in that $f_i(\boldsymbol{x} + \boldsymbol{y}) = f_i(\boldsymbol{x}) f_i(\boldsymbol{y}), \forall \boldsymbol{x}, \boldsymbol{y} \in \mathbb{F}_p^m$.

The following relationship ([32], observation 3.1) follows directly from the definition of γ and the fact that squaring in fields of characteristic 2 is a linear operation and that by definition of matching sets $(\boldsymbol{u}^{(i)}, \boldsymbol{v}^{(j)}) = (1 - \delta_{ij}) 2^r$ for some positive integer r (depending on i and j) . Thus

$$\begin{aligned} & f_i(\boldsymbol{x}) + f_i(\boldsymbol{x} + \boldsymbol{v}^{(j)}) + f_i(\boldsymbol{x} + \gamma \boldsymbol{v}^{(j)}) \\ &= g^{(\boldsymbol{u}^{(i)}, \boldsymbol{x})} + g^{(\boldsymbol{u}^{(i)}, (\boldsymbol{x} + \boldsymbol{v}^{(j)}))} + \boldsymbol{g}^{(\boldsymbol{u}^{(i)}, (\boldsymbol{x} + \gamma \boldsymbol{v}^{(j)})} \\ &= g^{(\boldsymbol{u}^{(i)}, \boldsymbol{x})} \left(1 + g^{(\boldsymbol{u}^{(i)}, \boldsymbol{v}^{(j)})} + g^{\gamma (\boldsymbol{u}^{(i)}, \boldsymbol{v}^{(j)})} \right) \\ &= \begin{cases} g^{(\boldsymbol{u}^{(i)}, \boldsymbol{x})} (1 + g + g^\gamma)^{2^r} = 0, & \text{if } i \neq j \\ g^{(\boldsymbol{u}^{(i)}, \boldsymbol{x})} (1 + 1 + 1) = g^{(\boldsymbol{u}^{(i)}, \boldsymbol{x})}, & \text{if } i = j . \end{cases} \end{aligned} \tag{8.4}$$

With these basic properties the codes, and local decoding algorithms for them, can be formed. The codes will be of length p^m with coordinates indexed by elements of $\mathbb{F}_p^m$ and coordinate values from $\mathbb{F}_{2^t}$ – it will encode information $\boldsymbol{\alpha} = (\alpha_1, \alpha_2, \ldots, \alpha_n) \in \mathbb{F}_{2^t}^n$ by forming the codewords

$$\begin{aligned} \mathscr{C} = \Big\{ \boldsymbol{c}_f = (f(\boldsymbol{a}_1), f(\boldsymbol{a}_2), \ldots, f(\boldsymbol{a}_{p^m})), \; f(x) \in \mathbb{F}_{2^t}[x] \Big\}, \\ f(\boldsymbol{a}) = \sum_{i=1}^{n} \alpha_i f_i(\boldsymbol{a}) \, , \; \boldsymbol{a} \in \mathbb{F}_p^m, \; \alpha_i \in \mathbb{F}_{2^t} \end{aligned}$$

for some ordering of the elements of $\mathbb{F}_p^m$. This is the LDC encoding of the information $\boldsymbol{\alpha} \in \mathbb{F}_{2^t}^m$.

The fact the code is a 3-query code follows directly from its construction. Suppose it is desired to determine the information symbol $\alpha_i \in \mathbb{F}_{2^t}$, recalling the matching sets $U = \{\boldsymbol{u}^{(j)}\}$, $V = \{\boldsymbol{v}^{(k)}\}$. Let $\boldsymbol{b} \in \mathbb{F}_p^m$ be an arbitrary coordinate position of the code. The code is queried in the three positions $\boldsymbol{b}, \boldsymbol{b}+\boldsymbol{v}^{(i)}$, and $\boldsymbol{b}+\gamma\boldsymbol{v}^{(i)}$ to obtain the values $f(\boldsymbol{b})$, $f(\boldsymbol{b}+\boldsymbol{v}^{(i)})$ and $f(\boldsymbol{b}+\gamma\boldsymbol{v}^{(i)})$ and compute the quantity

$$\begin{aligned} &g^{-(\boldsymbol{u}^{(i)},\boldsymbol{b})}\left(c_f(\boldsymbol{b}) + c_f(\boldsymbol{b}+\boldsymbol{v}^{(i)}) + c_f(\boldsymbol{b}+\gamma\boldsymbol{v}^{(i)})\right) \\ &= g^{-(\boldsymbol{u}^{(i)},\boldsymbol{b})}\sum_{j=1}^{n}\alpha_j\left(f_j(\boldsymbol{b}) + f_j(\boldsymbol{b}+\boldsymbol{v}^{(i)}) + f_j(\boldsymbol{b}+\gamma\boldsymbol{v}^{(i)})\right)\end{aligned}$$

is computed and by Equation 8.3 and the last lines of Equation 8.4 it follows that

$$g^{-(\boldsymbol{u}^{(i)},\boldsymbol{b})}\left(c_f(\boldsymbol{b}) + c_f(\boldsymbol{b}+\boldsymbol{v}^{(i)}) + c_f(\boldsymbol{b}+\gamma\boldsymbol{v}^{(i)})\right) = \alpha_i g^{-(\boldsymbol{u}^{(i)},\boldsymbol{b})} g^{(\boldsymbol{u}^{(i)},\boldsymbol{b})} = \alpha_i .$$

If the probability a coordinate queried is incorrect is δ, then the probability all three coordinates queried are correct is $(1-\delta)^3 \approx 1-3\delta$. Hence this system is a $(3,\delta,3\delta)$ LDC system as claimed, with a length complexity noted earlier of $m = \exp(n^{10^{-7}})$. Other LDC parameter sets are obtainable with similar considerations (e.g., [32, 41]).

Comments

The monographs [42, 43] and survey papers [44, 45] make excellent reading for the material of this and the next chapter, as do the research papers, e.g., [37, 38, 39, 40, 41], all the work of Yekhanin.

There is a natural connection with work on LDCs and PIR considered in Chapter 9 to the extent that a construction in one area yields one in the other. A brief outline of this connection is noted. To discuss the connection it is necessary to extend the LDC discussion to the q-ary case. Further comments on the connection can be found in the Comments section of Chapter 9.

It will be noted that an r-server q-ary PIR protocol $\mathcal{P}$ is a triplet of algorithms $\mathcal{Q}, \mathcal{A}, \mathcal{C}$ that, respectively, query, answer and compute information from servers. It is assumed in a PIR (uncoded) scheme that each server contains the same n q-ary symbols $\boldsymbol{x} = \{x_1, x_2, \ldots, x_n\}$ and that the user $\mathcal{U}$ wishes to retrieve a particular symbol x_i by querying the r servers (using r here for the number of queries as q is the field size) and computing the information received back, without revealing to any of the servers the particular symbol of interest. The query algorithm $\mathcal{Q}$ generates a random string $rand$ from which r queries $q_1, q_2, \ldots, q_r$ are generated and transmitted to the respective servers. The servers respond with answers a_j and the computation $\mathcal{C}$, on inputs $a_1, a_2, \ldots, a_r$, computes x_i. For such a PIR protocol the answer should be correct with probability one and that, for the servers to glean no information as to the symbol of interest, the queries should be uniformly distributed for all the servers and for any symbol of interest.

To obtain an r-server PIR protocol from a smooth q-ary r query LDC suppose ([19], [42], lemma 7.2) $\mathscr{C}$ is a q-ary r query LDC that is perfectly smooth (uniformly distributed queries over all codeword symbols) that encodes k symbol-long messages to codewords of length N. Each server contains the codeword corresponding to the original information symbols. The user constructs the queries $q_1, q_2, \ldots, q_r$ (the set of codeword positions the LDC queries) and the j-th server responds with a symbol $C(q_j)$ (a computation of the symbols that query contained). From the responses the user is able to compute the required symbol x_i via the LDC construction process. The privacy of this PIR protocol follows from the assumed uniform distribution of the LDC queries. The correctness of the PIR protocol follows from the assumed properties of the LDC code protocol. Further comments on the PIR/LDC connection are noted in the next chapter.

Only the basic approaches to LDCs and their connection to PIR have been discussed. The literature on these and numerous related topics is extensive and deep. The numerous interesting works on this connection include [11, 12, 17, 29, 35].

References

[1] Babai, L., Fortnow, L., Levin, L.A., and Szegedy, M. 1991. Checking computations in polylogarithmic time. Pages 21–32 of: *Proceedings of the Twenty-Third Annual ACM Symposium on Theory of Computing*. STOC '91. ACM, New York.

[2] Ben-Sasson, E., Maatouk, G., and Shpilka, A. 2011. Symmetric LDPC codes are not necessarily locally testable. *Proceedings of the Annual IEEE Conference on Computational Complexity*, 06.

[3] Cameron, P.J. 1994. *Combinatorics: topics, techniques, algorithms*. Cambridge University Press, Cambridge.

[4] Chandran, N., Kanukurthi, B., and Ostrovsky, R. 2014 (February). Locally updatable and locally decodable codes. Pages 489–514: *Theory of Cryptography Conference*.

[5] Dvir, Z., and Hu, G. 2013. Matching-vector families and LDCs over large modulo. Pages 513–526 of: *Approximation, randomization, and combinatorial optimization*. Lecture Notes in Computer Science, vol. 8096. Springer, Heidelberg.

[6] Dvir, Z., and Shpilka, A. 2006. Locally decodable codes with two queries and polynomial identity testing for depth 3 circuits. *SIAM J. Comput.*, **36**(5), 1404–1434.

[7] Dvir, Z., Gopalan, P., and Yekhanin, S. 2010 (October). Matching vector codes. Pages 705–714 of: *2010 IEEE 51st Annual Symposium on Foundations of Computer Science*.

[8] Dvir, Z., Gopalan, P., and Yekhanin, S. 2011. Matching vector codes. *SIAM J. Comput.*, **40**(4), 1154–1178.

[9] Efremenko, K. 2009. 3-Query locally decodable codes of subexponential length. Pages 39–44 of: *Proceedings of the Forty-First Annual ACM Symposium on Theory of Computing*. STOC '09. ACM, New York.

[10] Efremenko, K. 2012. 3-Query locally decodable codes of subexponential length. *SIAM J. Comput.*, **41**(6), 1694–1703.

[11] Gasarch, W. 2004. A survey on private information retrieval. *Bull. EATCS*, **82**(72–107), 113.

[12] Goldreich, O., Karloff, H., Schulman, L.J., and Trevisan, L. 2006. Lower bounds for linear locally decodable codes and private information retrieval. *Comput. Complex.*, **15**(3), 263–296.

[13] Gopalan, P. 2009. A note on Efremenko's Locally Decodable Codes. *Electron. Colloquium Comput. Complex.*, **16**, 69.

[14] Grolmusz, V. 2000. Superpolynomial size set-systems with restricted intersections mod 6 and explicit Ramsey graphs. *Combinatorica*, **20**(1), 71–85.

[15] Grolmusz, V. 2002. Constructing set systems with prescribed intersection sizes. *J. Algorithms*, **44**(2), 321 – 337.

[16] Guo, A., and Sudan, M. 2012. New affine-invariant codes from lifting. *CoRR*, abs/1208.5413.

[17] Hielscher, E. 2007. *A survey of locally decodable codes and private information retrieval schemes*.

[18] Holzbaur, L., Polyanskaya, R., Polyanskii, N., and Vorobyev, I. 2020. Lifted Reed-Solomon codes with application to batch codes. Pages 634–639 of: *2020 IEEE International Symposium on Information Theory (ISIT)*.

[19] Katz, J., and Trevisan, L. 2000. On the efficiency of local decoding procedures for error-correcting codes. Pages 80–86 of: *Proceedings of the Thirty-Second Annual ACM Symposium on Theory of Computing*. STOC '00. ACM, New York.

[20] Kaufman, T., and Litsyn, S. 2005 (October). Almost orthogonal linear codes are locally testable. Pages 317–326 of: *46th Annual IEEE Symposium on Foundations of Computer Science (FOCS'05)*.

[21] Kaufman, T., and Viderman, M. 2010. *Locally testable vs. locally decodable codes*. Springer, Berlin, Heidelberg.

[22] Kedlaya, K.S., and Yekhanin, S. 2008. Locally decodable codes from nice subsets of finite fields and prime factors of Mersenne numbers. Pages 175–186 of: *Twenty-Third Annual IEEE Conference on Computational Complexity*. IEEE Computer Society, Los Alamitos, CA.

[23] Kedlaya, K.S., and Yekhanin, S. 2008/09. Locally decodable codes from nice subsets of finite fields and prime factors of Mersenne numbers. *SIAM J. Comput.*, **38**(5), 1952–1969.

[24] Kerenidis, I., and de Wolf, R. 2004. Exponential lower bound for 2-query locally decodable codes via a quantum argument. *J. Comput. System Sci.*, **69**(3), 395–420.

[25] Kopparty, S., and Saraf, S. 2017. Local testing and decoding of high-rate error-correcting codes. *Electron. Colloquium Comput. Complex.*, **24**, 126.

[26] Kopparty, S., Saraf, S., and Yekhanin, S. 2011. High-rate codes with sublinear-time decoding. Pages 167–176 of: *Proceedings of the Forty-Third Annual ACM Symposium on Theory of Computing*. STOC '11. ACM, New York.

[27] Kopparty, S., Saraf, S., and Yekhanin, S. 2014. High-rate codes with sublinear-time decoding. *J. ACM*, **61**(5), Art. 28, 20.
[28] Li, R., and Wootters, M. 2020. Lifted multiplicity codes and the disjoint repair group property. *IEEE Trans. Inform. Theory*, 1–1.
[29] Obata, K. 2002. Optimal lower bounds for 2-query locally decodable linear codes. Pages 39–50 of: *Randomization and approximation techniques in computer science*. Lecture Notes in Computer Science, vol. 2483. Springer, Berlin.
[30] Polyanskii, N., and Vorobyev, I. 2019. Constructions of batch codes via finite geometry. Pages 360–364 of: *2019 IEEE International Symposium on Information Theory (ISIT)*.
[31] Polyanskii, N., and Vorobyev, I. 2019 (October). Trivariate lifted codes with disjoint repair groups. Pages 64–68 of: *2019 XVI International Symposium "Problems of Redundancy in Information and Control Systems" (REDUNDANCY)*.
[32] Raghavendra, P. 2007. A note on Yekhanin's locally decodable codes. *Electron. Colloquium Comput. Complex.*, **14**(01).
[33] Saraf, S., and Yekhanin, S. 2011. Noisy interpolation of sparse polynomials, and applications. Pages 86–92 of: *26th Annual IEEE Conference on Computational Complexity*. IEEE Computer Society, Los Alamitos, CA.
[34] Sudan, M. 1995. *Efficient checking of polynomials and proofs and the hardness of approximation problems*. Lecture Notes in Computer Science, vol. 1001. Springer-Verlag, Berlin; Association for Computing Machinery (ACM), New York.
[35] Wehner, S., and de Wolf, R. 2005. *Improved lower bounds for locally decodable codes and private information retrieval*. Springer, Berlin, Heidelberg.
[36] Woodruff, David. 2007. New lower bounds for general locally decodable codes. *Electron. Colloquium Comput. Complex.*, **14**(01).
[37] Woodruff, D., and Yekhanin, S. 2007. A geometric approach to information-theoretic private information retrieval. *SIAM J. Comput.*, **37**(4), 1046–1056.
[38] Yekhanin, S. 2006. New locally decodable codes and private information retrieval schemes. *Electron. Colloquium Comput. Complex.*, **13**.
[39] Yekhanin, S. 2007. *Locally decodable codes and private information retrieval schemes*. PhD thesis, Massachusetts Institute of Technology, Cambridge, MA.
[40] Yekhanin, S. 2007. Towards 3-query locally decodable codes of subexponential length. Pages 266–274 of: *STOC '07: Proceedings of the 39th Annual ACM Symposium on Theory of Computing*. ACM, New York.
[41] Yekhanin, S. 2008. Towards 3-query locally decodable codes of subexponential length. *J. ACM*, **55**(1), Art. 1, 16.
[42] Yekhanin, S. 2010. Locally decodable codes. *Found. Trends Theor. Comput. Sci.*, **6**(3), front matter, 139–255.
[43] Yekhanin, S. 2010. *Locally decodable codes and private information retrieval schemes*, 1st ed. Springer, Berlin, Heidelberg.
[44] Yekhanin, S. 2011. Locally decodable codes: a brief survey. Pages 273–282 of: *Coding and cryptology*. Lecture Notes in Computer Science, vol. 6639. Springer, Heidelberg.
[45] Yekhanin, S. 2014. Codes with local decoding procedures. Pages 683–697 of: *Proceedings of the International Congress of Mathematicians – Seoul 2014. Vol. IV*. Kyung Moon Sa, Seoul.

9
Private Information Retrieval

Chapters 6, 7, 8, 10 and this chapter all deal with some aspect of information storage and retrieval. This chapter introduces the notion of a private information retrieval (PIR) protocol which allows a user to download information from a server, without the server being aware of the particular information of interest. It was noted in Chapter 8 that this notion is intimately connected with that of locally decodable codes (LDCs) in that any smooth LDC yields a PIR scheme. Some additional aspects of this connection are noted later in the chapter.

The elements of PIR protocols are introduced. As with previous chapters, efficient and practical schemes that could achieve these goals are of interest. Current approaches and results on these concepts are given in the next two sections, divided into single-server and multiple-server cases. The multiple-server case often involves replicating the database leading to inefficiency of data storage. The final section studies the recent ideas of [19] for PIR protocols that use coding of the databases to reduce storage requirements.

The basic notions of PIR were first introduced in [14, 15]. In the basic model, a file of n bits $x_1, x_2, \ldots, x_n$ is replicated on k servers (databases)

$$\mathcal{DB}_1,\ \mathcal{DB}_2, \ldots,\ \mathcal{DB}_k.$$

A user $\mathcal{U}$ wants to retrieve a bit, say x_i without the servers being aware which bit the user is interested in. Further, only one-round protocols are of interest in this discussion, where the user sends queries to each of the databases and computes the information of interest from the responses. It is assumed the servers are unable to collude. With only a single server (database), to ensure information-theoretic privacy a trivial solution is for the server to download the entire database and it can be shown [15] that this trivial solution is in fact the only possible one. However, in this case, one can achieve *computational security*, meaning there is no polynomial computation on the user-server information to discover the bit of interest. While seemingly counterintuitive, this can be achieved using public key cryptographic techniques. Such systems are considered in the next section. It is not clear how much current interest there is in this single-server case. Yet the fact it is possible seems worthy of some attention. It requires certain concepts from computational number theory, of which an outline is provided in the next section.

In the multiserver case, it is assumed there are k servers, each holding identical copies of the database. The total amount of communication between the user and the servers is of interest and the servers do not collude. The seminal work of this area is [14] (and the full journal paper [15]). For privacy the protocol must ensure that communication between servers and user is uniformly distributed to divulge no information as to the index i of the bit of interest. Furthermore, for a set $S \subseteq [n] = \{1, 2, \ldots, n\}$ and an index $i \in [n]$ define the notation

$$S \oplus i \triangleq \begin{cases} S \cup \{i\} & \text{if } i \notin S \\ S \backslash \{i\} & \text{if } i \in S. \end{cases} \tag{9.1}$$

This use of the $\oplus$ operator between a set and set element is for the remainder of the chapter while as a binary operator between bits it retains the usual definition. Both meanings are used and it will be clear from the context which is applicable.

For the two-server case, $k = 2$, the following simple example [14] illustrates a technique to achieve download privacy. The servers $\mathcal{DB}_1$ and $\mathcal{DB}_2$ each contain the bitstring $\boldsymbol{x} \in \mathbb{F}_2^n$ of length n. The user $\mathcal{U}$ wishes to retrieve the bit x_i without either server being aware of this fact or chooses a subset S of $[1, n]$ at random (choosing each element of $[1, n]$ for inclusion in the set with probability $1/2$). The set is transmitted to $\mathcal{DB}_1$. The user computes $S \oplus i$ and transmits it to $\mathcal{DB}_2$. The server $\mathcal{DB}_1$ computes $y_1 = \oplus_{j \in S} x_j$ (mod 2 sum) which is transmitted to $\mathcal{U}$ and $\mathcal{DB}_2$ computes $y_2 = \oplus_{j \in S \oplus i} x_j$ which is also transmitted to $\mathcal{U}$. The user computes $y_1 \oplus y_2 = x_i$. The complexity of this protocol is $O(n)$ since each server receives a list of $O(n)$ bits and each server returns one bit to the user. Neither server gleans any information on the bit of interest to the user. Although simple, the protocol is instructive and suggestive for further development.

A more formal description of a PIR protocol is given, similar to that in [5, 14, 15, 39]

Definition 9.1 A k-server PIR protocol includes the following:

(i) k servers $\mathcal{DB}_1, \mathcal{DB}_2, \ldots, \mathcal{DB}_k$, each containing an identical copy of the n bits $\boldsymbol{x} \in \{0, 1\}^n$.
(ii) A user $\mathcal{U}$ requires $x_i \in \boldsymbol{x}$ in a manner such that no information on i is leaked.

The protocol is a triple of algorithms $(Q, \mathcal{A}, C)$, a query algorithm Q that generates queries to send to a subset of the servers, an answer algorithm $\mathcal{A}$ used by the servers to generate answers to respond to user queries and a reconstruction algorithm C used by the user to construct the required bit from the responses received. Thus

(iii) $\mathcal{U}$ generates random bits with a uniform distribution from which queries $q_1, q_2, \ldots, q_k$ are generated, sending q_j to server $\mathcal{DB}_j$.
(iv) Server $\mathcal{DB}_j$ computes a response $a_j = \mathcal{A}(j), j = 1, 2, \ldots, k$ which is transmitted to the user $\mathcal{U}$.
(v) From the responses, the user computes $C(a_1, a_2, \ldots, a_k)$ and with probability 1 over the random string r, this is equal to the required bit x_i.

The protocol has the properties:

(i) (Correctness) for any $\boldsymbol{x} \in \{0,1\}^n$ and $i \in [n]$ the algorithm produces the correct value of the bit x_i with probability 1.
(ii) (Privacy) each server learns no information on the index i – the distributions of the queries are the same for any requested bit.

Interest is restricted to *one-round* protocols where the user determines the bit of interest from a computation involving the responses to their queries. While in practice interest is in small values k (fewer servers and less replication of the data), the general case is also of interest. The *communication complexity* of the protocol is the total number of bits sent in the queries and answers. For a given number of servers, ways of minimizing the total amount of communication are sought.

The next section considers protocols for the single-server case. As has been noted, the single-server protocol involves computational security rather than the stronger information-theoretic security. It is somewhat counterintuitive to imagine retrieving a bit from a server without the server being aware of some information on the bit in question, in the case of a single server. The techniques involve notions drawn from public-key cryptography which are typically computationally expensive. For a reader with no background in computational number theory, the section might seem onerous and since the single server is of limited interest, that section can be skipped. However, the techniques of such protocols are interesting and appear to be of potential value to pursue.

Section 9.2 considers the case of multiple servers and the aim is to minimize the complexity (amount) of communications while preserving information-theoretic security. It is assumed here the servers (databases) do not collude, although some works define t-privacy where up to t servers are allowed to collude. It is also assumed the multiple servers contain identical copies of the information of n bits, $\boldsymbol{x} \in \mathbb{F}_2^n$, and a user wishes to receive a single bit x_i with no server obtaining any information on which bit is of interest.

This multiple-server model typically has the significant disadvantage of having the information replicated across all servers, leading to storage inefficiency. The recent work of Fazeli et al. [19] addresses this problem and considers coding techniques that allow the total storage across all servers to approach n, the original size of the database, while preserving information-theoretic privacy. Section 9.3 discusses this approach.

As noted there is a strong connection between PIR protocols and LDCs and indeed an instance of the latter with smooth queries yields a k-server PIR protocol. This connection was briefly discussed in the previous chapter and will be noted upon further in the Comments section of this chapter.

9.1 The Single-Server Case

It was shown in [14] that at least $\Omega(n)$ bits of communication are required to securely retrieve a single bit from a single ($k = 1$) database in the information-theoretic model (information-theoretic security). This is on the order of the trivial solution of the database downloading the entire file to the user for receiving a request for a single bit and is clearly not of interest. The situation is similar to the nonexistence of a $(1, \delta, \epsilon)$ LDC noted earlier, shown by [24]. However, it has been shown [25] that an interesting (and unexpected) solution for this single-server case can be achieved in the computationally secure case [5]. While computationally expensive, this is an interesting aspect of the subject. It requires some elements from computational number theory which is the basis of many public key cryptosystem algorithms. An overview of the requirements is given.

Much of the prequantum world of public-key cryptography depended on either the difficulty of factoring a large integer which is the product of two primes or the difficulty of finding discrete logarithms in either a finite field or in the group of an elliptic curve. Less well used for this purpose is the *quadratic residuosity problem* which will be described and used for the single-server PIR case. For this section let p, q denote distinct odd large primes and $n = pq$ their product. The problem to be described will be trivial to solve for a user who knows this factorization and computationally infeasible for one who does not. The term "computationally infeasible" will mean beyond the means of a nonquantum computer in feasible time.

Some preliminary notions and facts from elementary number theory are needed. An excellent source for these is the handbook [27]. No proofs are given here. Let $\mathbb{Z}_p$ be the set of integers modulo the prime p and $\mathbb{Z}_p^*$ be the set of nonzero integers modulo p. It forms a multiplicative subgroup of $\mathbb{Z}_p$ of order $(p-1)$.

An integer $a \in \mathbb{Z}_p^*$ is called a *quadratic residue* (QR) mod p if there exists an integer x such that the equation

$$x^2 \equiv a \pmod{p}$$

is satisfied and if no such x exists a is called a *quadratic nonresidue* (QNR) mod p. The zero element is not in $\mathbb{Z}_p$ or $\mathbb{Z}_p^*$ and is neither a QR nor a QNR. The set of QRs modulo p is denoted Q_p and the NQRs modulo p denoted $\overline{Q}_p$. It is easy to establish that

$$|Q_p| = (p-1)/2 = |\overline{Q}_p| \ .$$

For a prime p the *Legendre symbol* is defined as

$$\left(\frac{a}{p}\right) = \begin{cases} 0, \text{ if } & p \mid a \\ 1, \text{ if } & a \in Q_p \\ -1, \text{ if } & a \in \overline{Q}_p. \end{cases}$$

The Legendre symbol enjoys many properties that allow its efficient computation. A generalization of this symbol is the Jacobi symbol to be defined shortly.

Consider the situation for the equation for a general positive integer n

$$x^2 \equiv a \pmod{n} \tag{9.2}$$

and define Q_n so that $a \in Q_n$ if the equation has solutions modulo n and $a \in \overline{Q}_n$ if it does not. It can be shown that for n the product of two primes $n = pq$,

$$|Q_n| = (p-1)(q-1)/4 \quad \text{and} \quad |\overline{Q}_n| = 3(p-1)(q-1)/4$$

and if $a \in Q_n$ the equation has four solutions.

In general if $n = \prod_{i=1}^{k} p_i^{e_i}$, then

$$|\mathbb{Z}_n^*| = \phi(n) \triangleq \prod_{i=1}^{k} (p_i - 1) p_i^{e_i - 1}$$

where $\mathbb{Z}_n$ is the set of integers mod n and $\mathbb{Z}_n^*$ the set of elements relatively prime to n. If $a \in Q_n$, Equation 9.2 has 2^k solutions, where k is the number of distinct prime divisors of n and where $\phi(n)$ is the *Euler Totient function*, $|\mathbb{Z}_n^*|$.

Similar to the Legendre symbol for prime moduli one can define the *Jacobi symbol* for the composite case using the Legendre symbol. In particular for $(a,n) = 1$ and $n = \prod_{i=1}^{k} p_i^{e_i}$ define the Jacobi symbol as

$$\left(\frac{a}{n}\right) \triangleq \prod_{i=1}^{k} \left(\frac{a}{p_i}\right)^{e_i},$$

a product of powers of Legendre symbols. If n is a prime the Jacobi symbol is the Legendre symbol. The Jacobi symbol enjoys many properties (section 2.4.5 of [27], true for arbitrary positive n, not just for products of two primes):

$$\begin{aligned}
&(i)\ \left(\tfrac{a}{n}\right) = 0, 1, \text{ or } -1 \quad (\text{and is 0 iff } \gcd(a,n) \neq 1),\\
&(ii)\ \left(\tfrac{ab}{n}\right) = \left(\tfrac{a}{n}\right)\left(\tfrac{b}{n}\right) \quad (\text{hence } \left(\tfrac{a^2}{n}\right) = 1),\\
&(iii)\ \left(\tfrac{a}{mn}\right) = \left(\tfrac{a}{m}\right)\left(\tfrac{a}{n}\right),\\
&(iv)\ \text{if } a \equiv b \pmod{n}, \text{ then } \left(\tfrac{a}{n}\right) = \left(\tfrac{b}{n}\right) \quad \text{and } \left(\tfrac{a}{n}\right) = \left(\tfrac{a \ (\text{mod } n)}{n}\right),\\
&(v)\ \left(\tfrac{a}{n}\right) = \left(\tfrac{n}{a}\right)(-1)^{(a-1)(n-1)/4}.
\end{aligned}$$

With these properties one can determine the Jacobi symbols easily (without knowing the factorization of n, an important property for the sequel).

Define the sets (for arbitrary n)

$$J_n = \left\{ a \in \mathbb{Z}_n^* \mid \left(\frac{a}{n}\right) = 1 \right\} \quad \text{and} \quad \widetilde{Q}_n = J_n \backslash Q_n.$$

Note that $Q_n \subset J_n$. Thus for an element $a \in J_n \backslash Q_n$ the equation $x^2 \equiv a$ mod n has no solutions $\pmod{n}$ while for $a \in Q_n$ (i.e., a is a QR mod n) it has 2^k solutions for n a product of powers of k distinct primes (and hence four solutions if $n = pq$). Elements in $J_n \backslash Q_n$ are referred to [27] as *pseudoprimes* since they are not squares yet have a Jacobi symbol of 1.

For the remainder of the section assume $n = pq$, a product of two large primes. It follows that $a \in \mathbb{Q}_n$ iff $a \in \mathbb{Q}_p$ and $a \in Q_q$. For a given element $a \in \mathbb{Z}_n^*$ and not knowing the factorization of n it is easy to determine if $a \in J_n$ (via the formulae for the Jacobi symbols noted earlier) but computationally difficult (termed computationally infeasible in the cryptographic context) to determine if $a \in Q_n$. On the other hand, knowing the factorization of n it is simple to determine if $a \in Q_n$ since one can easily determine the Legendre symbols mod p and mod q. It is this fact that is used in the following single-server PIR protocol.

From the above discussion two problems are defined ([27], p. 99):

Definition 9.2

(i) The *quadratic residuosity problem* (QRP) is: given $a \in J_n$, decide if $a \in Q_n$.

(ii) The *square root modulo n problem* (SQROOT) is: given a composite integer n and $a \in Q_n$ find a square root of $a \pmod{n}$.

Conjecture 9.3 *For $n = pq$ the QRP problem is as difficult as the problem of factoring n.*

It is clear the SQROOT problem for $n = pq$ is computationally equivalent to factoring.

Example 9.4 Let $n = 5 \cdot 7 = 35$. The set of QRs mod 35 is computed as $Q_{35} = \{1,4,9,11,16,29\}$ and, for $a \in Q_{35}$ the equation $x^2 \equiv a \pmod{35}$ has four solutions. Similarly the set $\widetilde{Q}_{35} = J_{35} \backslash Q_{35} = \{6,9,10,12,15,18\}$ is such that for $a \in \widetilde{Q}_{35}$ the equation $x^2 \equiv a \pmod{35}$ has no solutions.

These facts form the basis of certain public key cryptographic protocols. For the single-server PIR example to be described it is assumed the primes p and q are chosen so that the integer $n = pq$ is computationally infeasible to factor (a standard assumption, e.g., in implementing an RSA public key cryptosystem). Thus the QRP problem mod n is assumed computationally infeasible.

For the basic PIR single-server scheme of [25] assume the file $\boldsymbol{x}$ on st bits is written on the server as an $s \times t$ array of bits, $M = (m_{ij})$, an array of s rows and t columns. A user wants to retrieve the bit m_{ab} from the server with privacy. The user generates the primes p, $q, n = pq$ and t integers $y_1, y_2, \ldots, y_t$ such that y_b is a QNR and the remaining $(t-1)$ y_j are QRs. These t integers are sent to the server along with n. The server computes the s integers z_i as follows:

$$w_{r,j} = \begin{cases} y_j^2 & \text{if } m_{rj} = 0 \\ y_j & \text{if } m_{rj} = 1 \end{cases}$$

and then

$$z_r = \prod_{j=1}^{t} w_{rj}, \; r = 1,2,\ldots,s.$$

It is clear that for $j \neq b$, w_{rj} is a QR for all r while for $j = b$, w_{rb} is a QR iff $m_{rb} = 0$ and a QNR otherwise. Thus z_r is a QR iff $m_{rb} = 0$ and a QNR otherwise.

The server sends the integers $z_1, z_2, \ldots, z_s$ to the user. The only variable of interest to the user is z_a – if it is a QR, then m_{ab} is 0 and otherwise it is a 1. Since the factorization of n is known to the user it is easy for the user to determine if it is a QR or not.

If n, the product of two primes, is k bits and if one chooses $s = t = \sqrt{n}$, the communication complexity of this scheme is

$$(2\sqrt{n} + 1)k$$

bits or $O(n^{1/2})$ for a constant number of servers. For the argument that the scheme is private see [25]. The argument is essentially to show that if for two indices i and i' one is able to distinguish the queries and responses, one can create an argument that some information is leaked concerning the quadratic residuosity of a generated parameter, in contradiction to the assumptions stated above.

A further protocol, based on a recursion technique, is described in [25].

Thus the single database case under computational security, rather than the information-theoretic case used in the following sections, can be made relatively efficient in term of communications.

Other aspects of single-server PIR systems (under computational security) are surveyed in [28] and [5] including their relationships to several cryptographic primitives such as collision-resistant hash functions, oblivious transfer, public key systems and homomorphic encryption. A single-server PIR system with constant communication complexity is given in [10, 21].

9.2 The Multiple-Server Case

The case of multiple servers (databases) and information-theoretic security is considered. Several interesting techniques are discussed as much for their novelty and inventiveness as for their effectiveness. The current state of affairs is broader than these techniques might indicate yet these systems remain of interest. More recent approaches will be noted later in the section.

Recall the model of interest in the case of multiple databases (as defined in Definition 9.1) is that the k databases, $\mathcal{DB}_1, \mathcal{DB}_2, \ldots, \mathcal{DB}_k$ contain the same string of n bits, $\boldsymbol{x} = \{x_1, x_2, \ldots, x_n\}$ and they do not communicate. A user $\mathcal{U}$ sends queries to each of the databases, and, depending on the query, each database computes a response which is returned to the user, from which the user computes the (single) bit of interest. No database gains any information as to which bit is of interest to the user (i.e., the protocol is information-theoretically private). Only one-round protocols are considered.

The first scheme is an extension of the two-database scheme described in the introduction to this chapter – that scheme will correspond to the case of $d = 1$ in the more general discrete d-dimensional cube encountered in other chapters [14].

Consider the case of $k = 2^d$ databases, for some positive integer d, each database containing $n = \ell^d$ bits for some positive integer ℓ. These are viewed as d-dimensional cubes, with the i-th bit, x_i, of the database $\boldsymbol{x} \in \mathbb{F}_2^n$ contained in cell $(i_1, i_2, \ldots, i_d)$ (say the ℓ-ary expansion of $i \in [n]$) of the cube. Thus the databases are labeled with binary d-tuples and the cells of each database by ℓ-ary d-tuples. Such a scenario is referred to as a subcube code in the literature and these appear as examples for numerous structures (e.g., see batch codes).

Suppose the user $\mathcal{U}$ wishes to retrieve the bit corresponding to position $(i_1, i_2, \dots, i_d)$ in the cube, $i_j \in [\ell]$. $\mathcal{U}$ chooses uniformly at random d subsets $S_1^0, S_2^0, \dots, S_d^0$, $S_i^0 \subseteq [\ell]$ (hence generally of differing sizes). As with the $d = 1$ case considered in the introduction, a second set of subsets is chosen by the user as $S_1^1 = S_1^0 \oplus i_1, S_2^1 = S_2^0 \oplus i_2, \dots, S_d^1 = S_d^0 \oplus i_d$ where the definition of "$\oplus$" of Equation 9.1 for these sets is recalled.

Consider the database $\mathcal{DB}_\sigma$ indexed by $\sigma = (\sigma_1, \sigma_2, \dots, \sigma_d)$, $\sigma_j \in \{0,1\}$. The user $\mathcal{U}$ sends to $\mathcal{DB}_\sigma = \mathcal{DB}_{\sigma_1,\sigma_2,\dots,\sigma_d}$ the sets $S_1^{\sigma_1}, S_2^{\sigma_2}, \dots, S_d^{\sigma_d}$. This is done for each of the $k = 2^d$ databases, each database receiving different sets depending on its binary d-tuple label.

The database $\mathcal{DB}_\sigma$ responds with the XOR of all the bits in the rectangular solid defined by the sets sent to it, i.e., by the bit

$$\bigoplus_{j_1 \in S_1^{\sigma_1}, \dots, j_d \in S_d^{\sigma_d}} x_{j_1,\dots,j_d}. \tag{9.3}$$

There are $|S_1^{\sigma_1}| \times \cdots \times |S_d^{\sigma_d}|$ bits in the XOR sum.

The user XORs the k bits received from all $k = 2^d$ servers to construct the bit x_i. The correctness of the scheme is clear from the construction and the privacy is assured by the uniform and random construction of the sets sent to each database. The complexity (number of bits transmitted) of the scheme is as follows: each set transmitted to the k databases consists of d bits (as an indicator vector for a subset of size d) and each database returns a single bit for a total of $k(\ell d + 1) \approx 2^d n^{1/d}$ bits (recall $k = 2^d$ and $n = \ell^d$). The following trivial example illustrates some of the issues involved with the computations.

Example 9.5 As above assume $k = 2^d$ servers, labeled by binary d-tuples, each containing the information $\boldsymbol{x} \in \{0,1\}^n, n = \ell^d$, $\boldsymbol{x}$ stored as a d-dimensional ℓ-ary array, $\ell > 4$. Suppose the sets S_i^0 are chosen as $S_i^0 = \{1\}, i = 1, 2, \dots, d$ and that it is wished to retrieve the data element $x_{4,4,\dots,4}$ from the servers, i.e., the element of $\boldsymbol{x}$ stored in cell $(4,4,\dots,4)$ of each array. By the above construction the sets $S_i^1 = S_i^0 \oplus \{4\} = \{1,4\}, i = 1, 2, \dots, d$. The set $S_1^{\sigma_1} \times S_2^{\sigma_2} \times \cdots \times S_d^{\sigma_d}$ is sent to database labeled $\sigma = (\sigma_1, \dots, \sigma_d) \in \{0,1\}^d$. Clearly the only database whose response includes $x_{4,4,\dots,4}$ is the one labeled with $(1, 1, \dots, 1)$ which received the set $\{1,4\}^d$.

Since all server responses are XORed together it remains to show that every other server location (than $(4,4,\dots,4)$) is included an even number of times (including 0) counting all server responses. A simple general proof of this seems not immediately available but, e.g., consider the number of servers that include the data element $x_{1,1,1,4,4,\ldots,4}$ XORed in its response. Clearly only the eight servers with addresses of the form $(a,b,c,1,1,\ldots,1), a,b,c \in \{0,1\}$ have this cell in its response. Any other server cell location can be treated in a similar manner.

A slight variation of this scheme is able to accommodate a smaller number of servers at the expense of greater communication complexity using the notion of a *covering code*. In the general case consider the vector space of n-tuples over $\mathbb{F}_q$, $V_n(q)$. An e-covering code of $V_n(q)$ is a set of codewords $C = \{c^1, c^2, \ldots, c^M, c_i \in \mathbb{F}_q^n\}$ such that every q-ary n-tuple of $V_n(q)$ is within Hamming distance e of at least one codeword. Said another way, the spheres of radius e around each codeword cover $V_n(q)$ or every element of $V_n(q)$ is in at least one sphere around some codeword. Perfect codes are an extreme or optimal case of covering codes, when they exist, in that each word in $V_n(q)$ is in exactly one sphere around the codewords.

Example 9.6 The code $C = \{000, 111\}$ is a 1-covering code for $V_3(2)$ since the spheres of radius 1 around each codeword exhaust $V_3(2)$, i.e., the two spheres are

$$S_1 = \{000, 001, 010, 100\} \quad S_2 = \{111, 110, 101, 011\}.$$

Only 1-covering binary codes are considered here. As above let the size of the database be $n = \ell^d$. A database $\mathcal{DB}_\sigma$ for $\sigma = (\sigma_1, \sigma_2, \ldots, \sigma_d) \in \{0,1\}^d$ receives the sets $S_1^{\sigma_1}, S_2^{\sigma_2}, \ldots, S_d^{\sigma_d}$ and responds as above in Equation 9.3. Let C be a binary 1-covering code

$$C = \left\{ \boldsymbol{c}^1, \boldsymbol{c}^2, \ldots, \boldsymbol{c}^k,\ \boldsymbol{c}^i \in \mathbb{F}_2^d \right\}.$$

The strategy will be as follows. Each server is assigned a codeword (hence the need for k codewords). Without loss of generality assume the all-zero d-tuple is the codeword $\boldsymbol{c}^1$ assigned to server labeled $00\ldots0$. The user sends $\mathcal{DB}_{0,0,0}$ the queries $S_1^0, S_2^0, \ldots, S_d^0$ and the server responds with the bit computed as the XOR of all the bits in the database subcube corresponding to all the indices in the set

$$S_1^0 \times S_2^0 \times \cdots \times S_d^0.$$

The user requires such information from all 2^d sets $S_1^{\sigma_1}, S_2^{\sigma_2}, \ldots, S_d^{\sigma_d}$ although now only k of these sets correspond to server assignments due to the use of the 1-covering code. However, the database $\mathcal{DB}_{0,0,\ldots,0}$ knows that if database

$\mathcal{DB}_{0,0,\dots,1}$ did exist, the information it would have received from the user would have been of the form

$$S_1^0 \times S_2^0 \times \cdots \times S_d^1$$

for S_d^1 of the form $S_d^0 \oplus j$ for some $j \in [\ell]$. Thus in place of $\mathcal{DB}_{0,0,\dots,1}$, the $\mathcal{DB}_{0,0,\dots,0}$ computes all of the bits corresponding to the ℓ sets

$$S_1^0 \times S_2^0 \times \cdots \times \left\{ S_d^0 \oplus j \right\}, \quad j = 1, 2, \dots, \ell$$

and transmits all ℓ bits to the user who knows which bit is the correct one they would have received had the corresponding server existed. The server $\mathcal{DB}_{0,0,\dots,0}$ repeats the process for each of the d missing servers corresponding to binary d-tuples distance 1 from its own codeword. Each server does the same and hence each of the k servers sends to the user a total of

$$(1 + d\ell)k$$

bits. However, if the code is not perfect some of the d-tuples are within distance 1 of more than one codeword and these should be eliminated. The total number of bits transmitted then is

$$k + (2^d - k)\ell$$

bits from the databases to the user and

$$d\ell k$$

bits from the user to the databases, for a total of (theorem 1 of [14])

$$k + (2^d + (d-1)k)\ell, \quad \ell = n^{1/d}.$$

Thus one can reduce the number of servers at the expense of increasing the number of bits transmitted from the servers via covering codes.

Another scheme of the influential paper ([14], section 4) is based on polynomial interpolation, a technique used for several other schemes. The basic idea is for the user to distribute partial information to the servers who compute partial information on the interpolation which is relayed back to the user who is able to complete the polynomial interpolation from the information provided.

An unusual construction from [41] is of interest. The protocol will yield a k-server PIR protocol with communication complexity of $O\big(k^3 n^{1/(2k-1)}\big)$ for a database of size n symbols (assumed from a finite field of size q, i.e., $\mathbb{F}_q$). The construction is sufficiently involved that it is not described here.

There are numerous other ingenious protocols to achieve the PIR property, including the work of Beimel et al. [6] which achieves a communication complexity of $n^{O\left(\log\log k / k \log k\right)}$ k for a k-server scheme, more efficient than previous schemes for $k \geq 3$. It is noted the techniques used also result in the construction of LDCs of length $\exp\left(n^{O\left(\frac{\log\log k}{k\log k}\right)}\right)$ which is shorter than previous constructions for all $k \geq 4$. The work is interesting for its use of multivariable polynomials. This approach is further developed in ([6], corollary 3.9) to yield a k-server PIR protocol for $k \geq 3$ with bit complexity $O(n^{2\log\log k/k\log k})$, a significant improvement for this number of servers.

The cases of $k = 2$ and $k = 3$ servers have received considerable attention. A 2-server protocol with communication complexity $n^{O\left(\sqrt{\log\log n/\log n}\right)}$ is described in [16, 17], a significant improvement in known 2-server protocols which previously had complexity $O(n^{1/3})$. The construction requires families of matching vectors, which gives further evidence of the relationship between PIR protocols and LDCs. The work of [35] develops tight bounds on the complexity of 2-server schemes that imply many of the known 2-server protocols are in fact close to optimal.

The equivalence of the notions of a smooth LDC and a PIR has been considered (indeed, this was noted in the original work of [24]). Thus given a smooth LDC one can construct an information-theoretic PIR and, given a PIR one can describe a smooth LDC, making the appropriate identification of the various parameters. Further aspects of this equivalence are noted in the Comments section. The relationships are further discussed in numerous works including [13, 22, 24, 35].

This equivalence has proved very useful in constructing new efficient PIRs. In particular it means finding lower bounds for information-theoretic PIRs is essentially equivalent to finding lower bounds on code lengths for smooth LDCs.

In particular, the breakthrough construction of 3-query LDC codes of [37] (see Chapter 8) of subexponential length, had length $N = \exp(O(1/\log\log n))$, a considerable improvement from the previously best-known $O(n^{1/2})$, and these smooth LDC codes can be used to construct PIR schemes with three servers.

Finally it is mentioned that many of the works on PIR protocols address the notion of t-private k-server PIR protocols in which the user's privacy is maintained under collusion of up to t servers. These are not considered in this work.

9.3 Coding for PIR Storage

Two obvious shortcomings of the PIR protocols described in the previous section include the large amount of storage required to implement them and the somewhat unrealistic assumption that a user wishes to retrieve only one bit. Replicating a large database across multiple servers is expensive in both storage costs and transmission costs and coded approaches to this problem are considered. The first work that considers the possibility of coding rather that replicating the database appears to be [32]. It is shown there (theorem 1) that using k servers containing R bits and any erasure code, then any PIR algorithm achieving information-theoretic privacy must download at least $R+1$ bits and must connect to at least $R+1$ servers which in a typical application would be large. The papers [11, 12] consider the optimal trade-off between storage cost and complexity and show that the optimal trade-off can be achieved with MDS codes and that this optimal trade-off depends on the number of records in the system. Augot et al. [3] use multiplicity codes considered earlier and a PIR scheme to achieve more efficient storage schemes than replication.

The significant breakthrough on this issue that is considered here is the work of Fazeli et al. [19] (the author is unaware if this important contribution was published in a journal) which shows that PIR protocols can be adapted to work with information on the servers *coded* in such a way that the total amount of storage can be made to approach the size of one copy of the database rather than k times this amount as required in the k-server protocols described in the previous section. This while maintaining privacy of the downloads. The work has inspired a number of important papers. A brief overview of the original work in [18, 19] is described here.

It will be helpful to elaborate on the informative example in ([19], example 2). Our treatment will be informal to focus on the ideas behind the coded system. As before a database $X = (x_1, \ldots, x_n)$ of n bits and a k-server PIR defined previously as the triple $\mathcal{P} = (Q, \mathcal{A}, C)$ which will allow the downloading of a single bit x_i with accuracy and privacy. As before, to invoke the protocol the user calls on Q to (randomly) generate k queries with query q_j sent to server $S_j, j = 1, 2, \ldots, k$. Each server uses routine $\mathcal{A}$ to generate answers to the queries, the answer from the j-th server designated as a_j. The user then invokes the reconstruction algorithm $C(a_1, a_2, \ldots, a_k)$ to generate the answer x_i. The PIR protocol to be described can be proved correct and private for which the reader is referred to the original paper [19]. Interest will be in showing how the storage capacity can be significantly reduced, in fact to not much more than the size of a single server, with the use of a coding strategy while maintaining privacy.

PIR Codes from Binary Generating Matrices

All the coded PIR constructions of this section can be cast as being from binary code generating matrices. The technique is best illustrated via an extended example. For the baseline uncoded system assume a 3-server PIR protocol is available, each server storing the entire database of n bits, a total storage of $3n$. Consider another base system with the same storage ($3n$) where it is assumed that each server is only able to store $n/4$ bits and hence 12 servers are needed. From the assumed $k = 3$-server PIR protocol available generate 3 queries (on $n/4$ bits). These queries can be sent to any of the 12 servers (any 3 queries in each round). The user receives answers back from the servers it has queried and determines the bit of interest from these answers. To use the protocol the user determines which quarter of the database the bit it is wished to recover lies in. Suppose it is in the first quarter – three of the servers contain this quarter and the user implements the 3-server protocol using these three servers. It will be assumed the PIR protocols are additive in that the response to a query corresponds to the sum of two database bits. Virtually all PIR protocols developed so far have this additive property. This is the uncoded baseline system. Its total storage, as before, is $3n$.

The aim of the following example (an expansion of example 2 in [19]) is to show how a PIR protocol with eight servers and total storage of $2n$ (compared to $3n$ for the original 3-server strategy or the 12-server strategy) can be used, thus demonstrating the advantage of coding.

Example 9.7 Divide the entire database x into four parts $\boldsymbol{X}_1, \boldsymbol{X}_2, \boldsymbol{X}_3$ and $\boldsymbol{X}_4$, each of size $n/4$ (as in the above 12-server model). Thus $\boldsymbol{X} = (\boldsymbol{X}_1, \boldsymbol{X}_2, \boldsymbol{X}_3, \boldsymbol{X}_4) = (x_1, x_2, \ldots, x_n)$ and

$$\boldsymbol{X}_i = \left(x_{(i-1)n/4+1}, x_{(i-1)n/4+2}, \ldots, x_{(i-1)n/4+n/4}\right), i = 1,2,3,4.$$

Rather than store these quarters on the servers they are first coded into pieces $\boldsymbol{C}_1, \boldsymbol{C}_2, \ldots, \boldsymbol{C}_8$, each of size $n/4$ according to the $(8,4)_2$ binary code:

$$\boldsymbol{C} = \boldsymbol{X} \cdot \boldsymbol{M}$$

$$(\boldsymbol{C}_1, \boldsymbol{C}_2, \ldots, \boldsymbol{C}_8) = (\boldsymbol{X}_1, \boldsymbol{X}_2, \ldots, \boldsymbol{X}_4) \begin{bmatrix} 1 & 0 & 0 & 0 & 1 & 0 & 0 & 1 \\ 0 & 1 & 0 & 0 & 1 & 1 & 0 & 0 \\ 0 & 0 & 1 & 0 & 0 & 1 & 1 & 0 \\ 0 & 0 & 0 & 1 & 0 & 0 & 1 & 1 \end{bmatrix}$$

The arithmetic is in $\mathbb{F}_2$ – thus $\boldsymbol{C}_5 = \boldsymbol{X}_1 \oplus \boldsymbol{X}_2$ which is of dimension $n/4$ and is stored on server 5.

This matrix has an interesting property, the *disjoint repair group* property which is described as follows (it was also used in LDCs from multiplicity codes discussed in Chapter 8). Let $\boldsymbol{e}^i, i = 1,2,3,4$ be column vectors, binary 4-tuples, the indicator vectors for 4-tuples. The matrix $\boldsymbol{M}$ is said to have the disjoint repair group property for a $k = 3$-PIR if, for each $i = 1,2,3,4$, there are three (in general k) disjoint sets of columns of the matrix $\boldsymbol{M}$, $R_3^{(i)}$ (different sets for each i) such that the columns in each set add to $\boldsymbol{e}^i, i = 1,2,3,4$. Thus if we label the column vectors of the matrix $\boldsymbol{M}$ by $\boldsymbol{m}^1, \boldsymbol{m}^2, \ldots, \boldsymbol{m}^8$ we have the following sets of disjoint repair groups:

$$\begin{array}{l} R_3^{(1)} = \big\{\{1\},\{2,5\},\{4,8\}\big\} : \boldsymbol{e}^1 = \boldsymbol{m}^1 = \boldsymbol{m}^2 \oplus \boldsymbol{m}^5 = \boldsymbol{m}^4 \oplus \boldsymbol{m}^8 = (1,0,0,0)^t \\ R_3^{(2)} = \big\{\{2\},\{1,5\},\{3,6\}\big\} : \boldsymbol{e}^2 = \boldsymbol{m}^2 = \boldsymbol{m}^1 \oplus \boldsymbol{m}^5 = \boldsymbol{m}^3 \oplus \boldsymbol{m}^6 = (0,1,0,0)^t \\ R_3^{(3)} = \big\{\{3\},\{2,6\},\{4,7\}\big\} : \boldsymbol{e}^3 = \boldsymbol{m}^3 = \boldsymbol{m}^2 \oplus \boldsymbol{m}^6 = \boldsymbol{m}^4 \oplus m^7 = (0,0,1,0)^t \\ R_3^{(4)} = \big\{\{4\},\{3,7\},\{1,8\}\big\} : \boldsymbol{e}^4 = \boldsymbol{m}^4 = \boldsymbol{m}^3 \oplus \boldsymbol{m}^7 = \boldsymbol{m}^1 \oplus \boldsymbol{m}^8 = (0,0,0,1)^t \end{array}$$

For general k, $R_k^{(j)}$ will consist of a set of k disjoint subsets of columns of the coding matrix $\boldsymbol{M}$ which each add up to the same indicator vector, $\boldsymbol{e}^j$ of dimension equal to the number of rows of $\boldsymbol{M}$. Denote the j-th subset of $R_k^{(i)}$ as $R_{k,j}^{(i)}$. The construction of such disjoint (repair) groups is of great interest in this and other chapters.

The crucial property of this matrix now is that the following inner products form independent assessments of the various information bits required:

$$\begin{aligned} X_1 &= \Big(X, \sum_{j \in R_{3,1}^{(1)}} \boldsymbol{m}^j\Big) = \Big(X, \sum_{j \in R_{3,2}^{(1)}} \boldsymbol{m}^j\Big) = \Big(X, \sum_{j \in R_{3,3}^{(1)}} \boldsymbol{m}^j\Big) \\ X_2 &= \Big(X, \sum_{j \in R_{3,1}^{(2)}} \boldsymbol{m}^j\Big) = \Big(X, \sum_{j \in R_{3,2}^{(2)}} \boldsymbol{m}^j\Big) = \Big(X, \sum_{j \in R_{3,3}^{(2)}} \boldsymbol{m}^j\Big) \\ X_3 &= \Big(X, \sum_{j \in R_{3,1}^{(3)}} \boldsymbol{m}^j\Big) = \Big(X, \sum_{j \in R_{3,2}^{(3)}} \boldsymbol{m}^j\Big) = \Big(X, \sum_{j \in R_{3,3}^{(3)}} \boldsymbol{m}^j\Big) \\ X_4 &= \Big(X, \sum_{j \in R_{3,1}^{(4)}} \boldsymbol{m}^j\Big) = \Big(X, \sum_{j \in R_{3,2}^{(4)}} \boldsymbol{m}^j\Big) = \Big(X, \sum_{j \in R_{3,3}^{(4)}} \boldsymbol{m}^j\Big). \end{aligned}$$

Consider the three sets of $R_3^{(1)}$ and the queries generated by the 3-query PIR, say (q_1, q_2, q_3). These queries are designed to retrieve a bit x_i from a database (of size $n/4$) and denote the responses from the three servers as (a_1, a_2, a_3). Suppose these queries are sent to the appropriate coded servers. Specifically suppose q_1 is sent to the first set of $R_{3,1}^{(1)}$, namely server 1 with response a_1. Send query q_2 to the coded servers S_2 and S_5 and denote the responses $a_2^{'}$ and $a_5^{'}$, respectively, and send q_3 to coded servers S_4 and S_8 and their responses $a_4^{'}$ and $a_8^{'}$, respectively. The primes indicate the responses are from a coded database. Queries may be sent to other databases but their responses, indicated with a $*$, are not used in the reconstruction. It is clear then because of the disjoint repair property discussed the three responses are

$$a_1, \quad a_2 = a_2^{'} \oplus a_5^{'} \quad \text{and} \quad a_3 = a_4^{'} \oplus a_8^{'}$$

because of the assumed additive property of the responses. Said another way, the sum of the parts of the databases stored in servers 2 and 5 is X_1 which is also the sum of the databases of servers 4 and 8. Thus by the coding of the disjoint repair property three identical databases $\boldsymbol{X}_1$ have been created to which the original 3-PIR protocol can be applied. This explains the use of the disjoint repair groups constructed. Similar comments lead to the other equations.

To recap, to retrieve a bit from the sub-database $\boldsymbol{X}_1$ one uses the 3-PIR to generate queries (q_1, q_2, q_3) and sends these to the coded servers as $(q_1, q_2, *, q_3, q_2, *, *, q_3)$ and receives responses $(a_1', a_2', *, a_4', a_5', *, *, a_8')$ from which are derived the responses $a_1 = a_1', a_2 = a_2' \oplus a_5',\ a_3 = a_4' \oplus a_8'$ from which the original reconstruction algorithm of the 3-PIR protocol can be used to determine the bit x_i. The paper [19] sends random queries from those of the servers not used to assist with satisfying the privacy condition in that responses from these other servers are not used in the reconstruction process. We use the $*$ to explain the "null" responses.

The summary of this process for the bit of interest in each of the four sub-databases is as follows for the 3-server queries (q_1, q_2, q_3) and coded server responses $(a_1', a_2', \ldots, a_8')$ used to create input to the 3-server PIR reconstruction:

x_i in	coded sever queries	inputs to the 3-PIR reconstruction
X_1	$(q_1, q_2, *, q_3, q_2, *, *, q_3)$	$a_1 = a_1',\ a_2 = a_2' \oplus a_5',\ a_3 = a_4' \oplus a_8'$
X_2	$(q_2, q_1, q_3, *, q_2, q_3, *, *)$	$a_1 = a_2',\ a_2 = a_1' \oplus a_5',\ a_3 = a_3' \oplus a_6'$
X_3	$(*, q_2, q_1, q_3, *, q_2, q_3, *)$	$a_1 = a_3',\ a_2 = a_2' \oplus a_6',\ a_3 = a_4' \oplus a_7'$
X_4	$(q_3, *, q_2, q_1, *, *, q_2, q_3)$	$a_1 = a_4',\ a_2 = a_3' \oplus a_7',\ a_3 = a_1' \oplus a_8'$

Note in particular the use of eight servers while using only a 3-server PIR protocol.

The example thus achieves the same storage/performance ($3n$ bits) of a 12-server uncoded PIR protocol while using only storage of $2n$ bits.

To generalize this instructive example to k servers it will be assumed a k-server PIR protocol is available and that the database $\boldsymbol{X}$ of n bits is divided into s sub-databases $\boldsymbol{X}_1, \boldsymbol{X}_2, \ldots, \boldsymbol{X}_s$, each containing n/s bits (in the above example $k = 3, s = 4$). There will be $m > k$ servers in the coded protocol.

A *k-server PIR code* is defined as follows ([19], definition 4):

Definition 9.8 An $s \times m$ binary matrix $\boldsymbol{G}$ has property A_k if, for each $i \in [s]$ there exists k disjoint subsets of columns of $\boldsymbol{G}$ that add to $\boldsymbol{e}^i$ (for each i) the indicator s-tuple of the i-th position. A binary linear $[m, s]$ code C will be called a *k-server PIR code* if there exists a generator matrix $\boldsymbol{G}$ for C with property A_k, i.e., for each $i \in [s]$ there exist k disjoint sets of columns of $\boldsymbol{G}$, $R_k^{(i)}, i = 1,2,\ldots,s$, $|R_k^{(i)}| = k$, such that

$$\boldsymbol{e}^j = \sum_{i \in R_{k,1}^{(j)}} \boldsymbol{m}^i = \cdots = \sum_{i \in R_{k,k}^{(j)}} \boldsymbol{m}^i, \quad j = 1,2,\ldots,s.$$

It is shown ([19], theorem 5) that the upload and download complexity of an $[m,s]$ k-server PIR code with each server containing n/s bits is m times that of the uncoded k-server PIR protocol with servers containing n/s bits. A lower bound is given in [31]. Aspects of the problem of constructing matrices with the disjoint repair groups property are explored further in [26, 29].

An interesting question is to determine for a given value of s, the number of pieces the original database is divided into (in essence the number of information symbols of the code), values for m and k.

A natural figure of merit for such coded PIR systems is the ratio of total storage among all servers to the size of the database. Define [19] the quantity $A(s,k)$ as the optimum value of m such that an $[m,s]$ k-server PIR code exists, which is the total number of coded bits of the system. Thus

$$A(s,k) = \text{total number of stored bits and} \quad \eta = A(s,k)/n = \text{storage efficiency}$$

where η is the ratio of total stored to original information bits, the storage overhead, is a measure of the efficiency of the coded scheme. It should be noted it is irrelevant whether a single bit or portion of the database is stored in each cell for these measures. For the examples it will be assumed a single bit is stored in each cell.

There remains the problem of constructing k-server PIR codes. The problem is addressed in the original manuscript [19] where the multidimensional cubic construction (a similar construction was used for uncoded PIR codes earlier in this chapter and batch codes in Chapter 10), Steiner systems and majority logic decodable codes are used to give simple constructions. Only the cubic construction is briefly described here.

The Cubic k-PIR Code Construction

For the derivation of a cubic $[m,s]$ k-server PIR code (recall: in general, m servers of which k will be contacted in the protocol, each containing n/s bits) the general cubic construction considers a discrete set of ℓ^{k-1} cells in $(k-1)$ dimensions, an $\ell \times \ell \times \cdots \times \ell, ((k-1)$-fold) array. For simplicity it will be assumed each cell contains a single bit of the database, i.e., $n = s$. The extension to n/s will be clear. The cube side notation of ℓ is used to be consistent with the cubic construction in other sections (σ is used in [19]).

Thus $n = s = \ell^{(k-1)}$. On each of the $(k-1)$ "faces" of the cube, each containing $\ell^{(k-2)}$ cells, a parity symbol is placed in each cell, giving a total of $(k-1)\ell^{(k-2)}$ parity-check symbols. The parity bit is the XOR of the contents of all cells corresponding to an addressing coordinate being constant. Thus the length of the code (information plus parity bits, also the total number of servers) is

$$m = \ell^{(k-1)} + (k-1)\ell^{(k-2)}.$$

The situations for $k = 3$ and $\ell = 3$ are shown in Figure 9.1.

It is a simple matter to translate this cube arrangement into an ordinary binary code via a generating matrix – as will be done for the following example. The geometric picture of the $(k-1)$-dimensional cube, however, is useful.

To use these cubic structures to generate the disjoint repair groups of columns of the matrix, consider an information bit in a particular cell and consider the information symbols and parity symbols obtained by running lines through this symbol (cell) along directions of the $(k-1)$ perpendicular axes of the cell, plus the singleton of the symbol in the cell itself. This generates k disjoint repair subsets of the columns of the $s \times m$ generating matrix G and these form the set of sets $R_k^{(x_i)}$ where the content of the cell is the symbol x_i.

For the k-server cubic codes we have

$$A(s,k) = \ell^{k-1} + (k-1)\ell^{k-2} \quad \text{and} \quad \eta = 1 + (k-1)/\ell.$$

This quantity goes to unity with ℓ for a fixed dimension k. Compare this with the replication schemes with a value equal to the number of servers. Numerous works consider the design of k-server coded schemes ([7, 8] and references cited there).

Example 9.9 Consider the case $\ell = 2, k = 3$ which is similar to the case of the previous example. The situation is shown in Figure 9.1(a) with the equivalent matrix G (b). The disjoint repair groups for each possible information bit readily follow. For example, for the information symbol x_4 in Figure 9.1 the three disjoint repair groups are

$$\{x_4, x_4 = x_1 \oplus x_7 \oplus p_1, x_4 = x_5 \oplus x_6 \oplus p_5\}.$$

Note that in the general construction k is the number of disjoint repair groups for each information bit and hence the scheme requires a k-PIR protocol and a total of m servers.

The work of Shah et al. [32], noted previously, apart from introducing the notion of erasure coding for PIR download schemes, also introduced schemes that considered a database of k records X, each record R bits in length and a

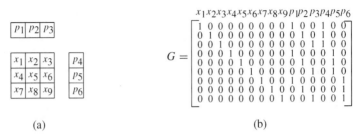

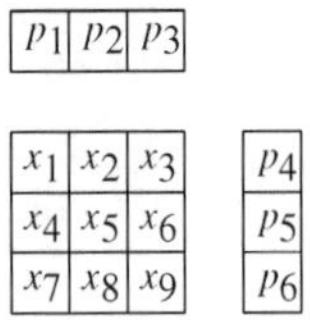

(a)

$$G = \begin{bmatrix} 1&0&0&0&0&0&0&0&0&1&0&0&1&0&0 \\ 0&1&0&0&0&0&0&0&0&0&1&0&1&0&0 \\ 0&0&1&0&0&0&0&0&0&0&0&1&1&0&0 \\ 0&0&0&1&0&0&0&0&0&1&0&0&0&1&0 \\ 0&0&0&0&1&0&0&0&0&0&1&0&0&1&0 \\ 0&0&0&0&0&1&0&0&0&0&0&1&0&1&0 \\ 0&0&0&0&0&0&1&0&0&1&0&0&0&0&1 \\ 0&0&0&0&0&0&0&1&0&0&1&0&0&0&1 \\ 0&0&0&0&0&0&0&0&1&0&0&1&0&0&1 \end{bmatrix}$$

(columns: $x_1\, x_2\, x_3\, x_4\, x_5\, x_6\, x_7\, x_8\, x_9\, p_1\, p_2\, p_3\, p_4\, p_5\, p_6$)

(b)

Figure 9.1 Subcube codes: (a) $k = 3$ (b) $\ell = 3$ associated generator matrix

user wishes to retrieve a record, with privacy and accuracy, rather than a single bit, as has been considered to this point. A very large number of works since that seminal work has furthered development of this model, e.g., [4, 9, 11, 12, 33, 34] but such interesting avenues are not pursued here.

The interesting relationships between LDCs and PIR codes have been noted earlier (and in the Comments section to follow) and the relationship to batch codes will be discussed in Chapter 10. The notion of disjoint recovery or repair sets appears in them all to some extent. What might be viewed as a precursor to these ideas is that of majority logic decodable codes of coding theory, a notion studied in the coding literature since the 1960s. For the binary majority logic decodable codes case, this involves determining parity-check vectors (codewords in the dual code) which each involve the bit x_i of interest but each is otherwise disjoint on remaining coordinate positions. If $2e+1$ such check vectors can be found and if e or fewer errors are made in transmission, then a majority of the check equations will yield the same correct answer. In essence the parity-check equations can be viewed as disjoint repair equations and so their application to the construction of coded PIR systems is clear.

PIR Codes from Multiplicity Codes

The role of disjoint recovery sets is clear in the previous generating matrix approach to PIR codes, where a recovery set is defined more formally in the definition below. Each column indicator vector can be described as a disjoint sum of generator matrix columns, each such column vector associated with a part of the database. The number of disjoint sums for the same indicator vector is k to allow a k-PIR scheme to be used on the same part of the database. This motivates the use of disjoint recovery sets below.

The work [1] shows how the multiplicity codes $C_{m,d,s,q}$ can be used to generate such recovery sets and hence yield PIR codes. Recall from Section 8.2 the following properties of multiplicity codes. For a database of n symbols from $\mathbb{F}_q$, the multiplicity code gives $N = q^m$ coded symbols over the alphabet $\Sigma = q^{\binom{m+s-1}{m}}$. As each coordinate position contains all Hasse derivates of order $< s$ of the m-variate polynomial of degree at most d over $\mathbb{F}_q$, the code is also thought of as an array of $\binom{m+s-1}{m} \times q^m$ $\mathbb{F}_q$ symbols. The code $C_{m,d,s,q}$ has parameters:

$$\text{length} = q^m, \quad \text{rate} = \frac{\binom{m+d}{m}}{\binom{m+s-1}{m} q^m}, \quad \text{normalized distance } \Delta = 1 - \frac{d}{sq}.$$

The definition of a PIR code is modified slightly:

Definition 9.10 ([1]) The code C will be called a k-PIR code, denoted by $[N,n,k]_q$ if for every information symbol $x_i, i \in [n]$ there exists k mutually disjoint sets $R_{i,1}, R_{i,2}, \ldots, R_{i,k}$ such that for all $j \in [k]$ x_i is a function of the symbols in $R_{i,j}$ Such sets are referred to as recovery sets for the code position i.

It is desired to minimize N, the number of coded symbols, for a given n and k, and such a value will be designated $N_{pir}(n,k)$ and schemes that give $\lim_{n\to\infty} N_{pir}(n,k)/n = 1$ are sought. A brief outline of the use of multiplicity codes [1] in the derivation of k-PIR codes is given.

An *interpolation set* for $f(\boldsymbol{x}) \in \mathbb{F}_q[x_1, \ldots, x_m] = \mathbb{F}_q[\boldsymbol{x}]$ is a set $R \subseteq \mathbb{F}_q^m$ such that if for every polynomial for which $g(\boldsymbol{x}) \in \mathbb{F}_q[\boldsymbol{x}]$, $f(\boldsymbol{x}) = g(\boldsymbol{x})\ \forall \boldsymbol{x} \in R$, then $f(\boldsymbol{x}) = g(\boldsymbol{x})\ \forall \boldsymbol{x} \in \mathbb{F}_q^m$. Consider sets of the form $Z = A_1 \times A_2 \times \cdots \times A_m$ with $|A_i| = d+1$, $A_i \subset \mathbb{F}_q$, a set of distinct elements and note that this is an interpolation set for $f \in \mathbb{F}_q[x_1, x_2, \ldots, x_m]$ of degree d ([1], lemma 10).

Consider a codeword $\boldsymbol{c}_f \in C_{m,d,s,q}$

$$\boldsymbol{c}_f = \left(f^{[<s]}(\boldsymbol{a}_1), f^{[<s]}(\boldsymbol{a}_2), \ldots, f^{[<s]}(\boldsymbol{a}_{q^m})\right)$$

for some ordering of the elements $\boldsymbol{a}_j$ of $\mathbb{F}_q^m$. If $\boldsymbol{A} \subseteq \mathbb{F}_q^m$ is an interpolation set for homogeneous polynomials of degree at most $s-1$, then for any $\boldsymbol{a} \in \mathbb{F}_q^m$

$$R_{\boldsymbol{a},\boldsymbol{v}} \stackrel{\Delta}{=} \left\{\boldsymbol{a} + \lambda \boldsymbol{v},\ \lambda \in \mathbb{F}_q^*\right\},\ \boldsymbol{v} \in \mathbb{F}_q^m$$

is a recovery set for the code coordinate position corresponding to $\boldsymbol{a} \in \mathbb{F}_q^m$. Thus it is to be shown how the codeword coordinate position $\boldsymbol{a}$ of $\boldsymbol{c}_f$, i.e., $f^{[<s]}(\boldsymbol{a})$, can be computed from the coordinate positions of the recovery set $R_{\boldsymbol{a},\boldsymbol{v}}$, i.e., to compute

$$f^{[<s]}(\boldsymbol{a}) \quad \text{from} \quad f^{[<s]}(\boldsymbol{u}),\ \boldsymbol{u} \in R_{\boldsymbol{a},\boldsymbol{v}}.$$

The argument is similar to the one for the discussion of multiplicity codes in Chapter 8. For a codeword polynomial $f(\boldsymbol{x}) \in \mathbb{F}_q[\boldsymbol{x}]$ of total degree at most d, define the univariate polynomial

$$h(\lambda) = f(\boldsymbol{a} + \lambda \boldsymbol{v}) \in \mathbb{F}_q[\lambda]$$

of degree at most d. Assume this polynomial is known for $\lambda \in \mathbb{F}_q^*$. Furthermore it can be shown [1] to be unique and can be used to determine $f^{[<s]}(\boldsymbol{a})$ since

$$h(0) = f(\boldsymbol{a}).$$

It follows that from the values of $f^{[<s]}(\boldsymbol{b}), \boldsymbol{b} \in R_{\boldsymbol{a},\boldsymbol{v}}$ the value of the codeword coordinate position $\boldsymbol{a}$, i.e., $f^{[<s]}(\boldsymbol{a})$ can be determined. Thus any interpolation set for a homogeneous polynomial of degree less than s leads to a recovery set for a codeword position. For details see ([1], theorem 14).

Furthermore ([1], theorem 16, corollary 17) it can be shown there are $\left\lfloor \frac{q}{s} \right\rfloor^{m-1}$ mutually disjoint such recovery sets and hence the multiplicity code $C_{m,d,s,q}$ yields a k-PIR code with

$$N = q^m, \; n = \frac{\binom{d+m}{m}}{\binom{m+s-1}{m}}, \quad k = \left\lfloor \frac{q}{s} \right\rfloor^{m-1}, \quad \text{and symbol field size } \Sigma = q^{\binom{m+s-1}{m}}$$

where N is the number of coded symbols and each database contains n symbols. (Note: The roles of the parameters m and s here are reversed from those of [1].)

Comments

The intimate connection between the construction of LDCs and PIR schemes was noted in Chapter 8. As a more concrete relationship consider the following ([38], lemma 7.2). Suppose there is available a q-ary smooth LDC with r queries that encodes a database $\boldsymbol{x} \in \mathbb{F}_q^k$ to a codeword $\boldsymbol{y} \in \mathbb{F}_q^n$. The codeword is stored on each of the r servers, $S_1, S_2, \ldots, S_r$. The PIR scheme then uses the r queries of the LDC to query each of the servers to receive answers $\mathcal{A}(\boldsymbol{y}, j), j = 1, 2, \ldots, r$ from which the desired database element $x_i \in \mathbb{F}_q$ is recovered via the LDC decoding algorithm. The privacy of the resulting PIR scheme derives from the assumed smoothness of the LDC code.

As further evidence of this connection the following two results are quoted, the first a result of the 3-query construction of LDCs of Yekhanin introduced in Chapter 8:

Theorem 9.11 ([30], theorem 5,3) *Let $p = 2^t - 1$ be a fixed Mersenne prime. For every positive integer n there exists a 3-server PIR protocol of length $O(n^{1/t})$ and answers of length t. Specifically, for every positive integer n there exists a 3-server PIR protocol with communication complexity $O(n^{1/32582658})$.*

In the reverse direction the following lemma is a generalization of a result of [22]:

Lemma 9.12 ([23], lemma 4.1) *Assume the existence of a PIR scheme with k servers, database size n, query size t, answer size a, and recovery probability at least* $1/2+\epsilon$. *Then there exists a* $(k,k+1,\epsilon)$*-smooth LDC code* $C\colon \{0,1\}^n \longrightarrow (\{0,1\}^a)^m$, *with* $m \le (k^2+k)\cdot 2^t$. *Further:*

(i) If the PIR scheme is linear, then the LDC C is also linear.
(ii) If in the PIR scheme the user only uses k predetermined bits from each answer a_j, *then this is also true for the decoding algorithm of C.*

This lemma uses the slightly different definition of an LDC noted earlier in that the code is (q,c,ϵ)-smooth if, in the decoding, the probability the algorithm queries a particular coordinate is upper-bounded by c/m, the meaning of the other parameters the same.

Research into techniques for the distributed storage and private retrieval of information into servers with various types of restrictions has developed rapidly over the past decade. The chapters of this volume on coding for distributed storage, PIR, LDCs and batch codes are strongly related and simply different aspects of the storage and retrieval problem. These four short chapters do not capture the full scope of work in progress on these important problems but hope to have provided a glimpse into the models and direction the work is taking.

The monographs [38] and [39] make excellent reading and cover both LDCs and PIR schemes and their relationships. In addition there are excellent survey papers such as [2], [20], [23] and [40]. Yekhanin's doctoral thesis at MIT [36] is also interesting reading.

References

[1] Asi, H., and Yaakobi, E. 2019. Nearly optimal constructions of PIR and batch codes. *IEEE Trans. Inform. Theory*, **65**(2), 947–964.

[2] Asonov, D. 2001. Private information retrieval – an overview and current trends. https://api.semanticscholar.org/CorpusID:14698164.

[3] Augot, D., Levy-dit Vehel, F., and Shikfa, A. 2014. A storage-efficient and robust private information retrieval scheme allowing few servers. Pages 222–239 of: *Cryptology and network security*. Lecture Notes in Computer Science, vol. 8813. Springer, Cham.

[4] Banawan, K., and Ulukus, S. 2018. The capacity of private information retrieval from coded databases. *IEEE Trans. Inform. Theory*, **64**(3), 1945–1956.

[5] Beimel, A. 2008. Private information retrieval: a primer. Ben-Gurion University.

[6] Beimel, A., Ishai, Y., Kushilevitz, E., and Raymond, J.F. 2002. Breaking the O(n1(2k-1)/) barrier for information-theoretic Private Information Retrieval. Pages 261–270 of: *The 43rd Annual IEEE Symposium on Foundations of Computer Science, 2002. Proceedings.*

[7] Blackburn, S.R., and Etzion, T. 2016. PIR array codes with optimal PIR rates. *CoRR*, abs/1609.07070.

[8] Blackburn, S.R., Etzion, T., and Paterson, M.B. 2017 (June). PIR schemes with small download complexity and low storage requirements. Pages 146–150 of: *2017 IEEE International Symposium on Information Theory (ISIT).*

[9] Blackburn, S.R., Etzion, T., and Paterson, M.B. 2020. PIR schemes with small download complexity and low storage requirements. *IEEE Trans. Inform. Theory*, **66**(1), 557–571.

[10] Cachin, C., Micali, S., and Stadler, M. 1999. *Computationally private information retrieval with polylogarithmic communication.* Springer, Berlin, Heidelberg.

[11] Chan, T.H., Ho, S.-W., and Yamamoto, H. 2014. Private information retrieval for coded storage. *CoRR*, abs/1410.5489.

[12] Chan, T.H., Ho, S., and Yamamoto, H. 2015 (June). Private information retrieval for coded storage. Pages 2842–2846 of: *2015 IEEE International Symposium on Information Theory (ISIT).*

[13] Chee, Y.M., Feng, T., Ling, S., Wang, H., and Zhang, Liang, F. 2013. Query-efficient locally decodable codes of subexponential length. *Comput. Complex.*, **22**(1), 159–189.

[14] Chor, B., Goldreich, O., Kushilevitz, E., and Sudan, M. 1995. Private information retrieval. Pages 41– of: *Proceedings of the 36th Annual Symposium on Foundations of Computer Science.* FOCS '95. IEEE Computer Society, Washington, DC.

[15] Chor, B., Kushilevitz, E., Goldreich, O., and Sudan, M. 1998. Private information retrieval. *J. ACM*, **45**(6), 965–981.

[16] Dvir, Z., and Gopi, S. 2015. 2-Server PIR with sub-polynomial communication. Pages 577–584 of: *STOC '15: Proceedings of the 2015 ACM Symposium on Theory of Computing.* ACM, New York.

[17] Dvir, Z., and Gopi, S. 2016. 2-Server PIR with subpolynomial communication. *J. ACM*, **63**(4), Art. 39, 15.

[18] Fazeli, A., Vardy, A., and Yaakobi, E. 2015 (June). Codes for distributed PIR with low storage overhead. Pages 2852–2856 of: *2015 IEEE International Symposium on Information Theory (ISIT).*

[19] Fazeli, A., Vardy, A., and Yaakobi, E. 2015. PIR with low storage overhead: coding instead of replication. *CoRR*, abs/1505.06241.

[20] Gasarch, W. 2004. A survey on private information retrieval. *Bull. EATCS*, **82**(72–107), 113.

[21] Gentry, C., and Ramzan, Z. 2005. *Single-database private information retrieval with constant communication rate.* Springer, Berlin, Heidelberg.

[22] Goldreich, O., Karloff, H., Schulman, L.J., and Trevisan, L. 2006. Lower bounds for linear locally decodable codes and private information retrieval. *Comput. Complex.*, **15**(3), 263–296.

[23] Hielscher, E. 2007. A survey of locally decodable codes and private information retrieval schemes. https://citeseerx.ist.psu.edu/document?repid=rep1&type=pdf&doi=f0e9a3406eeb7facef0c0ce55e0a3c67cb28877d.

[24] Katz, J., and Trevisan, L. 2000. On the efficiency of local decoding procedures for error-correcting codes. Pages 80–86 of: *Proceedings of the Thirty-Second Annual ACM Symposium on Theory of Computing*. STOC '00. ACM, New York.

[25] Kushilevitz, E., and Ostrovsky, R. 1997. Replication is not needed: single database, computationally-private information retrieval. Pages 364 of: *Proceedings of the 38th Annual Symposium on Foundations of Computer Science*. FOCS '97. IEEE Computer Society, Washington, DC.

[26] Li, R., and Wootters, M. 2019. Lifted multiplicity codes and the disjoint repair group property. Art. No. 38, 18 of: *Approximation, randomization, and combinatorial optimization: Algorithms and techniques*. LIPIcs – Leibniz International Proceedings in Informatics, vol. 145. Schloss Dagstuhl. Leibniz-Zentrum für Informatik, Wadern.

[27] Menezes, A.J., van Oorschot, P.C., and Vanstone, S.A. 1997. *Handbook of applied cryptography*. CRC Press Series on Discrete Mathematics and Its Applications. CRC Press, Boca Raton, FL.

[28] Ostrovsky, R., and Skeith, W.E. 2007. *A survey of single-database private information retrieval: techniques and applications*. Springer, Berlin, Heidelberg.

[29] Polyanskii, N., and Vorobyev, I. 2019. Constructions of batch codes via finite geometry. Pages 360–364 of: *2019 IEEE International Symposium on Information Theory (ISIT)*.

[30] Raghavendra, P. 2007. A note on Yekhanin's locally decodable codes. *Electron. Colloquium Comput. Complex.*, **14**(01).

[31] Rao, S., and Vardy, A. 2016. Lower bound on the redundancy of PIR codes. *CoRR*, abs/1605.01869.

[32] Shah, N.B., Rashmi, K.V., and Ramchandran, K. 2014 (June). One extra bit of download ensures perfectly private information retrieval. Pages 856–860 of: *2014 IEEE International Symposium on Information Theory*.

[33] Sun, H., and Jafar, S.A. 2017. The capacity of private information retrieval. *IEEE Trans. Inform. Theory*, **63**(7), 4075–4088.

[34] Sun, H., and Jafar, S.A. 2019. The capacity of symmetric private information retrieval. *IEEE Trans. Inform. Theory*, **65**(1), 322–329.

[35] Wehner, S., and de Wolf, R. 2005. *Improved lower bounds for locally decodable codes and private information retrieval*. Springer, Berlin, Heidelberg.

[36] Yekhanin, S. 2007. *Locally decodable codes and private information retrieval schemes*. Ph.D. thesis, Massachusetts Institute of Technology, Cambridge, MA. AAI0819886.

[37] Yekhanin, S. 2007. Towards 3-query locally decodable codes of subexponential length. Pages 266–274 of: *STOC '07: Proceedings of the 39th Annual ACM Symposium on Theory of Computing*. ACM, New York.

[38] Yekhanin, S. 2010. Locally decodable codes. *Found. Trends Theor. Comput. Sci.*, **6**(3), front matter, 139–255 (2012).

[39] Yekhanin, S. 2010. *Locally decodable codes and private information retrieval schemes*, 1st ed. Springer-Verlag, Berlin, Heidelberg.

[40] Yekhanin, S. 2011. Locally decodable codes: a brief survey. Pages 273–282 of: *Coding and cryptology*. Lecture Notes in Computer Science, vol. 6639. Springer, Heidelberg.

[41] Yuval, I., and Kushilevitz, E. 1999. Improved upper bounds on information-theoretic private information retrieval (extended abstract). Pages 79–88 of: *Annual ACM Symposium on Theory of Computing (Atlanta, GA, 1999)*. ACM, New York.

10
Batch Codes

Previous chapters have addressed various topics related to the distributed storage and transmission of data on networks of computers. These have included: network coding where problems related to the efficient transmission of information through a network of servers were considered; locally decodable and locally repairable codes capable of reconstructing a codeword with errors or erasures by querying a few coordinate positions; coding for distributed storage so the contents of a server can be restored from adjacent servers and private information retrieval (PIR) that considers how information can be coded so a user is able to download information in privacy. This chapter considers coding information on servers in such a manner that the amount of information downloaded from any particular server is limited regardless of the information requested – referred to as *batch codes*. The literature on this type of coding and its relationships with the other types of coding considered is perhaps less developed than the other aspects of information storage considered. The potential value of such a system is nonetheless of interest. A brief glimpse of the topic is given.

The next section introduces the notion of batch codes and considers some of their basic properties. Section 10.2 considers an important class of batch codes, the *combinatorial batch codes* (CBCs) which involve no coding of the original data, only the storage of carefully designed subsets in a manner that achieves the goal of limited download from any server. Just as there are natural connections between PIR protocols and locally decodable codes (LDCs), the final Section 10.3 considers relations between the three types of coding.

10.1 Batch Codes

The data/server model will be much as before. A database $\boldsymbol{x} \in \Sigma^n$ is assumed to be n symbols over an alphabet Σ, often (but not necessarily) assumed to be binary, $\Sigma = \{0,1\}$. There are m servers (often referred to as buckets in the literature) whose contents are possibly coded versions of subsets of $\boldsymbol{x}$, i.e., there is a coding function C such that

$$C\colon \Sigma^n \longrightarrow \left(\Sigma^*\right)^m$$

where Σ^* indicates a string of arbitrary length over Σ (**not** the set of nonzero symbols as used before) – thus the map C yields a set of m strings, each of some length and each stored on one of the m servers.

The initial problem of interest will be to design an encoding function such that k users decide independently one symbol (each) they wish to recover from the servers in such a way that no server is required to download more than t symbols to satisfy the users, i.e., the users choose subscripts $i_1, i_2, \ldots, i_k$ so that the decoding algorithm provides $x_{i_1}, x_{i_2}, \ldots, x_{i_k}$ to the k users, downloading no more than t bits from any server. There are two commonly considered cases – where the k indices are distinct and where repetition is allowed. The first case might arise when a single user is requesting the symbols and the second where k users are requesting them and there is no coordination between the users and each user determines a distinct way to download the bit of interest. The second case will be referred to as a *multiset batch code*. (The standard definition of the term multiset allows for repeated elements.) When distinct elements are required, it will be referred to simply as a batch code and when repetition of the requests is allowed it will be referred to explicitly as a multiset batch code.

An interesting subcase is that of *replication-based* batch codes [14] where, rather than coding, subsets of the original database are stored directly on the servers to satisfy the download requirement. This reduces the situation to a purely combinatorial problem and such codes are referred to as CBCs which will be discussed in the next section.

The seminal paper [14] lays out many of the basic properties of batch codes.

Definition 10.1 An (n, N, k, m, t) batch code over an alphabet Σ is defined by an encoding function $C\colon \Sigma^n \longrightarrow \left(\Sigma^*\right)^m$ (mapping a database Σ^n of n symbols over an alphabet Σ to m nonempty strings, each of some arbitrary length stored on the m servers) and a decoding algorithm A such that

(i) the total number of bits (length) stored in all servers is N;

(ii) for any database $\boldsymbol{x} \in \Sigma^n$ and set (distinct elements) $\{i_1, i_2, \ldots, i_k\} \subseteq [n]$ the algorithm

$$A(C(\boldsymbol{x}), i_1, i_2, \ldots, i_k) = (x_{i_1}, x_{i_2}, \ldots, x_{i_k})$$

retrieves the k requested elements and A receives at most t symbols from each server (positions determined by indices).

It will be assumed that batch codes are systematic meaning each of the elements of the database is included uncoded on some server. Further, the term (n, N, k, m) standard batch code refers to an $(n, N, k, m, 1)$ batch code over $\Sigma = \{0, 1\}$, i.e., $t = 1$ with distinct queries. The notion is easily extended to the case $t > 1$.

With minor modification:

Definition 10.2 An (n, N, k, m) *multiset batch code* is an (n, N, k, m) standard batch code (i.e., $t = 1$) such that for any multiset $\{i_1, i_2, \ldots, i_k\} \in [n]$ (not necessarily distinct) there is a partition of the servers such that $S_1, S_2, \ldots, S_k \subseteq [m]$ such that each item $x_{i_j}, j \in [k]$ can be recovered by reading at most one symbol from each server in $S_j, j = 1, 2, \ldots, k$.

It is important to note that in the multiset case, when users request the same element from the database, distinct download estimates of that element are required. Equivalently the algorithm is unaware there is repetition among the elements requested.

Definition 10.3 A *primitive batch code* is an (n, N, k, m) batch code where each server contains a single symbol (hence $N = m$).

The following lemma ([14], lemma 2.4) gives some obvious relations between batch codes with various parameter sets. Only two of the statements are demonstrated:

Lemma 10.4 *The following statements hold for both batch and multiset batch codes:*

- *(i)* *An (n, N, k, m, t) batch code implies an (n, tN, k, tm) batch code (thus downloading at most a single element from a server).*
- *(ii)* *An (n, N, k, m) batch code implies an (n, N, tk, t) code and an $(n, N, k\lceil m/t \rceil, t)$ code.*
- *(iii)* *An (n, N, k, m) (binary) batch code implies an (n, N, k, m) code over $\Sigma = \{0, 1\}^w$ for arbitrary w.*
- *(iv)* *An (n, N, k, m) batch code over $\Sigma = \{0, 1\}^w$ implies a (wn, wN, k, wm) code over $\Sigma = \{0, 1\}$.*

To see (i), if the m servers and their contents are replicated t times, then rather than downloading t items from the original server, a single item can be downloaded from each of the t replicated servers, proving the assertion. To see the first part of (ii) consider an (n, N, k, m) batch code and consider expanding the query set from k indices to tk, say in t batches of k. Each batch of k queries uses the original $t = 1$ batch code, downloading at most one element from each server. Repeating the process for each of the t batches of queries means that the number downloaded from a server is at most t and hence the overall effect is that of an (n, N, tk, m, t) batch code. The remaining arguments are similar.

In a similar vein, consider the following lemma for multiset batch codes (referred to as the "Gadget Lemma" in [14], lemma 3.3, [18], lemma 1):

Lemma 10.5 *Let $C = (n, N, k, m)$ be a multiset batch code ($t = 1$). Then for any positive integer r there is an (rn, rN, k, m) multiset batch code C_r.*

Proof: Let $\boldsymbol{x} = \{x_1, x_2, \ldots, x_n\}$ denote the database for the multiset batch code C. Define the $r \times n$ data array $\boldsymbol{x}' = (x_{ij},\ i = 1, 2, \ldots, r,\ j = 1, 2, \ldots, n)$, $x_{ij} = x_j, j = 1, 2, \ldots, n$. In the batch code C replace the element x_j in all servers combining it, possibly in coded form, with the r-tuple $x_{ij}, i = 1, 2, \ldots, r$ (the j-th column of the data array $\boldsymbol{x}'$). It follows from the fact that any k-multiset drawn from the data array $\boldsymbol{x}'$ (of size rn) can be obtained by querying the same servers as in the code C and the multiset batch code property follows. ■

A simple example of a batch code from [14] is instructive. Suppose $\boldsymbol{x}$ is a database of n symbols and it is desired to download 2 symbols from servers with $t = 1$. This requires at least two servers each containing the entire database with a total storage of $2n$ symbols. An alternative is to consider a three-server case by dividing the database into two halves, say L and R and computing $L \oplus R$, each of the three servers containing $n/2$ symbols (and each stored on a separate server). Thus $m = 3$, $k = 2$ and we want $t = 1$. Consider x_{i_1}, x_{i_2} a pair of items requested by a user, downloading at most one item from a server. If the items are on different servers, they can be downloaded successfully. If they reside on the same server (say L), then one item (say i_1) can be downloaded from L. The other item can be obtained from downloading some item i_3 from R and $i_2 \oplus i_3$ from $L \oplus R$ to obtain i_2, thus achieving the download with one item from each server. Thus an $(n, 3n/2, 2, 3)$ multiset batch code. This is an example of the *subcube batch codes* to be considered shortly.

Another simple scheme is also interesting ([14], lemma 3.4). Suppose the database size is n and the batch code is to be primitive ($N = m$). The number of servers is $m = n + 1$ with a single distinct symbol stored on each of the first n servers and $x_1 \oplus x_2 \oplus \cdots \oplus x_n$ stored on the $(n+1)$-st. Two symbols are to be read with at most one downloaded per server. If $\{i_1, i_2\}$ are such that $i_1 \neq i_2$, the symbols can be read directly from the associated servers (among servers 1 to n). If $i_1 = i_2$, then x_{i_1} can be read from the i_1-th server and the other i_1 the XOR of symbols from all other servers – leading to an $(n, n + 1, 2, n + 1)$ primitive multiset batch code. Refer to this code as C_*.

It is of interest to consider using existing batch codes to construct others and in this the following composition lemma ([14], lemma 3.5), stated without proof, is useful:

Lemma 10.6 *Let C_1 be an (n_1,N_1,k_1,m_1) batch code and C_2 an (n_2,N_2,k_2,m_2) such that $N_1 = m_1n_2$. Then there is an (n,N,k,m) batch code with $n = n_1, N = m_1N_2, k = k_1k_2$, and $m = m_1m_2$. Moreover, if C_1 and C_2 are multiset batch codes, then so is C and if the servers of C_2 are of equal size, then so also are those of C. Denote $C = C_1 \otimes C_2$.*

The literature on batch codes contains a number of construction techniques based on a variety of combinatorial structures. For the remainder of this section only a few types are considered: the subcube batch codes, batch codes obtained from linear error-correcting codes, batch codes derivable from bipartite graphs and batch codes from multiplicity codes. While the constructions discussed are typical, they are by no means exhaustive of the techniques available. The original paper on batch codes [14] considered several such constructions.

Subcube Codes

Subcube codes [14] have previously been described for LDC and PIR codes and their construction for (multiset) batch codes is on very similar lines. Consider the 3×3 square array with parity checks on rows and columns and on checks as illustrated in Figure 10.1(a). The 3×3 array contains the nine elements of the database. There are three row checks and three column checks. There is an overall parity check (which is the XOR of all nine elements in the database). It is easy to check that any four elements of the database can be recovered by downloading a single element from the appropriate cells. Recall that in the multiset case four distinct downloads must be obtained, corresponding to four users requesting the same database element.

Example 10.7 To obtain the multiset x_2,x_2,x_2,x_2 from the $(9, N = m = 16, k = 4)$ primitive multiset batch code of Figure 10.1(a) one can download the sets as follows:

$$\{x_2\}$$
$$\{x_1 \oplus x_2 \oplus x_3, x_1, x_3\}$$
$$\{x_2 \oplus x_5 \oplus x_8, x_5, x_8\}$$
$$\left\{\bigoplus_{i=1}^{9} x_i, x_4 \oplus x_5 \oplus x_6, x_7 \oplus x_8 \oplus x_9, x_1 \oplus x_4 \oplus x_7, x_4, x_7, x_3 \oplus x_6 \oplus x_9, x_6, x_9\right\}$$

Adding the elements within a set yields an independent estimate of x_2. Notice that no cell is downloaded from more than once.

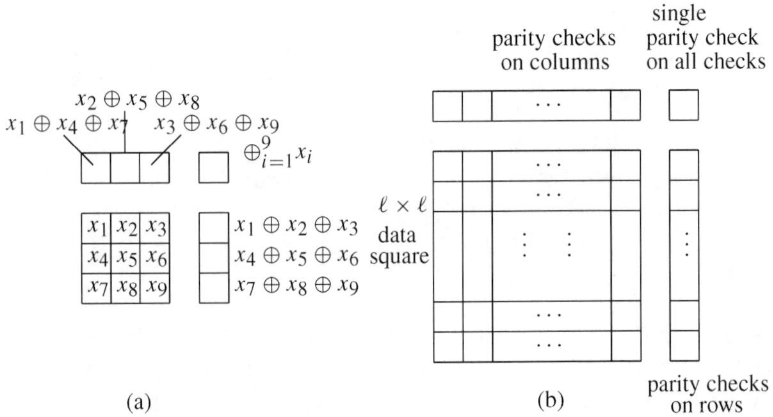

Figure 10.1 Subcube codes: (a) $d = 2$ (b) $\ell = 2$ arbitrary ℓ

The example is generalized to $\ell \times \ell$ database cells as shown in Figure 10.1(b) and it is straightforward to verify that this gives an $(\ell^2, N = m = (\ell + 1)^2, k = 4)$ primitive multiset batch code. The check on parity-checks cell contains an overall parity check.

This two-dimensional case codes can be generalized to higher dimensions. For example in three dimensions ($d = 3$) one can form an $\ell \times \ell \times \ell$ cube of cells containing the database. One could form a $1 \times \ell \times \ell$ array on the right-hand side of parity check on rows of the database. Similarly one can form arrays of cells of parity checks on front to back rows and also of columns to form a total of $3\ell^2$ parity-check cells. Parities on the $3 \times \ell \times \ell$ "side" parities can be formed and finally one could form a single cell containing the XOR of all ℓ^3 database elements. One can verify that such is a primitive multiset batch code with parameters $(n = \ell^3, N = m = (\ell + 1)^3, k = 2^3)$.

The extension to higher dimensions is immediate to yield, in dimension d, an $(n = \ell^d, N = m = (\ell + 1)^d, k = 2^d)$ primitive multiset batch code ([14], lemma 3.6).

Batch Codes from Linear Codes

The notions of using LDCs, in particular smooth codes, and their relation to batch codes, were noted in [14] and are discussed further in Section 10.3. The purpose of this subsection is to consider more direct uses of the properties of linear block codes. The principles of such an approach were laid down in the work of [16]. There is a conflict in notation between the areas of error-correcting codes and their application to batch codes. We adjust the notation of other works to that introduced so far in this chapter.

Definition 10.8 The alphabet $\Sigma = \mathbb{F}_q$ is considered (rather than the binary considered so far). A primitive multiset batch code will be referred to as a *linear* batch code if the contents of all servers are linear combinations over $\mathbb{F}_q$ of database elements.

Only primitive multiset batch codes will be considered (one $\mathbb{F}_q$ element per server or bucket). For the finite field $\mathbb{F}_q$ let $\boldsymbol{x} = \{x_1, x_2, \ldots, x_n\} \in \mathbb{F}_q^n$ be the database and let the contents of the m servers be $\boldsymbol{y} = \{y_1, y_2, \ldots, y_m\} \in \mathbb{F}_q^m$ consisting of linear combinations over $\mathbb{F}_q$ of the database elements, stored in the $N = m$ servers, one $\mathbb{F}_q$ element per server. In particular let

$$y_i = \sum_{j=1}^{n} g_{j,i} x_j, \ i = 1,2,\ldots,m, \ j = 1,2,\ldots,n \quad x_i, y_j, g_{i,j} \in \mathbb{F}_q$$

and form the $n \times m$ matrix $\boldsymbol{G} = (g_{j,i})$, i.e., $\boldsymbol{y} = \boldsymbol{x}\boldsymbol{G}$. Let $\boldsymbol{g}_{(i)} \in \mathbb{F}_q^m$ denote the i-th row of $\boldsymbol{G}, i = 1,2,\ldots,n$ and $\boldsymbol{g}^{(j)}$ the j-th column, $j = 1,2,\ldots,m$.

Denote the i-th indicator column vector $\boldsymbol{e}^t_{(i)} \in \mathbb{F}_q^n$ with a one in the i-th position and zeros elsewhere. The matrix $\boldsymbol{G}$ will define an $(n, N = m, k, m)_q$ multiset batch code (over $\mathbb{F}_q$) if *any* multiset of database items $\{x_{i_1}, x_{i_2}, \ldots, x_{i_k}\}$ can be retrieved from the contents of the servers in k disjoint ways. Suppose from $\boldsymbol{G}$ it is possible to form k disjoint sets of columns of $\boldsymbol{G}$, $S_j, j = 1,2,\ldots,k$ such that for any such database subset (multiset) $\{i_1, i_2, \ldots, i_k\}$ it is possible to form k distinct sums of columns such that

$$\boldsymbol{e}^t_{(i_j)} = \sum_{\ell \in S_j} \alpha_\ell \boldsymbol{g}^{(\ell)}, j = 1,2,\ldots,k,$$

where $\boldsymbol{e}^t_{(i_j)}$ is an indicator vector (weight one) and all such indicator vectors can be formed. It follows the query set can be obtained as

$$x_{i_j} = \left(\boldsymbol{x}, \boldsymbol{e}^t_{(i_j)}\right) \quad \text{(inner product)}.$$

Note the matrix G must have rows of weight at least k since the columns of G can be partitioned to form $e^t_{(j)}$ from k distinct subsets of columns and similarly, G must be of full rank since any of the indicator column vectors $\boldsymbol{e}^t_{(j)}, j = 1,2,\ldots,n$ can be formed. Numerous other properties of the code are given [16]. The success of this method depends on the ability to form the column subsets required.

Example 10.9 An example of such codes from binary Hamming codes [2] is adapted. Consider a binary Hamming code $(2^\ell - 1, 2^\ell - 1 - \ell, 3)_2$ with $(2^\ell - 1 - \ell) \times (2^\ell - 1)$ generator matrix $\boldsymbol{G}$ and $\ell \times (2^\ell - 1)$ parity-check matrix $\boldsymbol{H}$. Assume $\boldsymbol{G}$ is in systematic form. A binary nonprimitive batch code with the following parameters is constructed with: $n = 2^\ell - 1 - \ell$ binary database elements total storage of $N = 2^\ell - 1$ bits, $k = 2$ queries and $m = 2^{\ell-1}$ servers. Let $\boldsymbol{x} \in \mathbb{F}_2^n$ be the n binary database elements, and let $\boldsymbol{y} = \boldsymbol{x}\boldsymbol{G} \in \mathbb{F}_2^N$ be the corresponding codeword, and $\boldsymbol{y}\boldsymbol{H}^t = \boldsymbol{0}_N$, the all-zero word of length N. All m servers will contain 2 bits of the codeword y except one which will contain a single element. For convenience label the columns of $\boldsymbol{H}$ by $\boldsymbol{h}_i, i = 1, 2, \ldots, 2^\ell - 1$ and associate the i-position of $\boldsymbol{y}$, y_i, with column $\boldsymbol{h}_i$. The columns of $\boldsymbol{H}$ consist of all $2^\ell - 1$ nonzero binary ℓ-tuples. If $\boldsymbol{h}_i \oplus \boldsymbol{h}_j = \boldsymbol{1}_\ell$, the all-ones ℓ-tuple, then bits y_i and y_j are placed on the same server. The columns of $\boldsymbol{H}$ divide into $2^{\ell-1} - 1$ pairs that add to $\boldsymbol{1}_\ell$ plus the all-ones ℓ-tuple. The single element corresponding to the all-ones column of $\boldsymbol{H}$ is placed into the last server. Notice wlog it can be assumed the code is systematic and database elements of $\boldsymbol{x}$ appear explicitly. The database elements also appear as coded on other servers.

To show that this arrangement allows a batch code with $k = 2$ and $t = 1$, argue as follows. It is desired to retrieve y_i and y_j, $i \neq j$. If they reside on separate servers, they can be retrieved. If they are on the same server, it is argued as follows. Suppose y_i is downloaded. Let $\boldsymbol{h}_j$ be the binary ℓ-tuple column of $\boldsymbol{H}$ associated with y_j. It will always be possible (usually in many different ways) to represent $\boldsymbol{h}_j$ as the sum of two other columns say $\boldsymbol{h}_r$ and $\boldsymbol{h}_s$, $\boldsymbol{h}_j = \boldsymbol{h}_r \oplus \boldsymbol{h}_s$, corresponding to bits y_s and y_r with neither $\boldsymbol{h}_r$ nor $\boldsymbol{h}_s$ equal to $\boldsymbol{h}_j$. These clearly reside on distinct servers and y_r and y_s can be downloaded and added to give $y_j = y_r \oplus y_s$, resulting in obtaining y_i and y_j with $t = 1$ when they are on the same server.

Other techniques for obtaining batch codes from linear codes, GRM codes in particular are discussed in [2] as well in [14].

In general, it can be shown that the generator matrix for an $(n, N, k, m = N)$ primitive batch code yields a binary error-correcting code of length N, dimension n and minimum distance $\geq k$. It is noted the converse is not true in that the generator matrix of an error-correcting code may not yield a batch code.

Batch Codes from Bipartite Graphs

The use of unbalanced bipartite expander graphs in constructing good batch codes was noted in [14] (a balanced bipartite graph has an equal number of left and right vertices). This approach is able to take advantage of the vast literature on such graphs but tends to produce batch codes that are not optimal. The more combinatorial/constructive approach in [18] is considered here.

Consider a bipartite graph $\mathcal{G}(V_1, V_2, \mathcal{E})$ with a left set of vertices V_1, $|V_1| = n$ and a right set, V_2, $|V_2| = N - n$ with edge set $\mathcal{E}$ containing edges only between the two sets. It is shown how, under certain circumstances, such a graph can yield an (n, N, k, m) primitive multiset batch code (theorem 1 and lemma 2 of [18]). The n left vertices are associated with the database symbols and assumed binary for this discussion. The $N - n$ right vertices are associated with parity checks on the left vertices and the total N symbols will be associated with the contents with the $N = m$ servers, one per server (i.e., a primitive multiset code). The resulting code is systematic. (This is not a Tanner graph of a code as defined in Chapter 3 as the left vertices are only the database bits and do not include parity bits.)

The notion of *repair group* in the context of the graph is considered. The left vertices, associated with the database bits, are labeled $x_1, x_2, \ldots, x_n$ and the right vertices with the parity checks $c_{n+1}, c_{n+2}, \ldots, c_N$. There is an edge between x_i and c_j if x_i participates in the check c_j and the edge is denoted $(x_i, c_j) \in \mathcal{E}$. As in Chapter 3 denote the set of neighbors of the left node x_i as N_{x_i}, and of the check node c_j by N_{c_j}. Further, assuming $(x_i, c_j) \in \mathcal{E}$ is an edge of the graph, denote by N_{x_i, c_j} (resp. N_{c_j, x_i}) as the set of neighbors of x_i *except* c_j (resp. as the set of neighbors of c_j except x_i). Let $n_{x_i} = |N_{x_i}|$ denote the degree of left node x_i and similarly n_{c_j}, $|N_{c_j}|$ for right nodes.

Recall that, identifying data bits and check bits with their nodes,

$$c_j = \bigoplus_{x_k \in N_{c_j}} x_k,$$

i.e., the value of the check symbol c_j is the mod 2 sum of the connected left data symbols.

It follows directly from this observation that a repair group for the database symbol x_i can be formed for each $c_j \in N_{x_j}$ as the mod 2 sum of the data symbols in the set

$$R_{i,j} = \left\{ x_k \in N_{c_j, x_i} \right\},$$

i.e., the set of all left nodes participating in the formation of the check symbol c_j except the symbol x_i itself. Clearly $R_{i,j}$ is a repair group for the data symbol x_i since it is the mod 2 sum of its variables that gives x_i, i.e., for $(x_i, c_j) \in \mathcal{E}$

$$\text{since} \quad c_j = \bigoplus_{x_k \in \mathbb{N}_{c_j}} x_k \quad \text{then } x_i = c_j \bigoplus_{x_k \in N_{c_j, x_i}} x_i.$$

This is true for each set $R_{i,j}$, $c_j \in N_{x_i}$.

The problem might arise that these repair groups are not disjoint since only one download is permitted per server. Before discussing this issue, recall that for a subset of vertices of a general graph $\mathcal{G}$ with vertex set V, for any subset of vertices $U \subseteq V$ the *induced subgraph* $\mathcal{G}(U) = \tilde{\mathcal{G}}$ is the graph that contains all the edges of $(v_i, v_j) \in \mathcal{E}$ with $v_i, v_j \in U$.

The conditions and parameters of theorem 1 and lemma 2 of [18] guarantee that the above repair groups are in fact disjoint resulting in the primitive multiset batch code being (n, N, k, m) and these conditions are as follows. Returning to bipartite graphs, suppose there exists a subset of check nodes, $\tilde{V_2} \subset V_2$ such that the induced subgraph $\tilde{\mathcal{G}}(V_1, \tilde{V_2})$ (consisting of all edges between V_1 and $\tilde{V_2}$) with the property that

(i) all vertices in V_1 have degree at least k;
(ii) the bipartite induced graph has girth at least 8.

Then the code $C = (n, N, k, m = N)$ is a primitive multiset batch code.

The proof of this statement shows some surprising elements. Suppose there is a subset of check nodes $\tilde{V_2}$ satisfying the above conditions. Then it is shown the k repair groups are indeed disjoint and, further, that for $i \neq j, i, j = 1, 2, \ldots, n$ any one of the disjoint repair groups for x_i has common elements with at most one of the disjoint repair groups for x_j.

The proof of this interesting result is intricate. While not proven here the following observations are noted. The requirement (i) above ensures that each data symbol vertex has at least k repair groups. It is first established that, under the conditions stated, the repair groups associated with a data symbol vertex are disjoint. Assuming the contrary, it can be shown to imply a graph cycle of length 4 which is a contradiction. To show that repair groups for x_i have common symbols with at most one of the repair groups of x_j, $i \neq j$, again assume the opposite. It is shown this implies either a cycle of length 4 or 6 which is again a contradiction to the assumption of girth 8. To complete the proof of the theorem, an algorithm is given ([18], figure 4) that assigns k queries to k disjoint sets of data symbols.

The remainder of the paper [18] is concerned largely with construction techniques for suitable large girth bipartite graphs, a problem that has received attention in the mathematical literature in the sense there are many papers on the construction of bipartite graphs with large girth. Numerous constructions of suitable graphs are given that result in a wide variety of batch code parameters.

Batch Codes from Multiplicity Codes

It is shown [1] that the multiplicity codes defined in Chapter 8 can also be used to define a class of (primitive) batch codes (as they were used in Chapter 9 to define a class of k-PIR codes). A slightly modified definition of batch code is used:

Definition 10.10 A code C will be called a k-batch code, denoted $(N,n,k)_q$ if for every multiset request of k symbols, $\{i_1,i_2,\ldots,i_k\}$, there exist k mutually disjoint subsets $R_{i_1},R_{i_2},\ldots,R_{i_k}$ of $[N]$ such that for all $j \in [k]$, x_{i_j} is a function of the symbols in R_{i_j}.

Here it is assumed a database of n symbols from $\mathbb{F}_q$ is encoded into N symbols over $\mathbb{F}_q$ such that the $(N,n)_q$ code is linear and a single coded symbol is stored on each server.

Recall from Chapter 8 the multiplicity code $C_{m,d,s,q}$ using m-variable polynomials of degree at most d and order s-evaluations over $\mathbb{F}_q$ has parameters

$$\text{length} = q^m, \quad \text{rate} = \frac{\binom{m+d}{m}}{\binom{m+s-1}{m}q^m}, \quad \text{normalized distance } \Delta = 1 - \frac{d}{sq}.$$

It is shown in [1] that the multiplicity code $C_{m,d,s,q}$ for $d \leq s(q - ks^{m-1} - 2)$ and $k \leq \lfloor q/s \rfloor^{m-1}$ is a k batch code with q^m servers, a database of length $n = \binom{d+m}{m}$ and symbol alphabet of size $q^{\binom{m+s-1}{m}}$ in that k disjoint recovery sets can be generated in the same manner as for the PIR codes, basically by considering lines in the vector space $\mathbb{F}_q^m$ that intersect in a single point, representing codeword coordinate positions. The code is of length $N = q^m$ and a single $\mathbb{F}_q$ element is stored on each server. By the definition of disjoint recovery sets any multiset of $k \leq \lfloor q/s \rfloor^{m-1}$ requests can be satisfied with no server downloading more than a single element.

The constructions of batch codes described here are but a few on the many available. Describing the range of combinatorial techniques brought to bear on the problem would require a separate volume.

10.2 Combinatorial Batch Codes

A special type of batch code, the CBC has attracted attention in the literature. In the original work [14] such a code was referred to as replication coding, where the information stored on the servers is a subset of the database x, i.e., the stored information is not coded at all. Furthermore, only the case of $t = 1$, where only one symbol can be read from a server is generally considered and the set of k queries is distinct (although multiset combinatorial codes have been considered [24]). These conditions tend to cast the situation as a purely combinatorial problem allowing standard techniques to be used.

The following definition (from [17]) specializes the definition of a CBC in a convenient form:

Definition 10.11 An (n,N,k,t) CBC is a set system $(\boldsymbol{x},\mathcal{B})$, where $\boldsymbol{x} = \{x_1,x_2,\ldots,x_n\}$ (the database) is a set of n symbols (also called items, often bits), $\mathcal{B} = \{B_1,B_2,\ldots,B_m\}$ is a collection of m subsets of $\boldsymbol{x}$ stored on m servers, $\mathcal{S}$, where B_i is the set of elements of the database $\boldsymbol{x}$ stored on server i, with a total storage of $N = \sum_{B_i \in \mathcal{B}} \mid B_i \mid$, such that for each k-subset of (distinct) subscripts $\{x_{i_1},x_{i_2},\ldots,x_{i_k}\} \subset \boldsymbol{x}$ there exists a subset $C_i \subseteq B_i$, where $|C_i| \leq t, i = 1,2,\ldots,m$ such that

$$\left\{x_{i_1},x_{i_2},\ldots,x_{i_k}\right\} \subset \bigcup_{i=1}^{m} C_i.$$

Such a set system could as well be described by the sets $\mathcal{B}^* = \{B_1^*,B_2^*,\ldots,B_n^*\}$ where B_i^* is the set of servers storing element x_i of the database $\boldsymbol{x}$ and $\mathcal{S}$ the set of servers. Such a pair of sets $(\mathcal{S},\mathcal{B}^*)$ is referred to as the *dual set system*. In the set system $(\boldsymbol{x},\mathcal{B})$ the set $\boldsymbol{x}$ is referred to as the set of points and $\mathcal{B}$ the set of blocks while in the dual set system $\mathcal{S}$ is the set of servers and $\mathcal{B}^*$ the set of blocks.

Since consideration is restricted to the case $t = 1$ for CBCs, they will be designated as (n,N,k,m) CBCs. Much of the work on such codes is centered on the problem of determining codes, for given parameters n,k and m, that have the lowest storage requirements N. The least such value of N for given parameters is denoted $N(n,k,m)$ and the corresponding batch code with this storage is referred to as *optimal*. A problem of interest is to determine this value for various ranges of parameters.

Interest in this section is restricted to a few of the many examples of parameter sets where exact descriptions of optimal CBC batch codes exist as well as to introduce a few of the combinatorial structures used to obtain optimal codes. Only a few of the many bounds on $N(n,k,m)$ will be noted. The standard batch code model is assumed – distinct queries – although the multiset version will be noted later.

Consider first some simple examples [17] of optimal CBCs, batch codes that achieve the minimum amount of storage for given n,k,m. If the number of servers is equal to the size of the database n, then clearly one symbol can be placed in each server and

$$N(n,k,n) = n.$$

If the number of servers m is equal to the query size, k, then we must have ([17], theorem 2.2)

$$N(n,k,k) = kn - k(k-1),\ n \geq k.$$

For the database $\boldsymbol{x} = \{x_1, x_2, \ldots, x_n\}$ a code that achieves this bound places $x_1, x_2, \ldots, x_k$ into the k servers, one symbol per server. (The argument is valid for any distinct k elements.) It places all the remaining symbols $x_{k+1}, \ldots, x_n$ into each of the servers for a total of $k(n-k+1)$ storage. To see that this is optimal, consider the query for the symbols $x_1, x_2, \ldots, x_k$. These symbols must appear in different servers (by the batch code assumption). Since $t = 1$ the server containing x_1 say, does not contain another copy of x_1. However, it must contain each of the symbols $x_{k+1}, \ldots, x_n$ since the query $x_j, x_2, \ldots, x_k$ must be satisfied for each $x_j, j = k+1, \ldots, n$ and can only be satisfied if the server contains all these elements. Similarly for the other servers.

A slight variation of this argument ([17], theorem 2.5) shows that

$$N(m+1,k,m) = m+k.$$

In this case the optimal code [4, 6] places items $x_1, x_2, \ldots, x_m$ in separate servers, one symbol per server (any m distinct items). Note the database is $\{x_1, x_2, \ldots, x_m, x_{m+1}\}$. It then places k copies of x_{m+1} in any k of the servers (one copy per server chosen). A query of the form $x_1, \ldots, x_{k-1}, x_{m+1}$ can be satisfied since x_{m+1} is contained in at least one of the servers not containing the first $k-1$ symbols.

The determination of $N(m+2,k,m)$ is surprisingly more difficult. It was established in [6] (and in [8]) that for all $m \geq k \geq 1$:

(i) if $m+1-k \geq \lceil \sqrt{k+1} \rceil$, then $N(m+2,k,m) = m+k-2+\left\lceil 2\sqrt{k+1} \right\rceil$;

(ii) if $m+1-k < \lceil \sqrt{k+1} \rceil$, then $N(m+2,k,m) = 2m-2+\left\lceil 1+\frac{k+1}{m+1-k} \right\rceil$.

In addition it is shown that

$$\begin{aligned} N(m+2,k,m) &= N(m+1,k,m-1)+1 \\ &= m+k-2+\left\lceil 2\sqrt{k+1} \right\rceil \quad \text{for all } m > k+\sqrt{k}. \end{aligned}$$

More generally [8],

$$\begin{aligned} N(n,k,m) &\leq N(n-1,k,m-1)+1 \quad \text{for all} \quad n \geq m \geq k+1 \geq 2 \\ N(n,k,m) &\leq 2m+(n-m-1)\left\lfloor \frac{k}{\frac{m-k}{n-m-1}+1} \right\rfloor - b \quad \text{for all } n \geq m+2 \\ b &= \text{remainder } (m-k) \pmod{m-k+1}. \end{aligned}$$

An interesting general result of [17] is outlined. Suppose $n \geq (k-1)\binom{m}{k-1}$. Consider creating an $m \times n$ incidence matrix A of a dual set system with rows indexed with the m servers and columns indexed with the n database elements. Let the first $(k-1)\binom{m}{k-1}$ columns of A each contain $(k-1)$ copies of each possible server subset of size $(k-1)$ – meaning that each particular column with $(k-1)$ ones appears $(k-1)$ times for each of the $\binom{m}{k-1}$ possible server subsets of size $(k-1)$. The remaining $n-(k-1)\binom{m}{k-1}$ columns of A are filled with any k ones arbitrarily. The total number of elements stored is

$$\begin{aligned} N &= (k-1)^2\binom{m}{k-1} + k\left(n-(k-1)\binom{m}{k-1}\right) \\ &= kn-(k-1)\binom{m}{k-1} \end{aligned} \tag{10.1}$$

which is also obtained by noting that each column contains either k or $k-1$ ones and the number containing $(k-1)$ is $(k-1)\binom{m}{k-1}$. This construction clearly represents an (n,N,k,m) CBC since, by construction, for any k elements of the database it will be possible to find k distinct servers from which to download an element. The argument will be made more convincing later in the section through an argument involving systems of distinct representatives and Hall's conditions on set systems and transversals.

The following results of [17] are shown without proof:

Theorem 10.12

$$\begin{aligned} &(i) \quad \textit{For } n \geq (k-1)\binom{m}{k}, \quad N(n,k,m) \leq kn-(k-1)\binom{m}{k-1}. \\ &(ii) \quad \textit{For } n \geq (k-1)\binom{m}{k-1}, \quad N(n,k,m) = kn-(k-1)\binom{m}{k-1}. \end{aligned}$$

The following result has also been shown by several authors [4, 7, 19]:

$$\text{For all } \binom{m}{k-2} \leq n \leq (k-1)\binom{m}{k-1},$$

$$N(n,k,m) = (k-1)n - \left\lfloor \frac{(k-1)\binom{m}{k-1}-n}{m-k+1} \right\rfloor.$$

Also [19]

(i) If $n \geq (k-1)\binom{m}{k-1}-m+k,$ then $N(n,k,m) = n(k-1).$

(ii) If $n \geq (k-1)\binom{m}{k-1}-m+k-1,$ then $N(n,k,m) = n(k-1)-1.$

Numerous other works are available for certain parameter sets [9, 15, 20]. Also a complete enumeration [7] of $N(n,3,m)$ and $N(n,4,m)$ is given in [7]. It remains an open problem to determine $N(n,k,m)$ for $n \leq \binom{m}{k-2}$.

Recall from Definition 10.11 that the dual set system $(\mathcal{S},\mathcal{B}^*)$ where $\mathcal{S} = \{S_1,S_2,\ldots,S_m\}$ is the set of m servers and $\mathcal{B}^* = \{B_1^*,B_2^*,\ldots,B_n^*\}$ is the set of blocks, where B_i^* represents the set of servers that database element x_i resides on. Such a system is entirely equivalent to the $m \times n$ incidence matrix $A = (a_{i,j})$:

$$a_{i,j} = \begin{cases} 1 & \text{if } \ x_j \in S_i \\ 0 & \text{otherwise} \end{cases}$$

where the matrix rows represent servers and columns the database elements.

Suppose the incidence matrix A has the property that for any k columns (corresponding to k database elements) there exist k rows (corresponding to servers) such that the $k \times k$ submatrix has the property that there are 1's in each of the k rows that appear in distinct columns. If for any k columns there are k rows with this property, it is described as having a *k-transversal.* Clearly the dual set system having a k-transversal for any set of k columns is equivalent to a CBC (n, N, k, m).

These comments are closely related to the notion of a system of distinct representatives (SDRs) defined as follows:

Definition 10.13 A set system on $\boldsymbol{x}$ is a set of subsets of the set $\boldsymbol{x}$, $\mathcal{A} = \{A_1, A_2, \ldots, A_m\}, A_i \subseteq \boldsymbol{x}, i = 1, 2, \ldots, m$. The set system has an SDR if there exists an m-tuple of elements $\{x_1, x_2, \ldots, x_m\}$ such that $x_i \in A_i, i \in [m]$ and for $i \neq j$, $x_i \neq x_j$, i.e., the representatives of the sets x_i are distinct.

The following Hall's theorem (sometimes referred to as Hall's marriage theorem [5]) gives conditions for the existence of an SDR:

Theorem 10.14 *A set system $\mathcal{A} = \{A_i, A_2, \ldots, A_m\}$ has an SDR iff for any set $S \subseteq [m]$*

$$|\bigcup_{i \in S} A_i| \geq |S|.$$

The reference to "marriage theorem" follows if one interprets the m sets as the sets of boys m girls know, then having an SDR is equivalent to each girl knowing at least one distinct boy.

For the study of CBCs a slightly weaker form of the theorem is sufficient [4, 17]: Given the dual set system $(\mathcal{S}, \mathcal{B}^*)$ a CBC (n, N, k, m) will exist iff for any k subset of $\mathcal{B}^*$, $\{\mathcal{B}^*_{i_1}, \ldots, \mathcal{B}^*_{i_k}\}$ it is true that

$$\left| \bigcup_{1 \leq j \leq r} \mathcal{B}^*_{i_j} \right| \geq r \quad \text{for all } 0 \leq r \leq k. \qquad \text{HC1}$$

An equivalent form of this condition that is sometimes more convenient to apply is that any r-element subset of $\mathcal{S}$ contains at most r sets of $\mathcal{B}^*$ for $0 \leq r \leq k-1$ as otherwise Hall's condition would be violated, i.e., $\mathcal{B}^*$ has an SDR iff

$$\text{Any } S \subset \mathcal{S},\ |S| = r, \text{ contains at most } r \text{ sets of } \mathcal{B}^* \text{ for all } r \leq m, \quad 0 \leq r \leq k-1. \qquad \text{HC2}$$

These conditions on sets will be used in the construction of CBCs [4, 17].

The relevance of SDRs to the batch coding problem is as follows. If the dual set system $(\mathcal{S}, \mathcal{B}^*)$ is an (n, N, k, m) CBC, then any k sets of $\mathcal{B}^*$ must have an

SDR, i.e., by definition, to retrieve the query set $x_{i_1}, x_{i_2}, \ldots, x_{i_k}$ the code finds servers $S_{i_1}, S_{i_2}, \ldots, S_{i_k}, x_{i_j}$ in S_{i_j}, i.e., $S_{i_j} \in B^*_{i_j}$ as required.

Other aspects of CBCs are briefly noted. The notion of a *c-uniform CBC* or *fixed-rate* CBC was introduced in [17] as follows:

Definition 10.15 A c-uniform CBC is one in which each database element is contained in exactly c servers. Alternatively it is one whose $m \times n$ incidence matrix has each column containing c $1's$.

Clearly the total number of stored elements is cn and results in an (n, cn, k, m) CBC. The issue of interest for such systems is the maximum value of n for which such a system exists which is denoted by $n(m, c, k)$. The uniformity imposed on the problem allows a variety of combinatorial techniques to be employed in their construction.

A great many results on such systems are available. For example the following bound on $n(m, c, k)$ [17] can be shown: Consider an $\binom{m}{k-1} \times n$ matrix $M = (M_{ij})$ with the rows labeled by all possible $(k-1)$ sets of the m servers and the columns labeled with the blocks of the dual system (which are blocks of servers containing a given database element). Let $M_{ij} = 1$ if the block of column j is contained in the row corresponding to a set of $(k-1)$ severs. By the modified Hall's condition HC2 any k blocks of the dual system must contain at least k elements and hence each row of the matrix M can contain at most $(k-1)$ $1's$. Since by the c-uniformity assumption each block of the dual system contains c elements (the c servers the corresponding database element is stored on) and each such set is contained in $\binom{m-c}{k-1-c}$ sets and hence each column of M contains this many $1's$. Thus

$$n \binom{m-c}{k-1-c} \leq (k-1) \binom{m}{k-1}$$

which is manipulated to

$$n(m, c, k) \leq \frac{(k-1)\binom{m}{c}}{\binom{k-1}{c}}.$$

A simple construction [17] that led to Equation 10.1 can also be used for c-uniform CBCs. Consider the following $m \times n$ incidence matrix of a c-uniform CBC with m servers and a database of n elements. Each of the possible $\binom{m}{c}$ columns with c $1's$ is replicated c times. It is seen this represents an $(n = c\binom{m}{c}, N = c^2\binom{m}{c}, k = c+1, m)$ CBC. There are a great many constructions of such uniform CBCs in the literature that take advantage of the many combinatorial regular structures and readers are referred to [3, 4, 17, 19, 22] for other results.

This section has focused on the condition that $t = 1$. For the extension to $t > 1$ readers are referred to [9, 10, 19]. In addition the CBC codes studied here are of the standard type (distinct download elements). The case of multiset CBCs has been considered in [24, 25].

Hall's theorem is used in many works on constructing CBCs. In [9, 10] a modified form of Hall's theorem (a (k,t) Hall condition) is used to consider construction of CBCs with $t > 1$.

The following is also shown ([9], proposition 12): For every set of positive integers n,k,m and t if $m = \lceil k/t \rceil$ and $n \geq tm$, then $N(n,k,m,t) = mn - tm(m-1)$ which gives an interesting generalization of a theorem of [17].

Theorem 13: If $m \geq \lceil k/t \rceil$ and $n > t\left(\left\lceil \frac{k}{t} \right\rceil - 1\right)\binom{m}{\lceil k/t \rceil - 1}$, then

$$N(n,k,m,t) = n\lceil k/t \rceil - t\left(\left\lceil \frac{k}{t} \right\rceil - 1\right)\binom{m}{\lceil k/t \rceil - 1}.$$

The notions of transversals and SDRs are used in [6] to find optimal CBCs for $n = m + 1$ and $m + 2$ as already noted. Similarly, these notions are used in [22] to construct transversal designs and from such designs, c-uniform CBCs for $c \sim \sqrt{k}$ – as noted above they were previously only known for $c = 2, k - 1$ or $k - 2$.

10.3 The Relationship of Batch Codes to LDCs and PIR Codes

As mentioned, there are strong relationships between the notions of LDCs, PIR codes and batch codes. All notions deal with aspects of retrieving data from storage according to different scenarios. A few comments on these relationships are in order.

The notion of recovery sets for information bits, central to PIR codes and batch codes, was already present in the definition of one-step majority logic decodable binary codes from the 1960s. As noted previously, for such codes it is possible to formulate $2e + 1$ parity checks with the property that each check contains the information bit of interest and whose other check bits are disjoint with other check equations. This is another version of disjoint repair groups. Thus if e or fewer errors are made in transmission, a majority of the check equations applied to the received word are correct.

That a smooth LDC can also be viewed as a batch code [14] is evident in the following sense. For a smooth LDC code one can recover an information symbol by querying up to some fixed number of codeword positions. If the symbols of the LDC codeword of length N are placed in N servers, and the k queries of the batch code are processed serially as an LDC code, one obtains from the responses a batch code. Of course there is no notion of errors in the batch code scenario.

Other relationships between batch, PIR and LDC codes are noted in [23] and [13]. They are also illustrated in the work [1] where, as has been described, several classes of batch and PIR codes are constructed from multiplicity and array codes.

An (r, δ, ϵ) LDC has been defined (8.3) as one that produces an estimate of an information symbol by making at most r queries to codeword symbols and yields a correct answer with probability at least $1-\epsilon$ if the fraction of codeword errors is at most δ. Other versions such as locally repairable codes produce estimates of the codeword coordinate position (rather than just information coordinates) which is similar to the case of locally correctable codes (LCCs). The notion of *locality* and *availability* has also been considered [23]:

Definition 10.16 The code C has locality $r \geq 1$ and availability $\alpha \geq 1$ if for any $\boldsymbol{y} \in C$ any symbol in the codeword can be reconstructed by using any of α disjoint sets of symbols, each of size at most r.

The notion of a *k-server PIR code* introduced in [12], discussed in Chapter 9, involves querying *coded* databases and involved the construction of an $s \times m$ binary matrix $\boldsymbol{G}$ which has property A_k if, for each $i \in [s]$, there exists k disjoint subsets of columns of $\boldsymbol{G}$ that add to $\boldsymbol{e}_{(i)}$ the indicator s-tuple of the i-th position. A binary linear $[m, s]$ code C will be called a *k-server PIR code* if there exists a generator matrix $\boldsymbol{G}$ for C with property A_k, i.e., for each $i \in [s]$ there exist k disjoint sets of columns of $\boldsymbol{G}$, $R_j^{(i)}, i = 1, 2, \ldots, s,\ j = 1, 2, \ldots, k$, such that for each indicator vector

$$\boldsymbol{e}_{(i)} = \sum_{j \in R_1^{(i)}} m_j = \cdots = \sum_{j \in R_k^{(i)}} m_j\,, \quad i = 1, 2, \ldots, s.$$

The notion of disjoint repair groups is common to locally decodable, PIR and batch codes.

The idea of a distributed storage code is somewhat different in that they seek to be able to reconstruct the contents of a failed server in an efficient manner (in terms of downloads) from adjacent servers.

Various relationships exist between these types of codes. For example ([23], corollary 1) a linear systematic code is an LRC code with locality r and availability $\alpha = t - 1$ of information symbols iff it is a PIR code with queries of size t and size of reconstruction sets at most r. Additional references on this problem include [11] and [21].

Comments

Of the coding systems considered in this volume, the notions of PIR and batch codes are perhaps the least developed. However, given the importance of privacy and server efficiency problems in today's information storage and transmission environment, it is not too difficult to imagine that a major development in either area could have a far-reaching impact on commercial systems.

References

[1] Asi, H., and Yaakobi, E. 2019. Nearly optimal constructions of PIR and batch codes. *IEEE Trans. Inform. Theory*, **65**(2), 947–964.

[2] Baumbaugh, T., Diaz, Y., Friesenhahn, S., Manganiello, F., and Vetter, A. 2018. Batch codes from Hamming and Reed-Muller codes. *J. Algebra Comb. Discrete Struct. Appl.*, **5**(3), 153–165.

[3] Bhattacharya, S. 2015. Derandomized construction of combinatorial batch codes. *CoRR*, abs/1502.02472.

[4] Bhattacharya, S., Ruj, S., and Roy, B. 2012. Combinatorial batch codes: a lower bound and optimal constructions. *Adv. Math. Commun.*, **6**(2), 165–174.

[5] Bollobás, B. 1986. *Combinatorics*. Cambridge University Press, Cambridge.

[6] Brualdi, R.A., Kiernan, K.P., Meyer, S.A., and Schroeder, M.W. 2010. Combinatorial batch codes and transversal matroids. *Adv. Math. Commun.*, **4**(3), 419–431.

[7] Bujtás, C., and Tuza, Z. 2011. Optimal batch codes: many items or low retrieval requirement. *Adv. Math. Commun.*, **5**(3), 529–541.

[8] Bujtás, C., and Tuza, Z. 2011. Optimal combinatorial batch codes derived from dual systems. *Miskolc Math. Notes*, **12**(1), 11–23.

[9] Bujtás, C., and Tuza, Z. 2012. Relaxations of Hall's condition: optimal batch codes with multiple queries. *Appl. Anal. Discrete Math.*, **6**(1), 72–81.

[10] Bujts, C., and Tuza, Z. 2011. Combinatorial batch codes: extremal problems under Hall-type conditions. *Electron. Notes Discrete Math.*, **38**, 201–206. The Sixth European Conference on Combinatorics, Graph Theory and Applications, EuroComb 2011.

[11] Diffie, W., and Hellman, M. 1976. New directions in cryptography. *IEEE Trans. Inform. Theory*, **22**(6), 644–654.

[12] Fazeli, A., Vardy, A., and Yaakobi, E. 2015. PIR with low storage overhead: coding instead of replication. *CoRR*, abs/1505.06241.

[13] Henry, R. 2016. Polynomial batch codes for efficient IT-PIR. *Proc. Priv. Enhancing Technol.*, **2016**(02).

[14] Ishai, Y., Kushilevitz, E., Ostrovsky, R., and Sahai, A. 2004. Batch codes and their applications. Page 262–271 of: *Proceedings of the Thirty-Sixth Annual ACM Symposium on Theory of Computing*. STOC '04. Association for Computing Machinery, New York.

[15] Jia, D., and Gengsheng, Z. 2019. Some optimal combinatorial batch codes with $k = 5$. *Discrete Appl. Math.*, **262**, 127–137.

[16] Lipmaa, H., and Skachek, V. 2015. Linear batch codes. Pages 245–253 of: *Coding theory and applications*. The CIM Series in Mathematical Sciences, vol. 3. Springer, Cham.

[17] Paterson, M.B., Stinson, D.R., and Wei, R. 2009. Combinatorial batch codes. *Adv. Math. Commun.*, **3**(1), 13–27.

[18] Rawat, A.S., Song, Z., Dimakis, A.G., and Gál, A. 2016. Batch codes through dense graphs without short cycles. *IEEE Trans. Inform. Theory*, **62**(4), 1592–1604.

[19] Ruj, S., and Roy, B.K. 2008. More on combinatorial batch codes. *CoRR*, abs/0809.3357.

[20] Shen, Y., Jia, D., and Gengsheng, Z. 2018. The results on optimal values of some combinatorial batch codes. *Adv. Math. Commun.*, **12**(11), 681–690.

[21] Shor, P.W. 1994. Algorithms for quantum computation: discrete logarithms and factoring. Pages 124–134 of: *35th Annual Symposium on Foundations of Computer Science (Santa Fe, NM, 1994)*. IEEE Computer Society Press, Los Alamitos, CA.

[22] Silberstein, N., and Gál, A. 2016. Optimal combinatorial batch codes based on block designs. *Des. Codes Cryptogr.*, **78**(2), 409–424.

[23] Skachek, V. 2016. Batch and PIR codes and their connections to locally repairable codes. *CoRR*, abs/1611.09914.

[24] Zhang, H., Yaakobi, E., and Silberstein, N. 2017. Multiset combinatorial batch codes. Pages 2183–2187 of: *2017 IEEE International Symposium on Information Theory (ISIT)*.

[25] Zhang, H., Yaakobi, E., and Silberstein, N. 2018. Multiset combinatorial batch codes. *Des. Codes Cryptogr.*, **86**(11), 2645–2660.

11
Expander Codes

Expander graphs, which informally are graphs for which subsets of vertices have a large number of neighbors, have been of interest to numerous areas of mathematics and computer science. The paper/monograph [20] is an excellent reference for such graphs. In the context of coding, the notion of graph expansion arose in the search for codes that could be encoded and decoded with algorithms that have linear complexity in code length.

The aim of this chapter is to understand how the notion of graph expansion leads to the interesting class of expander codes which have low decoder complexity and excellent performance. The literature on the subject of codes and expander graphs is large and only the fundamentals are discussed.

The next section introduces basic graph notions and defines graph expansion along with the use of eigenvalues of the graph adjacency matrix as an aid in determining graph expansion properties. The important notion of a Ramanujan graph is noted but, except for an example, is not pursued. Many of the expander code constructions use a particular technique (and variants) due to Tanner [30], and all such variants will be referred to here as Tanner codes. This construction and its properties are discussed in Section 11.2. The final Section 11.3 gives a discussion of the seminal work of Sipser and Spielman [29] on expander codes as well as some related work.

11.1 Graphs, Eigenvalues and Expansion

All graphs are assumed to be finite, connected, undirected and with no multiple edges or self-loops. A graph $\mathcal{G}$ with vertex set V and edge set E will be denoted $\mathcal{G} = (V, E)$. The work will be concerned largely with two classes of graphs, the d-regular graphs and (d_v, d_c)-regular bipartite graphs. A few of their basic properties are noted.

The *degree* of a vertex $v \in V$ is the number of edges incident with v. The graph $\mathcal{G}$ will be denoted *d-regular* if all vertices have degree d. This use of the parameter d is unfortunate in that it is also used conventionally for code minimum distance. The convention followed in this chapter (as in other chapters) is that d will denote vertex degree while any code minimum distance will be denoted explicitly as $d_{\min}$.

The *diameter* of a graph is the maximum distance (edge distance – the number of edges traversed) between any two distinct vertices and the *girth* of a graph is the length of the shortest cycle in the graph.

As with previous chapters the (d_v, d_c)-regular bipartite graphs will be of interest here with variable nodes of degree d_v and check nodes of degree d_c, referred to as biregular. If there are n variable nodes, then the number of edges is $|E| = nd_v = md_c$ where there are m check nodes.

A graph $\mathcal{G} = (V, E)$, $|V| = v$, has a $v \times v$ *adjacency matrix* $\boldsymbol{A} = (a_{ij})$ where $a_{ij} = 1$ if $(v_i, v_j) \in E$ and zero otherwise. The set of eigenvalues of $\boldsymbol{A}$, $\{\lambda_0 \geq \lambda_1 \geq \cdots \geq \lambda_{n-1}\}$, is referred to as the *spectrum* of the graph. Since the adjacency matrix is real and symmetric, the eigenvalues are real.

There is a wealth of information on graph properties and a few are noted.

Theorem 11.1 ([12], theorem 1.3.14, p. 15) *For a connected graph $\mathcal{G}$ the following are equivalent:*

(i) $\mathcal{G}$ is bipartite;
(ii) if λ is an eigenvalue of $\mathcal{G}$, then so is $-\lambda$ with the same multiplicity;
(iii) if $\rho(A)$ is the maximum eigenvalue of A, then $-\rho(A)$ is also an eigenvalue.

Furthermore ([12], p. 56) if $\mathcal{E}(\lambda)$ is an eigenspace of the adjacency matrix of a bipartite graph corresponding to the eigenvalue λ, then the dimension of $\mathcal{E}(\lambda)$ is the same as that of $\mathcal{E}(-\lambda)$. Additionally a bipartite graph can be characterized ([14], p. 15) as a graph that contains no odd cycles (i.e., a graph is bipartite iff it contains no odd cycles) and any graph of diameter d_1 that contains a cycle has girth $g \leq 2d_1 + 1$. A graph with diameter d_1 must have at least $d_1 + 1$ distinct eigenvalues [17].

Let $\mathcal{G}$ be a d-regular graph with n vertices and spectrum and $\{\lambda_0 \geq \lambda_1 \geq \cdots \geq \lambda_{n-1}\}$. The following theorem summarizes some useful properties of such graphs:

Theorem 11.2 ([20], p. 16) *Let $\mathcal{G}$ be a finite d-regular graph with n vertices with eigenvalues $\lambda_0 \geq \lambda_1 \geq \cdots \geq \lambda_{n-1}$. Then*

(i) $\lambda_0 = d$;
(ii) $\mathcal{G}$ *is connected iff* $\lambda_0 > \lambda_1$;
(iii) $\mathcal{G}$ *is bipartite iff* $\lambda_0 = -\lambda_{n-1}$.

For a (d_v, d_c)-regular bipartite graph [21] the maximum and minimum eigenvalues are $\pm\sqrt{d_v d_c}$. In fact [5] a nontrivial (at least one edge) graph is bipartite iff its spectrum is symmetric about the origin (as noted above) and a connected d-regular graph is bipartite iff $-d$ is an eigenvalue.

An interesting characterization of regular graphs [12] is that a graph is regular iff $\sum_{i=0}^{n-1} \lambda_i^2 = n\lambda_0$ (and if the graph is not connected, the multiplicity of λ_0 is the number of its components).

The girth of d-regular graphs is of interest. For a fixed integer $d \geq 3$ let $\mathcal{G}_i = (V_i, E_i)_{\geq 1}$ be a family of d-regular graphs of increasing order $| V_i |= v_i$ with girth g_i. Such a family is said to have *large girth* if

$$g_i \geq \gamma \log_{d-1}(v_i)$$

for some constant γ. It is known that $\gamma \leq 2$ although the largest such constant for any family of graphs is 4/3 and the family of *Ramanujan* graphs, to be introduced shortly, asymptotically achieves this bound [18].

For coding applications, the parity-check matrix of a linear code of length n with m independent parity checks, **H** is the $m \times n$ matrix indicating the connection between m constraint (check) and n codeword (variable) positions. The related $(n+m) \times (n+m)$ adjacency matrix **A** of the bipartite graph, which includes both the left vertices of the Tanner graph (variables of the code) and the right vertices (constraints of the code), is of the form

$$\boldsymbol{A} = \begin{bmatrix} \boldsymbol{O} & \boldsymbol{H}^T \\ \boldsymbol{H} & \boldsymbol{O} \end{bmatrix}. \tag{11.1}$$

This is a matrix whose associated graph, in particular whose graph expansion properties, will be of interest. In the case of a (d_v, d_c)-regular bipartite graph, the rows of the $m \times n$ matrix $\boldsymbol{H}$ have weight d_c and columns have weight d_v. The matrix $\boldsymbol{H}$ can be viewed as an incidence matrix of a bipartite graph in the sense that row i and column j contains a 1 if there is an edge between constraint node i and variable node j.

Note that

$$\boldsymbol{A}^2 = \begin{bmatrix} \boldsymbol{H}^T\boldsymbol{H} & \boldsymbol{O} \\ \boldsymbol{O} & \boldsymbol{H}\boldsymbol{H}^T \end{bmatrix} \tag{11.2}$$

and if $\boldsymbol{x}$ is an eigenvector of $\boldsymbol{H}^T\boldsymbol{H}$ with eigenvalue $\lambda \neq 0$, then $\boldsymbol{x}' = \boldsymbol{H}\boldsymbol{x}$ is an eigenvector of $\boldsymbol{H}^T$ with the same eigenvalue. Thus $\boldsymbol{H}^T\boldsymbol{H}$ and $\boldsymbol{H}\boldsymbol{H}^T$ have the same eigenvalues (possibly with different multiplicities). It is not difficult to show that if the eigenvalues of the matrix $\boldsymbol{H}$ in Equation 11.1 are $\{\mu_i,\ i = 0,1,\dots,n-1\}$, then the eigenvalues of $\boldsymbol{H}^T\boldsymbol{H}$ are $\{\mu_i^2,\ i = 0,1,\dots,n-1\}$ and similarly the eigenvalues of $\boldsymbol{A}$ are the square roots of those of $\boldsymbol{H}^T\boldsymbol{H}$.

Graph Expansion and the Second Eigenvalue

The notion of graph expansion has played an important role in many aspects of computer science and mathematics. Their appearance in coding theory arises from the construction of asymptotically good codes ([4] and, most notably, [29]). A standard reference for expander graphs is [20].

Let $\mathcal{G} = (V,E)$ be a graph with vertex set V and edge set E. As noted, the two classes of graphs of most interest for present purposes are the set of d-regular graphs and the set of (d_v,d_c)-regular bipartite graphs. There is a connection between these two classes, the *edge-vertex graph*, that will arise sufficiently often that the following definition is useful:

Definition 11.3 Let $\mathcal{G} = (V,E)$ be a d-regular graph with n vertices. The bipartite graph is formed with left set of vertices the $nd/2$ edges of $\mathcal{G}$ and right set of n vertices the vertices of $\mathcal{G}$, with an edge between vertices on the left and right if there is an edge in $\mathcal{G}$ joining the vertices on the right, i.e., if the vertex on the right is an endpoint of the edge on the left. It is a $(2,d)$-regular bipartite graph referred to as the *edge-vertex graph of d-regular graph $\mathcal{G}$* and denoted $\mathcal{G}_{ev}$.

From a coding theory point of view one might expect expander bipartite graphs to be interesting in that having numerous variable/check edges might lead to numerous estimates to decode a codeword symbol/variable node of interest.

In the literature the formal definition of expansion tends to vary slightly. With minor variations the notation of [20] will be followed. For a graph $\mathcal{G}$ (connected, undirected, no multiple edges, or loops, not necessarily bipartite) let S,T be subsets of vertices V and $E(S,T)$ the set of all edges between S and T. The *edge boundary* of a set $S \subset V$ is defined as

$$\partial^* S = E(S,\bar{S}),$$

the set of all edges between S and $\bar{S}$. Less restrictively, one can define

$$\partial S = \bigcup_{v\in S} \partial v, \quad \partial v = \{u \in V \mid (u,v) \in E\}.$$

Note that $\partial v = N_v$, the set of neighbors of v in previous chapters.

Definition 11.4 The graph $\mathcal{G} = (V, E)$ is called an (α, ϵ) expander if for all sets of vertices $S \subset V$ of size $|S| \leq \alpha \, |V|$, $\alpha \leq 1/2$, the set ∂S is of size at least $\epsilon \, |S|$. The parameter ϵ is called the *expansion* parameter of the graph where

$$\epsilon = \min_{\substack{S,\ |S| \leq \alpha |V| \\ \alpha \leq 1/2}} \frac{\partial S}{|S|}.$$

In some works, ∂S is replaced with $\partial^* S$ in this definition. Often interest in expansion is restricted to d-regular graphs. For (d_v, d_c)-regular bipartite graphs the notion is the same except that only expansion of variable nodes is of interest and all neighbor vertices are distinct from the variable nodes – and the distinction disappears:

Definition 11.5 The (d_v, d_c)-regular bipartite graph $\mathcal{G} = (N \cup M, E)$ is called an (α, ϵ) expander if for all sets of vertices $S \subset N$ of size $|S| \leq \alpha \, |N|$, $\alpha \leq 1/2$, the set ∂S is of size at least $\epsilon \, |S|$. The parameter ϵ is called the *expansion* parameter of the graph where

$$\epsilon = \min_{\substack{S,\ |S| \leq \alpha |N| \\ \epsilon \leq 1/2}} \frac{\partial S}{|S|},$$

i.e., only expansion of variable node sets is of interest. Such a code can be referred to an $(n, d_v, d_c, \alpha, \epsilon)$ expander graph or code. For some applications the expansion parameter will be denoted as $c(\alpha)$.

For a given graph $\mathcal{G}$, it appears to be a complex task to determine its expansion property in the sense that the expansion of all vertex subsets of size at most half the vertices should be found. In a surprising result the expansion of graphs is found to be related to the size of the second largest eigenvalue of the graph adjacency matrix.

To discuss this, let $\mathcal{G}$ be a d-regular graph with n vertices and spectrum $\{\lambda_0 = d > \lambda_1 \geq \cdots \geq \lambda_{n-1}\}$. Define the parameter λ^* as

$$\lambda^* = \lambda^*(\mathcal{G}) = \max_{|\lambda_i| \neq d} \, |\lambda_i| \, .$$

For obvious reasons this parameter $\lambda^* = \lambda_1$ is referred to as the *second eigenvalue* of $\mathcal{G}$. The quantity $d - \lambda^*$ for d-regular graphs, referred to as the *spectral gap* of the graph, can be related to the expansion of the graph, as discovered for finite graphs independently by Tanner [31] for (d_v, d_c)-regular bipartite graphs and Alon and Milman [3]. The result of Tanner, translated to our terminology is: let $\mathcal{G}$ be a (d_v, d_c)-regular bipartite graph with second eigenvalue λ^*. Then $\mathcal{G}$ is an (α, ϵ) expander where ([31], theorem 2.1):

$$\epsilon \geq \frac{d_v{}^2}{[\alpha(d_v d_c - \lambda^{*2}) + \lambda^{*2}]}. \tag{11.3}$$

Note that in the Tanner work, λ_2 is the second eigenvalue of $\boldsymbol{MM}^t$, where $\boldsymbol{M}$ is the incidence matrix of variable to constraint vertices (corresponding to the matrix $\boldsymbol{H}$ in Equation 11.3 and hence λ_2 is the square of an eigenvalue of the corresponding matrix $\boldsymbol{H}$).

For a d-regular graph:

Theorem 11.6 ([20], theorem 2.4) *For a d-regular graph $\mathcal{G}$ the expansion parameter ϵ is bounded by*

$$\frac{d - \lambda^*}{2} \leq \epsilon \leq \sqrt{2d(d - \lambda^*)}.$$

This relates the spectral gap $d - \lambda^*$ to expansion. Indeed it can be shown that ([20], theorem 6.1) for any d-regular graph of size n and diameter Δ there exists a constant c such that

$$\lambda^* \geq 2\sqrt{(d-1)}\left(1 - \frac{c}{\Delta^2}\right).$$

Thus d-regular graphs for which the spectral gap is large (λ^* small) will have good expansion.

The edge-vertex graph of a d-regular graph plays a prominent role in the work of Sipser and Spielman [29], as will be discussed in the next section. Its expansion factor in terms of the expansion factor of the d-regular graph $\mathcal{G}$ from which (Definition 11.3) it is derived is given in the next section (see Proposition 11.12). As pointed out in the work of [1], this is a special case of the more general construction of using all paths of length k for the left-hand side of the bipartite graph and vertices for the right-hand side of $\mathcal{G}$.

Let there be a (d_v, d_c)-regular bipartite graph with n variable vertices and $m = nd_v/d_c$ constraint vertices with r_{d_v,d_c} the rank of its adjacency matrix and $\lambda_{\max} = \lambda_0$ and λ^* be the maximum and second eigenvalue, respectively. Then ([19], corollary 1):

(i) $\lambda_{\max} = \sqrt{d_v d_c}$ and

$$\lambda^* \geq \left(\frac{nd_v - d_v d_c}{r_{d_v,d_c} - 1}\right)^{1/2}$$

with equality iff the eigenvalues are $\pm\lambda_{\max}$ with multiplicity 1, $\pm\lambda$ with multiplicity $(r_{d_v,d_c} - 1)$ and 0 with multiplicity $m - r_{d_v,d_c}$.

(ii)

$$\lambda^* \geq \left(\frac{nd_v - d_v d_c}{n-1} \right)^{1/2}$$

with equality if the eigenvalues are $\pm\lambda_{\max}$ with multiplicity 1, $\pm\lambda$ with multiplicity $(n-1)$ and 0 with multiplicity $n-m$.

Ramanujan Graphs

The following result will motivate the definition of Ramanujan graphs:

Theorem 11.7 ([13], theorem 1.3.1) *Let $\{\mathcal{G}_i\}$ be a family of d-regular graphs such that for $|V_i| = v_i \to +\infty$ as $i \to +\infty$, then*

$$\liminf_{i \to +\infty} \lambda^* \{\mathcal{G}_i\} \geq 2\sqrt{d-1}.$$

Notice the bound asymptotically limits the size of the spectral gap. In addition ([13], theorem 1.3.1), if the girth of such a family tends to $+\infty$ with i, then the smallest eigenvalue μ is upper bounded by $\limsup_{i \to +\infty} \mu(\mathcal{G}_i) \leq -2\sqrt{d-1}$.

Ramanujan graphs are extremal in terms of this bound. While our interest in them for this work is limited, they appear sufficiently often in the coding literature, as do Cayley graphs, that they are defined here.

Definition 11.8 (i) A *Ramanujan graph* is a d-regular graph satisfying, for every nontrivial eigenvalue λ,

$$|\lambda| \leq 2\sqrt{d-1} \tag{11.4}$$

and in particular this is true of λ^*.

(ii) A (d_v, d_c)-regular bipartite graph has maximum/minimum eigenvalues [19, 25] of $\pm\sqrt{d_v d_c}$ and is defined to be Ramanujan if

$$\lambda^*(\mathcal{G}) \leq \sqrt{d_v - 1} + \sqrt{d_c - 1}, \tag{11.5}$$

a natural extension of the regular graph case by setting $d_v = d_c$.

It was recently ([25], theorem 5.6) established that there exists an infinite sequence of (d_v, d_c)-regular bipartite graphs for all $d_v, d_c \geq 3$. It has also been established [25] that there exists (d_v, d_c)-regular Ramanujan bipartite graphs of every degree and number of vertices.

Thus the spectral gap of d-regular graphs is maximized when the second eigenvalue achieves the bound of Equation 11.4 and hence the expansion parameter (via spectral gap) ϵ of the graph is maximized for Ramanujan graphs. From the result of Tanner in Equation 11.3 the largest spectral gap for (d_v, d_c)-regular bipartite graphs is also achieved when the second eigenvalue achieves the bound in Equation 11.5.

The second eigenvalue λ^* has become a useful criteria for determining if a given graph is a good expander. The *second eigenvalue problem* for graphs has typically meant the determination of graphs with "small" second eigenvalue or large spectral gap.

The construction of graphs that meet the above bounds has been of great interest in graph theory, computer science and mathematics. Until recently the construction of the most famous examples of such graphs involved a certain amount of number theory, group theory, quaternion algebra and the transformation group $PSL_2(q)$ to construct a celebrated class of graphs labeled $X^{p,q}$. They are Cayley graphs (see below) and the situation is summarized as:

Theorem 11.9 ([13], theorem 4.2.2) *Let p,q be distinct odd primes with $q > 2\sqrt{p}$. The graphs $X^{p,q}$ are $(p+1)$-regular graphs which are connected and Ramanujan. Moreover*

(i) If $\left(\frac{p}{q}\right) = 1$ (Legendre symbol), then $X^{p,q}$ is a nonbipartite graph with $q(q^2-1)/2$ vertices satisfying the girth estimate

$$g(X^{p,q}) \geq 2\log_p q.$$

(ii) If $\left(\frac{p}{q}\right) = -1$, then $X^{p,q}$ is a bipartite graph with $q(q^2-1)$ vertices, satisfying

$$g(X^{p,q}) \geq 4\log_p q - \log_p 4.$$

That the class of graphs constructed has second eigenvalue satisfying the Ramanujan condition is inherent in the construction. The graphs were first constructed in the work of Margulis [26] and later by Gabber and Galil [16] and also considered in the works [24] and [13].

Morgenstern [27] extended the above by giving a class of Ramanujan graphs that are $(q+1)$-regular for all prime powers q, some bipartite and some not. Importantly and impressively the recent work of Marcus et al. [25] has shown the existence of bipartite Ramanujan graphs of every degree and every number of vertices using the notion of interlacing families of polynomials (sets of polynomials whose roots interlace).

The notion of a Cayley graph is sufficiently common in coding literature that a definition [17] may be of value:

Definition 11.10 Let C be a subset of a group G such that if $g \in C$, then $g^{-1} \in G$ (closed under inverses) and does not contain the identity. The vertices of the Cayley graph of G, $\mathcal{G}$, are the elements of G and (g,h) is an edge in $\mathcal{G}$ iff $gh^{-1} \in C$. The graph is undirected with no loops.

11.2 Tanner Codes

The notion of a Tanner code introduced in [30] will be helpful for the expander codes to be introduced in the next section. A brief discussion of a Tanner code and some of their variants, all of which will be referred to as Tanner codes, is given here. Although expressions for the minimum distances of these codes will involve the second eigenvalue, a discussion of the role of expansion in the definition and properties of expander codes is postponed to the next section.

A related concept to a Tanner code is that of a product code whose definition is recalled. A product code of two binary linear codes $C_1 = (n_1,k_1,d_1)_2$ and $C_2 = (n_2,k_2,d_2)_2$ is a collection of $n_1 \times n_2$ arrays such that each column is a codeword in C_1 and each row is a codeword in C_2. One might view the construction as starting with a $k_1 \times k_2$ array of information bits and completing each row and column using the appropriate code and then computing parity checks on parity checks, noting that because of linearity the two possible ways of computing these parity-on-parity checks yield the same result. The resulting product code is an $(n_1n_2,k_1k_2,d_1d_2)_2$ linear code.

Such codes can be decoded by first decoding rows and then decoding columns. It can be shown (e.g., [23]) that this technique is able to correct any pattern of $d_1d_2/4$ errors. With a more sophisticated algorithm it is possible to decode up to $(d_1d_2-1)/2$ errors.

From this product code interpretation, a few variations of a Tanner code will be described, all of which will be termed Tanner codes, although the original work of Tanner [30] used only the first variation. It will be convenient to use notation that overlaps and use of any particular variation will be resolved by sufficient description.

In their most general terms each of them will involve a (d_v,d_c)-regular bipartite graph $\mathcal{G}$ with disjoint vertex sets $V = N \cup M$, $|N| = n$, $|M| = m$, with the vertices of N of degree d_v and those of M of degree d_c – and hence with edge set E, $|E| = nd_v = md_c$. Code minimum distance will be explicitly stated as $d_{\min}$.

Let C be a $(d_c, rd_c, \delta d_c)_2$ linear subcode of length d_c, rate r and relative distance δ. Denote by $\boldsymbol{x} = (x_1,x_2,\ldots,x_n) \in \mathbb{F}_2^n$ and associate the bit x_i with variable node $v_i, i = 1,2,\ldots,n$. For a subset $S = \{i_1,\ldots,i_s\} \subseteq N, |S| = s$ let

$$\boldsymbol{x}|_s = \Big(x\big(v_{i_1}\big),\ldots,x\big(v_{i_s}\big)\Big),$$

i.e., the projection of the binary n-tuple $\boldsymbol{x}$ onto the set S. Let $N_c \subseteq N$ set of degree d_c variable neighbors of the constraint node $c \in M$ in the graph $\mathcal{G}$. Consider the code

$$\mathscr{C}_1 = \left\{ \boldsymbol{x} \in \mathbb{F}_2^n \mid \boldsymbol{x}_{|N_c} \in C \text{ for all } c \in M \right\},$$

i.e., the d_c codeword bits associated with the neighbors of each constraint node $c \in M$ form a codeword in the code C, for some fixed ordering of the variable nodes. The binary code $\mathscr{C}_1$ is of length n and its rate and distance are of interest. It is noted [30] that the properties of the code $\mathscr{C}_1$ will depend on the labeling of the variable node neighbors of the constraint node with the codeword bits, i.e., how the codeword bits are assigned to the d_c neighbors. Each variable node is included in d_v parity-check equations of code $\mathscr{C}$ and each constraint node contributes $d_c - d_c r$ check equations and hence a total of $m(d_c - d_c r)$ equations. To see this one might imagine constructing a parity- check matrix for the code $\mathscr{C}_1$ of length n. In the worst case, for each codeword assigned to the d_c variable neighbors of a constraint code, one might have the $d_c - d_c r$ check equations of a codeword of the code C of length d_c. Hence a bound for the overall rate of the Tanner code is given by

$$R \geq \frac{n - md_c(1-r)}{n} = 1 - (1-r)\frac{d_c m}{n} = 1 - (1-r)c, \quad \text{as } n_v = md_c. \tag{11.6}$$

It will be shown [29] in the next section that if the graph $\mathcal{G}$ is an (α, ϵ) expander, then the minimum distance of the code (for a sufficiently large ϵ) will be at least αn (since a simple decoding algorithm will be given that can correct $\alpha n/2$ errors). This code will be denoted $\mathscr{C}_1(\mathcal{G}, C)$ and is a linear $(n, R \geq (1-(1-r)d_v, d_{min} \geq \alpha n)_2$ code.

Variations on the above code construction technique might be considered. Suppose $\mathcal{G}$ is a d-regular graph or a (d_v, d_c)-regular bipartite graph and $C = (d_c, rd_c, \delta d_c)_2$ a binary linear code of length d_c – as above. One might populate the edges of either of these graphs with codeword bits so that in the d-regular graph the bits on the edges incident with a given vertex (assuming some fixed edge ordering for the codeword bits) contain a codeword of C (for $d_c = d$) or, in the case of the (d_v, d_c)-regular bipartite graph, the edges incident with a constraint node, contain a codeword of C of length d_c. Notice the difference of this model with that of the former where variable nodes were identified with codeword bits rather than edges. Thus in these models the length of the code is either $nd/2$ for the d-regular graph or md_c for the (d_v, d_c)-regular bipartite graph. This model will be designated $\mathscr{C}_2(\mathcal{G}, C)$ where $\mathcal{G}$ is either the d-regular or (d_v, d_c)-regular bipartite. Any confusion will be resolved by designating the model and graph before use.

More generally, one might consider a (d_v, d_c)-regular bipartite graph and two codes, $C_1 = (d_v, r_1 d_v, d_1 = \delta_1 d_v)_2$ and $C_2 = (d_c, r_2 d_c, d_2 = \delta_2 d_c)_2$ and assign codeword bits to all edges so that edges emanating from variable nodes form a codeword of C_1 and those from constraint nodes a codeword of C_2. This Tanner code is denoted $\mathscr{C}_3(\mathcal{G}, C_1, C_2)$. Note that the encoding is not trivial in that it requires codeword bits on edges serve the two purposes of being a bit in a codeword of C_1 as well as one of C_2.

All of these versions have been considered in the literature.

The remainder of the section gives a sketch of work of Janwa and Lal [21, 22] which:

(i) gives a lower bound on the expansion of the graph $(n, 2, d, \alpha, c(\alpha))$ where $\mathcal{G}_{ev}$ is the edge-vertex graph of a d-regular graph $\mathcal{G}$ with n vertices and second eigenvalue λ^*;
(ii) establishes a lower bound on the minimum distance of the code $\mathscr{C}_2(\mathcal{G}_{ev}, C)$ where $\mathcal{G}$ is a (d_v, d_c)-regular bipartite graph [21];
(iii) establishes a lower bound on the minimum distance of the code $\mathscr{C}_3(\mathcal{G}, C_1, C_2)$ where $\mathcal{G}$ is a (d_v, d_c)-regular bipartite graph [21].

The techniques used in both cases are of interest although the proofs are not considered here.

Expansion Properties of $\mathcal{G}_{ev}$ from the Properties of $\mathcal{G}$

Let $\mathcal{G}$ be a d-regular graph with n vertices and second eigenvalue $\lambda^*_{\mathcal{G}}$, and let $\boldsymbol{M}$ be its edge-vertex incidence matrix (of dimension $e \times n$). The edge-vertex graph $\mathcal{G}_{ev}$ of $\mathcal{G}$ then has adjacency matrix

$$\boldsymbol{A}_{\mathcal{G}_{ev}} = \begin{bmatrix} \mathbf{O} & \boldsymbol{M} \\ \boldsymbol{M}^t & \mathbf{O} \end{bmatrix}.$$

The $n \times n$ adjacency matrix of $\mathcal{G}$ is seen to satisfy

$$\boldsymbol{A}_{\mathcal{G}} + d\boldsymbol{I} = \boldsymbol{M}^t \boldsymbol{M}$$

and note that

$$\boldsymbol{A}_{\mathcal{G}_{ev}} \boldsymbol{A}^t_{\mathcal{G}_{ev}} = \begin{bmatrix} \boldsymbol{M}\boldsymbol{M}^t & \mathbf{O} \\ \mathbf{O} & \boldsymbol{M}^t\boldsymbol{M} \end{bmatrix}.$$

The eigenvalues of $\boldsymbol{M}\boldsymbol{M}^t$ are the same (possibly different multiplicities) as those of $\boldsymbol{M}^t\boldsymbol{M}$ and the eigenvalues of $\boldsymbol{A}_{\mathcal{G}_{ev}}$ are the square roots of those of $\boldsymbol{A}_{\mathcal{G}} + d\boldsymbol{I}$. It follows easily then that

$$\lambda_{0,\mathcal{G}_{ev}} = \sqrt{2d} \quad \text{and} \quad \lambda^*_{\mathcal{G}_{ev}} = \sqrt{d + \lambda^*_{\mathcal{G}}}.$$

These can be applied to the expansion result of Tanner (see Equation 11.3) which is recalled, for a (d_v, d_c)-regular bipartite graph as:

$$\epsilon = c(\alpha) \geq \frac{{d_v}^2}{\left[\alpha\left(d_v d_c - \lambda^{*2}_{\mathcal{G}_{ev}}\right) + \lambda^{*2}_{\mathcal{G}_{ev}}\right]},$$

which for the case in hand (for the $(2,d)$-regular bipartite graph $\mathcal{G}_{ev}$) gives:

Proposition 11.11 ([22], proposition 2.1) *Let $\mathcal{G}$ be a d-regular graph with n vertices and second eigenvalue $\lambda^*_{\mathcal{G}}$. Then its edge-vertex graph $\mathcal{G}_{ev}$ is an $(nd/2, 2, d, \alpha, c(\alpha))$ expander graph with maximum eigenvalue and second eigenvalue as in the above equations. The graph expansion factor is given by*

$$c(\alpha) \geq \frac{4}{\alpha\left(d - \lambda^*_{\mathcal{G}}\right) + \left(d + \lambda^*_{\mathcal{G}}\right)}. \tag{11.7}$$

Better Expansion with Alon and Chung Edge Estimate

The expansion coefficient $c(\alpha)$ above can be improved upon using the celebrated result of Alon and Chung [2] on the number of edges in an induced subgraph. Since the Alon and Chung result makes frequent appearances in research on expander codes, a sketch of the expansion coefficient improvement is given.

Let $\mathcal{G}$ be a d-regular graph with n vertices, second eigenvalue λ^*, S a subset of $|S| = \gamma n$ vertices and $e(S)$ the number of edges in the subgraph induced by S. Then

$$\left|e(S) - \frac{1}{2} d\gamma^2 n\right| \leq \frac{1}{2}\lambda^* \gamma(1-\gamma)n. \tag{11.8}$$

It is noted that $\frac{1}{2}d\gamma^2 n$ is approximately the expected number of edges one would expect from a subgraph of αn vertices of a d-regular graph. This interesting result [22] can be used to improve the expansion coefficient $c(\alpha)$ to:

Proposition 11.12 ([22], theorem 2.2) *Let $\mathcal{G}$ be a d-regular graph with n vertices and second eigenvalue $\lambda^*_{\mathcal{G}}$. Then the edge-vertex graph $\mathcal{G}_{ev}$ is an $(nd/2, 2, d, \alpha, c(\alpha))$ expander graph with graph expansion factor of*

$$c(\alpha) \geq \frac{4}{\lambda^* + \sqrt{(\lambda^*)^2 + 4\alpha d(d - \lambda^*)}}. \tag{11.9}$$

It is straightforward to show that this result is indeed an improvement over that of the previous proposition, i.e.,

$$c(\alpha) \geq \frac{4}{\lambda^* + \sqrt{(\lambda^*)^2 + 4\alpha d(d - \lambda^*)}} \geq \frac{4}{\alpha(d - \lambda^*) + d + \lambda^*}.$$

It is clear that in some cases the size of the vertex expansion set size αn and expansion coefficient $c(\alpha)$ can be traded off. In the above theorem concerning the expander graph $\mathcal{G}_{ev}$ edge-vertex graph of $\mathcal{G}$, a d-regular graph $\mathcal{G}$ on n vertices with second eigenvalue λ^*, $\mathcal{G}_{ev}$ an $(nd/2, 2, d, \alpha, c(\alpha))$ expander graph, if one chooses for the code $\mathscr{C} = (\mathcal{G}_{ev}, C)$, a code $C = (d, rd, \delta d)_2$ and

$$\alpha' = \frac{\delta(d\delta - \lambda^*)}{(d - \lambda^*)}$$

then

$$\begin{aligned} c(\alpha') &\geq \frac{4}{\lambda^* + \sqrt{(\lambda^*)^2 + 4d\alpha'(d - \lambda^*)}} \\ &\geq \frac{4}{\lambda^* + \sqrt{(\lambda^*)^2 + 4d\frac{\delta(d\alpha - \lambda^*)}{(d - \lambda^*)}(d - \lambda^*)}} \\ &\geq \frac{4}{\lambda^* + \sqrt{(\lambda^* - 2d\delta)^2}} \\ &\geq \frac{2}{d\delta}. \end{aligned}$$

This computation results in an expander graph with parameters

$$\left(nd/2, 2, d, \frac{\delta(d\delta - \lambda^*)}{(d - \lambda^*)}, \frac{2}{d\delta}\right). \tag{11.10}$$

This result will be of interest in the next section where, as for a previous code, it will be shown that in fact the corresponding code to this expander graph has a minimum distance of $\dfrac{\delta(d\delta - \lambda^*)}{(d - \lambda^*)}$ and rate (Equation 17.2 with $c = 2$) of $2r - 1$. This is in fact the code used in [29] which contained the minimum distance bound

$$\left[\frac{(d\delta - \lambda^*)}{(d - \lambda^*)}\right]^2$$

referred to here as $\mathscr{C}_2(\mathcal{G}_{ev}, C)$ for C a $(d, rd, \delta d)_2$ code, and this bound can be shown to be inferior to the previous bound above.

For the item (iii) mentioned above, it is of interest to determine the minimum distance of the code $\mathscr{C}_3 = C(\mathcal{G}, C_1, C_2)$ following the interesting argument of [21], where $\mathcal{G}$ is a (d_v, d_c)-regular bipartite graph with vertices $N \cup M$, $|N| = n$, $|M| = m$ and $C_1 = (d_v, r_1 d_v, d_1 = \delta_1 d_v)_2$ and $C_2 = (d_c, r_2 d_c, d_2 = \delta_2 d_c)_2$ two binary linear codes. Again assuming a fixed edge assignment to codeword bit positions, label the graph edges incident with variable nodes with codewords of C_1 and those incident with constraint nodes with codewords of C_2 (a single codeword bit per edge). Clearly $\mathscr{C}_3$ is a binary linear code of length $\mathcal{N} = nd_v$ (number of edges in the bipartite graph). The rate of the code can be bounded using the previous argument, i.e., each of the n variable nodes would add at most $d_v - d_v r_1$ equations to the parity-check matrix of $\mathscr{C}$ and each of the constraint nodes $d_c - d_c r_2$ for a total of $n(d_v - d_v r_1) + m(d_c - d_c r_2)$. The code rate R is then bounded as

$$R \geq \Big[\mathcal{N} - (n(d_v - d_v r_1) + m(d_c - d_c r_2))\Big]/\mathcal{N} = r_1 + r_2 - 1.$$

The minimum distance, as above, is derived via estimates of edge counts ([21], theorem 3.1) in the following manner.

As an extension of this edge count notion of Alon and Chung used for the previous case, for the codes of interest here, an estimate of the quantity $e(S,T)$, the number of edges from a set $S \subseteq N$ to $T \subseteq M$ in a (d_v, d_c)-regular bipartite graph with disjoint vertex sets N and M is required. Such edge density estimates were also considered in [33] for the case of balanced bipartite graphs $d_v = d_c$ (i.e., regular bipartite). The improvement of those estimates of [22] is used. Assume the (d_v, d_c)-regular bipartite graph $\mathcal{G}$ (i.e., its adjacency matrix) has second eigenvalue λ^*. Then it is shown ([22], equation 1) that

$$\begin{aligned} \left| e(S,T) - \frac{d_c \mid S \mid \mid T \mid}{n} \right| &\leq \tfrac{\lambda^*}{2}\left(\mid S \mid + \mid T \mid \; - \frac{\mid S \mid^2}{n} - \frac{\mid T \mid^2}{m} \right) \\ &\leq \tfrac{\lambda^*}{2}\Big(|S| + |T| \Big). \end{aligned} \tag{11.11}$$

As noted, this is a generalization of the edge count given in [33] for the d-regular bipartite graph (all nodes, variable and constraint, of degree d). This edge count estimate is used in the proof of the following:

Theorem 11.13 ([21], theorem 3.1) *Let $\mathcal{G}$ be a (d_v, d_c)-regular bipartite graph with n variable nodes and second eigenvalue λ^* and let $\mathscr{C} = C(\mathcal{G}, C_1, C_2)$ be the code described above, assuming $d_1 \geq d_2 > \lambda^*/2$. Then $\mathscr{C}$ is an $(\mathcal{N} = nd_v, R \geq r_1 + r_2 - 1, d_{\min})_2$ where*

$$d_{\min} \geq \frac{n}{d}\left[d_1 d_2 - \frac{\lambda^*}{2}(d_1 + d_2) \right].$$

To decode these codes [21] one can first decode the codewords in C_1 associated with the variable nodes – since they are independent this can be done in parallel. Then the codewords of C_2 associated with the constraint nodes can be decoded in parallel, and so on, iterating until there is no change. It can be shown ([21], theorem 5.2) this algorithm will converge to the correct codeword if the weight of the error word is less than

$$\alpha\frac{\delta_1}{2}\left(\frac{\delta_2}{2}-\frac{\lambda^*}{d_c}\right)nd_v.$$

In some works regular bipartite graphs are of interest, i.e., both sets of vertices have the same degree (hence the same number of vertices on each side of the graph). Consider [7, 33] a d_v-regular bipartite graph $\mathcal{G}$ with vertex sets N, M such that $|N|=|M|=n$, all vertices of degree d_v. The graph is d_v-regular. The graph has nc edges and the code will consist of binary codewords of length $N = nc$ derived from bits assigned to the graph edges. Consider a subcode C a linear $(d_v, rd_v, d_1 = \delta_1 d_v)_2$. The edges from each node of the graph, both left and right sets of nodes, will be assigned a codeword from C, for some fixed ordering of the edges for each node to allow the codeword bits of C to be assigned to the edges, as in the code $\mathscr{C}_3(\mathcal{G}, C, C)$. Thus the set of edges incident with each node, variable and check, contains a codeword of C (again according to some fixed assignments of codeword bits to edges).

As previously, a simple decoding algorithm for this code [33] is to decode each of the subcode words on the edges emanating from the N variable nodes. Since the corresponding codewords are disjoint this can be done in parallel. Then decode the codewords on edges from all constraint node codewords, again in parallel. Continuing in this fashion it is shown in ([33], theorem 6) that, for λ^* the second eigenvalue of the adjacency matrix of $\mathcal{G}$ and $d_1 \geq 3\lambda^*$, as long as the error vector (of length nd_v) $\boldsymbol{e}$ is such that

$$w(\boldsymbol{e}) \leq \alpha\frac{\delta_1}{2}\left(\frac{\delta_1}{2}-\frac{\lambda^*}{c}\right)nd_v$$

for some $\alpha < 1$, the above algorithm will converge to the transmitted codeword, implying the minimum distance of this Tanner code is at least twice this amount. Furthermore the decoding algorithm takes a number of steps that is logarithmic in the number of edges of the graph nd_v.

It is noted [33] that the above decoding algorithm reduces to that of the product codes mentioned if $d_v = m$ and $\mathcal{G}$ is the complete bipartite graph.

It is interesting to observe [29] the following decoding was proposed for the code $(\mathcal{G}, C)$ where the codeword associated with the n variable nodes is such that the set of neighbors of each constraint node supports a codeword of the subcode C. Consider a constraint node c and the set of its d_c variable neighbors forming a word – a d_c-tuple. If this word differs from a codeword of C in at most $\delta d_c/4$ places, then send a "flip" message to each of those variables that differ. Then in parallel flip all variables that received at least one flip message. It is shown [29] that for this algorithm to decode successfully it is necessary that the fraction of variables in error α must be such that

$$\alpha < \delta^2(1/3 - 4\lambda^*/\delta d_c)/16 < \delta^2/48.$$

Techniques introduced in [33] showed how this fraction can be reduced to $\delta^2/4$, a factor of 12 improvement.

11.3 Expander Graphs and Their Codes

The notion of graph expansion was introduced in the first section and aspects of it in terms of the second eigenvalue of a graph were considered in the previous section. The explicit use of graph expansion to determine codes that have very efficient decoding algorithms is considered here.

The seminal work of [29] is first considered. As before, let $\mathcal{G}$ be a (d_v, d_c)-regular bipartite graph with n variable and $m = nd_v/d_c$ check nodes and nd_v edges. Let $C = (d_c, rd_c, \delta d_c)_2$, $r > (d_v - 1)/d_v$, be a binary linear code. Let the variable neighbors of check node c_i be $N_{c_i} = \{v_{i_1}, \ldots, v_{i_{d_c}}\}$ labeled in an appropriate manner. The expander code $\mathscr{C} = C(\mathcal{G}, C)$ (previously $\mathscr{C}_1$) then is a binary code of length n with codeword $\boldsymbol{x} = (x_1, x_2, \ldots, x_n)$ such that for each check node $c_i, i = 1, 2, \ldots m$, $\boldsymbol{x}_{|N_{c_i}}$, the projection of $\boldsymbol{x}$ onto the neighbors of c_i, is a codeword in C.

The following ([29], theorem 7) gives the crucial property of graph expansion to code properties:

Theorem 11.14 *Let $\mathcal{G}$ be an $(n, d_v, d_c, \alpha, d_v/d_c\delta)$ expander (d_v, d_c)-regular bipartite graph and $C = (d_c, r > (d_v - 1)/d_v, d_{\min} > \delta d_c)_2$ binary linear code. Then the binary linear code $\mathscr{C} = (\mathcal{G}, C)$ has rate $> d_v r - (d_v - 1)$ and minimum relative distance at least α.*

An informal argument [29] is used to illustrate the proof. To see the minimum distance of the code is at least αn, suppose the variables support a codeword (of $\mathscr{C}$) of weight at most αn and let X, $\mid X \mid = x$ be the nonzero variable positions, i.e., supporting 1s. There are $d_v x$ edges leaving these variable positions and, by the expansion properties of the graph, these edges lead to at least $(d_v/d_c\delta)x$ constraint nodes. The average number of edges on these constraint nodes is therefore at most $d_v x \big/ \big((d_v/d_c\delta)x\big) = d_c\delta$. There must be at least one constraint node c^* with fewer than $d_c\delta$ edges leaving it (in the original graph the other edges leaving c^* are attached to variable positions with values of 0). By definition the edges leaving c^* support a codeword of C of weight at least $d_c\delta$ and the variable nodes of it supporting 1s are in X. But there are fewer than $d_c\delta$ 1s in it so it cannot be a codeword of C and hence the supposed codeword cannot be of weight less than αn.

Consider the following simple decoding algorithm for the code $\mathscr{C}(\mathcal{G}, C)$ above. Assume the graph has a received codeword with at most $e = \alpha n/2$ errors in it. A constraint node is said to satisfied if it has a value 0, i.e., the mod 2 sum of attached variable values is 0. By "flipping" a value (either variable or constraint) is meant changing 1 to 0 and 0 to 1.

Algorithm 11.15

(i) If there is a variable node with more unsatisfied constraints than satisfied, flip that variable value and all attached constraint values.
(ii) Continue until no such variable node exists.

It seems remarkable that it is possible to show this simple algorithm will correct up to $\alpha n/2$ errors in a received word ([28], [29], theorem 10) and hence the code has minimum distance at least αn, as has been referenced previously. For the argument consider the code $\mathscr{C}(\mathcal{G}, C)$ for a subcode C of all even-weight codewords and (c,d)-regular bipartite expander graph $\mathcal{G} = (n, d_v, d_c, \alpha, 3d_v/4)$. However, such a code is equivalent to a code where the constraint vertices are simple parity checks since in that case if the variable neighbors supported a codeword of even weight its mod 2 sum would be zero. Equivalently the argument applies to any code where the expander (d_v, d_c)-regular bipartite graph corresponds to an $m \times n$ parity-check matrix where the m rows correspond to the right-hand vertices (constraint or parity checks) and the n columns the variable vertices.

Refer to the variable positions supporting errors as "corrupt positions." Assume initially there are $e \leq \alpha n/2$ corrupt variable positions and that by definition of the algorithm e is strictly decreasing as the algorithm progresses. It is required to show that there will exist variable positions attached to more unsatisfied constraints than satisfied ones until there are no unsatisfied constraint positions, i.e., all constraint positions have values of 0 and the corresponding variable positions form a codeword. At any iteration of the algorithm let u be the number of unsatisfied constraint positions and s the number of satisfied constraint positions attached to at least one corrupt variable position (i.e., the constraint position is attached to an even number of corrupt variable positions). Since the expansion factor of the graph is $3d_v/4$, by the expansion properties of the graph

$$u + s > (3d_v/4)e.$$

Since each satisfied constraint position attached to a corrupt position is attached to at least two of them it follows that

$$u + 2s \leq d_v e$$

and these two equations imply that

$$u > d_v e/2.$$

This means that more than half of the check neighbors of a corrupt variable position connect to an unsatisfied check position. Thus there must be at least one corrupt variable attached to more unsatisfied constraint variables than satisfied. However, it could be there is an uncorrupted variable position also attached to more unsatisfied than satisfied check positions. If this position is chosen to be flipped, it will increase the number of corrupted positions e – making matters worse. Regardless, it has been shown that under the conditions noted there will be a variable position with more unsatisfied constraints than satisfied (whether it is corrupted or not). It could be that as e increases it reaches a point where $e = \alpha n$ and the corrupt positions form a codeword. We show this cannot occur. Suppose it does and that $e = \alpha n$. Recall that u is strictly decreasing. If $e = \alpha n$, then by the formula above

$$u > d_v \alpha n/2$$

but this cannot occur as the algorithm started with $u = \alpha n/2$ and is always less than this quantity. Thus the flipping algorithm is capable of correcting at least $\alpha n/2$ errors and hence the code distance is at least αn.

What has been shown is that graph expansion can be used in the formulation of good codes and efficient and simple decoding algorithms. In particular if the coding graph is an $(n, d_v, d_c, \alpha, \epsilon)$ expander, then as long as the expansion coefficient ϵ is sufficient, the relative distance of the resulting code $\mathscr{C}$ is α. This is the fundamental relationship established in [29] on which much work on expander codes is based.

Consider the code $\mathscr{C}(\mathcal{G}_{ev}, C)$ where $\mathcal{G}$ is a d-regular graph with n vertices, with second eigenvalue λ^*, $\mathcal{G}_{ev}$ its edge-vertex graph and C a binary linear code of length d, rate r and minimum relative distance δ and $\mathcal{G}$ an $(n, d_v, d_c, \alpha, \epsilon = d_v(\alpha))$ expander graph. As shown in the previous section if α is chosen as

$$\frac{\delta(d_c\delta - \lambda^*)}{(d_c - \lambda^*)},$$

then this is a lower bound on the minimum distance of the code and the expansion factor is

$$c(\alpha) = \frac{2}{d_c \delta}.$$

Using the Ramanujan graphs of Theorem 11.9 and good codes which are known to exist (by Varshamov–Gilbert bound) the following theorem can be shown ([29], theorem 19, improvement by [33], theorem 1A):

Theorem 11.16 *For all δ_0 such that $1 - 2H(\delta_0) > 0, H(\cdot)$ the binary entropy function, there exists a polynomial-time constructible family of expander codes of length n and rate $1-2H(\delta_0)$ and minimum relative distance arbitrarily close to δ_0^2 in which any $\alpha < \delta_0^2/4$ fraction of errors can be corrected with a circuit of size $O(n \log n)$ and depth $O(\log n)$.*

It is mentioned that [4] also uses the notion of expander graphs to construct asymptotically good codes (including nonbinary) but relies on the pseudorandom properties such graphs tend to have for the results.

The work of Sipser and Spielman [29] established the importance of the notion of graph expansion in the construction of codes and decoding algorithms of linear complexity. It has generated a very large literature of significant results. Due to space only a few of these research directions and results will be noted.

Note that for a (d_v, d_c)-regular bipartite graph the largest expansion possible would be c and this would be realized only if the corresponding constraint vertices for a given set of variable vertices would be distinct. Thus many works refer to $(\alpha, (1 - \epsilon)d_v)$ expanders with ϵ corresponding to the backoff from the maximum possible expansion. It is crucial to observe many of the results achieve excellent distance properties and linear decoding complexity even for codes over small fields.

While the following result from [15] refers to a particular type of decoding, linear programming or LP decoding, not discussed here, it is the expansion/distance properties of the codes that are of interest.

Theorem 11.17 ([15], theorem 1) *Let C be an LDPC code with length n and rate at least $1 - m/n$ described by a Tanner graph $\mathcal{G}$ with n variable nodes, m check nodes and regular left degree d_v. Suppose $\mathcal{G}$ is an $(\alpha, \delta d_v)$ expander, where $\delta > 2/3 + 1/(3d_v)$ and δd_v is an integer. Then the LP decoder succeeds, as long as at most $\frac{3\delta-2}{2\delta-1}(\alpha n - 1)$ bits are flipped by the channel.*

Similarly from [32]:

Theorem 11.18 ([32], theorem 3.1) *Let $d_v, d_c, \epsilon, \delta$ be constants and let $C \subset \mathbb{F}_2^n$ be a $(d_v, d_c, \epsilon, \delta)$ code. Assume that $\epsilon > 1/2$ and $\epsilon d_v + h - d_v > 0$ where $h = \lceil (2\epsilon - 1)c \rceil$. Then C is decodable in linear time from $((\epsilon d_v + h - d_v)/h)\lfloor \delta n \rfloor$ errors. Moreover C is decodable in logarithmic time on a linear number of processors from $((\epsilon d_v + h - d_v)/h)\lfloor \delta n \rfloor$ errors.*

Theorem 1.1 of [11] shows that the best minimum distance one might expect of such codes derived from a (c,d)-regular expander code is about $\alpha n/2\epsilon$ in the sense that:

Theorem 11.19 ([11], theorem 1.1) *Given any (d_v, d_c)-regular bipartite graph with expansion parameters $(\alpha, (1-\epsilon)\, d_v)$ and corresponding code $\mathscr{C}$, then the minimum distance of the code is at least $\frac{\alpha}{2\epsilon} n - O_\epsilon(1)$ and for any constant $\eta > 0$ there exists an $(\alpha, (1-\epsilon)\, d_v)$ expander graph which has distance at most $\left(\frac{\alpha}{2\epsilon} + \eta\right)$ and whose corresponding code has minimum distance at most $\left(\frac{\alpha}{2\epsilon} + \eta\right)n$.*

Actually the work requires only that the left vertices are d_v-regular and that the right vertices are parity checks on their respective variable nodes. Furthermore under certain conditions these codes will have a linear-time decoding algorithm:

Theorem 11.20 ([11], theorem 1.2) *Given any constants $\alpha, \eta > 0$ and $0 < \epsilon < 1/4$ there exists a linear-time decoding algorithm that will decode up to $\left(\frac{3\alpha}{16\epsilon} - \eta\right)n$ adversarial errors for any expander code associated with an $(\alpha, (1-\epsilon)d_v)$ expander graph.*

Since the minimum distance of the code is $\alpha n/16\epsilon$, the linear complexity algorithm cannot correct all errors the code is capable of.

The following decoding algorithm [11, 32] is of interest. Let $\mathscr{C}$ be the code associated with the expander regular bipartite graph. For $\boldsymbol{c} \in \mathscr{C}$ the transmitted codeword let $\boldsymbol{y} \in \mathbb{F}_2^n$ be the received word with errors at locations $F \subset N$ and number of errors $|F|$. Recall the constraint neighbors of variable node $v_i \in N$ is denoted N_{v_i}. Let Δ be a real number, a threshold parameter. L will be a subset of N and R a subset of M (the left and right vertices).

Algorithm 11.21 Input: $\boldsymbol{y}, \Delta$

1. $L \leftarrow \phi$
2. $R \leftarrow$ unsatisfied parity checks of $\boldsymbol{y}$
3. $L \leftarrow \phi$
4. $h \leftarrow (1 - 2\Delta)d_v$ (a threshold)
5. **while** $\exists\, v_i \in N \backslash L$ *st* $|N_{v_i} \cap R| \geq h$ **do**
6. $\quad L \leftarrow L \cup v_i$

7. $R \leftarrow R \cup N_{v_i}$
8. **end while**
9. **return** L
10. **end**

Note that in Step 5 if the algorithm finds a variable node attached to more than h unsatisfied parity checks, that node is added to the set of suspicious (possibly corrupted) variable nodes.

Under certain conditions it can be shown the set F of all error positions in the received codeword is contained in the set L after the algorithm completes. One can then treat the set F as erasures which can be corrected up to the minimum distance of the code (twice the number of correctable errors). Numerous variations/improvements to this basic algorithm are discussed in [11]. It is also noted there that from an $(\alpha, (1-\epsilon)\,d_v)$ expander one can trade-off parameters to give a $(k\alpha, (1-k\epsilon)\,d_v)$ expander as long as $1 > k\epsilon$, a useful tool for some constructions.

Numerous other aspects of the properties of expander codes have been considered in the literature, including their error probability and error exponent curves [6, 7, 8, 9] where it is shown, among numerous other properties, they achieve capacity on the BSC with a linear decoding algorithm and error probability that decreases exponentially with code length n. Burshtein [10] shows that expander-type arguments can be used to show that the message-passing algorithms (as considered, e.g., in Chapter 3) can actually be improved with the use of expander-type arguments to achieve the correction of a linear number of errors in a codeword.

Comments

The simple notion of graph expansion, well established as a topic of interest in the combinatorial, graph theoretic and computer science communities, has found a very interesting application in coding theory. The achievement of codes with excellent distance properties and simple linear-time decoding algorithms, using this notion, is surprising. The notion of expansion, briefly introduced in this chapter, will continue to be a topic of great interest in coding theory.

References

[1] Ajtai, M., Komlos, J., and Szemeredi, E. 1987. Deterministic simulation in LOGSPACE. Pages 132–140 of: *Proceedings of the Nineteenth Annual ACM Symposium on Theory of Computing*. STOC '87. ACM, New York.

[2] Alon, N., and Chung, F.R.K. 2006. Explicit construction of linear sized tolerant networks. *Discrete Math.*, **306**, 1068–1071.

[3] Alon, N., and Milman, V.D. 1984 (October). Eigenvalues, expanders and superconcentrators. Pages 320–322 of: *25th Annual Symposium on Foundations of Computer Science, 1984.*

[4] Alon, N., Bruck, J., Naor, J., Naor, M., and Roth, R.M. 1992. Construction of asymptotically good low-rate error-correcting codes through pseudo-random graphs. *IEEE Trans. Inform. Theory*, **38**(2), 509–516.

[5] Asratian, A.S., Denley, T.M.J., and Häggkvist, R. 1998. *Bipartite graphs and their applications*. Cambridge Tracts in Mathematics, vol. 131. Cambridge University Press, Cambridge.

[6] Barg, A., and Mazumdar, A. 2011. On the number of errors correctable with codes on graphs. *IEEE Trans. Inform. Theory*, **57**(2), 910–919.

[7] Barg, A., and Zémor, G. 2002. Error exponents of expander codes. *IEEE Trans. Inform. Theory*, **48**(6), 1725–1729.

[8] Barg, A., and Zémor, G. 2004. Error exponents of expander codes under linear-complexity decoding. *SIAM J. Discrete Math.*, **17**(3), 426–445.

[9] Barg, A., and Zémor, G. 2006. Distance properties of expander codes. *IEEE Trans. Inform. Theory*, **52**(1), 78–90.

[10] Burshtein, D., and Miller, G. 2001. Expander graph arguments for message-passing algorithms. *IEEE Trans. Inform. Theory*, **47**(2), 782–790.

[11] Xue, C., Cheng, K., Li, X., and Ouyang, M. 2021. Improved decoding of expander codes. *CoRR*, abs/2111.07629.

[12] Cvetković, D., Rowlinson, P., and Simić, S. 1997. *Eigenspaces of graphs*. Encyclopedia of Mathematics and Its Applications, vol. 66. Cambridge University Press, Cambridge.

[13] Davidoff, G., Sarnak, P., and Valette, A. 2003. *Elementary number theory, group theory, and Ramanujan graphs*. London Mathematical Society Student Texts, vol. 55. Cambridge University Press, Cambridge.

[14] Diestel, R. 2018. *Graph theory*, 5th ed. Graduate Texts in Mathematics, vol. 173. Springer, Berlin.

[15] Feldman, J., Malkin, T., Servedio, R.A., Stein, C., and Wainwright, M.J. 2007. LP decoding corrects a constant fraction of errors. *IEEE Trans. Inform. Theory*, **53**(1), 82–89.

[16] Gabber, O., and Galil, Z. 1981. Explicit constructions of linear-sized superconcentrators. *J. Comput. System Sci.*, **22**(3), 407–420.

[17] Godsil, C., and Royle, G. 2001. *Algebraic graph theory*. Graduate Texts in Mathematics, vol. 207. Springer-Verlag, New York.

[18] Halford, T.R., and Chugg, K.M. 2006. An algorithm for counting short cycles in bipartite graphs. *IEEE Trans. Inform. Theory*, **52**(1), 287–292.

[19] Høholdt, T., and Janwa, H. 2012. Eigenvalues and expansion of bipartite graphs. *Des. Codes Cryptogr.*, **65**(3), 259–273.

[20] Hoory, S., Linial, N., and Wigderson, A. 2006. Expander graphs and their applications. *Bull. Amer. Math. Soc. (N.S.)*, **43**(4), 439–561.

[21] Janwa, H., and Lal, A.K. 2003. On Tanner codes: minimum distance and decoding. *Appl. Algebra Engrg. Comm. Comput.*, **13**(5), 335–347.

[22] Janwa, H., and Lal, A.K. 2004. On expanders graphs: parameters and applications. *CoRR*, cs.IT/0406048.

[23] Justesen, J., and Høholdt, T. 2004. *A course in error-correcting codes*. EMS Textbooks in Mathematics. European Mathematical Society (EMS), Zürich.

[24] Lubotzky, A., Phillips, R., and Sarnak, P. 1988. Ramanujan graphs. *Combinatorica*, **8**(3), 261–277.

[25] Marcus, A.W., Spielman, D.A., and Srivastava, N. 2015. Interlacing families IV: bipartite Ramanujan graphs of all sizes. Pages 1358–1377 of: *2015 IEEE 56th Annual Symposium on Foundations of Computer Science—FOCS '15*. IEEE Computer Society, Los Alamitos, CA.

[26] Margulis, G.A. 1988. Explicit group-theoretic constructions of combinatorial schemes and their applications in the construction of expanders and concentrators. *Problemy Peredachi Informatsii*, **24**(1), 51–60.

[27] Morgenstern, M. 1994. Existence and explicit constructions of $q + 1$ regular Ramanujan graphs for every prime power q. *J. Combin. Theory Ser. B*, **62**(1), 44–62.

[28] Richardson, T., and Urbanke, R. 2008. *Modern coding theory*. Cambridge University Press, Cambridge.

[29] Sipser, M., and Spielman, D.A. 1996. Expander codes. *IEEE Trans. Inform. Theory*, **42**(6, part 1), 1710–1722.

[30] Tanner, R.M. 1981. A recursive approach to low complexity codes. *IEEE Trans. Inform. Theory*, **27**(5), 533–547.

[31] Tanner, R.M. 1984. Explicit concentrators from generalized N-gons. *SIAM J. Algebra. Discrete Meth.*, **5**(3), 287–293.

[32] Viderman, M. 2013. Linear-time decoding of regular expander codes. *ACM Trans. Comput. Theory*, **5**(3), 10:1–10:25.

[33] Zémor, G. 2001. On expander codes. *IEEE Trans. Inform. Theory*, **47**(2), 835–837.

12
Rank-Metric and Subspace Codes

Algebraic coding theory usually takes place in the vector space of n-tuples over a finite field with the Hamming metric where the distance between two n-tuples is the number of coordinate positions where they differ. In essence one tries to find subsets of the space (codes) with the property that any two elements of the subset are at least some minimum distance apart. This can be viewed as finding a packing of spheres of a given radius in the Hamming space.

Similar notions can be found in many other structures with different metrics – a familiar one is Euclidean n-space with the usual Euclidean distance – a structure where the sphere packing problem has been of considerable mathematical interest as well as practical interest for the construction of codes for the Gaussian channel. Another is the real line with absolute difference between numbers as a metric.

This chapter considers two other spaces with metrics that are perhaps less familiar. The first is the case of rank-metric codes as subsets of matrices of a given shape (not necessarily square) over a finite field with norm/distance defined by matrix rank rather than Hamming distance. They have been of independent interest in coding theory and of interest for their relationship to space-time codes. The second is the case of subspace codes as subsets of the set of all subspaces of a vector space over a finite field with a certain metric. They arise naturally in the subject of network codes for file transmission over a network. This subject might well have been included in the chapter on network codes (Chapter 5) but is included here because of its relation to rank-metric codes. The structure and properties of both rank-metric codes and subspace codes has generated an enormous amount of research using sophisticated mathematical techniques. Our treatment of these interesting topics is limited to an introduction of the basic ideas.

The first instance of rank-metric codes appeared in the work of Delsarte [5] in the language of bilinear forms and association schemes. The explicit notion of a rank-metric code as such, first appeared in the work of Gabidulin [16], work which has had a profound impact on subsequent research in the area.

Rank-metric codes are defined and some of their basic properties established in the next section. Section 12.2 describes several important constructions of rank-metric codes. The final Section 12.3 introduces the notion of subspace codes and considers their relation to rank-metric codes.

The amount of work on the rank-metric codes in the past decade is extensive with connections to network coding and to several areas in combinatorics and other areas of discrete mathematics. The seminal contributions of [5, 16, 27] have led to enormous interest in the coding community and the subject continues to be an active area of research.

Two metrics are used in this chapter, matrix rank and subspace dimension. The notation d for the distance metric will be for both since it will be clear from the context whether matrices or subspaces are of interest.

12.1 Basic Properties of Rank-Metric Codes

Let $\mathbb{F}_q$ and $\mathbb{F}_{q^n}$ be the finite fields with q and q^n elements, respectively, and $V_n(q)$ the vector space of n-tuples over $\mathbb{F}_q$ which is also denoted as $\mathbb{F}_q^n$ and there is a natural association of $V_n(q)$ with $\mathbb{F}_{q^n}$. Similarly denote by $\boldsymbol{M}_{k\times m}(q)$ the space of $k \times m$ matrices over $\mathbb{F}_q$ viewed as a vector space of shape $k \times m$ over $\mathbb{F}_q$, as the term dimension will have another meaning. Denote the all-zero vector in $V_n(q)$ as $\mathbf{0}_n$ and the all-one vector as $\mathbf{1}_n$. The usual inner product on $V_n(q)$ is denoted

$$(\boldsymbol{x},\boldsymbol{y}) = \sum_{i=1}^{n} x_i y_i \in \mathbb{F}_q,\ \boldsymbol{x},\boldsymbol{y} \in V_n(q).$$

Recall that a *norm* N on $V_n(q)$ satisfies the properties

(i) $N(\boldsymbol{x}) \geq 0\ \forall \boldsymbol{x} \in V_n(q)$;
(ii) $N(\boldsymbol{x}) = 0 \iff \boldsymbol{x} = \mathbf{0}_n \in V_n(q)$;
(iii) $N(\boldsymbol{x}+\boldsymbol{y}) \leq N(\boldsymbol{x}) + N(\boldsymbol{y})\ \forall \boldsymbol{x},\boldsymbol{y} \in V_n(q)$.

Similarly a *distance* function $d(\boldsymbol{x},\boldsymbol{y})$ on $V_n(q)$ satisfies the axioms

(i) $d(\boldsymbol{x},\boldsymbol{y}) \geq 0\ \forall \boldsymbol{x},\boldsymbol{y} \in V_n(q)$;
(ii) $d(\boldsymbol{x},\boldsymbol{y}) = 0 \iff \boldsymbol{x} = \boldsymbol{y}$;
(iii) $d(\boldsymbol{x},\boldsymbol{y}) \leq d(\boldsymbol{x},\boldsymbol{z}) + d(\boldsymbol{z},\boldsymbol{y})\ \forall \boldsymbol{x},\boldsymbol{y},\boldsymbol{z} \in V_n(q)$ (triangle inequality).

The norm function can be used to define a distance function on the space.

The space $\boldsymbol{M}_{k\times m}(q)$ is endowed with an inner product, the *trace product* as follows. Let $\boldsymbol{A},\boldsymbol{B} \in \boldsymbol{M}_{k\times m}(q)$ and denote their i-th columns as $\boldsymbol{A}_i, \boldsymbol{B}_i \in \mathbb{F}_q^k, i = 1,2,\ldots,m$, respectively. An inner product on $\boldsymbol{M}_{k\times m}(q)$ is defined as

$$\langle \boldsymbol{A},\boldsymbol{B}\rangle = \mathrm{Tr}(\boldsymbol{A}\boldsymbol{B}^t), \quad \boldsymbol{A},\boldsymbol{B} \in \boldsymbol{M}_{k\times m}(q), \tag{12.1}$$

where Tr indicates the matrix trace, the sum of the diagonal elements. It follows from the above that

$$\langle \boldsymbol{A},\boldsymbol{B}\rangle = \sum_{j=1}^{m} (\boldsymbol{A}_i,\boldsymbol{B}_i) = \sum_{i=1}^{m}\sum_{j=1}^{k} a_{ij}b_{ij} \tag{12.2}$$

where $(\boldsymbol{x},\boldsymbol{y})$ is the usual inner product of vectors. It is easy to verify that this matrix inner product is an inner product, i.e., it is bilinear, symmetric and nondegenerate on the space $\boldsymbol{M}_{k\times m}(q)$. For one-dimensional matrices the definition is the usual one for vectors.

Denote by rank($\boldsymbol{A}$) the rank of a matrix $\boldsymbol{A}$ in $\boldsymbol{M}_{k\times m}(q)$, assumed to be the rank over $\mathbb{F}_q$ unless otherwise specified. Recall that the rank is also the dimension of the row space generated by the rows of the matrix A which is the same as the dimension of the column space generated by the columns of A. Since

$$\left|\,\mathrm{rank}(\boldsymbol{A}) - \mathrm{rank}(\boldsymbol{B})\,\right| \leq \mathrm{rank}(\boldsymbol{A}+\boldsymbol{B}) \leq \mathrm{rank}(\boldsymbol{A}) + \mathrm{rank}(\boldsymbol{B})$$

the rank satisfies the axioms for a norm function and in particular gives a valid distance function on $\boldsymbol{M}_{k\times m}(q)$, i.e., that

$$d(\boldsymbol{A},\boldsymbol{B}) = \mathrm{rank}(\boldsymbol{A}-\boldsymbol{B}), \quad \boldsymbol{A},\boldsymbol{B} \in \boldsymbol{M}_{k\times m}(q) \tag{12.3}$$

satisfies the axioms of a distance function follows readily from the arguments used to prove the previous inequality. The following informal argument is given. Let $C(\boldsymbol{X})$ and $R(\boldsymbol{X})$ denote the column and row space, respectively of the matrix $\boldsymbol{X} \in \boldsymbol{M}_{k\times m}(q)$. Note that for two matrices $\boldsymbol{A}$ and $\boldsymbol{B}$ of the same shape $C(\boldsymbol{A}+\boldsymbol{B}) \subseteq C(\boldsymbol{A}) \cup C(\boldsymbol{B})$ and let $c = \dim(C(\boldsymbol{A}) \cap C(\boldsymbol{B}))$ and hence

$$\mathrm{rank}(\boldsymbol{A}+\boldsymbol{B}) \leq \dim(C(\boldsymbol{A}))+\dim(C(\boldsymbol{B}))-\dim(C(\boldsymbol{A})\cap C(\boldsymbol{B})) = \mathrm{rank}(\boldsymbol{A})+\mathrm{rank}(\boldsymbol{B})-c$$

and similarly for $R(\boldsymbol{A})$ and $R(\boldsymbol{B})$ and $r = \dim(R(\boldsymbol{A}) \cap R(\boldsymbol{B})$. It follows that

$$\mathrm{rank}(\boldsymbol{A}+\boldsymbol{B}) \leq \mathrm{rank}(\boldsymbol{A})+\mathrm{rank}(\boldsymbol{B})-\max(c,r) \quad \text{or} \quad \mathrm{rank}(\boldsymbol{A}-\boldsymbol{B}) \leq \mathrm{rank}(\boldsymbol{A})+\mathrm{rank}(\boldsymbol{B})$$

and these relations can be used to show

$$d(\boldsymbol{A},\boldsymbol{C}) = \mathrm{rank}(\boldsymbol{A}-\boldsymbol{C}) \leq d(\boldsymbol{A},\boldsymbol{B}) + d(\boldsymbol{B},\boldsymbol{C}) = \mathrm{rank}(\boldsymbol{A}-\boldsymbol{B}) + \mathrm{rank}(\boldsymbol{B}-\boldsymbol{C})$$

(noting that rank($-\boldsymbol{B}$) = rank($\boldsymbol{B}$)).

Definition 12.1 A *rank-metric code* $\mathscr{C}$ of shape $k \times m$ will be a subset $\mathscr{C}$ of the set $\boldsymbol{M}_{k\times m}(q)$ of $k \times m$ matrices over $\mathbb{F}_q$. The minimum distance of the code is

$$d_{\min} = \min_{\boldsymbol{A},\boldsymbol{B}\in\mathscr{C},\boldsymbol{A}\neq\boldsymbol{B}} d(\boldsymbol{A}-\boldsymbol{B}) = \min_{\boldsymbol{A},\boldsymbol{B}\in\mathscr{C},\boldsymbol{A}\neq\boldsymbol{B}} \operatorname{rank}(\boldsymbol{A}-\boldsymbol{B}).$$

The number of matrices in the code will be called its size. In the case of a linear rank-metric code, the set $\mathscr{C}$ is a linear space over $\mathbb{F}_q$ and the minimum distance is the rank of the minimum rank nonzero matrix in the code subspace. If the size of such a linear code is q^K, it will be designated as a $(k \times m, K, d_{\dim})_q$ rank-metric code, $1 \leq K \leq km$, $|\mathscr{C}| = q^K$ and K will be the *dimension* of the code. For nonlinear codes the actual size (number of codewords M) will be used and referred to as a $(k \times m, M, d_{\min})_q$ code.

The definition attempts to reflect that in use for ordinary algebraic codes. The notation used here is not standard in the literature. It is emphasized the term "dimension K" is restricted to give the dimension of a linear rank-metric code as q^K, i.e., the set of matrices forms a vector space over $\mathbb{F}_q$ of size q^K and dimension K. In general nonlinear sets of matrices are also of interest.

Definition 12.2 In the case a rank-metric code contains only matrices of the same rank, it is referred to as a *constant-rank code* (CRC).

Definition 12.3 The *rank distribution* of a rank-metric code $\mathscr{C}$ is the set of nonnegative integers $|A_i(C)|$ where

$$A_i(\mathscr{C}) = \left|\{\boldsymbol{A} \in \mathscr{C} \mid \operatorname{rank}(\boldsymbol{A}) = i\}\right|, \; i = 0, 1, \ldots, \min(k, m)$$

and if $\mathscr{C}$ is linear, the *minimum distance* d of the code is the smallest $i > 0$ such that $A_i(\mathscr{C}) \neq 0$.

In analogy with ordinary algebraic codes, one can define a *distance distribution* [21] of rank-metric codes. In the case of linear codes the two concepts of rank and distance distribution are the same.

The reference [21] is an excellent collection of chapters on network codes and subspace designs by researchers. Various notions of the *equivalence* of rank-metric codes have arisen ([21], chapter by Otal and Özbudak) and [3]:

Definition 12.4 Rank-metric codes $\mathscr{C}, \mathscr{C}' \subseteq \boldsymbol{M}_{k\times m}(q)$ are *equivalent* if there exist $\boldsymbol{X} \in GL_k(q), \boldsymbol{Y} \in GL_m(q)$ ($GL_k(q)$ the general linear group of $k \times k$ nonsingular matrices over $\mathbb{F}_q$) and $\boldsymbol{Z} \in \boldsymbol{M}_{k\times m}(q)$ such that

$$\mathscr{C}' = \boldsymbol{X}\mathscr{C}^{\sigma}\boldsymbol{Y} + \boldsymbol{Z}, \text{ when } k \neq m$$
$$\mathscr{C}' = \boldsymbol{X}\mathscr{C}^{\sigma}\boldsymbol{Y} + \boldsymbol{Z}, \text{ or } \mathscr{C}' = \boldsymbol{X}(\mathscr{C}^{t})^{\sigma} + \boldsymbol{Z} \text{ when } k = m$$

where σ is an isometry of $\boldsymbol{M}_{k \times m}(q)$ (a bijective map that preserves rank) acting on the entries of $C \in \mathscr{C}$ and the superscript t indicates matrix transpose. If both $\mathscr{C}$ and $\mathscr{C}'$ are closed under addition (additive codes), then $\boldsymbol{Z}$ must be the zero matrix, and if they are both linear codes, then σ can be taken as the identity map.

Certain types of rank-metric codes have been investigated, initiated in the works of Delsarte [5] and Gabidulin [16], respectively. Although the next section gives constructions of these rank-metric codes, it is convenient to discuss the definitions and some of their properties here before giving their constructions. Reference [34] contains an excellent view of this material.

Definition 12.5 A *Delsarte rank-metric* code of shape $k \times m$ over $\mathbb{F}_q$ is an $\mathbb{F}_q$ linear subspace $\mathscr{C} \subseteq \boldsymbol{M}_{k \times m}(q)$. The *dual* code of $\mathscr{C}$ is

$$\mathscr{C}^{\perp} = \{\boldsymbol{A} \in \boldsymbol{M}_{k \times m}(q) \mid \langle \boldsymbol{A}, \boldsymbol{B} \rangle = 0 \, \forall \, \boldsymbol{B} \in \mathscr{C}\}.$$

Gabidulin [16] introduced the notion of the rank of a vector over a subfield as follows:

Definition 12.6 Let $\mathbb{F}_{q^m}$ be an extension field of $\mathbb{F}_q$ of degree m. A *Gabidulin rank-metric* $\mathscr{C}$ code of length k over $\mathbb{F}_{q^m}$ is an $\mathbb{F}_{q^m}$ linear subspace $\mathscr{C} \subseteq \mathbb{F}_{q^m}^k$. The *rank* of a vector $\boldsymbol{\alpha} = (\alpha_1, \alpha_2, \ldots, \alpha_k) \in \mathbb{F}_{q^m}^k$ is the $\mathbb{F}_q$ dimension of the subspace spanned by $\alpha_1, \ldots, \alpha_k$, denoted $\langle \alpha_1, \ldots, \alpha_k \rangle$. The dual of a Gabidulin code $\mathscr{C}$ is

$$\mathscr{C}^{\perp} = \left\{ \boldsymbol{\beta} \in \mathbb{F}_{q^m}^k \mid (\boldsymbol{\alpha}, \boldsymbol{\beta}) = 0 \, \forall \, \boldsymbol{\alpha} \in \mathscr{C} \right\}.$$

The above definitions of Delsarte and Gabidulin codes are slightly different from the codes defined in the original works ([5] and [16], respectively). At times it is convenient to refer to $m \times k$ matrices rather than $k \times m$ and the equivalence is clear.

Note that both Gabidulin and Delsarte rank codes are linear over their respective fields. It is clear that a Gabidulin code can be viewed as a Delsarte code by using a basis of $\mathbb{F}_{q^m}$ over $\mathbb{F}_q$ and noting that linearity over $\mathbb{F}_{q^m}$ implies linearity over $\mathbb{F}_q$. To this end let $\mathcal{G} = \{\gamma_1, \ldots, \gamma_m\}$ be a basis of $\mathbb{F}_{q^m}$ over $\mathbb{F}_q$. For a codeword (vector) in the Gabidulin code, $\boldsymbol{\alpha} \in \mathscr{C} \subseteq \mathbb{F}_{q^m}^k$, $\boldsymbol{\alpha} = (\alpha_1, \ldots, \alpha_k)$, $\alpha_j \in \mathbb{F}_{q^m}$, associate an $m \times k$ matrix over $\mathbb{F}_q$ $\boldsymbol{M}_{\mathcal{G}}(\boldsymbol{\alpha})$ by defining its j-th column as the expansion of α_j with respect to the basis vectors of $\mathcal{G}$, i.e.,

$$\alpha_j = \sum_{i=1}^{m} \boldsymbol{M}_{\mathcal{G}}(\alpha)_{ij} \gamma_i \quad \alpha \in \mathscr{C}, \; j = 1, 2, \ldots, k$$

and the Delsarte code associated with the Gabidulin code $\mathscr{C} \subseteq \mathbb{F}_{q^m}^k$ is

$$C_{\mathcal{G}}(\mathscr{C}) = \{\boldsymbol{M}_{\mathcal{G}}(\alpha) \mid \alpha \in \mathscr{C}\} \subseteq M_{m\times k}(q),\ \boldsymbol{M}_{\mathcal{G}}(\alpha) = \left(\boldsymbol{M}_{\mathcal{G}}(\alpha)_{ij}\right).$$

It is clear that for every $\mathbb{F}_q$ basis of $\mathbb{F}_{q^m}$, $\mathcal{G}$, the mapping

$$\begin{aligned} \phi\colon \mathbb{F}_{q^m}^k &\longrightarrow \boldsymbol{M}_{m\times k}(q) \\ v &\mapsto \boldsymbol{M}_{\mathcal{G}}(v) \end{aligned}$$

is an $\mathbb{F}_q$-linear bijective isometry. Further if $\mathscr{C}$ is a Gabidulin code (at times also referred to as a vector rank-metric code), then the corresponding rank-metric code $C_{\mathcal{G}}(\mathscr{C})$ has the same cardinality, rank distribution and minimum distance ([21], chapter by Gorla and Ravagnani). By this process it is clear that any Gabidulin code can be viewed as a Delsarte code. However, since Gabidulin codes are $\mathbb{F}_{q^m}$ subspaces while Delsarte codes are $\mathbb{F}_q$ subspaces, the converse is clearly not true. It is noted [34] that for a Gabidulin code $\mathscr{C} \subseteq \mathbb{F}_{q^m}^k$ and associated Delsarte code $C_{\mathcal{G}}(\mathscr{C})$,

$$\dim_{\mathbb{F}_q} C_{\mathcal{G}}(\mathscr{C}) = m \dim_{\mathbb{F}_{q^m}}(\mathscr{C})$$

and that the rank distribution of $C_{\mathcal{G}}(C)$ is the same as that of $\mathscr{C}$. Thus a Delsarte code C with $m \nmid \dim_{\mathbb{F}_q}(C)$ cannot arise from a Gabidulin code.

It is noted that in general

$$C_{\mathcal{G}}(\mathscr{C}^{\perp}) \neq C_{\mathcal{G}}(\mathscr{C})^{\perp}.$$

However, if $\mathcal{G} = \{\gamma_i\}$ and $\mathcal{G}' = \{\beta_j\}$ are *dual* (trace dual, orthogonal or complementary – see Chapter 1) bases (i.e., $\mathrm{Tr}(\gamma_i, \beta_j) = \delta_{ij}$), then ([21], chapter by Gorla and Ravagnani, theorem 1, [34], theorem 21)

$$C_{\mathcal{G}'}(\mathscr{C}^{\perp}) = C_{\mathcal{G}}(\mathscr{C})^{\perp}.$$

For clarity and since the definition of a rank-metric code in the literature is not consistent, the following definitions are restated:

(i) A rank-metric code is a subset of $\boldsymbol{M}_{k\times m}(q)$.
(ii) A vector rank-metric code is a subset of $\mathbb{F}_{q^m}^k$.
(iii) A Delsarte rank-metric code is an $\mathbb{F}_q$ linear subspace of $\boldsymbol{M}_{k\times m}(q)$.
(iv) A Gabidulin rank-metric code is an $\mathbb{F}_{q^m}$ subspace of $\mathbb{F}_{q^m}^k$.

Recall from Definition 12.1 a (nonlinear) rank-metric code with M $k \times m$ matrices over $\mathbb{F}_q$ and minimum distance $d_{\min}$ is denoted $\mathscr{C} = (k\times m, M, d_{\min})_q$ where $k \times m$ is the code shape and M the size of the code (number of codewords) and if the code is linear (over $\mathbb{F}_q$) and $M = q^K$ denoted a $(k \times m, K, d_{\min})_q$ code.

The notion of a *maximum-rank distance* (MRD) code is of importance and follows from the following bound, reminiscent of the Singleton bound in algebraic coding theory. It was first proved in [5]:

Theorem 12.7 *Let $\mathscr{C} \subseteq \mathbf{M}_{k\times m}(q)$ be a nonzero (not necessarily linear) rank-metric code with minimum rank distance $d_{\min}$. Then for (wlog) $k \leq m$*

$$|\mathscr{C}| \leq \min \left\{ q^{m(k-d_{\min}+1)}, q^{k(m-d_{\min}+1)} \right\}. \qquad \text{Singleton bound}$$

Rank metric codes that meet this bound are called MRD codes.

Proof: From $\mathscr{C}$ obtain the set of matrices $\mathscr{C}'$ in $\boldsymbol{M}_{(k-d_{\min}+1)\times m}$ by deleting the bottom $(d_{min} - 1)$ rows of the matrices of $\mathscr{C}$. The matrices of $\mathscr{C}'$ must be distinct as otherwise a subtraction would yield an all-zero matrix implying a nonzero matrix in $\mathscr{C}$ had rank less than $d_{\min}$ contrary to assumption. Since the number of such matrices is at most $q^{m(k-d_{\min}+1)}$ the result follows. ■

As noted, both the Delsarte and Gabidulin codes in the original works [5, 16] were specific constructions to be given in the next section. In a sense the MRD rank-metric codes can be viewed as analogs of the Reed–Solomon codes (or more generally maximum distance separable (MDS) codes) of algebraic coding theory since they achieve similar upper bounds on their size. To obtain further bounds, much as for block codes, define the notion of spheres and balls of radius i in $\boldsymbol{M}_{k\times m}(q)$ with center $\boldsymbol{Y}$ as

$$\mathcal{S}_i(\boldsymbol{Y}) \overset{\Delta}{=} \{\boldsymbol{X} \in \boldsymbol{M}_{k\times m}(q) \mid \operatorname{rank}(\boldsymbol{X} - \boldsymbol{Y}) = i\} \quad \text{and} \quad \mathcal{B}_t = \bigcup_{i=0}^{t} \mathcal{S}_i.$$

It is clear the quantities are independent of the center chosen. Define the quantities

$$s_t = |\mathcal{S}_t| \quad \text{and} \quad b_t = |\mathcal{B}_t|$$

where in these expressions the dependence on q, k and m is understood. Let $\mathscr{N}_q(k \times m, r)$ denote the set of $k \times m$ matrices over $\mathbb{F}_q$ of rank r and denote by $N_q(k \times m, r) = |\mathscr{N}_q(k \times m, r)|$. As shown in Appendix A this quantity is evaluated as

$$N_q(k \times m, r) = \prod_{i=0}^{r-1} (q^m - q^i) \frac{(q^{k-i} - 1)}{(q^{i+1} - 1)}. \tag{12.4}$$

This can be expressed in terms of Gaussian coefficients. Recall that the number of subspaces of a vector space of dimension k in $V_n(q)$ is given by the Gaussian coefficient (see Appendix A) which is:

$$\begin{aligned}\begin{bmatrix} n \\ k \end{bmatrix}_q &= \frac{(q^n-1)(q^n-q)(q^n-q^2)\cdots(q^n-q^{k-1}-1)}{(q^k-1)(q^k-q)\cdots(q^k-q^{k-1})} \\ &= \frac{(q^n-1)(q^{n-1}-1)\cdots(q^{n-k+1}-1)}{(q^k-1)(q^{k-1}-1)\cdots(q-1)}.\end{aligned} \tag{12.5}$$

(It is understood that for $n < 0, k < 0$ or $k > n$ the value is 0 and for $n = 0$ and $k \geq 0$ the value is 1.) These coefficients have interesting and useful properties, a few of which are noted in Appendix A. Thus

$$N_q(k \times m, r) = q^{\binom{r}{2}} \begin{bmatrix} m \\ r \end{bmatrix}_q \begin{bmatrix} k \\ r \end{bmatrix}_q \prod_{i=0}^{r} (q^i - 1). \tag{12.6}$$

The quantities of interest in the above computations are s_t and b_t. Bounds on s_t and b_t can be obtained ([30], proposition 1):

$$\begin{aligned} q^{(k+m-2)t-t^2} &\leq s_t \leq q^{(k+m+1)t-t^2} \\ q^{(k+m-2)t-t^2} &\leq b_t \leq q^{(k+m+1)t-t^2+1} \end{aligned}$$

The *sphere packing bound* on $(k, M, d_{\min})_{q^m}$ codes in the rank metric then is, for $t = \lfloor (d_{\min} - 1)/2 \rfloor$

$$M \cdot |\mathcal{B}_t| = M b_t \leq q^{km} \qquad \text{Sphere packing bound}$$

and it is observed ([30], proposition 2) that no perfect codes exist in the rank metric (i.e., equality cannot be obtained for any set of parameters in the above inequality). The proof of this is relatively simple following from the form of the equations for the sizes of s_t and b_t given above. The corresponding statement for algebraic coding theory, that all perfect codes are known (essentially two Golay codes, Hamming codes, binary repetition codes and some nonlinear codes the same parameters as Hamming codes) was a difficult result achieved after many years of research.

The following proposition is viewed as an analog of the Varshamov–Gilbert (VG) bound of algebraic coding theory:

Proposition 12.8 ([30], proposition 3) *Let $k, m, M, d_{\min}$ be positive integers, $t = \lfloor (d_{\min} - 1)/2 \rfloor$. If*

$$M \times |\mathcal{B}_{d-1}| < q^{mk} \qquad \text{Varshamov–Gilbert bound}$$

then a $(k, M+1, d_{\min})_{q^m}$ vector rank code exists.

In the next section MRD codes for all allowable parameters will be constructed and so the question of obtaining further bounds for non-MRD codes seem less compelling than is the case in algebraic coding theory.

As with algebraic block codes that meet the Singleton bound, MRD codes have interesting mathematical structure and have been studied widely. Many aspects of their construction and properties are considered in ([21], chapter by Otal and Özbudak). These include known constructions of linear, nonlinear additive, nonadditive (but closed under multiplication by a scalar) and nonadditive (and not closed under multiplication by a scalar) MRD codes. A brief overview of a small part of this work is considered in the next section.

The following properties of such codes are straightforward consequences of the definitions. Let $\mathscr{C}$ be a Delsarte rank-metric code, a linear subspace of $\boldsymbol{M}_{k\times m}(q)$ over $\mathbb{F}_q$. Recall that

$$\mathscr{C}^{\perp} = \{\boldsymbol{A} \in \boldsymbol{M}_{k\times m}(q) \mid \langle \boldsymbol{A}, \boldsymbol{B}\rangle = 0 \, \forall \boldsymbol{B} \in \mathscr{C}\}$$

noting the inner product on $\boldsymbol{M}_{k\times m}(q)$. Clearly

$$\dim_{\mathbb{F}_q}(\mathscr{C}) + \dim_{\mathbb{F}_q}(\mathscr{C}^{\perp}) = km$$

(where dim here refers to the dimension of $\mathscr{C}$ as a subspace of $\boldsymbol{M}_{k\times m}(q)$) and

$$(\mathscr{C}^{\perp})^{\perp} = \mathscr{C}.$$

Further properties of the dual space are contained in lemma 5 of [34] (the journal version of [33]) including that for two subspaces $\mathscr{C}$ and $\mathscr{D}$

$$(\mathscr{C} \cap \mathscr{D})^{\perp} = \mathscr{C}^{\perp} + \mathscr{D}^{\perp} \quad \text{and} \quad (\mathscr{C} + \mathscr{D})^{\perp} = \mathscr{C}^{\perp} \cap \mathscr{D}^{\perp}.$$

The addition of spaces here refers to the smallest subspace containing both spaces.

In analogy with the linear MDS codes of algebraic coding theory it was shown for both Delsarte codes ([5], theorem. 5.6) and Gabidulin codes ([16], theorem 3) that:

Theorem 12.9 *A linear rank-metric code $\mathscr{C}$ is MRD $\iff$ $\mathscr{C}^{\perp}$ is MRD.*

Furthermore in both cases it is established that the rank distribution of a linear MRD is entirely determined by its parameters [5, 16] much like the weight distribution for MDS codes of algebraic coding theory. The expression for the rank distribution of a linear MRD code is omitted. It is shown in [2] this is true of any (linear or nonlinear) MRD code.

The following proposition ([21], chapter by Gorla and Ravagnani, proposition 3) summarizes a further property:

Proposition 12.10 *Let $\mathscr{C} \subseteq \mathbf{M}_{k\times m}(q)$ be a linear rank-metric code with $1 \leq \dim(\mathscr{C}) \leq km - 1$ and $k \leq m$. The following are equivalent:*

(i) $\mathscr{C}$ *is MRD.*
(ii) $\mathscr{C}^{\perp}$ *is MRD.*
(iii) $d_{\min}(\mathscr{C}) + d_{\min}(\mathscr{C}^{\perp}) = k + 2$.

It is also easy to show that ([2], proposition 39) for $\mathscr{C} \subseteq \mathbb{F}_{q^m}^k$, a Gabidulin code of minimum distance $d_{\min}$ (i.e., rank) and dual minimum distance $d_{\min}^{\perp}$ one of the following two statements holds:

(i) $\mathscr{C}$ is MRD and $d_{\min} + d_{\min}^{\perp} = k + 2$.
(ii) $d_{\min} + d_{\min}^{\perp} \leq k$.

Given that codes in the rank metric share many properties with codes in the Hamming metric it is perhaps not surprising that MacWilliams identities can be formulated for such codes. The following theorem is analogous to the familiar MacWilliams identities for algebraic coding theory:

Theorem 12.11 ([34], theorem 31) *Let* $\mathscr{C} \subseteq \mathbf{M}_{k\times m}(q)$ *be a linear rank-metric code. Let* $\{A_i\}_{i\in\mathbb{N}}$ *and* $\{B_j\}_{j\in\mathbb{N}}$ *be the rank distributions of* $\mathscr{C}$ *and* $\mathscr{C}^{\perp}$*, respectively. For any integer* $0 \leq \nu \leq k$ *we have*

$$\sum_{i=0}^{k-\nu} A_i \begin{bmatrix} k-i \\ \nu \end{bmatrix}_q = \frac{|\mathscr{C}|}{q^{m\nu}} \sum_{j=0}^{\nu} B_j \begin{bmatrix} k-j \\ \nu-j \end{bmatrix}_q, \qquad \nu = 0, 1, 2, \ldots, k.$$

It follows directly from this theorem that the rank distribution of a linear rank-metric MRD code $\mathscr{C} \subseteq \boldsymbol{M}_{k\times m}(q)$ depends only on k, m and the minimum rank of the code. Indeed by Theorem 12.11 and Proposition 12.10 if $d_{\min} = \min\operatorname{rank}(\mathscr{C})$, then $\mathscr{C}^{\perp}$ has minimum rank distance $k - d_{\min} + 2$ and hence the above identities yield ([34], corollary 44), for $0 \leq \nu \leq k - d_{\min}$

$$\begin{bmatrix} k \\ \nu \end{bmatrix}_q + \sum_{i=d}^{k-\nu} A_i \begin{bmatrix} k-i \\ \nu \end{bmatrix}_q = \frac{|\mathscr{C}|}{q^{m\nu}} \begin{bmatrix} k \\ \nu \end{bmatrix}_q, \qquad \nu = 0, 1, 2, \ldots, k - d_{\min}.$$

This system of $k - d_{\min} + 1$ equations with $k - d_{\min} + 1$ unknowns is seen to be upper triangular with ones on the diagonal and hence easily solvable and the above statement is verified.

Finally it is noted that both Delsarte [5] and Gabidulin [16] gave expressions for the rank distributions of their MRD codes. An impressive recent result of ([21], chapter by Gorla and Ravagnani, theorem 8) gives the following expression for the rank distribution of a not necessarily linear MRD code $\mathscr{C}$ with $|\mathscr{C}| \geq 2$, $\mathbf{0} \in \mathscr{C}$, $k \leq m$, with $\min\operatorname{rank}(\mathscr{C}) = d_{\min}$ and $A_0 = 1, A_i = 0$ for $1 \leq i \leq d_{\min} - 1$ and

$$A_i = \sum_{u=0}^{d-1} (-1)^{i-u} q^{\binom{i-u}{2}} \begin{bmatrix} k \\ i \end{bmatrix}_q \begin{bmatrix} i \\ u \end{bmatrix}_q + \sum_{u=d_{\min}}^{i} (-1)^{i-u} q^{\binom{i-u}{2} + m(u-d_{\min}+1)} \begin{bmatrix} k \\ i \end{bmatrix}_q \begin{bmatrix} i \\ u \end{bmatrix}_q,$$

for $d_{\min} \leq i \leq k$.

The next section discusses constructions of certain classes of rank-metric codes, including the Delsarte and Gabidulin MRD codes, and shows, in particular, that such codes exist for all allowable parameter sets.

The section is concluded with a brief discussion of the notion of an *anticode*, first appearing in the work of Delsarte [4] and discussed in ([21], chapter by Gorla and Ravagnani) and in ([26], chapter by Kschischang).

Definition 12.12 A *rank-metric* Δ *anticode* is a nonempty subset $\mathcal{A} \subseteq \boldsymbol{M}_{k\times m}(q)$ such that $d(\boldsymbol{A},\boldsymbol{B}) \leq \Delta \ \forall \ \boldsymbol{A},\boldsymbol{B} \in \mathcal{A}$ and it is a linear anticode if it is an $\mathbb{F}_q$ linear subspace of $\boldsymbol{M}_{k\times m}(q)$.

The equivalent notion for algebraic codes over $\mathbb{F}_q$ with the Hamming metric would lead to an optimal (maximal size) anticode being a sphere of diameter Δ in the Hamming space. This is not the case for codes in the rank metric – specifically a sphere of diameter d with the rank metric is an anticode but not an optimal anticode (see below for a definition of optimal for this case). Most importantly we have the following ([21], chapter by Gorla and Ravagnani, theorem 9):

Theorem 12.13 *Let $\mathcal{A}$ be a Δ anticode in $\mathbf{M}_{k\times m}(q)$ with $0 \leq \Delta \leq k-1$. Then $|\mathcal{A}| \leq q^{m\Delta}$ and the following conditions are equivalent:*

1. $|\mathcal{A}| = q^{m\Delta}$.
2. $\mathcal{A} + \mathcal{C} = \mathbf{M}_{k\times m}(q)$ *for some MRD code* $\mathcal{C}$ *with* $d_{\min}(\mathcal{C}) = \Delta + 1$.
3. $\mathcal{A} + \mathcal{C} = \mathbf{M}_{k\times m}(q)$ *for all MRD codes* $\mathcal{C}$ *with* $d_{\min}(\mathcal{C}) = \Delta + 1$

where

$$\mathcal{A} + \mathcal{C} = \left\{\mathbf{X} + \mathbf{Y} \mid \mathbf{X} \in \mathcal{A}, \mathbf{Y} \in \mathcal{C}\right\}.$$

A Δ anticode is said to be *optimal* if it attains the above bound, $|\mathcal{A}| = q^{m\Delta}$. The same work ([21], chapter by Gorla and Ravagnani, proposition 4) also contains the result:

Proposition 12.14 *Let $\mathcal{A} \subseteq \mathbf{M}_{k\times m}(q)$ be a linear rank-metric code with* $\dim(\mathcal{C}) = m\Delta$ *and $0 \leq \Delta \leq k-1$. The following are equivalent:*

1. *$\mathcal{A}$ is an optimal Δ anticode.*
2. $\mathcal{A} \cap \mathcal{C} = \{0\}$ *for all nonzero MRD linear codes* $\mathcal{C} \subseteq \mathbf{M}_{k\times m}(q)$ *with* $\dim(\mathcal{C}) = \Delta + 1$.

Finally it is noted that the dual of an optimal linear Δ anticode in $\boldsymbol{M}_{k\times m}(q)$ is an optimal linear $k - \Delta$ anticode.

12.2 Constructions of MRD Rank-Metric Codes

The previous section has considered general properties of rank-metric codes and in particular linear rank-metric codes. The original constructions of linear MRD codes of Delsarte [5] and Gabidulin [16] are described here along with proof they are MRD. Other more recent constructions of MRD codes are also considered. The notation of these works is adapted to ours as required.

It has been commented on that since MRD codes meet the rank-metric Singleton bound and, as is about to be shown, they exist for all allowable sets of parameters, there seems less motivation to search for other classes of codes. A question that has been of interest is whether there are MRD codes that are not Gabidulin or Delsarte codes. This has been answered in the affirmative, an example of which is the class of *twisted Gabidulin* (TG) codes discussed later in this section.

The Delsarte MRD codes are first constructed. The original work of Delsarte [5] is in the setting of bilinear forms and association schemes. The matrix approach used here is based on that approach. Without loss of generality, assuming $k \leq m$, the construction will develop a linear space of $q^{m(k-d+1)}$ $m \times k$ matrices each over $\mathbb{F}_q$, each nonzero matrix of rank at least d (hence MRD). Let $\mathrm{Tr}_{q^m|q}$ denote the trace function of $\mathbb{F}_{q^m}$ over $\mathbb{F}_q$. Let $\{\omega_1, \omega_2, \ldots, \omega_m\}$ be a basis of $\mathbb{F}_{q^m}$ over $\mathbb{F}_q$. Let $\{\mu_1, \ldots, \mu_k\}$ be a set of k elements from $\mathbb{F}_{q^m}$ linearly independent over $\mathbb{F}_q$. Let $\boldsymbol{u} = \{u_0, u_1, \ldots, u_{k-d}\}$ be an arbitrary set of $(k-d+1)$ elements from $\mathbb{F}_{q^m}$ for some integer d (which will be the code minimum distance), $1 < d < k$. Define the set of $m \times k$ matrices

$$\begin{aligned}\mathscr{C} = \Big\{ &\boldsymbol{M}(u) = (\boldsymbol{M}(u)_{ij}) \mid \boldsymbol{M}(u)_{ij} = \mathrm{Tr}_{q^m|q}\Big(\sum_{\ell=0}^{k-d} u_\ell \omega_i \mu_j^{q^\ell} \Big), \\ &i = 1, 2, \ldots, m,\ j = 1, 2, \ldots, k \Big\}.\end{aligned} \tag{12.7}$$

Since the set of $(k-d+1)$ elements of $\boldsymbol{u}$, $u_i \in \mathbb{F}_{q^m}$, can be chosen in $q^{m(k-d+1)}$ ways (including the all-zero one), this is the number of matrices (codewords) generated. To see that each nonzero matrix in the code has rank at least d consider the number of solutions to the matrix equation

$$\boldsymbol{M}(u)\,\boldsymbol{c}^t = 0, \quad \boldsymbol{c} = (c_1, \ldots, c_k) \in \mathbb{F}_q^k. \tag{12.8}$$

Equivalently

$$\begin{aligned}\sum_{j=1}^{k} \boldsymbol{M}(u)_{ij} &= \sum_{j=1}^{k} \mathrm{Tr}_{q^m|q}\Big(\sum_{\ell=0}^{k-d} u_\ell \omega_i \mu_j^{q^\ell} \Big) c_j \\ &= \mathrm{Tr}_{q^m|q}\Big(\omega_i \sum_{\ell=0}^{k-d} u_\ell \Big(\sum_{j=1}^{k} c_j \mu_j \Big)^{q^\ell} \Big) = 0, \quad i = 1, 2, \ldots, m,\end{aligned}$$

since $c_j \in \mathbb{F}_q$. The polynomial

$$p(x) = \sum_{\ell=0}^{k-d} u_\ell x^{q^\ell}, \ u_\ell \in \mathbb{F}_{q^m}$$

is a linearized or q-polynomial (see Appendix A) over $\mathbb{F}_{q^m}$ and its set of roots forms a vector space of some dimension over $\mathbb{F}_q$. Since the result of the Trace operation is zero for all i if and only if $p(\sum_{j=0}^{m} c_j \mu_j) = 0$ (see Chapter 1) the equation has at most q^{k-d} solutions and the solution space of Equation 12.8 is of dimension at most $k - d$ and the matrices $\boldsymbol{M}(u)$ have rank at least d for each of the $q^{m(k-d+1)}$ possible values of $\boldsymbol{u} \in \mathbb{F}_{q^m}^{k-d+1}$. The transposes of the matrices are of shape $k \times m$ and have the same properties. It is clear that such a construction yields MRD rank-metric codes for all allowable parameter sets. Also, as noted, the rank distribution of such a code is known.

To describe *Gabidulin MRD codes* consider the following construction of $m \times k$ matrices over $\mathbb{F}_{q^m}$. Let $g_1, g_2, \ldots, g_k \in \mathbb{F}_{q^m}$ be linearly independent over $\mathbb{F}_q, k \le m$. First, define the $K \times k$ matrix over $\mathbb{F}_{q^m}$:

$$\boldsymbol{M}_{K\times k} = \begin{bmatrix} g_1 & g_2 & \cdots & g_k \\ g_1^q & g_2^q & \cdots & g_k^q \\ \vdots & \vdots & \vdots & \vdots \\ g_1^{q^{K-1}} & g_2^{q^{K-1}} & \cdots & g_k^{q^{K-1}} \end{bmatrix}$$

and the matrix is of rank $K < k$ over $\mathbb{F}_{q^m}$ ([29], lemma 3.51 and appendix A) – since any square submatrix has nonzero determinant. Thus its row space over $\mathbb{F}_{q^m}$ contains q^{mK} vectors. The Gabidulin code $\mathscr{C} \subseteq \mathbb{F}_{q^m}^k$ is the row space over $\mathbb{F}_{q^m}$ of the matrix $M_{K\times k}$ containing q^{mK} codewords/vectors.

As previously, the rank of a vector $\boldsymbol{a} = (a_1, \ldots, a_k)$ in $\mathbb{F}_{q^m}^k$ is the dimension over $\mathbb{F}_q$ of the space spanned by $a_1, \ldots, a_k$. Equivalently expand each element of $\boldsymbol{a} \in \mathbb{F}_{q^m}$ into a column vector of dimension m over $\mathbb{F}_q$ to form an $m \times k$ matrix over $\mathbb{F}_q$. Thus each codeword vector over $\mathbb{F}_{q^m}$ of the Gabidulin code is associated to such a matrix and the rank of a vector is the rank of the associated matrix over $\mathbb{F}_q$.

To determine the rank distance of the Gabidulin code, the rank of the associated matrix of a codeword is determined by proceeding as follows. Consider a codeword $\boldsymbol{f}$ of the Gabidulin code $\mathscr{C}$ which can be expressed $\boldsymbol{f} = (f(g_1), f(g_2), \ldots, f(g_k)) \in \mathbb{F}_{q^m}^k$ where

$$f(x) = \sum_{i=0}^{K-1} f_i x^{q^i} \in \mathbb{F}_{q^m}[x], \text{ i.e., } f_i \in \mathbb{F}_{q^m}$$

which is a q-polynomial. Let $\{\eta_1, \ldots, \eta_m\}$ be a basis of $\mathbb{F}_{q^m}$ over $\mathbb{F}_q$ and express $f(g_i) = \sum_{j=1}^{m} f(g_i)_j \, \eta_j$ where $f(g_i)_j$ is the coefficient of η_j in the expansion $f(g_i)$ in the η basis. The codeword $\boldsymbol{f}$ can be expressed as the matrix equation

$$\boldsymbol{f} = (\eta_1, \ldots, \eta_m) \cdot \begin{bmatrix} f(g_1)_1 & f(g_2)_1 & \cdots & f(g_k)_1 \\ f(g_1)_2 & f(g_2)_2 & \cdots & f(g_k)_2 \\ \vdots & \vdots & \vdots & \vdots \\ f(g_1)_m & f(g_2)_m & \cdots & f(g_k)_m \end{bmatrix} = \boldsymbol{\eta} \cdot \boldsymbol{f}_{\eta}$$

where $\boldsymbol{\eta}$ is the row vector of basis elements and $\boldsymbol{f}_{\eta}$ is the $m \times k$ matrix corresponding to the Gabidulin codeword $\boldsymbol{f}$. That $\boldsymbol{f}_{\eta}$ has rank at least $k - K + 1$ follows from a similar argument as used in the Delsarte codes.

Consider the following. Let $\boldsymbol{f} = (f(g_1), f(g_2), \ldots, f(g_k)) \in \mathbb{F}_{q^m}^k$ be a codeword in the Gabidulin code and it is wished to determine the rank of the word $\boldsymbol{f}$ over $\mathbb{F}_q$. Equivalently, determine the null-space of the codeword over $\mathbb{F}_q$, i.e., determine the number of vectors $\boldsymbol{c} = (c_1, c_2, \ldots, c_k) \in \mathbb{F}_q^k$ orthogonal to $\boldsymbol{f}$. This is equivalent to determining the number of zeros of the equation:

$$\begin{aligned} \sum_{\ell=1}^{k} c_\ell \left(\sum_{j=0}^{K-1} f_j g_\ell^{q^j} \right) &= 0, \quad c_\ell \in \mathbb{F}_q,\ f_j \in \mathbb{F}_{q^m} \\ &= \sum_{j=0}^{K-1} f_j \left(\sum_{\ell=1}^{k} c_\ell g_\ell^{q^k} \right) \\ &= \sum_{j=0}^{K-1} f_j \left(\sum_{\ell=1}^{k} c_\ell g_\ell \right)^{q^j}. \end{aligned}$$

Since $\sum_{j=0}^{K-1} f_j x^{q^j}$ is a linearized polynomial its roots form a linear space with at most q^{K-1} elements. Thus the null-space of the $m \times k$ matrix over $\mathbb{F}_q$ is at most of dimension $K - 1$ and hence the rank of $\boldsymbol{f}_{\eta}$ is at least $k - K + 1$. Thus the Gabidulin code is equivalent to q^K $m \times k$ matrices over $\mathbb{F}_q$, each of rank at least $k - K + 1$. It follows MRD code.

Inherent in the previous arguments is the fact that a linear space of certain q-polynomials over $\mathbb{F}_{q^m}$ corresponds to a linear MRD code. Also it is noted that a vector rank-metric code is equivalent to a matrix rank-metric code (although the vector spaces involved may differ) in the sense that a vector rank-metric code over $\mathbb{F}_q^m$ of length k leads to an $m \times k$ rank-metric matrix code by representing elements in $\mathbb{F}_{q^m}$ as vectors over $\mathbb{F}_q$ as above.

Many of the more recent MRD codes found are also based on subsets of q-polynomials. A recent classification of the codes is given in ([21], chapter by

Otal and Özbudak). References are given there to recent constructions of linear MRD codes, nonlinear additive MRD codes, nonadditive (but closed under scalar multiplication) MRD codes, nonadditive and not closed under scalar multiplication codes. Morrison [32] examines notions of equivalence of such matrix codes. A few of the constructions of these codes are briefly noted. The class of *generalized Gabidulin codes* was formulated in [28] (see also [36] and [24]) by considering the vector space of linearized q-polynomials, for a fixed s and K,

$$\mathcal{G}_{k,s} = \left\{ f_0 + f_1 x^{q^s} + \cdots + f_{K-1} x^{q^{s(K-1)}} \mid f_i \in \mathbb{F}_{q^m} \right\}.$$

That these are linear MRD codes with shape $m \times k$ (where k is as in the construction of the usual Gabidulin codes) and minimum rank distance $(k - K + 1)$ is easily established. Defining $[i] \equiv q^i$ for the following matrix (in conflict with its usual meaning) the generator matrix for these generalized Gabidulin codes is

$$\boldsymbol{M}_{K \times k} = \begin{bmatrix} g_1^{s[0]} & g_2^{s[0]} & \cdots & g_k^{s[0]} \\ g_1^{s[1]} & g_2^{s[1]} & \cdots & g_k^{s[1]} \\ \vdots & \vdots & \vdots & \vdots \\ g_1^{s[K-1]} & g_2^{s[K-1]} & \cdots & g_k^{s[K-1]} \end{bmatrix}$$

for elements $g_1, g_2, \ldots, g_k$ in $\mathbb{F}_{q^m}$, linearly independent over $\mathbb{F}_q$ and $K < k$. The Gabidulin codes correspond to the space $\mathcal{G}_{k,1}$ (s=1) and it is known ([36]) that there exist codes in $\mathcal{G}_{k,s}$ that are not equivalent to Gabidulin codes. It is shown in [28] also ([24], proposition 2.9) that if $\mathscr{C}$ is a generalized Gabidulin code of dimension K over $\mathbb{F}_{q^m}$, then $\mathscr{C}^\perp \subseteq \mathbb{F}_{q^m}^k$ is a generalized Gabidulin code of dimension $k - K$ and these codes are MRD.

A class of TG codes is formulated in [36] by denoting the set of q-polynomials

$$\mathcal{H}_K(\eta, K) = \left\{ f_0 x + f_1 x^q + \cdots + f_{K-1} x^{q^{K-1}} + \eta f_0^{q^h} x^{q^K} \mid f_i \in \mathbb{F}_{q^m} \right\}$$

where $\eta \in \mathbb{F}_{q^m}$ is such that $N(\eta) \neq (-1)^{mK}$ where $N(x) = x^{(q^m-1)/(q-1)}$ is the norm function of $\mathbb{F}_{q^m}$ over $\mathbb{F}_q$ and $\gcd(m, s) = 1$. That these are linear MRD codes of dimension mK (q^{mK} matrices) and minimum distance $m - K + 1$ is established in [36]. It is also shown that some of these codes are not equivalent to Gabidulin codes.

Generalized TG codes are also discussed in ([21], chapter by Otal and Özbudak, theorem 4) and in [36]. Numerous other aspects of rank-metric codes have been considered in the extensive literature.

A class of rank-metric codes that will be of interest in the next section on subspace codes are those of *constant rank,* i.e., each codeword (vector or matrix) will have the same rank. These can be obtained from more general codes as subsets of constant rank and are generally nonlinear.

12.3 Subspace Codes

This section considers another related type of coding where the codewords are subspaces of a vector space according to a distance metric defined on such spaces. The application of these codes to network coding is noted. Clearly part of the section could have been placed in Chapter 5. It is included here due to its association with the use of rank-metric codes in their construction. The first part of this section introduces the notion of subspace codes and the remainder gives some constructions using the previously discussed rank-metric codes. The application of subspace codes to network coding is briefly noted.

The basic notions of subspaces of a vector space over the finite field $\mathbb{F}_q$ are recalled. Let $\mathcal{P}_q(n)$ denote the set of all subspaces of the vector space of dimension n over $\mathbb{F}_q$, $\mathbb{F}_q^n \sim V_n(q)$, and $\mathcal{G}_q(n,k)$ denote the set of all k-dimensional subspaces of the space $\mathbb{F}_q^n$, referred to as the *Grassmannian of order* k of $\mathbb{F}_q^n$. As noted previously the cardinality of this Grassmannian is

$$|\mathcal{G}_q(n,k)| = \begin{bmatrix} n \\ k \end{bmatrix}_q . \tag{12.9}$$

The following simple bound can be useful ([17, 27] and Appendix A):

$$q^{k(n-k)} \leq \begin{bmatrix} n \\ k \end{bmatrix}_q \leq K_q^{-1} q^{k(n-k)} \tag{12.10}$$

where $K_q = \prod_{i=1}^{\infty}(1 - q^{-i})$. The constant K_q^{-1} is monotonically decreasing with q with a maximum of 3.4594 at $q = 2$ and a value 1.0711 at $q = 16$. This follows directly from the definition of the Gaussian coefficient.

Consider now the problem of defining codes on $\mathcal{P}_q(n)$ for which a metric is required. For rank-metric codes the matrix rank was used for distance. For subspaces the equivalent notion is that of dimension of certain subspaces.

For any two subspaces of $V_n(q)$, U and V denote $U + V$ as the smallest subspace containing both U and V. Then

$$\dim(U + V) = \dim(U) + \dim(V) - \dim(U \cap V).$$

It is easy to verify that the subspace distance defined by

$$d_S(U,V) = \dim(U) + \dim(V) - 2\dim(U \cap V) \tag{12.11}$$

is a metric on subspaces of $V_n(q)$, the "subspace distance" between subspaces U and V. The similarity with the definition of the rank metric for matrices is immediate. It also follows that

$$d_S(U,V) = 2\dim(U+V) - \dim(U) - \dim(V). \tag{12.12}$$

For a subspace $U \subseteq V_n(q)$ the dual space is

$$U^\perp = \{\boldsymbol{v} \in V_n(q) \mid (\boldsymbol{u},\boldsymbol{v}) = 0,\ \forall \boldsymbol{u} \in U\}$$

where $(\boldsymbol{u},\boldsymbol{v})$ is the usual inner product on $V_n(q)$.

The following identities were noted earlier (e.g., [27]):

$$(U+V)^\perp = U^\perp \cap V^\perp \quad \text{and} \quad (U \cap V)^\perp = U^\perp + V^\perp.$$

and hence

$$d_S(U^\perp, V^\perp) = d_S(U,V).$$

Definition 12.15 A *subspace code* in $V_n(q)$, $\mathscr{C} \subseteq \mathcal{P}_q(n)$, is a collection of subspaces of $V_n(q)$. The minimum distance of $\mathscr{C}$ is

$$d_{S,\min}(\mathscr{C}) = \min_{U,V \in C, U \neq V} d_S(U,V).$$

Such a code is designated a $\mathscr{C} = (n,\ |\mathscr{C}|,\ d_{S,\min}(\mathscr{C}))_q$ code. Further define $d_{S.\max}(\mathscr{C}) = \max_{U \in \mathscr{C}} \dim(U)$, a maximum dimension of codewords. If all the words of the code have the same dimension ($\mathscr{C} \subseteq \mathcal{G}_q(n,k)$ for some k), it is referred to as a *constant-dimension code* (CDC), and designated as a $\mathscr{C} = (n,\ |\mathscr{C}|, k, d_{S,\min}(\mathscr{C}))_q$ CDC code. Denote by $A_q^{SC}(n, d_{S,\min})$ and $A_q^{CDC}(n,k,d_{S,\min})$ the maximum number of codewords (subspaces) in these two types of codes, respectively.

From the previous comments it is immediate that if $\mathscr{C} = (n, M, k, d_{S,\min})_q$ is a constant-dimension subspace code of subspaces of dimension k, then $\mathscr{C}^\perp$ is a constant-dimension $(n, M, n-k, d_{S,\min})_q$ subspace code since the dual of a subspace of dimension k is one of dimension $n-k$.

Another type of metric on subspaces is introduced in [39] suitable for an adversarial type of network channel. For network coding a collection of packets is identified with the subspace they generate, the reason for the interest in subspace coding. It is shown that in this adversarial case a more appropriate metric is the *injection metric* defined as follows:

Definition 12.16 The *injection distance* between two subspaces $U, V \in \mathcal{P}_q(n)$ is defined as

$$\begin{aligned} d_I(U,V) &\triangleq \max\{\dim(U), \dim(V)\} - \dim(U \cap V) \\ &= \dim(U+V) - \min\{\dim(U), \dim(V)\} \\ &= \tfrac{1}{2} d_S(U,V) + \tfrac{1}{2}\left|\dim(U) - \dim(V)\right| . \end{aligned}$$

That this is in fact a metric is shown in ([39], theorem 22). If an *injection* is defined as the deletion of a packet at a node and the insertion of another (orthogonal) packet (a packet outside the space of information packets received at that node), then the injection metric between spaces U and V may be viewed as the number of injections required to modify the space U to become the space V. Example 12.17 attempts to demonstrate the relationships in the above definition.

Example 12.17 Suppose $U, V \in \mathcal{G}_q(n,k)$. In this case the subspaces have constant dimension and

$$d_I(UV) = \frac{1}{2} d_S(U,V)$$

(the subspace distance for subspaces of the same dimension is always even) and for this case the two metrics are equivalent. As in Figure 12.1

$$d_S(U,V) = \dim U + \dim V - 2\dim U \cap V = k + k - 2(k-i) = 2i$$

and

$$d_I(U,V) = i.$$

Since $\dim(U \backslash U \cap V) = i$ it clearly takes i insertions and deletions to transform U to V.

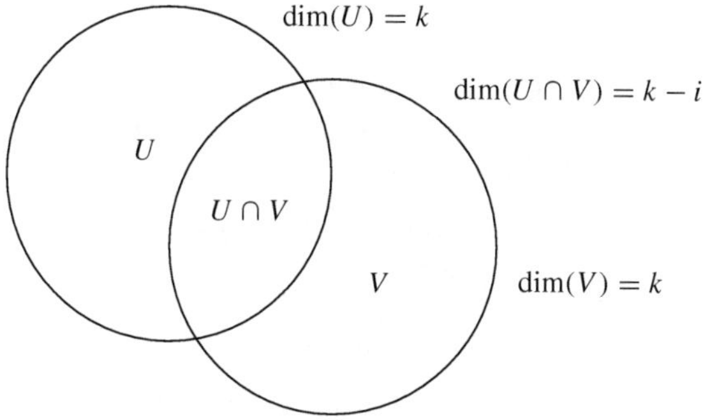

Figure 12.1 Subspace enumeration, $d_S(U,V) = 2i$, equal dimension spaces

The relationship between CRC (constant rank) matrix codes and CDC (constant dimension) subspace codes is interesting and is pursued in [18] using the injection distance. A few key results from that work are described adapted to our terminology.

For the remainder of the section interest will be on the bounding and construction of CDC codes in $\mathcal{P}_q(n)$ of dimension k (i.e., in $\mathcal{G}_q(n,k)$). Recall from Definition 12.15 the maximum size of such a code with minimum distance $d_{S,\min}$ is denoted $A_q^{CDC}(n,k,d_{S,\min})$ (minimum distance d and subspace dimension k).

To discuss the analogs of the usual algebraic coding bounds for subspaces, and the rank distance coding bounds, it is of interest to determine the number of subspaces at distance $2i$ from a fixed subspace U in $\mathcal{G}_q(n,k)$. Define [27] a ball of radius $2i$ and a sphere of radius $2t$ around the subspace U as (recalling such distances are even):

$$\begin{aligned} B_q(n,k,i) &= \{V \in \mathcal{G}_q(n,k) \mid d_S(U,V) = 2i\}, \quad b_q(n,k,i) = |B_q(n,k,i)| \\ S_q(n,k,t) &= \{V \in \mathcal{G}_q(n,k) \mid d_S(U,V) \le 2t\}, \quad s_q(n,k,t) = |S_q(n,k,t)| \\ &= \cup_{i \le t} B_q(n,k,i). \end{aligned} \tag{12.13}$$

It is clear the enumeration is independent of the "center," U, although allowing subspaces of other dimensions would negate this statement [27]. To enumerate this quantity, consider a fixed subspace $U \in \mathcal{G}_q(n,k)$ of dimension k and a fixed – for the moment – subspace of U, W, of dimension $k-i$. There are $\begin{bmatrix} k \\ k-i \end{bmatrix}_q = \begin{bmatrix} k \\ i \end{bmatrix}_q$ ways of choosing W in U. Consider the number of ways of adding points (linearly independent) to W in the exterior of U in such a way to expand $U \cap V$ into a k space $V \in \mathcal{G}_q(n,k)$. This would result in

$$d_S(U,V) = \dim(U) + \dim(V) - 2\dim(U \cap (V) = 2k - 2(k-i) = 2i.$$

The number of distinct ways this can be done using standard arguments is:

$$\begin{aligned} &\frac{(q^n - q^k)(q^n - q^{k+1})\cdots(q^n - q^{k+(i-1)})}{(q^k - q^{k-i})(q^k - q^{k-i+1})\cdots(q^k - q^{k-1})} \\ &= \frac{q^{k+(k+1)+\cdots+(k+(i-1))}}{q^{(k-i)+(k-i+1)+\cdots+(k-1)}} \begin{bmatrix} n-k \\ i \end{bmatrix}_q \\ &= q^{i^2} \begin{bmatrix} n-k \\ i \end{bmatrix}_q \end{aligned} \tag{12.14}$$

where the denominator of the first expression represents the number of ways of choosing independent vectors to complete a $(k-i)$ space to a k space.

Since the number of ways in choosing the intersection subspace $U \cap V$ in U is $\begin{bmatrix} k \\ k-i \end{bmatrix}_q = \begin{bmatrix} k \\ i \end{bmatrix}_q$ we have

$$s_q(n,k,t) = \sum_{i=0}^{t} q^{i^2} \begin{bmatrix} k \\ i \end{bmatrix}_q \begin{bmatrix} n-k \\ i \end{bmatrix}_q . \tag{12.15}$$

The simplest bound [11, 27] for the size of a CDC code of distance $2t$ is the *sphere packing bound*, the analog of that in algebraic coding theory,

$$\begin{aligned} A_q^{CDC}(n,k,d_S = 2t) &\leq \begin{bmatrix} n \\ k \end{bmatrix}_q \Big/ s_q(n,k,t) \\ &\leq \begin{bmatrix} n \\ k \end{bmatrix}_q \Big/ \left(\sum_{i=0}^{t} q^{i^2} \begin{bmatrix} k \\ i \end{bmatrix}_q \begin{bmatrix} n-k \\ i \end{bmatrix}_q \right). \end{aligned} \tag{12.16}$$

Sphere packing bound

With similar reasoning for a fixed $V \in \mathcal{G}_q(n,k)$ the number of $W \in \mathcal{G}_q(n,j)$ which intersect V in a subspace of dimension ℓ is given by ([26], chapter by Kschischang):

$$N_q(n,k,j,\ell) = q^{(k-\ell)(j-\ell)} \begin{bmatrix} k \\ \ell \end{bmatrix}_q \begin{bmatrix} n-k \\ j-\ell \end{bmatrix}_q$$

and the result of Equation 12.14 follows immediately from this.

Other bounds on the sizes of subspace codes are noted. The analog of the Varshamov–Gilbert (VG) bound of algebraic coding theory for constant-dimension subspace code is argued in a very similar manner. Consider constructing a subspace code of minimum distance $d_{S,\min}$ in the following way. Choose an element of $\mathcal{G}_q(n,k)$ as an initial codeword and surround it with a sphere of radius $d_{S,\min} - 1$. Continue selecting codewords (subspaces) that are external to any of the spheres. If at some point the number of codewords chosen in this manner is $M - 1$ and

$$(M-1) \cdot s_q(n,k,d_{S,\min} - 1) < \begin{bmatrix} n \\ k \end{bmatrix}_q ,$$

then another codeword can be added. Thus if $\mathscr{C}$ is an optimal code (with maximum number of codewords $A_q^{CDC}(n,d_{S,\min},k)$) with distance $d_{S,\min}$, then we must have

$$A_q^{CDC}(n,k,d_{S,\min}) \geq \begin{bmatrix} n \\ k \end{bmatrix}_q \Big/ s_q(n,k,d_{S,\min} - 1), \quad \text{Varshamov–Gilbert bound}$$

since otherwise another codeword could be added.

For another type of bound an interesting puncturing technique is described [27] that leads to an analog of the Singleton bound. Consider the CDC subspace code $\mathscr{C} = (n, M, k, d_{S,\min})_{q.} \subseteq \mathcal{G}_q(n,k)$. Let S be a subspace of $V_n(q)$ of dimension $n-1$. For a positive integer k define a *stochastic erasure operator* $\mathcal{H}_k(U)$ that operates on the subspaces $\mathcal{P}_q(n)$ such that if:

$$\mathcal{H}_k(U) = \begin{cases} \text{random subspace of } U \text{ of dimension } k \\ \text{if } \dim(U) > kU \quad \text{else.} \end{cases} \tag{12.17}$$

This operator is motivated by network coding considerations as in [27, 38]. This interesting work is not pursued here.

Define the code

$$\mathscr{C}' = \{U' = \mathcal{H}_{k-1}(U) \mid U \in \mathscr{C}\}.$$

It is shown that $\mathscr{C}' = (n-1, M', k-1, d' \geq d_{S,\min} - 2\}_q$ is a CDC code. To see this let $U' = \mathcal{H}_{k-1}(U)$ and $V' = \mathcal{H}_{k-1}(V)$ for two distinct codewords U, V of $\mathscr{C}$. Since

$$2\dim(U' \cap V') \leq 2\dim(U \cap V) = 2k - d_{S,\min},$$

by the definition of minimum distance (d_S) in C it follows that

$$d_S(U', V') = 2(k-1) - 2\dim(U' \cap V') \geq d_{S,\min} - 2.$$

Thus, while the codewords of the code $\mathscr{C}'$ may not be unique, due to the random erasure operator, the parameters of the code are $M' = M$ and $d' \geq d_{S,\min} - 2$, giving the parameters stated.

Suppose in the CDC code $\mathscr{C} = (n, M, k, d_{S,\min})_q$ the minimum subspace distance is $d_{S,\min} = 2\gamma + 2$. Applying this puncturing operation to the code C δ times yields a code $C''(n-\gamma, M, k-\gamma, d'' \geq 2)_q$. Since the codewords are unique, this implies that

$$M \leq \begin{bmatrix} n-\gamma \\ k-\delta \end{bmatrix}_q = \begin{bmatrix} n-\delta \\ n-k \end{bmatrix}_q.$$

We could also apply the puncturing to the dual code $C^{\perp} = (n, M, n-k, d)_q$ (same minimum distance) and hence

$$A_q^{CDC}(n, k, d_{S,\min} = 2\gamma + 2) \leq \begin{bmatrix} n-\gamma \\ \max\{k, n-k\} \end{bmatrix}_q, \quad \text{Singleton bound} \tag{12.18}$$

which is the desired bound ([11], equation 3, [27], theorem 9), regarded as an analog of the Singleton bound for CDC codes.

Using the inequality of Equation 12.10 gives another version of the inequality (assume $k \geq n/2$):

$$A_q^{CDC}(n,k,d = 2\gamma + 2) \leq K_q^{-1} q^{k(n-k-\gamma)}. \qquad (12.19)$$

Another approach gives a further interesting bound. As with rank-metric codes the following notion of a subspace anticode is of interest, first raised in the work of Delsarte on association schemes [4]. Delsarte's work was concerned with bounding the size of cliques in association schemes.

To introduce the notion of an anticode bound in this subspace environment consider first the notion applied to binary codes and define an anticode $\mathcal{A}_H(d-1)$ to be a binary set of n-tuples such that the Hamming distance between any two $\boldsymbol{x},\boldsymbol{y} \in \mathcal{A}_H(d-1)$ is at most $d-1$. A sphere of radius $(d-1)/2$ in the Hamming metric is an optimal anticode. Suppose C is an $(n,M,d)_2$ binary code and consider the array of n-tuples whose M rows consist of the words of the anticode added to a word of the code. It is clear that the words of this array are distinct and hence

$$M \cdot |\mathcal{A}_H(d-1)| \leq 2^n$$

which is viewed as an anticode bound for binary codes.

A similar situation occurs for CDC codes in $\mathcal{G}_q(n,k)$ although it is not justified here. As noted in the previous section, while spheres are optimal anticodes in Hamming spaces they are not in $V_n(q)$ and indeed [11] are far from optimal and the maximal size of such sets is, as discussed below, found by Frankl and Wilson [15].

Definition 12.18 A *CDC anticode* $\mathcal{A}$ in $\mathcal{G}_q(n,k)$ of diameter 2γ is any subset of $\mathcal{G}(n,k)$ such that $d_S(U,V) \leq 2\gamma$ for all $U,V \in \mathcal{A}$.

Thus any two subspaces in $\mathcal{A}$ intersect in a subspace of dimension at least $k-\gamma$. Fankl and Wilson [15] (see also [23]) determined the maximum size of anticodes in $\mathcal{G}_q(n,k)$ as follows. The set of subspaces of $V_n(q)$ which contain a fixed subspace of dimension $k-\gamma+1$ is an anticode of diameter $2\gamma-2$. This set is of size

$$\begin{bmatrix} n-k+\gamma-1 \\ \gamma-1 \end{bmatrix}_q.$$

Similarly the set of subspaces of $\mathcal{G}_q(n,k)$ contained in a fixed subspace of dimension $k+\gamma-1$ is also an anticode of dimension $2\gamma-2$ with at least

$$\begin{bmatrix} k+\delta-1 \\ k \end{bmatrix}_q$$

subspaces (see [15], theorem 1). These anticodes are the largest possible in $\mathcal{G}_q(n,k)$ whose elements intersect in at least a subspace of dimension $k-\gamma$ and is of size

$$\max\left\{\begin{bmatrix} n-k+\gamma-1 \\ \gamma-1 \end{bmatrix}_q, \begin{bmatrix} k+\gamma-1 \\ k \end{bmatrix}_q\right\}$$

and hence are of diameter $2(\gamma-1)$.

It is convenient to define γ so that the code minimum distance is $d_{S,\min} = 2\gamma+2$. For this the Delsarte anticode bound ([4], theorem 3.9, p. 32) then ([11], theorem 1, [26], theorem 1.6.2) is the size of $\mathcal{G}_q(n,k)$ divided by the size of the largest anticode or

$$A_q^{CDC}(n,k,d_{S,\min}) = 2\gamma+2) \leq |\mathcal{G}_q(n,k)| \Big/ |\mathcal{A}(\gamma)| = \begin{bmatrix} n \\ k \end{bmatrix}_q \Big/ \begin{bmatrix} n-k+\gamma \\ \gamma \end{bmatrix}_q .$$

Anticode bound

This bound was first obtained in [40] in the context of authentication codes. It was observed in [41] that this bound is always better than the Singleton bound (Equation 12.18) for nontrivial CDCs.

Another such bound is given in ([11], theorem 2) as:

Theorem 12.19

$$\mathcal{A}_q^{CDC}(n,k,d = 2\gamma+2) \leq \begin{bmatrix} n \\ k-\gamma \end{bmatrix}_q \Big/ \begin{bmatrix} k \\ k-\gamma \end{bmatrix}_q .$$

Proof: Suppose $\mathscr{C}$ is an optimal CDC $\mathscr{C}^{CDC}(n,k,d_{S,\min} = 2\gamma+2)$ code with $M = |\mathscr{C}|$. The proof is by enumeration of $k-\gamma$ subspaces. Note first that any two codewords U and V cannot have a $k-\gamma$ subspace of $V_n(q)$ in common as otherwise, if they had such a subspace in common, then

$$d_S(U,V) \leq 2k - 2(k-\gamma) = 2\gamma < d,$$

a contradiction. The total number of $k-\gamma$ subspaces in $V_n(q)$ is $\begin{bmatrix} n \\ k-\gamma \end{bmatrix}_q$.

On the other hand each codeword contains $\begin{bmatrix} k \\ k-\gamma \end{bmatrix}_q$ such subspaces and by the previous comment these are distinct and hence

$$M \cdot \begin{bmatrix} k \\ k-\gamma \end{bmatrix}_q \leq \begin{bmatrix} n \\ k-\gamma \end{bmatrix}_q$$

as was required to show. ■

There are numerous other bounds on CDC codes, including an analog of the Johnson bound of algebraic coding theory [41]. The chapter on Network Codes in the volume [26] gives an excellent discussion on bounds (and indeed on the whole subject of network coding). The work of [11] also discusses bounds and gives improvements of many of them, as does [23].

It is interesting to note that, as was the case for rank-metric codes, there are no perfect subspace codes (not just no perfect CDC codes) ([11], theorem 16).

With the above bound on the size of CDCs for a given distance and dimension, the construction of subspace codes is considered. An interesting construction of such codes uses the rank-metric codes of the previous two sections as follows.

Denote by $\mathscr{C}^{RM}$ a rank-metric code of shape $k \times m$ with minimum distance d_R and $\langle \boldsymbol{A} \rangle$ the row space of the matrix $\boldsymbol{A} \in \mathscr{C}_R$. Let $\boldsymbol{AB}$ denote the matrix which is the catenation (not multiplication) of compatible matrices $\boldsymbol{A}$ and $\boldsymbol{B}$ (i.e., the columns of $\boldsymbol{AB}$ are those of $\boldsymbol{A}$ followed by those of $\boldsymbol{B}$). Consider

$$\mathscr{C}_S = \left\{ \langle \boldsymbol{IA} \rangle \mid \boldsymbol{A} \in C^{RM} \right\}$$

a collection of row spaces generated by the rank metric matrices of $\mathscr{C}^{RM}$ with $\boldsymbol{I}$ a dimension-compatible identity matrix. Denote the minimum subspace distance of $\mathscr{C}_S$ by d_S and the minimum rank-metric distance of C^{RM} by $d_{R,\min}$. Further let

$$\langle \boldsymbol{AB} \rangle \quad \text{and} \quad \left\langle \begin{matrix} \boldsymbol{A} \\ \boldsymbol{B} \end{matrix} \right\rangle$$

be the row spaces of the matrix $\boldsymbol{A}$ catenated with $\boldsymbol{B}$ horizontally and vertically (rows of $\boldsymbol{B}$ added to those of $\boldsymbol{A}$), respectively, assuming dimension compatibility. Recall that the rank of a matrix is the dimension of its row space (or column space). In the proof of the following proposition it will be convenient to denote the corresponding matrices as

$$\left[\boldsymbol{AB} \right] \quad \text{and} \quad \begin{bmatrix} \boldsymbol{A} \\ \boldsymbol{B} \end{bmatrix}.$$

For any matrix $\boldsymbol{A}$, $\dim \langle \boldsymbol{A} \rangle = \text{rank} \left[\boldsymbol{A} \right]$. Then ([38], proposition 4):

Proposition 12.20 *With the notation above for subspaces*

$$d_S = 2 d_R.$$

Proof: For $\boldsymbol{A}, \boldsymbol{B} \in \mathscr{C}_R$ let $\boldsymbol{U} = \langle \boldsymbol{IA} \rangle$ and $\boldsymbol{V} = \langle \boldsymbol{IB} \rangle$ and note that $\dim \boldsymbol{U} = \dim \boldsymbol{V} = k$. From Equation 12.12

$$\begin{aligned} d_S(\boldsymbol{U},\boldsymbol{V}) &= 2\dim(\boldsymbol{U}+\boldsymbol{V}) - 2k \\ &= 2\text{rank}\begin{bmatrix} \boldsymbol{I} & \boldsymbol{A} \\ \boldsymbol{I} & \boldsymbol{B} \end{bmatrix} - 2k \\ &= 2\text{rank}\begin{bmatrix} \boldsymbol{I} & \boldsymbol{A} \\ 0 & \boldsymbol{B}-\boldsymbol{A} \end{bmatrix} - 2k \\ &= 2\text{rank}\begin{bmatrix} \boldsymbol{B}-\boldsymbol{A} \end{bmatrix} = 2d_R(\boldsymbol{A},\boldsymbol{B}) \end{aligned}$$

and the result is immediate. ■

This construction of a subspace code from a rank-matrix code is referred to as *lifting*. Thus an MRD rank-metric code yields a lifted subspace code with constant dimension and whose distance distribution is easily derivable from that of the MRD code. Such codes exist for all allowable parameter sets.

Suppose $\mathscr{C}^{MRD}$ is an MRD rank-metric code in $\boldsymbol{M}_{k\times m}(q)$ with minimum rank distance d_R and number of codewords (matrices) equal to the Singleton bound:

$$M = q^{m(k-d_R+1)}.$$

The above lifting process then yields a CDC code with M subspaces in $V_n(q)$ of dimension k and minimum distance $2d_R$ – hence a $\mathscr{C}^{CDC} = C(n = k + m, M = q^{m(k-d_R+1)}, k, 2d_R)_q$ code. It is argued in [25] that these MRD-lifted subspace codes are asymptotically optimal in the following sense. Using the Singleton bound for CDC codes with minimum distance $2\gamma + 2$ of Equation 12.18 and the q-coefficient approximation of Equation 12.10 yields the upper and lower bounds

$$q^{(n-k)(k-\gamma+1)} \leq A_q^{CDC}(n,k,d_S) \leq \begin{bmatrix} n \\ k-\gamma+1 \end{bmatrix}_q \leq K_q^{-1} q^{(n-k)(k-\gamma+1)}$$

but the codes lifted from an MRD rank-metric code have

$$q^{(n-k)(k-\gamma+1)}$$

subspaces which is within a factor of K_q^{-1} of the above upper bound and hence it is argued the lifted MRD codes are asymptotically optimal.

The work of Gadouleau and Yan [18] improved on the lifted construction as follows. Recall from Definition 12.15 $A_q^{CDC}(n,k,d_S)$ is the maximum number of subspaces of $V_n(q)$ of dimension k in $V_n(q)$ with distance d_S. For an arbitrary matrix $\boldsymbol{A} \in \boldsymbol{M}_{k\times m}(q)$ let $RSP(\boldsymbol{A})$ and $CSP(\boldsymbol{A})$ denote its row and column spaces, respectively. For a rank-metric code $\mathscr{C} \subseteq \boldsymbol{M}_{k\times m}(q)$ define subspace codes in the natural manner:

$$\begin{aligned} CSP(\mathscr{C}) &\triangleq \{U \in \mathcal{P}_q(k) \mid \exists \boldsymbol{A} \in \mathscr{C}, CSP(A) = U\} \\ RSP(\mathscr{C}) &\triangleq \{V \in \mathcal{P}_q(m) \mid \exists \boldsymbol{B} \in \mathscr{C}, RSP(B) = V\}. \end{aligned}$$

To derive CDC subspace codes from CRC rank-matrix codes the following results are noted:

Proposition 12.21 ([18], propositions 1 and 2) *Let $\mathscr{C}$ be a CRC of rank r and minimum rank distance d_R in $\mathbf{M}_{k\times m}(q)$*

(i) Then the subspace codes

$$RSP(C) \subseteq \mathcal{G}_q(m,r) \text{ and } CSP(\mathscr{C}) \subseteq \mathcal{G}_q(k,r)$$

have minimum distance at least $d_R - r$;

(ii) Suppose $\mathscr{C}$ has minimum rank distance $d+r$, $(1 \leq d \leq r)$ in $\mathbf{M}_{k\times m}(q)$. Then $RSP(\mathscr{C}) \subseteq \mathcal{G}_q(m,r)$ and $CSP(\mathscr{C}) \subseteq \mathcal{G}_q(k,r)$ have cardinality $|\mathscr{C}|$ and their minimum injection distances satisfy

$$d_I(CSP(\mathscr{C})) + d_I(RSP(\mathscr{C})) \leq d + r.$$

Interestingly, a reverse statement is also obtained in ([18], proposition 3) in that if $\mathcal{M}$ is a CDC in $\mathcal{G}_q(k,r)$ and $\mathcal{N}$ a CDC in $\mathcal{G}_q(m,r)$ such that $|\mathcal{M}|=|\mathcal{N}|$. Then there will exist a constant rank-metric code $\mathscr{C} \subseteq \boldsymbol{M}_{k\times m}(q)$ with constant rank r and cardinality $|\mathcal{M}|$ such that

$$CSP(\mathscr{C}) = \mathcal{M} \quad \text{and} \quad RSP(\mathscr{C}) = \mathcal{N}$$

and the minimum injection distances satisfy

$$d_I(\mathcal{N}) + d_I(\mathcal{M}) \leq d_R \leq \min\{d_I(\mathcal{N}), d_I(\mathcal{M})\} + r.$$

Theorem 2 of [18] gives conditions on the code parameters for the row spaces of an optimal constant rank, rank-metric code to give an optimal CDC. As a consequence it shows that if $\mathscr{C}$ is a constant rank-metric code in $\boldsymbol{M}_{k\times m}(q)$ with constant-rank codewords of rank r and minimum distance $d+r$ for $2r \leq k \leq m$ and $1 \leq d \leq r$, then for $m \geq m_0 = (k-r)(r-d+1)$ or $d = r$ the row space code of C, $RSP(C)$, is an optimal CDC in $\mathcal{G}_q(n,r)$. Thus finding optimal CRCs with sufficiently many rows, under the conditions stated gives optimal CDC codes. It seems that in general the rank-metric code problem yields more tractable approaches for the construction of subspace codes.

There is an impressive literature on the construction and analysis of subspace codes, in particular those of constant dimension, and this section is only a brief indication of that work. A variety of combinatorial structures have been applied to the problems and the area has attracted significant interest. The notions of Ferrer's diagrams, graph matchings and Grassman graphs have been of interest in these constructions including the works [6, 7, 8, 9, 10, 11, 13, 13, 14, 14, 19, 19, 20, 22, 31, 35, 37, 42]. The notion of a q-analog of a Steiner system is particularly interesting, consisting of a set of

k-dimensional subspaces S such that each t-dimensional subspace of $V_n(q)$ is contained in exactly one subspace of S [1, 6, 7, 12].

There is a natural association of subspace codes and network coding explored in [27, 38] among many other works. Consider a system where packets are transmitted on the network and at each node a random linear combination of the arriving packets is forwarded. Additive erroneous packets may be included at any of the intermediate nodes as the packets progress through the network and information packets may be lost at any node due to network impairments. Thus the spaces associated with a set of packets and the network operations of errors, including packet insertions and deletions, can be viewed via the subspace codes introduced here. The reader is referred to the references noted for an account of this interesting area.

Comments

The chapter has given a brief overview of rank-metric and subspace codes and their relationships. While such concepts of coding may be in unfamiliar settings, such as sets of matrices and sets of subspaces, the results obtained are of great interest, drawing for their development on geometry and combinatorics. The application of subspace codes to network coding is of particular interest and the survey article [25] is an excellent source for this area.

References

[1] Arias, F., de la Cruz, J., Rosenthal, J., and Willems, W. 2018. On q-Steiner systems from rank metric codes. *Discrete Math.*, **341**(10), 2729–2734.

[2] Cruz, J., Gorla, E., Lopéz, H., and Ravagnani, A. 2015 (October). Rank distribution of Delsarte codes. *CoRR*. arXiv:1510.00108v1.

[3] Cruz, J., Gorla, E., López, H.H., and Ravagnani, A. 2018. Weight distribution of rank-metric codes. *Des. Codes Cryptogr.*, **86**(1), 1–16.

[4] Delsarte, P. 1973. *An algebraic approach to the association schemes of coding theory*. Philips Research Reports Supplements. N.V. Philips' Gloeilampenfabrieken.

[5] Delsarte, P. 1978. Bilinear forms over a finite field, with applications to coding theory. *J. Combin. Theory Ser. A*, **25**(3), 226–241.

[6] Etzion, T. 2014. Covering of subspaces by subspaces. *Des. Codes Cryptogr.*, **72**(2), 405–421.

[7] Etzion, T., and Raviv, N. 2015. Equidistant codes in the Grassmannian. *Discrete Appl. Math.*, **186**, 87–97.

[8] Etzion, T., and Silberstein, N. 2009. Error-correcting codes in projective spaces via rank-metric codes and Ferrers diagrams. *IEEE Trans. Inform. Theory*, **55**(7), 2909–2919.

[9] Etzion, T., and Silberstein, N. 2013. Codes and designs related to lifted MRD codes. *IEEE Trans. Inform. Theory*, **59**(2), 1004–1017.

[10] Etzion, T., and Vardy, A. 2008 (July). Error-correcting codes in projective space. Pages 871–875 of: *2008 IEEE International Symposium on Information Theory*.

[11] Etzion, T., and Vardy, A. 2011. Error-correcting codes in projective space. *IEEE Trans. Inform. Theory*, **57**(2), 1165–1173.

[12] Etzion, T., and Vardy, A. 2011. On q-analogs of Steiner systems and covering designs. *Adv. Math. Commun.*, **5**, 161.

[13] Etzion, T., and Zhang, H. 2019. Grassmannian codes with new distance measures for network coding. *IEEE Trans. Inform. Theory*, **65**(7), 4131–4142.

[14] Etzion, T., Gorla, E., Ravagnani, A., and Wachter-Zeh, A. 2016. Optimal Ferrers diagram rank-metric codes. *IEEE Trans. Inform. Theory*, **62**(4), 1616–1630.

[15] Frankl, P., and Wilson, R.M. 1986. The Erdös-Ko-Rado theorem for vector spaces. *J. Combin. Theory Ser. A*, **43**(2), 228–236.

[16] Gabidulin, È. M. 1985. Theory of rank codes with minimum rank distance. *Problemy Peredachi Informatsii*, **21**(1), 1–12.

[17] Gadouleau, M., and Yan, Z. 2008. On the decoder error probability of bounded rank-distance decoders for maximum rank distance codes. *IEEE Trans. Inform. Theory*, **54**(7), 3202–3206.

[18] Gadouleau, M., and Yan, Z. 2010. Constant-rank codes and their connection to constant-dimension codes. *IEEE Trans. Inform. Theory*, **56**(7), 3207–3216.

[19] Gadouleau, M., and Yan, Z. 2010. Packing and covering properties of subspace codes for error control in random linear network coding. *IEEE Trans. Inform. Theory*, **56**(5), 2097–2108.

[20] Gorla, E., and Ravagnani, A. 2014. Partial spreads in random network coding. *Finite Fields Appl.*, **26**, 104–115.

[21] Greferath, M., Pavčević, M.O., Silberstein, N., and Vázquez-Castro, M.A. (eds.). 2018. *Network coding and subspace designs*. Signals and Communication Technology. Springer, Cham.

[22] Heinlein, D. 2019. New LMRD code bounds for constant dimension codes and improved constructions. *IEEE Trans. Inform. Theory*, **65**(8), 4822–4830.

[23] Heinlein, D., and Kurz, S. 2017. Asymptotic bounds for the sizes of constant dimension codes and an improved lower bound. Pages 163–191 of: *Coding theory and applications*. Lecture Notes in Computer Science, vol. 10495. Springer, Cham.

[24] Horlemann-Trautmann, A.-L., and Marshall, K. 2017. New criteria for MRD and Gabidulin codes and some rank-metric code constructions. *Adv. Math. Commun.*, **11**(3), 533–548.

[25] Huffman, W.C., Kim, J.-L., and Solé, P. (eds.). 2020. *Network coding in a concise encyclopedia of coding theory*. Discrete Mathematics and Its Applications (Boca Raton). CRC Press, Boca Raton, FL.

[26] Huffman, W.C., Kim, J.L., and Solé, P. (eds.). 2021. *A concise encyclopedia of coding theory*. CRC Press, Boca Raton, FL.

[27] Koetter, R., and Kschischang, F.R. 2008. Coding for errors and erasures in random network coding. *IEEE Trans. Inform. Theory*, **54**(8), 3579–3591.

[28] Kshevetskiy, A., and Gabidulin, E. 2005 (September). The new construction of rank codes. Pages 2105–2108 of: *Proceedings International Symposium on Information Theory, 2005. ISIT 2005.*

[29] Lidl, R., and Niederreiter, H. 1997. *Finite fields*, 2nd ed. Encyclopedia of Mathematics and Its Applications, vol. 20. Cambridge University Press, Cambridge.

[30] Loidreau, Pierre. 2006. Properties of codes in rank metric. *CoRR*, abs/cs/0610057.

[31] Martin, W.J., and Zhu, X.J. 1995. Anticodes for the Grassmann and bilinear forms graphs. *Des. Codes Cryptogr.*, **6**(1), 73–79.

[32] Morrison, K. 2014. Equivalence for rank-metric and matrix codes and automorphism groups of Gabidulin codes. *IEEE Trans. Inform. Theory*, **60**(11), 7035–7046.

[33] Ravagnani, A. 2014. Rank-metric codes and their MacWilliams identities. *arXiv:1410.11338v1.*

[34] Ravagnani, A. 2016. Rank-metric codes and their duality theory. *Des. Codes Cryptogr.*, **80**(1), 197–216.

[35] Schwartz, M., and Etzion, T. 2002. Codes and anticodes in the Grassman graph. *J. Combin. Theory Ser. A*, **97**(1), 27–42.

[36] Sheekey, J. 2016. A new family of linear maximum rank distance codes. *Adv. Math. Commun.*, **10**(3), 475–488.

[37] Silberstein, N., and Trautmann, A. 2015. Subspace codes based on graph matchings, Ferrers diagrams, and pending blocks. *IEEE Trans. Inform. Theory*, **61**(7), 3937–3953.

[38] Silva, D., Kschischang, F.R., and Koetter, R. 2008. A rank-metric approach to error control in random network coding. *IEEE Trans. Inform. Theory*, **54**(9), 3951–3967.

[39] Silva, D., Zeng, W., and Kschischang, F.R. 2009 (June). Sparse network coding with overlapping classes. Pages 74–79 of: *2009 Workshop on Network Coding, Theory, and Applications.*

[40] Wang, H., Xing, C., and Safavi-Naini, R. 2003. Linear authentication codes: bounds and constructions. *IEEE Trans. Inform. Theory*, **49**(4), 866–872.

[41] Xia, S.-T., and Fu, F.-W. 2009. Johnson type bounds on constant dimension codes. *Des. Codes Cryptogr.*, **50**(2), 163–172.

[42] Xu, L., and Chen, H. 2018. New constant-dimension subspace codes from maximum rank distance codes. *IEEE Trans. Inform. Theory*, **64**(9), 6315–6319.

13
List Decoding

The two commonly assumed error models that are encountered in the literature are reviewed. The first model is the random error model which can be viewed in terms of the binary symmetric channel (BSC) as discussed in several earlier chapters. Each bit transmitted on the channel has a probability p of being received in error, where p is the channel crossover probability. The number of errors in n bit transmissions then is binomially distributed with parameters n and p with an average number of errors in a block of n bits of pn. As usual, denote by an $(n,M,d)_2$ binary code with M codewords of length n and minimum distance d. The performance of an $(n,M,d)_2$ binary code with bounded distance decoding is of interest. In this model a received word is decoded to a codeword only if it falls within a sphere of radius $e = \lfloor \frac{d-1}{2} \rfloor$ about a codeword – otherwise an error is declared. The probability of correct reception then is the probability that no more than e errors are made in transmission where

$$P_{\text{correct}} = \sum_{j \le e} \binom{n}{j} p^j (1-p)^{n-j},$$

and $P_{\text{error}} = 1 - P_{\text{correct}}$. This is referred to as the random or Shannon error model [9].

Another channel model, referred to as the Hamming or adversarial model, specifies the maximum number of errors that are made in transmission. The adversary is free to choose the positions of the errors to insure maximum disruption to communications and the choice may depend on the particular codeword chosen for transmission. The error probability then is dependent on the code properties and the number of errors assumed.

The implications of the two models are quite different in the following sense. Consider the Hamming error model and let $A_2(n,d)$ denote the maximum number of binary codewords of length n with minimum Hamming distance d between any two codewords. It is argued that the maximum fraction of errors that can be tolerated with such a Hamming model is $p < 1/4$. For $p > 1/4$ the number of errors will be of the order pn requiring a code with minimum distance $d \geq 2pn \gtrsim n/2$. If $2d > n$, by the Plotkin bound (e.g., [20]) the maximum number of codewords with minimum distance d is bounded by

$$A_2(n,d) \leq 2\left\lfloor \frac{d}{2d-n} \right\rfloor \sim \frac{1}{1-n/2d}$$

and hence asymptotically for fixed d/n, the rate of the code

$$\frac{1}{n} \lim_{n\to\infty} \log_2 A_2(n,d) \longrightarrow 0.$$

Thus if $p > 1/4$ error-free communications at a positive rate are not possible in the Hamming model.

On the other hand, in the Shannon model, the capacity of the BSC is (see Chapter 1)

$$C_{BSC} = 1 - H_2(p) = 1 + p\log_2 p + (1-p)\log_2(1-p)$$

where $H_2(\cdot)$ is the binary entropy function, and this function is > 0 for all $0 < p < 1/2$ implying that asymptotically there are codes that will perform well for this fraction. The difference of the two models is of interest.

To introduce list decoding, consider decoding a code $C = (n,M,d)_q$, $d = 2e+1$. A codeword $\boldsymbol{c} \in C$ is transmitted on a DMC and $\boldsymbol{y} = \boldsymbol{c} + \boldsymbol{n} \in \mathbb{F}_q^n$ is received for some noise word $\boldsymbol{n} \in \mathbb{F}_q^n$. For unique (bounded distance) decoding, consider surrounding the received word with a sphere of radius e. If the sphere with center the received word $\boldsymbol{y}$ and radius e, contains a single codeword, it is assumed that at most e errors were made in transmission and the received word is decoded to that codeword. This is maximum-likelihood decoding since the probabilities are monotonic with Hamming distance.

Suppose now the sphere around the received word $\boldsymbol{y}$ is expanded to one of radius $\epsilon > e$. As the radius of the sphere ϵ increases around $\boldsymbol{y}$, it is likely that more codewords will lie in this sphere. Suppose, for a given ϵ there are L codewords in the sphere. Rather than unique decoding the list decoder produces a list of L possible codewords within distance ϵ of the received word. The relationship of the size of the list L and the decoding radius ϵ is of interest as well as algorithms to produce the list, central issues for list decoding.

The subject of list decoding is explored in detail in the works of Guruswami, including monographs [7, 8] and excellent survey papers on various aspects of the subject. The treatment of this chapter is more limited.

Definition 13.1 For a code $C \subset F_q^n$, $|C|= M$, and any $\boldsymbol{y} \in \mathbb{F}_q^n$ define $\Delta(\boldsymbol{y}, \epsilon)$ as the set of codewords within distance ϵ of the received word $\boldsymbol{y}$ or:

$$\Delta(\boldsymbol{y}, \epsilon) = \{\boldsymbol{c} \in C \mid d(\boldsymbol{y}, \boldsymbol{c}) \leq \epsilon\}.$$

The code $C = (n, M, d)_q$ is said to be an (n, ϵ, L) code if for any $\boldsymbol{y} \in \mathbb{F}_q^n$

$$|\Delta(\boldsymbol{y}, \epsilon)| \leq L.$$

Decoding is deemed successful if the transmitted codeword is in the list.

Thus a code is (n, ϵ, L) iff no sphere of radius ϵ around *any* $\boldsymbol{y} \in \mathbb{F}_2^n$ contains more than L codewords. Equivalently a code is (n, ϵ, L) if the spheres of radius ϵ around codewords in C contains each point of $\mathbb{F}_q^n$ at most L times. Unique bounded distance decoding can be viewed as $(n, e, 1)$ list decoding for a code with minimum distance $d = 2e + 1$.

As before, define the sphere of radius ϵ about $\boldsymbol{y} \in \mathbb{F}_q^n$ as:

Definition 13.2 The sphere of radius ϵ about $\boldsymbol{y} \in \mathbb{F}_q^n$ is the set of points

$$\begin{aligned} S_{n,q}(\boldsymbol{y}, \epsilon) &= \{\boldsymbol{x} \in \mathbb{F}_{q^n} \mid d(\boldsymbol{x}, \boldsymbol{y}) \leq \epsilon\} \\ s_{n,q}(\boldsymbol{y}, \epsilon) &= |S_{n,q}(\boldsymbol{y}, \epsilon)| = \sum_{i=0}^{\epsilon} \binom{n}{i} (q-1)^i \end{aligned}$$

where the last equality follows since $s_{n,q}(\epsilon)$ is independent of the sphere center.

Some useful approximations for these functions can be formulated as follows. The size of the sphere, as $n \to \infty$ takes on an interesting form. Recall from Equation 1.13 the q-ary entropy function is defined as

$$H_q(x) = \begin{cases} x \log_q(q-1) - x \log_q x - (1-x)\log_q(1-x), & 0 < x \leq \theta = (q-1)/q, \\ 0, & x = 0, \end{cases}$$

an extension of the binary entropy function.

It can be shown [8, 23, 26] that

$$\lim_{n \longrightarrow \infty} (1/n) \log_q s_{n,q}(\lfloor pn \rfloor) = H_q(p)$$

where p denotes the probability of error in each symbol of the received word (and pn the expected number of errors in a word). Thus a good approximation to the size of a sphere of radius $\epsilon = pn$ in $\mathbb{F}_q^n$ is given by $q^{H_q(\epsilon/n)n}$ for $\epsilon < \theta$. Indeed [8] for any $\boldsymbol{y} \in \mathbb{F}_q^n$ it can be shown[1]

$$q^{H_q(p)n - o(n)} \leq s_{n,q}(pn) \leq q^{H_q(p)n} \quad \text{or} \quad s_{n,q}(pn) \approx q^{H_q(p)n}.$$

Using this estimate and a random coding argument, the following is shown:

[1] Recall that a function $f(n)$ is $o(g(n))$ if for all $k > 0\, \exists n_0\ st\ \forall n > n_0\ |f(n)| < kg(n)$.

Theorem 13.3 ([5], proposition 15), [8], theorem 3.5) *For integer $L \geq 2$ and every $p \in (0, \theta)$, $\theta = (q-1)/q$, there exists a family of (n, pn, L) list-decodable codes over $\mathbb{F}_q$ of rate R satisfying*

$$R \geq 1 - H_q(p) - 1/L.$$

Thus the largest rate a code can have and be (n, pn, L) list decodable, where p is the fraction of errors that can be tolerated, for polynomially sized lists, is approximately $1 - H_q(p)$. It is shown ([8], section 3) that codes with rate

$$R \geq 1 - H_q(p) \geq 1 - p - \gamma$$

are possible for list size $O(1/\gamma)$ and the challenge is to devise the construction of codes and decoding algorithms that can achieve these bounds. This is the challenge met in Sections 13.2 and 13.3 of this chapter.

Notice that this also implies that a code of rate R can be list decoded with a fraction of errors up to $1 - R$ with list decoding. For this reason this quantity can be viewed as the *list-decoding capacity*. Since $1 - R = (n-k)/n$ this would suggest that on average the number of codeword positions in error should be less than $(n - k)$ which would seem to be a requirement for correct decoding since any more errors would prohibit the production of all information symbols. Thus a fraction of errors in a received codeword $\leq (1 - R)$ is a requirement for successful list decoding. The size of the list is an important consideration in this since an exponential size list gives problems for practical considerations. Expressions for the size of the list for the various codes considered here are generally available but, beyond the fact they will be polynomially related to code length, will not be a priority to discuss here.

The next section considers mainly the work of Elias [5] which deals largely with the capability of a given binary code to be list decodable for a number of errors beyond its unique decoding bound. Said another way, given a binary code, as one increases the size of decoding spheres, what can one say about the size of decoding lists produced? This work has very much a combinatorial flavor although some information-theoretic considerations are included. This line of inquiry was not pursued much in subsequent literature and while not contributing to the goal of constructing capacity-achieving codes, it remains of interest for the principles it introduces.

The remainder of the chapter traces the remarkable work of a few researchers, namely Sudan, Guruswami and others, to construct new classes of codes that achieve capacity on the list-decoding channel, i.e., to construct codes and decoding algorithms capable of correcting a fraction $1 - R$ errors with a code of rate R producing a list of L possible codewords containing the transmitted codeword with high probability. A sketch of the incremental steps taken to achieve this is given here. The monograph [7] (based on the author's prize-winning thesis) is a more comprehensive discussion of this subject.

13.1 Combinatorics of List Decoding

The work of Elias [5] treated only binary codes and derived analogs to the generalized Hamming and Varshamov–Gilbert bounds for list decoding. Interest here is restricted to determining conditions under which a binary $C = (n, M, d = 2e + 1)_2$ code can also be an (n, ϵ, L) list decodable for parameters ϵ and L. From the previous section, a code $C = (n, M, d)_2$ is an (n, ϵ, L) code iff for any $\boldsymbol{y} \in \mathbb{F}_2^n$ the decoder produces a list of L possible codewords in any sphere of radius ϵ around any received word. Decoding is successful if the transmitted codeword is in the list. Equivalently, a code is an (n, ϵ, L) code iff no sphere around *any* $\boldsymbol{y} \in \mathbb{F}_2^n$ of radius ϵ contains more than L codewords.

The Johnson bound of coding gives an upper bound on $A_q(n,d)$, the largest number codewords in a (linear or nonlinear) code of length n and distance d over $\mathbb{F}_q$. Also $A_q(n,d,w)$ is the same for a constant weight w code (see [16], section 2.3). Define $\mathcal{A}_2(n,d,\epsilon)$ to be the maximum number of points that may be placed in an ϵ-sphere (sphere of radius ϵ) around any point in $\mathbb{F}_2^n$ with all points a distance at least d apart. It follows immediately that any $(n, M, d)_2$ code (or an $(n, e, 1)$ list-decodable code) is an $(n, \epsilon, L = \mathcal{A}_2(n,d,\epsilon))$ list-decodable code. The following bounds for $\mathcal{A}_2(n,d,\epsilon)$, obtained by combinatorial arguments, are shown in ([5], proposition 10). Their derivations are straightforward but intricate and are omitted.

Proposition 13.4

(a) If $n \geq d \geq 2\epsilon + 1$, then $\mathcal{A}_2(n,d,\epsilon) = 1$.

(b) If $\epsilon > 0$ and $n \geq 2\epsilon \geq d > 2\epsilon(n-\epsilon)/n$, then $n \geq 2$ and

$$2 \leq \mathcal{A}_2(n,d,\epsilon) \leq \left\lfloor \frac{nd}{nd - 2\epsilon(n-\epsilon)} \right\rfloor \leq \max\{2, n(n-1)/2\}.$$

(c) If d is odd, then $\mathcal{A}_2(n,d,\epsilon) = \mathcal{A}_2(n+1,d+1,\epsilon)$. Therefore for odd d if $\epsilon > 0$ and $n+1 \geq 2\epsilon \geq d+1 > 2\epsilon(n+1-\epsilon/(n+1))$, then $n \geq 1$ and

$$\mathcal{A}_2(n,d,\epsilon) \leq \left\lfloor \frac{(n+1)(d+1)}{(n+1)(d+1) - 2\epsilon(n+1-\epsilon)} \right\rfloor \leq \max\{2, n(n+1)/2\}.$$

(d) If d is odd and the conditions in the first line of (b) hold, then $\mathcal{A}_2(n,d,\epsilon) \leq n$.

Two interesting propositions follow from this one ([5], propositions 11,12). To discuss them consider L, p_L, p as fixed and let $\epsilon_L = np_L, \epsilon = np$ determine the sphere radii and let $n \longrightarrow \infty$. Then:

Proposition 13.5 *Let $L \geq 2$ and p_L and p satisfy*

$$\begin{aligned} p_L &= p(1-p)L/(1-L), \quad 0 \leq p \leq 1/2 \\ p &= \tfrac{1}{2}\Big(1 - \sqrt{1 - 4p_L(1-1/L)}\Big), \quad 0 \leq p_L \leq L/4(L-1). \end{aligned}$$

If np_L and np are integers (e.g., $p = j/k$ (i.e., rational) and $n = ik^2(L-1)$), then each $(n, np_L, 1)$ code is also an (n, np, L) code. More generally for any n let

$$e_L(n) = \lfloor (n+1)p_L \rfloor, \quad e(n) = \lfloor (n+1)p \rfloor.$$

Then if a code is $(n, e_L(n), 1)$ it is also $(n, e(n), L)$.

The asymptotic version of the proposition ([5], proposition 12), keeping p fixed and L increasing reads as follows:

Proposition 13.6 *Let p and p_∞ satisfy*

$$\begin{aligned} p_\infty &= p(1-p), \quad 0 \leq p \leq 1/2 \\ p &= \tfrac{1}{2}\Big(1 - \sqrt{1 - 4p_\infty}\Big), \quad 0 \leq p_\infty \leq 1/4. \end{aligned}$$

If np and np_∞ are integers (e.g., $p = i/k, n = ik^2$), then if a code is an $(n, np_\infty, 1)$ code it is also an (n, np, n) code and more generally if for any n

$$e_\infty(n) = \lceil np_\infty \rceil, \quad e(n) = \lfloor np \rfloor$$

then if a code is an $(n, e_\infty(n), 1)$ it is also an $(n, e(n), n)$ code.

An interesting aspect of this approach is to determine the trade-off between list size and error-correction capability, keeping in mind that with list decoding successful decoding means the transmitted codeword is in the decoded list. The culmination of this approach is the following proposition ([5], proposition 17) which follows directly from the above two propositions. Assume a binary code with minimum distance $d = 2e_1 + 1$ which is then an $(n, e_1, 1)$ (list decodable) code.

Proposition 13.7

(a) *Let C be an $(n, e_1, 1)$ code and $e \geq e_1 + 1$. Then C is also an (n, e, L) code where*

$$L = \left\lfloor \frac{(n+1)(e_1+1)}{e^2 - (n+1)(e - e_1 - 1)} \right\rfloor$$

if the denominator is positive.

(b) *In particular if C is $(n, e, 1)$ it is an $(n, (e+1), \lfloor (n+1)/(e+1) \rfloor)$ list-decodable code.*

Notice that item (b) of the proposition is obtained by replacing e with $e_1 + 1$ in part (a).

Numerous examples of this proposition are given in ([5], appendix A), including the binary perfect Hamming codes and the $(23, 12, 7)_2$ Golay code. In particular the above results show that a $(2^k - 1, 2^k - k, 3)_2$ Hamming code is a $(2^k - 1, 2, 2^{k-1})$ list-decodable code. Similarly the $(23, 12, 7)_2$ Golay code is a $(23, 4, 6)$ list-decodable code. These two cases can also be obtained by simple combinatorial arguments which are outlined in the following example.

Example 13.8 Recall the notion of a t-design, denoted a $\lambda - (v, k, t)$ design. It is a collection of k-element subsets (or blocks $\mathcal{B}$) of a v-element set, say $\mathcal{X}$, such that every t-element subset of $\mathcal{X}$ is contained in exactly λ subsets of $\mathcal{B}$. Let λ_s be the number of subsets containing a given s-element subset, $s = 1, 2, \ldots, t$, $\lambda_t = \lambda$. Then by elementary counting

$$\lambda_s = \lambda \binom{v-s}{t-s} \Big/ \binom{k-s}{t-s}. \tag{13.1}$$

There are numerous conditions known on the parameters for the existence of such designs (e.g., see [3]). Of special interest are the Steiner systems corresponding to $\lambda = 1$, often denoted $S(t, k, v)$.

It is known that the codewords of weight 3 in the binary Hamming $(2^k - 1, 2^k - k, 3)_2$ perfect code form a $S(2, 3, 2^k - 1)$ Steiner triple system. A point in $\mathbb{F}_2^{2^k-1}$ is either a codeword or at distance 1 from a codeword. Spheres of radius 1 around codewords exhaust the space (i.e., it is a perfect code). Spheres of radius 2 around any codeword contain no other codeword. Consider a sphere of radius 2 around a word of $\mathbb{F}_2^{2^k-1}$ of weight 1, say $\boldsymbol{u}$. By Equation 13.1 exactly

$$\lambda_1 = \binom{2^k - 1 - 1}{2 - 1} \Big/ \binom{2}{1} = 2^{k-1} - 1$$

nonzero codewords lie within distance 2 of u. Including the all-zero codeword shows the code is a $(2^k - 1, 2, 2^{k-1})$ list-decodable code, as shown by the above proposition.

Similarly, it is known the codewords of weight 8 in the extended $(24, 12, 8)_2$ Golay code form a $S(5, 8, 12)$ system, a remarkable combinatorial structure. The codewords of weight 8 in the extended (parity check added) code correspond to codewords of weight either 7 or 8 in the perfect $(23, 12, 7)_2$ code. Consider a word v of weight 4 in $\mathbb{F}_2^{23}$. By Equation 13.1 the number of codewords in the extended code of weight 8 in the sphere of radius 4 around v is

$$\lambda_4 = \binom{24-4}{1} \Big/ \binom{8-4}{1} = 5.$$

In the $(23,12,7)_2$ code this means, taking into account the all-zero codeword, the code is a $(23,4,6)$ list-decodable code.

Returning from the combinatorial interlude, it is clear the above proposition can be applied to many other classes of codes. For example, the $(31,6,15)_2$ BCH code is a $(31,7,1)$ unique decodable code and also a $(31,8,4)$ list-decodable code.

This combinatorial approach to the list-decoding problem does not seem to have been pursued in the literature. The following two sections describe the remarkable work on obtaining codes that achieve list-decoding capacity.

13.2 The Sudan and Guruswami–Sudan Algorithms for RS Codes

The previous section has considered certain aspects of the list decodability of known codes. This section and the next are concerned with the problem of constructing codes and decoding algorithms that will allow us to obtain codes of rate R that are list decodable up to capacity, i.e., a fraction of codeword errors of $1 - R$. As noted, this can be effective when the fraction of errors exceeds the unique decoding bound which for Reed–Solomon (RS) codes would be $(n-k+1)/2n \sim (1-R)/2$, half of the capacity.

The story will be how researchers were able to reach the list-decoding capacity of $1 - R$ in a series of incremental steps, often introducing novel and ingenious ideas. The progression will yield codes of rates $1 - \sqrt{2R}$ (Sudan), $1 - \sqrt{R}$ (Guruswami/Sudan and slight improvements by Parvaresh and Vardy) and finally $1 - R$ (Guruswami and Rudra).

Reed–Solomon codes and certain relatives will be considered in the remainder of this chapter although much of the material can be extended to more general classes of codes such as generalized RS codes, alternant and algebraic geometry codes. Let $x_1, x_2, \ldots, x_n$ be n distinct evaluation points in $\mathbb{F}_q$ – hence $n \leq q$ – and consider the RS code

$$C = \left\{ \boldsymbol{f} = (f(x_1), f(x_2), \ldots, f(x_n)) \mid f(x) \in \mathbb{F}_q[x],\ \deg f(x) \leq k-1 \right\}$$

which defines a linear $(n,k,d = n-k+1)_q$ MDS code. The notions of codeword as an n-tuple $\boldsymbol{f} = (f_1, f_2, \ldots, f_n),\ f_i = f(x_i)$ over $\mathbb{F}_q$ and codeword polynomial $f(x)$ are often identified. To encode the $\mathbb{F}_q$ information or message symbols $m_0, m_1, \ldots, m_{k-1}$ one might use the polynomial $f(x) = \sum_{i=0}^{k-1} m_i x^i$ to form the nonsystematic codeword by evaluation of f at n distinct points of $\mathbb{F}_q$.

The approaches of this and the next section may appear somewhat challenging in that it involves the construction of multivariable interpolation polynomials with properties such as multiple zeros on given interpolation sets. However, the approach is a straightforward generalization of the single-variable case and the results achieved are of great interest.

As a precursor to introducing the innovative work of Sudan [25] the Berlekamp–Welch decoding algorithm for RS codes (e.g., see [6, 21]) is recalled. Let $C = (n, k, d = n-k+1)_q$ be an RS code and denote $e = \lfloor \frac{n-k+1}{2} \rfloor$ – for convenience assume d is odd and $d = 2e + 1$. Suppose the codeword $\boldsymbol{f}$ is transmitted and $\boldsymbol{y} = \boldsymbol{f} + \boldsymbol{\eta}$, $y_i = f_i + \eta_i, i = 1, 2, \ldots, n$, $\eta_i \in \mathbb{F}_q$ received where $\boldsymbol{\eta} \in \mathbb{F}_q^n$ is a noise word. ($\boldsymbol{\eta}$ is used as $\boldsymbol{n}$ will have another use.) Consider the problem of finding two polynomials, $D(x)$ and $N(x)$ over $\mathbb{F}_q$ with the following properties:

$$(i)\ \deg D(x) \leq e,$$
$$(ii)\ \deg N(x) \leq e + k - 1,$$

satisfying the conditions:

$$(iii)\ N(x_i) = y_i D(x_i), \quad i = 1, 2, \ldots, n.$$

Denote by E the set of (unknown) error positions of the received word, $E = \{i \mid y_i \neq f_i\}$, and let $D(x)$ be the error locator (monic) polynomial whose zeros are at the (as yet unknown) locations of the errors

$$D(x) = \prod_{j \in E} (x - x_j).$$

The actual number of errors, $|E|$, is assumed to be $\leq e$. Let $N(x) = f(x)D(x)$ which is of degree at most $e + k - 1$. Clearly

$$N(x_i) = f(x_i)D(x_i) = f_i D(x_i), \quad i = 1, 2, \ldots, n,$$

since if $x_i \in E$ both sides are zero and if $x_i \notin E$, $y_i = f_i$ and they are equal by definition.

Let

$$N(x_i) = \sum_{j=0}^{e+k-1} n_j x_i^j = f_i D(x_i) = y_i \sum_{j=0}^{e} d_j x_i^j, i = 1, 2, \ldots, n,$$

since $y_i = f_i$, $i \notin E$. The number of unknowns n_i, d_j in this equation is at most $2e + k < n$ ($D(x)$ is monic) and so the solutions for them, given the

n values y_i, may be found by Gaussian elimination. Note that if $t < e$ errors occur, the appropriate coefficients will be determined as zero. The complexity of the decoding is $O(n^3)$. Other algorithms have been formulated which have lower complexity.

With $N(x)$ and $D(x)$ determined, by definition the codeword polynomial is $f(x) = N(x)/D(x)$ assuming they are both nonzero. Suppose the rank conditions on the Gaussian elimination are such that more than one pair of solutions is obtained. Suppose $\{N_1(x), D_1(x)\}$ and $\{N_2(x), D_2(x)\}$ are both solutions – and hence result in the same codeword. We would like to verify this: i.e., that we must have $N_1(x)/D_1(x) = N_2(x)/D_2(x)$ or $N_1(x)D_2(x) = N_2(x)D_1(x)$. Suppose for some coordinate position $y_i = 0$. Then by property (iii) above $N_1(x_i) = N_2(x_i) = 0$ and the condition is satisfied trivially. Suppose $y_i \neq 0$ then

$$N_1(x_i)N_2(x_i) = y_i D_1(x_i)N_2(x_i) = y_i N_1(x_i)D_2(x_i)$$

and hence $D_1(x_i)N_2(x_i) = D_2(x_i)N_1(x_i),\ i = 1,2,\ldots,n$. Since they are equal on $n > 2e + k$ distinct points they are equal as polynomials and

$$N_1(x)/D_1(x) = N_2(x)D_2(x)$$

and so multiple solutions to the Gaussian elimination yield the same codeword solution.

The approach of [25] was to notice that the above Berlekamp–Welch decoding algorithm can be formulated as a bivariate polynomial problem [9] as follows. Consider the bivariate polynomial

$$Q(x, y) = yD(x) - N(x) = (y - f(x))D(x) = (y - f(x) \prod_{i \in E} (x - x_i),$$
$$\deg D \leq e, \quad \deg N \leq e + k - 1$$

where E is the set of error positions and $D(x), N(x)$ are as before. Notice that

$$Q(x_i, y_i) = 0, \quad i = 1,2,\ldots,n.$$

Conversely, suppose we could formulate a bivariate polynomial $Q(x, y)$ such that $Q(x_i, y_i) = 0,\ i = 1,2,\ldots,n$. If this polynomial could be factored it would have $(y - f(x))$ as a factor and the code would be decoded. Thus the RS decoding problem is cast as a bivariate interpolation problem. This is the innovative approach of Sudan [25].

Note that rings of multivariate polynomials such as $\mathbb{F}_q[x_1, x_2, \ldots, x_n]$ have unique factorization (up to order and multiplication by scalars) [4] and factorization algorithms for such polynomials are available in the computer science literature.

The bounded distance decoder produces the unique codeword closest to the received word provided no more than $e = \lfloor\frac{(n-k)}{2}\rfloor$ errors have been made. The maximum fraction of errors that can be tolerated for unique decoding is $(1 - R)/2$. For list decoding, we would like to be able to list decode up to a fraction of $1-R$ of errors in which case a list of possible codewords is produced and that this fraction of $(n-k)/n$ errors is clearly the maximum possible. Thus the problem of interest is to construct codes and list decoding algorithms for them that are able to "correct" (in the list-decoding context) up to a fraction of $1 - R$ of received symbols in error. Equivalently interest is in finding all codewords within a radius ϵ as ϵ expands beyond the unique decoding bound e.

In the sequel it may be helpful to note the equivalence of the following two algorithms [13] stated without details:

Algorithm 13.9 (RS list decoding)

Input: A finite field $\mathbb{F}_q$, code $C = (n,k,d = n - k + 1)_q$, an n-tuple $\boldsymbol{y} \in \mathbb{F}_q^n$ and error parameter $\epsilon \leq n$

Output: All codewords $\boldsymbol{c} \in C$ such that $d(\boldsymbol{y},\boldsymbol{c}) \leq \epsilon$

Algorithm 13.10 (Polynomial reconstruction)

Input: Integers k,t and n points $\{(x_i, y_i)\}_{i=1}^n \in \mathbb{F}_q \times \mathbb{F}_q$

Output: All polynomials $f(x) \in \mathbb{F}_q[x]$, $\deg f \leq k - 1$ and $\Big|\{f(x_i) = y_i, i = 1,2,\ldots,n\}\Big| \geq t$

The polynomial reconstruction algorithm asks for all polynomials f of degree less than k for which $y_i = f(x_i)$ for at least t points and the reduction of the first algorithm to the second for $t = n - \epsilon$ is clear.

In the remainder of this chapter four algorithms are discussed which, in sequence, achieve error fractions for a code of rate R of $1 - \sqrt{2R}, 1 - \sqrt{R}$, a low-rate improvement of this last bound and finally, $1 - R$, the capacity of the list-decoding problem. The last two achievements are discussed in the next section.

The contribution of Sudan [25] is discussed as it introduced a radically new and unexpected way of viewing the decoding problem that altered the mindset of many coding theorists. The work utilizes bivariate polynomials $Q(x, y) \in \mathbb{F}_q[x, y]$ and weighted degrees such as

$$Q(x, y) = \sum_{i,j} q_{ij} x^i y^j, \; q_{ij} \in \mathbb{F}_q$$

which is a linear sum of the monomials $x^i y^j$. The (r,s)-weighted degree of this monomial is

$$ir + js$$

and the weighted degree of the polynomial $Q(x, y)$ is the maximum of the weighted degrees of its nonzero monomials. For the RS code of interest we will only be interested in the $(1, k-1)$-weighted degree of $Q(x, y)$ and the reason for this is in the degree of the univariate polynomial $Q(x, f(x))$ when a polynomial of degree at most $k-1$ is substituted for the y variable. This corresponds to the RS code $(n, k, d = n-k+1)_q$ code where, as before, codewords are identified with polynomials over $\mathbb{F}_q$ of degree at most $k-1$. For the remainder of the chapter denote $k_1 = k-1$. Recall the RS code coordinate positions are labeled with the distinct $x_i \in \mathbb{F}_q$, $i = 1, 2, \ldots, n$ and the received word is $\boldsymbol{y} = (y_1, y_2, \ldots, y_n)$.

Two simple arguments will be of importance in the discussion of this and the next section:

(i) the fact that a bivariate (or multivariate) polynomial $Q(x, y)$ will exist if the number of its coefficients is greater than the number of homogeneous conditions placed on these coefficients; and
(ii) the fact that if the number of zeros of the univariate polynomial $Q(x, f(x))$, $\deg f < k$ exceeds its degree, it must be the zero polynomial.

Of course the degree of $Q(x, f(x))$ is the $(1, k_1)$-weighted degree of $Q(x, y)$ for $\deg f < k$. In spite of the inefficiency, the argument will be repeated for each case.

The essence of the Sudan list-decoding algorithm for RS codes is established in the following:

Algorithm 13.11 (Sudan decoding of RS codes)

Input: $n, k, t \in \mathbf{Z}$, $\{(x_i, y_i)\}_1^n \in \mathbb{F}_q \times \mathbb{F}_q$
Integer parameters ℓ, m – to be determined for optimization

Output:
1. A bivariate function $Q(x, y) \in \mathbb{F}_q[x, y]$ with the properties: $Q(x_i, y_i)$ has $(1, k_1)$-weighted degree at most $m + \ell k_1$, $0 < m < \ell$, and is such that $Q(x_i, y_i) = 0$, $i = 1, 2, \ldots, n$ and is not identically zero
2. Find the irreducible factors of $Q(x, y)$
3. Output all those irreducible factors of the form $(y - f(x))$ for which it is true that $y_i = f(x_i)$ for at least t of the points (x_i, y_i).

The algorithm is straightforward although there are a number of variables that must be kept track of. The approach of [25] is outlined. Consider the bivariate polynomial

$$Q(x, y) = \sum_{j=0}^{\ell} \sum_{h=0}^{m+(\ell-j)k_1} q_{hj} x^h y^j$$

with a maximum y degree of ℓ and maximum weighted degree of $m + \ell k_1$. Notice the maximum x-degree is $m + \ell k_1$. The total number of coefficients in such a polynomial is

$$\sum_{j=0}^{\ell}(m + (\ell - j)k_1 + 1) = (m + 1)(\ell + 1) + k_1\binom{\ell + 1}{2}.$$

To ensure we can solve for the coefficients of such a nonzero bivariate polynomial we need this quantity to be greater than n since we are requiring the n homogeneous conditions that

$$Q(x_i, y_i) = 0,\ i = 1,2,\ldots,n, \qquad \text{Condition (*)} \qquad (13.2)$$

and for a solution it is necessary to have more variables than conditions, i.e., one can set up an $n \times (m+\ell k_1)$ homogeneous matrix equation which will have nonzero solutions iff $m + \ell k_1 > n$. In determining the polynomial $Q(x, y)$ the Condition $(*)$ can be satisfied using Gaussian elimination of complexity at most $O(n^3)$.

Notice that n, k are fixed parameters and we want to determine the parameters t, m and ℓ to determine the bivariate polynomial $Q(x, y)$ with the above properties to find **all** polynomials $f(x)$ that correspond to codewords in the RS code of distance at most $n - t$ from the received word y for as small a value of t as possible, given the other parameters. The set of such polynomials correspond to the decoded list.

It seems remarkable that we will be able to show that if $f(x)$ corresponds to *any* codeword ($\deg f < k$) whose evaluations agree with the n received values in at least t places, then if

$$m + \ell k_1 < t$$

then $(y - f(x))$ must appear as an irreducible factor of $Q(x, y)$. Consider the polynomial

$$p(x) = Q(x, f(x))$$

which is a univariate polynomial of degree in x at most $m + \ell k_1$ since the degree of $f(x)$ is at most k_1. In the received codeword places that are correct $y_i = f(x_i) = f_i$ and for at least t values of x_i by construction, $Q(x_i, f(x_i)) = 0$ for these t places. Recall $y_i = f_i + \eta_i$, $i = 1,2,\ldots,n$ and it is assumed $\eta_i = 0$ for at least t values. If the degree of this univariate polynomial is at most $m+\ell k_1 < t$, as assumed, the polynomial has more zeroes than its degree and hence $Q(x, f(x))$ must be the zero polynomial (although $Q(x, y)$ is a nonzero polynomial). Viewing $Q(x, y)$ as a polynomial in y with coefficients from the rational field $\mathbb{K}$ of $\mathbb{F}_q[x]$ (field of fractions), $Q(x, y) =$

$Q_x(y) = \sum_i h_i(x)y^i, h_i(x) \in \mathbb{K}$, the polynomial can be factored over $\mathbb{K}$. If it has a zero, say $\eta \in \mathbb{K}$, then $(y-\eta) \mid Q_x(y)$. Hence $(y - f(x)) \mid Q(x,y)$, since the form of Q has unit denominator. The ring $\mathbb{F}_q[x,y]$ has unique factorization since if R is a ring with unique factorization (unique factorization domain (UFD)), then so is the polynomial ring $R[x]$. This argument holds for any polynomial satisfying the stated conditions and thus produces *all* possible codewords within the stated distance from the received word. This number cannot exceed ℓ, the y-degree of $Q(x,y)$. It is possible that a factor obtained from the process outlined produces a polynomial for which $f(x_i) = r_i$ for fewer than t values in which case it is discarded.

The algorithm requires the factorization of the bivariate polynomial $Q(x,y)$. Efficient means to achieve this for univariate polynomials have been well explored and algorithms to extend these to bivariate polynomials are available [18] and are not considered here.

For the above argument leading to a codeword polynomial to be valid we need the conditions:

$$m + \ell k_1 < t \quad \text{and} \quad (m+1)(\ell+1) + k_1\binom{\ell+1}{2} > n. \qquad \text{Condition (**)} \tag{13.3}$$

The first relation will be used to conclude that a polynomial in a single variable with degree $m + \ell k_1 < t$ and t zeros, must be the zero polynomial. The second ensures that a nonzero bivariate polynomial with more than n coefficients (monomials) satisfying n homogeneous conditions, exists – as noted previously.

We seek to "optimize" the parameters by choosing favorable values of ℓ, t and m. For a given value of ℓ the second condition above gives

$$m > \frac{n+1-k_1\binom{\ell+1}{2}}{\ell+1} - 1 \tag{13.4}$$

and so by the first condition

$$t \geq m + \ell k_1 > \frac{n+1-k_1\binom{\ell+1}{2}}{\ell+1} - 1 + \ell k_1 + 1 = \frac{n+1}{\ell+1} + \frac{k_1 \ell}{2}.$$

We would like the algorithm to work for as few agreements (t) between the received word and codewords as possible (to be able to "correct" as many errors ($n-t$) as possible). By differentiating the right-hand side of this last equation the optimum value of ℓ is found as

$$\ell = \sqrt{\frac{2(n+1)}{k_1}} - 1.$$

and from Equation 13.4

$$m > \frac{n+1}{\ell+1} - \frac{k_1\ell}{2} - 1 = \sqrt{\frac{k_1 n + 1}{2}} - \sqrt{\frac{k_1 n + 1}{2}} + \frac{k_1}{2} - 1 = \frac{k_1}{2} - 1.$$

To summarize, for the parameter values

$$m = \left\lceil \frac{k_1}{2} \right\rceil - 1 \quad \text{and} \quad \ell = \left\lceil \sqrt{\frac{2(n+1)}{k_1}} \right\rceil - 1$$

one can check that the conditions of Equation 13.3 are satisfied and hence the algorithm will find all codewords within distance $n - t$ to y. From the above equations, the corresponding value of t to these values is

$$t \geq \sqrt{2(n+1)k_1} - \frac{k_1}{2} - 1. \tag{13.5}$$

The result of the above discussion ([25], theorem 5) is that the Algorithm 13.18 will find all such polynomials that agree with the received word (polynomial) in at least t positions as long as the conditions $(*)$ and $(**)$ are met for the parameters chosen.

Rather than the optimum values of the parameters, if one chooses instead the approximations, then

$$\ell = \sqrt{\frac{2n}{k_1}} - \frac{1}{2} \quad \text{and} \quad m = \frac{k_1}{2}$$

it is easy to verify (ignoring integrality concerns which can be taken into account) the conditions $(**)$ are satisfied as long as t is chosen so that

$$t \geq m + k_1\ell = \frac{k_1}{2} + \left(\sqrt{\frac{2n}{k_1}} - \frac{1}{2}\right)k_1 = \sqrt{2nk_1}. \tag{13.6}$$

Thus, asymptotically, the normalized number of errors that can be tolerated by this Sudan algorithm is $\beta_{SD} = (n-t)/n$ satisfies

$$\beta_{SD} \leq 1 - \frac{t}{n} \approx 1 - \frac{\sqrt{2nk_1}}{n} \to 1 - \sqrt{2R}. \tag{13.7}$$

It follows that Algorithm 13.18 can correct an RS code received codeword with a fraction of errors at most $1 - \sqrt{2R}$ with a list of codewords within a sphere of radius at most $n - t$ around the received word of size as noted above. The list size will be at most $\ell \approx \sqrt{2n/k} \approx \sqrt{2/R}$ since this is the y-degree of the bivariate polynomial and there can be no more factors of the required form.

It is of interest to note the improvement of the error fraction obtained compared to unique decoding which is able to correct a fraction

$$\beta_{UD} = \frac{n-k+1}{n} \rightarrow (1-R)/2. \tag{13.8}$$

It is verified that the smallest value of R for which this expression (Equation 13.7) equals $(1-R)/2$ is about $R = 0.17$. Thus for code rates below 0.17 the list-decoding algorithm, for some size list, will correct a greater fraction of errors than unique decoding at the cost of producing a list of possible transmitted codewords rather than a unique word.

The next innovation to improve the situation is that of Guruswami and Sudan [13] who modified the algorithm to allow for higher-rate codes. The key to this improvement is to allow higher-total degree bivariate polynomials and for each zero to have a higher order multiplicity than one. In essence the curve represented by the bivariate polynomial will intersect itself at a zero of multiplicity s, s times. The improvement will develop an algorithm that will produce a list of polynomials satisfying the distance criterion that will be able to tolerate a fraction of errors up to $1-\sqrt{R}$ rather than $1-\sqrt{2R}$ as above. While it is not immediately intuitive why having higher-order singularities will lead to improvements, the development is of interest. The issue is discussed in [9] and [17] (chapter of Kopparty). While somewhat technical, the ideas are straightforward extensions of previous material.

Suppose the bivariate polynomial

$$P(x,y) = \sum_i \sum_j p_{i,j} x^i y^j$$

has a zero of order s at the point $(a,b) \in \mathbb{F}_q \times \mathbb{F}_q$. Consider the translated polynomial

$$P^t(x,y) \triangleq P(x+a, y+b) \triangleq \sum_i \sum_j p_{i,j}(x+a)^i (y+b)^j \triangleq \sum_i \sum_j p^t_{i,j} x^i y^j$$

and it is easy to show [13] that the coefficients of the translated polynomial are

$$p^t_{i,j} = \sum_{i \le i'} \sum_{j \le j'} \binom{i'}{i} \binom{j'}{j} p_{i',j'} a^{i'-i} b^{j'-j}.$$

For $P(x,y)$ to have a zero of order s at the point (a,b) it can be shown that it is necessary and sufficient that the translated polynomial $P^t(x,y) = P(x+a, y+b)$ have the coefficients of all monomials $x^i y^j$ to be zero for all $i \ge 0, j \ge 0$

and $i+j < s$. The number of possible monomials of the form $x_i^u y_i^v,\ u+v \leq s$ is $\binom{s+1}{2}$ (as shown in Chapter 1). Hence the total number of conditions such a requirement imposes on the coefficients of $P(x,y)$ to have zeros of order s at each of the points (x_i, y_i), $i = 1, 2, \ldots, n$ is

$$n\binom{s+1}{2}. \tag{13.9}$$

This is the number of monomials of degree $< s$ and hence is the number of equations relating the linear sums of coefficients of $P(x,y)$ to zero, set up as a matrix equation.

With these definitions the procedure for list decoding of RS codes is discussed, repeating the previous discussion with zeros of multiplicity s rather than $s = 1$. As before suppose $\boldsymbol{f} = (f_1, f_2, \ldots, f_n)$ is a codeword in the $(n, k, n-k_1)_q$ RS code corresponding to the polynomial $f(x)$, $f_i = f(x_i)$, $i = 1, 2, \ldots, n$, and $\boldsymbol{y} = (y_1, y_2, \ldots, y_n)$ the received word in $\mathbb{F}_q^n, \boldsymbol{y} = \boldsymbol{f} + \boldsymbol{\eta}$. Denote the set of pairs

$$\mathcal{I} = \Big\{(x_1, y_1), (x_2, y_2), \ldots, (x_n, y_n)\Big\},\ (x_i, y_i) \in \mathbb{F}_q{}^2\ \forall i$$

as the *interpolation set* and let

$$t = \Big|\big\{i \mid y_i = f(x_i), i = 1, 2, \ldots, n\big\}\Big|$$

be the number of agreements between the codeword f and received word y. Define the bivariate polynomial $Q(x,y)$ where

$$Q(x,y) = \sum_{j=0}^{\ell} \sum_{h=0}^{(\ell-j)k_1} q_{i,h} x^h y^j$$

where ℓ is the y-degree of $Q(x,y)$ and all monomials have $(1, k_1 = k-1)$-weighted degree at most ℓk_1. Notice the slight difference from the previous case of Sudan decoding (no m – the argument is easily changed to allow a total weighted degree of $m + \ell k_1$ as before). The number of coefficients of the above bivariate polynomial is computed as

$$(\ell+1) + k_1\binom{\ell+1}{2}. \tag{13.10}$$

From the above discussion such a nonzero bivariate polynomial will exist and have zeros of order s on all points of the interpolation set if

$$(\ell+1) + k_1\binom{\ell+1}{2} > n\binom{s+1}{2},$$

i.e., if the number of coefficients exceeds the number of homogeneous conditions. Furthermore if $f(x) \in \mathbb{F}_q[x]$ is of degree at most k_1, then the univariate polynomial $Q(x, f(x))$ is of degree at most ℓk_1. If the number of agreements $y_i = f_i$ is at least t (hence $n - t$ or fewer errors in the received word), then $Q(x, f(x))$ has st zeros since each zero is of order s. If $st > \ell k_1$ (i.e., the number of zeros exceeds the weighted degree of $Q(x, y)$) it must be the zero polynomial. The crucial lemma that will verify the Guruswami–Sudan list-decoding algorithm then is (adapted from [12] and [13]):

Lemma 13.12 *With the notations above, let $Q(x, y) \in \mathbb{F}_q[x, y]$ be a bivariate polynomial such that for each $i = 1, 2, \ldots, n$ the pair (x_i, y_i) is a zero of multiplicity s and $Q(x, y)$ has $(1, k_1)$-weighted degree at most $D = \ell k_1$. Let $f(x) \in \mathbb{F}_q[x]$ be of degree at most k_1 with the property that $y_i = f(x_i)$ for at least t values of $i \in [n]$. If the following conditions (Equations 13.9 and 13.10) are met:*

$$(i)\;\; (\ell + 1) + k_1 \binom{\ell + 1}{2} > n \binom{s + 1}{2} \qquad \textit{and} \qquad (ii)\;\; ts > \ell k_1,$$

then $(y - f(x))$ divides $Q(x, y)$ for any codeword polynomial within distance $n - t$ of the received word y.

Proof: Consider the polynomial $g(x) = Q(x, f(x))$ which is of degree at most $D = \ell k_1$ as the $(1, k_1)$-weighted degree of $Q(x, y)$ is at most D. By definition, since (x_i, y_i) is a zero of multiplicity s of $Q(x, y)$ it follows that $Q(x + x_i, y + y_i)$ has no monomial terms of degree less than s. Consider a value of i for which $y_i = f(x_i) = f_i$ and note that $x \mid f(x + x_i) - y_i$ and for such a value of i, $x^s \mid Q(x + x_i, y + y_i)$ and hence $(x - x_i)^s \mid Q(x, y)$ and this is true for t values of x_i which are distinct. Hence the number of zeros is at least st and since by assumption $st > \ell k_1$ the implication is that $Q(x, y)$ is the zero polynomial. As in a previous argument, writing

$$Q(x, y) = Q_x(y) = \sum_i q_i(x) y^i, \quad q_i(x) \in \mathbb{F}_q(x)$$

where the coefficients are in the rational field $\mathbb{F}_q(x)$, the polynomial has a zero in $\mathbb{F}_q(x)$, say ρ and $(y - \rho) \mid Q(x, y)$ or if $\rho = f(x)$, $(y - f(x)) \mid Q(x, y)$. ■

Notice that the algorithm finds *all* the polynomials (codewords) within the required distance $n - t$. This result leads to the algorithm :

Algorithm 13.13 (Guruswami–Sudan decoding of RS codes)

Input: $n,k,t \in \mathbf{Z}$, $\{(x_i,y_i)\}_1^n \in \mathbb{F}_q \times \mathbb{F}_q$

Output: All polynomials $f(x)$, $\deg f \leq k_1$, within distance $n-t$ of $\mathbf{y}$

Output:
1. Compute parameters s, ℓ and k_1 satisfying the inequalities $st > \ell k_1$ and $ns(s+1)/2 < (\ell+1) + k_1\binom{\ell+1}{2}$
2. A nonzero bivariate function $Q(x,y)$ over $\mathbb{F}_q$ of weighted degree at most ℓk_1 with the property that (x_i,y_i) is a zero of multiplicity $s, i = 1,2,\ldots,n$, i.e., for each $i \in [n]$ $Q(x+x_i, y+y_i)$ has no monomials of degree less than s
3. Find the irreducible factors of $Q(x,y)$
4. Output all those irreducible factors of the form $(y - f(x))$ for which it is true that $y_i = f(x_i)$ for at least t of the pairs of points.

Any set of parameters n,k,s,ℓ and t which satisfy the conditions of Lemma 13.12 will lead to the RS code list-decoding algorithm to find all codewords of distance less than $(n-t)$ to the received word. The following set are suggested (adapted from Kopparty [17]). For the given RS $(n,k,n-k_1)_q$, $k_1 = k-1$ code let

$$\ell = \left\lfloor \frac{k}{k-1}\sqrt{nk} \right\rfloor, \quad s = k, \quad t = \left\lceil \sqrt{nk} \right\rceil,$$

and assume the fractions in the arguments of the floor/ceiling functions are not integral. Then the first condition of Lemma 13.12 gives

$$(\ell+1)+k_1\binom{\ell+1}{2} > \frac{k_1\ell^2}{2} > \frac{k_1}{2}\cdot\frac{k^2\cdot nk}{k_1^2} = \frac{nk^3}{2k_1} > n\cdot\binom{k+1}{2} = n\cdot\frac{k(k+1)}{2}$$

and the last inequality follows from observing that $k^3/k_1 > k(k+1)$ or $k^3 > k(k^2-1)$.

The second condition for the parameters gives

$$ts = \left\lceil \sqrt{nk} \right\rceil k > \ell k_1 = \left\lfloor \frac{k}{k_1}\sqrt{nk} \right\rfloor k_1 \sim k\sqrt{nk}$$

where the inequality relies on the floor function and noninteger values of its argument. Thus the maximum fraction of errors in an RS code of rate R that can be corrected by the Guruswami–Sudan algorithm is

$$\beta_{GSD} = 1 - \frac{t}{n} \rightarrow 1 - \sqrt{R} \quad \text{with list size} \quad \ell \approx n\sqrt{R}. \tag{13.11}$$

The parameter set used for these calculations leads to the following (note the y-degree of $Q(x, y)$ is ℓ which is also the maximum number of solutions obtainable, the maximum size of the list obtained):

Theorem 13.14 *For every RS $(n,k,d)_q$ code there is a list decoding algorithm that corrects up to $n - \sqrt{kn}$ errors (fractional rate of $1 - \sqrt{R}$) that runs in time polynomial in n and q and outputs a list of at most $\sqrt{nk}$ polynomials.*

Another set of parameters is used in [13] with the same result. This is a considerable improvement over $1 - \sqrt{2R}$ result [25] and the decoding algorithm applies also to certain classes of algebraic-geometric codes as well as generalized RS codes. It still falls short of the capacity bound $1 - R$ of the previous section, a problem that is addressed in the next section by considering the construction of certain classes of codes (rather than only RS codes).

13.3 On the Construction of Capacity-Achieving Codes

The elegant results of Guruswami and Sudan of the previous section gave the fraction of errors tolerable in list decoders of $1-\sqrt{R}$ which was the best known until the work of Parvaresh and Vardy [22]. The key idea of that work was to introduce the notion of multivariate interpolation (again with multiplicities of zeros) and to use codes other than RS codes (although they were later shown to be related to RS codes (see [12])). The work is interesting from a technical view and an outline of this work is discussed here. The work led to the more complete results of Guruswami and Rudra [15] and the rather simple yet effective notion of *folded Reed–Solomon* (FRS) codes. Our aim is a description of these important contributions rather than a detailed analysis for which the reader is referred to the original papers. Numerous improvements in the efficiencies of the various original algorithms have been proposed. The work is strongly related to the decoding of interleaved Reed–Solomon codes as represented by such works as [1, 2, 24, 27].

While the general theory to be considered is valid for M variable polynomials over a field, the essence of the results can be appreciated by considering only the trivariate (three variable, $M = 3$) case which simplifies the situation considerably without compromising the power of the theory. Only the trivariate case will be considered in this section with the occasional reference to the M variable case.

Before embarking on the coding application some simple properties of trivariate polynomials are considered, generalizing those of the bivariate case of the previous section.

As before, consider the interpolation set

$$\mathcal{I} \stackrel{\Delta}{=} \{(x_1, y_1, z_1), (x_2, y_2, z_2), \ldots, (x_n, y_n, z_n),\ x_i, y_i, z_i \in \mathbb{F}_q\} \tag{13.12}$$

where the x_i are distinct. A polynomial will be a linear sum over $\mathbb{F}_q$ of monomials of the form $x^i y^{j_1} z^{j_2}$ and the degree of such a monomial will be $i + j_1 + j_2$. As before, we will be interested in the $(1, k_1, k_1)$-weighted degree of such a monomial $i + j_1 k_1 + j_2 k_1$. (Recall that we will be interested in the degree of the univariate polynomial after substituting polynomials of degree at most k_1 for the variables y and z, relating to an RS code $(n, k, d = n - k + 1 = n - k_1)_q$.) The weighted degree of a trivariate polynomial is the largest weighted degree of any monomial it contains.

For the treatment of the M-variable case it is useful to define a *lexicographic ordering* or lex ordering [4], on monomials and this notion is described. Denote M variables $x_1, x_2, \ldots, x_M$ simply by $\boldsymbol{x}$ and the exponents $a_1, a_2, \ldots, a_M$ by $\boldsymbol{a}$ where $a_i \in \mathbb{Z}_{\geq 0}$ and $a \in (\mathbb{Z}_{\geq 0})^M$. The monomial $x_1^{a_1} x_2^{a_2} \cdots x_M^{a_M}$ is then denoted $\boldsymbol{x}^{\boldsymbol{a}}$. The lex ordering on the integer M-tuples $\boldsymbol{a}, \boldsymbol{b} \in (\mathbb{Z}_{\geq 0})^M$ is then that $\boldsymbol{a} \succ \boldsymbol{b}$ if in the vector difference $\boldsymbol{a} - \boldsymbol{b} \in \mathbb{Z}^M$ the leftmost nonzero entry is positive. This ordering induces a lex ordering on monomials via their exponents, i.e., $\boldsymbol{x}^{\boldsymbol{a}} \succ \boldsymbol{x}^{\boldsymbol{b}} \Leftrightarrow \boldsymbol{a} \succ \boldsymbol{b}$. Notice that the definition implies that $x_i \succ x_j \Leftrightarrow i < j, i, j \in \mathbb{Z}$.

From the definition of zeros of multivariate polynomials discussed in Appendix B, a polynomial $P(x, y, z) \in \mathbb{F}_q[x, y, z]$ has a zero of multiplicity s at the point (α, β, γ) if the translated polynomial $P(x + \alpha, y + \beta, z + \gamma)$ has no monomials of degree less than s and at least one such monomial of degree s.

The extension to zeros of general multivariable polynomials is clear. The number of monomials of degree less than s and weighted $(1, k_1, k_1)$ degree at most D will be of interest.

The number of monomials of degree less than s is

$$\left|\{a + b + c < s, a, b, c \in \mathbb{Z}_{\geq 0}\}\right|.$$

As noted in Chapter 1 in the discussion of RM codes, this number is

$$\binom{s+2}{3}.$$

and the total number of conditions on the coefficients of the monomials of degree less than s in the n coordinates is n times this number.

The total number of monomials of $(1, k_1, k_1)$-weighted degree at most D is slightly more involved. Let $D = m + \ell k_1$ with $0 \leq m < \ell$. The number of weighted terms of the polynomial

$$P(x, y, z) = \sum_{i,j,h} p_{i,j,h} x^i y^j z^h$$

or equivalently the number of positive integer triples $(i,j,h) : i + jk_1 + hk_1 \leq D$ is needed. Clearly $j + h \leq \ell$ or the bound D would be violated. Let $j + h = u$, then the exponent of the variable x can range over $[0, m + 1 + (\ell - u)k_1]$ and hence the enumeration is

$$\sum_{u=0}^{\ell}(m+1+(\ell-u)k_1)(u+1) = (m+1)\binom{\ell+2}{2} + k_1\binom{\ell+2}{3} \qquad (13.13)$$

where use has been made of the relation $\sum_{i=1}^{\ell} i^2 = \ell(\ell+1)(2\ell+1)/6$. The second term is the dominant one in this expression and note that using the approximation $D \sim \ell k_1$

$$k_1\binom{\ell+2}{3} \sim k_1\frac{\ell^3}{6} \sim \frac{D^3}{6k_1^2}. \qquad (13.14)$$

For an approximation to this expression, consider the three-dimensional figure formed by the three axes planes, $x \geq 0, y \geq 0, z \geq 0$ and the condition $x + yk_1 + zk_1 \leq D$, a form of skewed pyramid $\mathcal{P}$ in three dimensions as shown in Figure 13.1. Let $N_3(D)$ be the number of such weighted monomials. To each valid weighted triple (a,b,c) associate a cube of volume unity, rooted at (a,b,c). Then it is argued the volume of the set of all such cubes exceeds the volume of $\mathcal{P}$. But the volume of $\mathcal{P}$ is easily evaluated by noting that a horizontal plane at height x through the figure intersects the figure in a triangle of area $(D-x)^2/2k_1^2$ and integrating this from $x = 0$ to $x = D$ gives

$$\frac{1}{2k_1^2}\int_0^D (D-x)^2 dx = D^3/6k_1^2$$

which is also the approximation of Equation 13.14. This serves as a good and tractable lower bound to the number of monomials of interest. For analytical

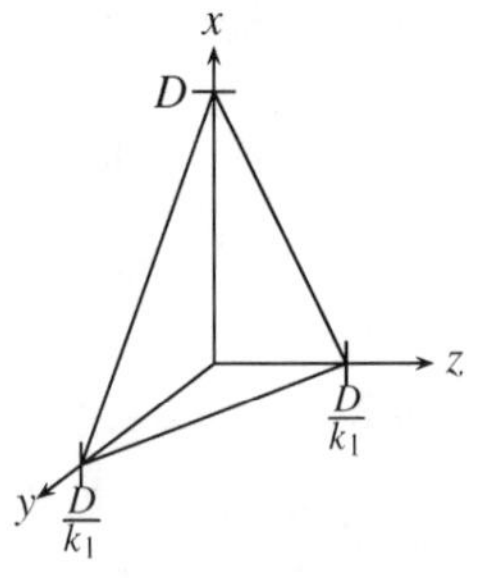

Figure 13.1 Pyramidal figure to approximate $N_3(D)$

purposes the approximation is easier to use than the exact expression above – although the exact expression is needed for some purposes.

As with the Guruswami–Sudan case, a nonzero trivariate polynomial $Q(x,y,z)$ is sought with zeros of order s at each of the n points of the interpolation set $\mathcal{I}$ of Equation 13.12. As before this requires that at each point of the set $\mathcal{I}$ the coefficients of each monomial of weighted $(1,k_1,k_1)$ degree $< s$ be zero giving a total number of conditions of $n\binom{s+2}{3}$.

In order for the required polynomial to exist it is necessary that the number of unknown coefficients in $Q_s(x,y,z)$ (the number of weighted $(1,k_1,k_1)$ monomials) exceed the number of conditions (the total number of monomials of (ordinary) degree less than s which evaluate to 0 in the translated polynomials). Hence we require (using the approximation to $N_3(D)$) that

$$N_3(D) = \frac{D^3}{6k_1^2} > n\frac{s(s+1)(s+2)}{6}. \tag{13.15}$$

Thus [22] the weighted degree of a polynomial that has zeros of degree s at each of the n points of the interpolation set $\mathcal{I}$ is at least

$$D \geq \left\lceil \sqrt[3]{nk_1^2 s(s+1)(s+2)} \right\rceil. \tag{13.16}$$

As long as the number of unknowns (coefficients of the monomials – as given by the left-hand side of Equation 13.15) is greater than this number, there will exist a nonzero polynomial $Q(x,y,z)$ with zeros of degree s at each of the n interpolation points of $\mathcal{I}$ of Equation 13.12. The importance of the $(1,k_1,k_1)$-weighted degree will be that when polynomials $f(x)$ and $g(x)$ of degree at most k_1 are substituted for the variables y and z, the univariate polynomial $Q(x, f(x), g(x))$ will be of degree at most D. Recall the similarity with the previous Guruswami–Sudan case.

The Parvaresh–Vardy Codes

With this background, an informal description of the Parvaresh–Vardy list-decoding algorithm is given. For the details and proof the reader is referred to [22]. The encoding process is first described. Let $f(x), g(x) \in \mathbb{F}_q[x]$ be two polynomials over $\mathbb{F}_q$ of degree at most k_1. The polynomial $f(x)$ is thought of as the "information polynomial," i.e., its coefficients carry the k information symbols over $\mathbb{F}_q$. The polynomial $g(x)$ will be derivable from $f(x)$ as

$$g(x) = f(x)^a \mod e(x)$$

where $e(x)$ is an arbitrary irreducible polynomial over $\mathbb{F}_q$ of degree k for a sufficiently large positive integer a. The value of a suggested ([22], equation 13) is $\lceil (sD+1)/k_1 \rceil$. Thus $g(x)$ represents redundant information which decreases the rate of the code but assists in the decoding process.

Let $\alpha \in \mathbb{F}_{q^2}$ be such that $\{1, \alpha\}$ is a basis for $\mathbb{F}_{q^2}$ over $\mathbb{F}_q$. For the n distinct points of x_i of $\mathbb{F}_q$ of the interpolation set $\mathcal{I}$ let the codeword over $\mathbb{F}_{q^2}$ corresponding to the polynomial $f(x)$ be $c = (c_1, c_2, \ldots, c_n)$ where

$$c_i = f(x_i) + \alpha g(x_i), \quad i = 1, 2, \ldots, n$$

where $g(x)$ is as in the previous equation. This is a subset of a Reed–Solomon code $C = (n, k, n-k+1)_{q^2}$ over $\mathbb{F}_{q^2}$ of length n dimension k over $\mathbb{F}_{q^2}$ and minimum distance $n-k+1$. To see this note that while both $f(x)$ and $g(x)$ are polynomials of degree at most k_1 over $\mathbb{F}_q$, the polynomial $f(x)+\alpha g(x)$ is also polynomial of degree at most k_1 over $\mathbb{F}_{q^2}$ and hence in the RS code. However, not all such polynomials are obtained with this construction and hence the set of such codewords is a subset of the code. It is generally not a linear code. Recall that the x_i are distinct elements of $\mathbb{F}_q$ – hence $n \leq q-1$ although the code is over $\mathbb{F}_{q^2}$. However, the rate of the subset C (nonlinear) code is $R = \log_{q^2} |C| / n = k/2n$ as each coördinate position of the codeword

$$\boldsymbol{c} = (c_1, c_2, \ldots, c_n) = (f(x_1)+\alpha g(x_1), f(x_2)+\alpha g(x_2), \ldots, f(x_n)+\alpha g(x_n))$$

is an element of $\mathbb{F}_{q^2}$.

It is convenient to designate the received word as a vector $\boldsymbol{v} = (v_1, v_2, \ldots, v_n)$, an n-tuple over $\mathbb{F}_{q^2}$. It is easily decomposed to two n-tuples over $\mathbb{F}_q$ via

$$v_i = y_i + \alpha z_i, \quad i = 1, 2, \ldots, n.$$

Given two polynomials $f(x)$, $g(x) \in \mathbb{F}_q[x]$ of degree at most k_1, used for encoding a word, for the interpolation set $\mathcal{I}$ define an agreement parameter:

$$t \triangleq \left| \left\{ i : f(x_i) = y_i \text{ and } g(x_i) = z_i \right\} \right|.$$

This will later be the number of correct code positions corresponding to a codeword generated by $f(x)$ compared to a received word $v_i = y_i + \alpha z_i$. (Notice: This definition of t is in contradiction to that of [22], which uses t for the number of error positions (disagreements), but conforms to our earlier use and later definitions.)

As before let $Q(x, y, z)$ be the interpolation polynomial corresponding to the interpolation set $\mathcal{I}$, of weighted $(1, k_1, k_1)$ degree D given by Equation 13.16. It has zeros of degree s at each of the n points of $\mathcal{I}$. Define the univariate polynomial

$$p(x) \triangleq Q(x, f(x), g(x)).$$

By definition, if t is the agreement parameter for these polynomials f, g, the number of zeros of $p(x)$ evaluated on a received word, is at least st, counting multiplicities of the t zeros. From the definition of the weighted degree of $Q(x,y,z)$, the degree of the univariate polynomial $p(x)$ is at most D bounded by the quantity in Equation 13.16 and

$$\begin{aligned} t &\geq \left\lceil \frac{1}{s}\left\{ \sqrt[3]{n(k_1)^2 s(s+1)(s+2)} + 1 \right\} \right\rceil, \quad k_1 = k-1 \\ &\geq \left\lceil \left\{ \sqrt[3]{n(k-1)^2(1+\frac{1}{s})(1+\frac{2}{s})} + \frac{1}{s} \right\} \right\rceil. \end{aligned} \tag{13.17}$$

In this case the weighted degree of $p(x) = Q(x, f(x), g(x))$ will be strictly less than st and the number of zeros of the polynomial is strictly greater than its degree and hence $p(x)$ is the zero polynomial.

In other words, if a given received n-tuple $v = (v_1, v_2, \ldots, v_n) \in \mathbb{F}_{q^2}^n$ differs from a codeword generated by the polynomials $f(x), g(x)$ in at most e positions where

$$e = n - t \leq \left\lfloor \left\{ n - \sqrt[3]{n(k-1)^2\left(1+\frac{1}{s}\right)\left(1+\frac{2}{s}\right)} - \frac{1}{s} \right\} \right\rfloor, \tag{13.18}$$

then $p(x) = Q(x, f(x), g(x))$ will have at least t zeros indicating positions where the given n-tuple and codeword agree.

A problem remains to actually determine the polynomials $f(x)$ and $g(x)$ from the interpolation polynomial $Q(x,y,z)$. This is slightly technical although not difficult. An overview of the technique is given from [22].

Recall the problem is to determine all codewords in a given code over $\mathbb{F}_{q^2}$ at distance at most $n-t$ from a received word $v \in \mathbb{F}_{q^2}^n$ whose i-th position is $v_i = y_i + \alpha z_i$. From this information the interpolation polynomial $Q_s(x,y,z)$ is found which has zeros of degree s on the interpolation set I and from this trivariate polynomial we are to determine a set of at most L codewords.

It is first noted ([22], lemma 5) that a polynomial $Q(x,y,z) \in \mathbb{F}_q[x,y,z]$ satisfies

$$Q(x, f(x), g(x)) \equiv 0$$

if and only if it belongs to the ideal in the ring $\mathbb{F}_q[x,y,z]$ generated by the elements $y - f(x)$ and $z - g(x)$, denoted by $\langle y - f(x)\rangle$ and $\langle z - g(x)\rangle$, respectively, i.e., iff there exists polynomials $a(x,y,z)$, $b(x,y,z)$ such that

$$Q(x,y,z) = a(x,y,z)(y - f(x)) + b(x,y,z)(z - g(x)).$$

In this setting an ideal is a subset of the ring $\mathbb{F}_q[x,y,z]$ which is closed under addition and multiplication by elements of $\mathbb{F}_q[x,y,z]$.

It will be convenient to consider the field

$$\mathbb{K} \triangleq \mathbb{F}_q[x]/\langle e(x)\rangle$$

for the irreducible polynomial $e(x)$ of degree k over $\mathbb{F}_q$, introduced earlier. The elements of $\mathbb{K}$ can be viewed as polynomials in x of degree less than k with multiplication in $\mathbb{K}$ taken modulo $e(x)$ and $|\mathbb{K}|= q^k$.

Clearly any trivariate polynomial such as $Q(x, y, z)$ can be written as

$$Q(x, y, z) = \sum_i \sum_j q_{ij}(x) y^i z^j, \quad q_{ij}(x) \in \mathbb{F}_q[x]$$

and if the polynomials are taken modulo $e(x)$ so that $q'_{ij}(x) \equiv q_{ij}(x) \mod e(x)$, then define

$$P(y, z) = \sum_i \sum_j q'_{ij}(x) y^i z^j \in \mathbb{K}[y, z].$$

It is not hard to establish that $P(y, z)$ is not the zero polynomial and that if β and $\gamma = \beta^a$ correspond to the polynomials $f(x)$ and $g(x) = f(x)^a$ in $\mathbb{K}$ and ([22], lemmas 7 and 8) if $Q(x, f(x), g(x)) \equiv 0$, then

$$P(\beta, \beta^a) = 0.$$

This implies that $\beta \equiv f(x) \in \mathbb{K}$ is a zero of the univariate polynomial

$$H(y) = P(y, y^a).$$

Thus finding the roots of $H(y)$ will yield a polynomial $f(x)$ corresponding to a codeword. Finding roots of univariate polynomials is a standard problem of finite fields for which there are many efficient algorithms. The algorithm then is as follows:

Algorithm 13.15 (Parvaresh–Vardy decoding of RS codes – trivariate case)

Input: $n, k, t \in \mathbf{Z}$,
received word $\boldsymbol{v} = (v_j = y_j + \alpha z_j)_1^n$
irreducible polynomial $e(x) \in \mathbb{F}_q[x]$, degree k,
positive integer a

Output: A list of codewords that agree with word $\boldsymbol{v}$ in at least t places

1. Compute parameters s and D satisfying the inequalities $st > D$ and $n\binom{s+2}{2} < N_3(D)$ (Equation 13.15)
2. Compute the interpolation polynomial $Q(x, y, z)$ over $\mathbb{F}_q$ of weighted $(1, k_1, k_1)$ degree at most D with the property that $(x_i, y_i, z_i) \in \mathcal{I}$ is a zero of multiplicity $s, i = 1, 2, \ldots, n$
3. Compute the polynomial $P(y, z) = Q(x, y, z) \mod e(x) \in \mathbb{K}[y, z]$
4. Compute the univariate polynomial $H(y) = P(y, y^a)$
5. Compute the roots of $H(y)$ in $\mathbb{K}$ as a polynomial $f(x)$ (and $y^a = g(x)$) for which it is true that $v_i = f(x_i) + \alpha g(x_i)$ for at least t of the points of $\mathcal{I}$.

The procedure outlined leads to the result ([22], theorem 10) that for a given received word $v \in \mathbb{F}_{q^2}^n$ and code rate $k/2n$ and zero parameter s, the algorithm outputs all codewords that differ from v in at most

$$e = \left\lfloor n - n\sqrt[3]{4R^2\left(1+\frac{1}{s}\right)\left(1+\frac{2}{s}\right)} - \frac{1}{s}\right\rfloor$$

coordinate positions and the size of the list of outputs is at most L^2 where

$$L = \left\lfloor s\sqrt[3]{\frac{\left(1+\frac{1}{s}\right)\left(1+\frac{2}{s}\right)}{2R}}\right\rfloor.$$

The factor of 4 in the equation for e arises from the rate of the code being $R = k/2n$. Thus asymptotically for this trivariate case, the code of rate R constructed is able to correct a fraction of errors of

$$1 - \sqrt[3]{4R^2} \sim 1 - (2R)^{2/3}.$$

The trivariate case discussed above contains all the ingredients for the general case which is briefly described. For the M-variate case ($M = 2$ for the trivariate case), the encoding process involves choosing a sequence of suitable integers $a_i, i = 1,2,\ldots,M-1$ and from the information polynomial $f(x) \in \mathbb{F}_q[x]$ generate polynomials

$$g_i(x) \equiv f(x)^{a_i} \mod e(x), \quad i = 1,2,\ldots,M-1$$

where, as before, $e(x)$ is an irreducible polynomial over $\mathbb{F}_q$ and the corresponding codeword to $f(x)$ is then

$$\boldsymbol{c} = (c_1,c_2,\ldots,c_n), \text{ where } c_j = f(x_j) + \sum_{i=1}^{M-1} \alpha_i g_i(x_j),\ j = 1,2,\ldots,n$$

where $\{1,\alpha_1,\alpha_2,\ldots,\alpha_{M-1}\}$ is a basis of $\mathbb{F}_{q^M}$ over $\mathbb{F}_q$. The rate of the code is then $R = k/Mn$. The addition of these related polynomials $g_i(x)$ decreases the code rate but improves the decoding procedure.

The remainder of the argument for the general multivariate case mirrors that of the trivariate case. The interpolation polynomial used is of the form

$$Q_s(x, f(x), g_1(x), \ldots, g_{M-1}(x)) \equiv 0$$

with each zero of multiplicity s. Proceeding as in the trivariate case, the general case yields the result:

Theorem 13.16 ([22], Theorem 1) *Let q be a power of a prime. Then for all positive integers $m, M, n \leq q$ and $k \leq n$ there is a code C of length n and rate $R = k/nM$ over $\mathbb{F}_{q^M}$ equipped with an encoder $\mathcal{E}$ and a decoder $\mathcal{D}$ that have* *the following properties. Given any vector* $\mathbf{v} \in \mathbb{F}_{q^M}^n$ *the decoder outputs the list of all codewords that differ from* $\mathbf{v}$ *in at most*

$$e = \left\lfloor n - n \sqrt[M+1]{M^M R^M \left(1 + \frac{1}{s}\right) \dots \left(1 + \frac{M}{s}\right)} - \frac{1}{s} \right\rfloor$$

positions. The size of the list is at most

$$L = \left\lceil m \sqrt[M+1]{\frac{\left(1 + \frac{1}{s}\right) \dots \left(1 + \frac{M}{s}\right)}{MR}} \right\rceil^M + o(1).$$

Moreover both the encoder and decoder run in time ($\mathbb{F}_q$ operations) bounded by a polynomial in n and s for every given $M \geq 1$.

Approximating the expression in the theorem for the fraction of errors tolerated by the algorithm to produce the decoding list yields:

$$\beta_{PVD} \to 1 - \sqrt[M+1]{M^M R^M} = 1 - (MR)^{M/M+1}.$$

This fraction of errors achievable for decoding improves on the Guruswami–Sudan bound of $1 - \sqrt{R}$ only for rates less than $1/16$ which, while it represented very significant progress, the general problem of achieving the full range of $1 - R$ remained open.

The Guruswami–Rudra Codes

The seminal work [11] introduced the notion of FRS codes and was successful in showing that, using a multivariable interpolation approach, with multiplicity of zeros, such codes could achieve list decoding in polynomial time up to a fraction of $1-R-\epsilon$ of adversarial (worst-case) errors. That work also considers the relationship between the Parvaresh–Vardy codes and the FRS codes. Rather than give an outline of that interesting and important work, the simpler and elegant approach of [10, 14] is considered in some detail.

The treatment here will follow these references with occasional adaptations of notation and results to suit our purpose. While the codes are not the best available in terms of list size and achievable rate, they are simpler to describe and, in a sense, representative of the approach.

Let γ be a primitive element of $\mathbb{F}_q$ and let $x_i = \gamma^i$, $i = 0, 1, 2, \dots, q-2 = n-1$. Consider the $(n, k, d)_q$ RS code

$$C = \left\{ (f(x_0), f(x_1), \dots, f(x_{n-1})), f \in \mathbb{F}_q[x],\ \deg f < k \right\}.$$

Let $m|n$, $N = n/m$. Codewords of this code will be designated $\boldsymbol{y} \in C$, $\boldsymbol{y} = (y_0, y_1, \dots, y_{n-1}) \in \mathbb{F}_q^n$.

The folded RS code then is a rearrangement of this RS code to the form:

$$C' = \left\{ \left(\begin{bmatrix} f(1) \\ f(x_1) \\ \vdots \\ f(x_{m-1}) \end{bmatrix}, \begin{bmatrix} f(x_m) \\ f(x_{m+1}) \\ \vdots \\ f(x_{2m-1}) \end{bmatrix}, \dots, \begin{bmatrix} f(x_{n-m}) \\ f(x_{n-m+1}) \\ \vdots \\ f(x_{n-1}) \end{bmatrix} \right), \deg f < k \right\}. \tag{13.19}$$

Denote this code as $FRS_q^{(m)}[n,k]$, an $(\frac{n}{m}, \frac{k}{m}, \frac{d}{m})_{q^m}$ code, also designated as C'. The code coordinates are viewed as column vectors of length m over $\mathbb{F}_q$ (elements of $\mathbb{F}_q^m$). The parameter m will be referred to as the *folding parameter* of the code. The rate of the code remains the same as the original RS code.

The reason such a construction might be of value lies in the fact that several errors might affect only a single or few coordinate positions in the folded RS code and thereby lead to improved performance.

Codewords of the code C will be designated as $\boldsymbol{y} = (y_0, y_1, \dots, y_{n-1}) \in \mathbb{F}_q^n, n = q-1$ and codewords of the code C' will be of the form of N-tuples of column vectors of length m over $\mathbb{F}_q$, i.e., as $\boldsymbol{z} = (z_0, z_1, \dots, z_{N-1}) \in (\mathbb{F}_q^m)^N$ where each coordinate position z_j is a column vector over $\mathbb{F}_q$, $z_j \in \mathbb{F}_q^m$. Thus

$$\boldsymbol{z} = (z_0, z_1, \dots, z_{N-1}) = \left(\begin{bmatrix} z_{0,0} \\ z_{0,1} \\ \vdots \\ z_{0,m-1} \end{bmatrix}, \begin{bmatrix} z_{1,0} \\ z_{1,1} \\ \vdots \\ z_{1,m-1} \end{bmatrix}, \dots, \begin{bmatrix} z_{N-1,0} \\ z_{N-1,1} \\ \vdots \\ z_{N-1,m-1} \end{bmatrix} \right) \in C'. \tag{13.20}$$

and it follows that $y_{jm+k} = z_{j,k}$. Refer to $\boldsymbol{y} \in C$ as the unfolded version of $\boldsymbol{z} \in C'$ and $\boldsymbol{z}$ as the folded version of $\boldsymbol{y}$.

It is the folded code C' that is of interest. As a slight variation of previous notation denote by $\boldsymbol{v} = (v_0, v_1, \dots, v_{N-1}), v_j \in \mathbb{F}_q^m$ a received word for some transmitted word $\boldsymbol{z} \in C'$. The received word $\boldsymbol{v}$ is said to have t agreements with a codeword $\boldsymbol{z} \in C'$ if t of the columns of $\boldsymbol{v}$ agree with the corresponding t columns of the transmitted codeword $\boldsymbol{z}$, i.e., the column m-tuples over $\mathbb{F}_q$ agree in all m positions.

Before proceeding, it is first noted that under certain conditions, the polynomial $e(x) = x^{q-1} - \gamma \in \mathbb{F}_q[x]$ (different from that of previous use) is irreducible over $\mathbb{F}_q$ (see [19], theorem 3.75 or [12], lemma 3.7). This polynomial is useful in the present context since for any polynomial $f(x) = \sum_i f_i x^i \in \mathbb{F}_q[x]$ it is true that

$$f(x)^q = \left(\sum_i f_i x^i\right)^q = \sum_i f_i^q x^{iq} = \sum_i f_i x^{iq} = f(x^q) = f(\gamma x), \mod e(x)$$

since in the field $\mathbb{K} = \mathbb{F}_q[x]/\langle e(x)\rangle$ it is true that $x^q = \gamma x$. Let $f(x) \in \mathbb{F}_q[x]$ be of degree at most k_1 which generates the codeword $\mathbf{y} = (y_0, y_1, \ldots, y_{n-1}) \in C$, an element of $\mathbb{F}_q^n$, $n = q - 1$. Further let $g(x) = f(\gamma x) = f(x)^q$. The definition of $g(x)$ may seem strange but it will be shown shortly to assist in the decoding of the code C'. Take as our interpolation set

$$\mathcal{I} = \left\{(\gamma^i, f(\gamma^i), g(\gamma^i) = f(\gamma^{i+1})), i = 0, 1, \ldots, n-1\right\}.$$

A polynomial $Q(x, y, z)$ is sought which has n zeros of order s at each point of the interpolation set $\mathcal{I}$. From previous considerations, the number of monomials in such a polynomial of weighted $(1, k_1, k_1)$ degree D is approximated by $D^3/6k_1^2$. The number of homogeneous conditions imposed by requiring zeros of order s at each point of the interpolation set is $n\binom{s+2}{3}$. Thus the nonzero polynomial $Q(x, y, z)$ satisfying these conditions will exist if its weighted degree satisfies

$$\frac{D^3}{6k_1^2} > n\binom{s+2}{3} \quad \text{or} \quad D > \left\lfloor \sqrt[3]{nk_1^2 s(s+1)(s+2)} \right\rfloor + 1.$$

Before discussing decoding the folded codes, consider the above interpolation polynomial written as $Q(x, y_1, y_2)$ which may be written as $Q(x, y_1, y_2) = \sum_{i,j} q_{ij}(x) y_1^i y_2^j$, $q_{ij}(x) \in \mathbb{F}_q[x]$ and denote

$$Q'(x, y_1, y_2) = \sum_{i,j} q'_{ij}(x) y_1^i y_2^j, \mod e(x)$$

where

$$q'_{ij}(x) = q_{ij}(x), \mod e(x).$$

The elements $q'_{ij}(x)$ can be thought of as elements of the field $\mathbb{K} = \mathbb{F}_q[x]/\langle e(x)\rangle$. It is clear that if $f(x)$ is a solution to $Q(x, f(x), f(\gamma x)) = 0$, then it is also a solution to $Q'(x, f(x), f(\gamma x)) = 0$. The problem is to find solutions to the equation

$$Q'(x, f(x), f(x)^q) \stackrel{\Delta}{=} P(f(x), f(x)^q) \in \mathbb{K}[x].$$

As with the previous codes, define

$$H(y) = P(y, y^q)$$

and it can be shown that this is a nonzero polynomial. The solution of this equation in $\mathbb{K}$ will be a polynomial of degree at most k_1 which will correspond to a codeword polynomial in the folded code C' within the specified distance to the received word $\boldsymbol{v}' \in (\mathbb{F}_q^m)^N$ and this polynomial will be a factor $(y - v(x))$ of $Q(x, y, y^q)$.

Suppose the received folded word is $\boldsymbol{v}' \in (\mathbb{F}_q{}^m)^N$ and let $\boldsymbol{v} \in \mathbb{F}_q^n$ be the unfolded version of this received word. Assume that t columns of the word v' are correct and assume column j of $\boldsymbol{v}', \boldsymbol{v}'_j \in \mathbb{F}_q^m$ is correct. Then

$$\boldsymbol{v}_j'^t = \left(v_{j,0}, v_{j,1}, \ldots, v_{j,m-1}\right) = \left(\gamma^{jm}, \gamma^{jm+1}, \ldots, \gamma^{jm+m-1}\right).$$

It follows (and this is the point of the folding process) that for a correct column the $(m-1)$ values

$$Q\left(\gamma^{jm+\ell}, f\left(\gamma^{jm+\ell}\right), f\left(\gamma^{jm+\ell+1}\right)\right) = 0, \quad \ell = 0, 1, \ldots, m-2$$

and thus as $Q(x, y, z)$ is evaluated over all adjacent pairs of the received word, each correct column contributes $s(m-1)$ zeros.

(Note: A slightly better argument, credited to Jørn Justesen in ([12], section 3.D), follows from the observation that an error in the j-th position of the FRS code can lead to *at most* $m + 1$ errors in the interpolating triples and this leads to slightly better parameter values for t.)

Thus if the total number of such zeros exceeds the weighted degree of $Q(x, y, z)$, then it is the zero polynomial and by the process noted above a codeword polynomial within distance at most $n - t$ is attained under the circumstances noted, i.e., if

$$t\,(m-1)s > D.$$

Hence t, the maximum number of error-free positions in the received word tolerated by the list-decoding algorithm, can be chosen as

$$t = \left\lfloor \sqrt[3]{\frac{k_1^2 n}{(m-1)^3}\left(1 + \frac{1}{s}\right)\left(1 + \frac{2}{s}\right)} + \frac{1}{(m-1)s} \right\rfloor + 1.$$

Since $n = mN$ this expression can be written as

$$\begin{aligned} t &= \left\lfloor \sqrt[3]{\tfrac{k_1^2 n}{(m-1)^3}\left(1 + \tfrac{1}{s}\right)\left(1 + \tfrac{2}{s}\right)} + \tfrac{1}{(m-1)s} \right\rfloor + 1 \\ &= \left\lfloor \sqrt[3]{\tfrac{k_1^2}{n^2}\tfrac{n^3}{(m-1)^3}\left(1 + \tfrac{1}{s}\right)\left(1 + \tfrac{2}{s}\right)} + \tfrac{1}{(m-1)s} \right\rfloor + 1. \end{aligned}$$

Asymptotically, using $(n/(m-1))^3 \approx (1+\epsilon)N^3$, this gives

$$t \approx N(1+\epsilon)R^{2/3}.$$

Thus the fractional error rate for the folded RS codes is

$$\frac{1}{N}(N-t) \approx 1-(1+\epsilon)R^{2/3}.$$

The conclusion of these arguments is the following [11, 12]:

Theorem 13.17 *For every $\epsilon > 0$ and rate $R, 0 < R < 1$ there is a family of m-FRS codes for $m = O(1/\epsilon)$ that have rate at least R and which can be list decoded up to a fraction $1-(1+\epsilon)R^{2/3}$ of errors in time polynomial in the block length and $1/\epsilon$.*

This development has been for the trivariate case. The M-variate case is commented on below. The details discussed above for the trivariate case are gathered in the algorithm:

Algorithm 13.18 (Guruswami–Rudra decoding of FRS codes – trivariate case)

Input: $n, k, d, \mathbb{F}_q, \gamma \in \mathbb{F}_q$ primitive, folding parameter m, $N = n/m$, irreducible polynomial $e(x) \in \mathbb{F}_q[x]$, degree $q-1$, s = zeros order, codeword polynomial $f(x) \in \mathbb{F}_q[x]$ degree $\leq k_1$, received word, $\boldsymbol{v} \in (\mathbb{F}_q^m)^N$, $\mathbb{K} = \mathbb{F}_q[x]/\langle e(x)\rangle$

Output: A list of codewords that agree with word v in at least t places

Algorithm:
1. Compute parameters t and D satisfying the inequalities $n\binom{s+2}{3} < D^3/6k_1^2$ and $t(m-1)s > D$
2. Compute the interpolation polynomial $Q(x,y,z)$ over $\mathbb{F}_q$ of weighted $(1,k_1,k_1)$ degree at most D with zeros of order s at the n points $(\gamma^i, f(\gamma^i), f(\gamma^{i+1})), i = 1, \ldots, n$
3. Compute the polynomial $P(y,z) = Q(x,y,z) \mod e(x)$, $P(y,z) \in \mathbb{K}[y,z]$
4. Compute the univariate polynomial $H(y) = P(y, y^a)$
5. Compute the roots of $H(y)$ in $\mathbb{K}$ as a polynomial $f(x)$ in x (and $y^a = g(x)$)
6. Output those polynomials $f(x)$ that agree with received word in at least t positions

As with the Parvaresh–Vardy codes, the M-variate case can be used for which the fraction of errors tolerated in the list-decoding algorithm can be shown to be

$$1-\left(1+\frac{M}{s}\right)\left(\frac{m}{m-M+1}\right)R^{M/(M+1)}$$

and asymptotically this tends to the $1 - R$ result promised. Thus for the Guruswami–Rudra decoding folded RS decoding, the fraction of errors that can be tolerated in an RS code of block length n is, asymptotically for large block lengths,

$$\beta_{GR} \rightarrow 1 - R. \tag{13.21}$$

For details and comments on the resulting list size, the references cited should be consulted.

Comments

The chapter has attempted to give a broad outline of the interesting development of the field of list decoding that have led to capacity-achieving codes. The work of Sudan and Guruswami and others has been exceptional in terms of the new techniques introduced and the results obtained. While many of the derivations considered are quite specialized and intricate, the impressive results obtained and the insight they give are surprising.

References

[1] Bartz, H., and Puchinger, S. 2021. Decoding of interleaved linearized Reed–Solomon codes with applications to network coding. *ArXiv*, abs/2101.05604.

[2] Bleichenbacher, D., Kiayias, A., and Yung, M. 2003. Decoding of interleaved Reed Solomon codes over noisy data. Pages 97–108 of: *Proceedings of the 30th International Colloquium on Automata, Languages and Programming (ICALP)*.

[3] Cameron, P.J., and van Lint, J.H. 1991. *Designs, graphs, codes and their links*. London Mathematical Society Student Texts, vol. 22. Cambridge University Press, Cambridge.

[4] Cox, D.A., Little, J., and O'Shea, D. 2015. *Ideals, varieties, and algorithms*. Fourth edn. Undergraduate Texts in Mathematics. Springer, Cham.

[5] Elias, P. 1991. Error-correcting codes for list decoding. *IEEE Trans. Inform. Theory*, **37**(1), 5–12.

[6] Gemmell, P., and Sudan, M. 1992. Highly resilient correctors for polynomials. *Inform. Process. Lett.*, **43**(4), 169–174.

[7] Guruswami, V. 2005. *List decoding of error-correcting codes*. Vol. 3282. Springer.

[8] Guruswami, V. 2006. Algorithmic results in list decoding. *Found. Trends Theor. Comput. Sci.*, **2**(2), 107–195.

[9] Guruswami, V. 2010. Bridging Shannon and Hamming: list error-correction with optimal rate. Pages 2648–2675 of: *Proceedings of the International Congress of Mathematicians. Volume IV*. Hindustan Book Agency, New Delhi.

[10] Guruswami, V. 2011. Linear-algebraic list decoding of folded Reed-Solomon codes. Pages 77–85 of: *26th Annual IEEE Conference on Computational Complexity*. IEEE Computer Society, Los Alamitos, CA.
[11] Guruswami, V., and Rudra, A. 2006. Explicit capacity-achieving list-decodable codes or decoding up to the Singleton bound using folded Reed-Solomon codes. Pages 1–10 of: *STOC '06: Proceedings of the 38th Annual ACM Symposium on Theory of Computing*. ACM, New York.
[12] Guruswami, V., and Rudra, A. 2008. Explicit codes achieving list decoding capacity: error-correction with optimal redundancy. *IEEE Trans. Inform. Theory*, **54**(1), 135–150.
[13] Guruswami, V., and Sudan, M. 1999. Improved decoding of Reed-Solomon and algebraic-geometry codes. *IEEE Trans. Inform. Theory*, **45**(6), 1757–1767.
[14] Guruswami, V., and Wang, C. 2013. Linear-algebraic list decoding for variants of Reed-Solomon codes. *IEEE Trans. Inform. Theory*, **59**(6), 3257–3268.
[15] Guruswami, V., Umans, C., and Vadhan, S. 2009. Unbalanced expanders and randomness extractors from Parvaresh–Vardy codes. *J. ACM*, **56**(4), 20:1–20:34.
[16] Huffman, W.C., and Pless, V. 2003. *Fundamentals of error-correcting codes*. Cambridge University Press, Cambridge.
[17] Huffman, W.C., Kim, J.L., and Solé, P. (eds.). 2021. *A concise encyclopedia of coding theory*. CRC Press, Boca Raton, FL.
[18] Kaltofen, E. 1992. Polynomial factorization 1987–1991. Pages 294–313 of: Simon, Imre (ed.), *LATIN '92*. Springer, Berlin, Heidelberg.
[19] Lidl, R., and Niederreiter, H. 1997. *Finite fields*, 2nd ed. Encyclopedia of Mathematics and Its Applications, vol. 20. Cambridge University Press, Cambridge.
[20] MacWilliams, F.J., and Sloane, N.J.A. 1977. *The theory of error-correcting codes. I and II*. North-Holland Mathematical Library, vol. 16. North-Holland, Amsterdam/New York/Oxford.
[21] Mullen, G.L. (ed.). 2013. *Handbook of finite fields*. Discrete Mathematics and Its Applications (Boca Raton). CRC Press, Boca Raton, FL.
[22] Parvaresh, F., and Vardy, A. 2005. Correcting errors beyond the Guruswami-Sudan radius in polynomial time. Pages 41 of: *Proceedings of the 46th Annual Symposium on Foundations of Computer Science*. FOCS '05. IEEE Computer Society, Washington, DC.
[23] Peterson, W.W. 1961. *Error-correcting codes*. The MIT Press, Cambridge, MA.
[24] Puchinger, S., and Rosenkilde né Nielsen, J. 2017. Decoding of interleaved Reed-Solomon codes using improved power decoding. Pages 356–360 of: *2017 IEEE International Symposium on Information Theory (ISIT)*.
[25] Sudan, M. 1997. Decoding of Reed–Solomon codes beyond the error-correction bound. *J. Complexity*, **13**(1), 180–193.
[26] van Lint, J.H. 1992. *Introduction to coding theory*, 2nd ed. Graduate Texts in Mathematics, vol. 86. Springer-Verlag, Berlin.
[27] Wachter-Zeh, A., Zeh, A., and Bossert, M. 2014. Decoding interleaved Reed–Solomon codes beyond their joint error-correcting capability. *Des. Codes Cryptogr.*, **71**(2), 261–281.

14

Sequence Sets with Low Correlation

The design of sequences for communication purposes such as ranging, synchronization, frequency-hopping spread spectrum, code-division multiple access (CDMA), certain cell phone systems and space communications has a long and interesting history. As with so many other of the areas covered in this volume, the mathematical techniques used to construct sequences with the desired properties tend to be challenging. While many of the results and techniques date back to the 1950s, the area continues to encompass a vigorous research community as more and more applications and variations are discovered. This chapter provides the basic information of the area.

14.1 Maximum-Length Feedback Shift Register Sequences

Consider an n-stage recursion over $\mathbb{F}_q$ (although generally interest will be in $\mathbb{F}_2$) as shown in Figure 14.1:

$$x_t = a_1 x_{t-1} + a_2 x_{t-2} + \cdots + a_n x_{t-n}, \quad t = n, n+1, \ldots, \quad a_i \in \mathbb{F}_q \quad (14.1)$$

with initial values in $\mathbb{F}_q$ given for the n cells, $x_{n-1}, x_{n-2}, \ldots, x_0$, referred to as the initial state of the shift register. This equation is referred to as the *characteristic equation* of the sequence it generates.

The operation of the shift register is that at each clock cycle t of the register a new feedback value x_t is formed, the contents of each of the n registers are shifted one cell to the right, the new value is fed into the leftmost cell and a value x_{t-n} is produced for output. The set of values $(x_{t-1}, x_{t-2}, \ldots, x_{t-n})$ is referred to as the *state* of the shift register at time t. If the state of the register is all-zeroes, then the register remains in that state always. We will be interested in sequences over $\mathbb{F}_q$, $x_t \in \mathbb{F}_q$ for all t. In order to analyze the system, however,

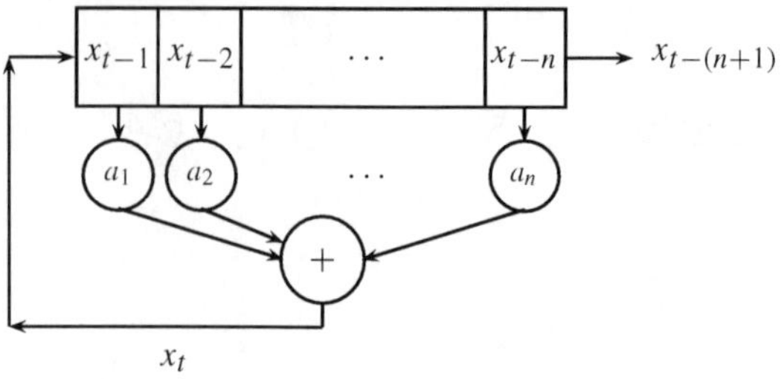

Figure 14.1 Linear-feedback shift register

we will also consider values of x_t in some extension field, say $\mathbb{F}_{q^n}$ of $\mathbb{F}_q$ as a means to an end.

Such sequences will be referred to as *linear-feedback shift register* (LFSR) sequences. The properties of such sequences are of interest in this chapter, especially the correlation and cross-correlation properties which play a crucial role in numerous synchronization, ranging, CDMA and frequency-hopping spread spectrum systems. One could also replace the linear sum of the cell contents with an arbitrary Boolean function to form the feedback variable, to give a nonlinear-feedback shift register but the properties of such sequences are considerably more difficult to analyze and are not considered here.

For sequences over $\mathbb{F}_q$ there are $q^n - 1$ possible nonzero states of the register. If the register is in the all-zero state, it remains there as the clock cycles and produces only zeros as output. If the sequence produced by the shift register is such that

$$x_t = x_{t+N} \text{ for all } t$$

for the smallest N, the sequence has period N. Clearly the maximum period of such a shift register is $q^n - 1$ and this is achieved if and only if the state of the register cycles through all $q^n - 1$ nonzero states in each cycle, assuming a sequence over $\mathbb{F}_q$. Such sequences are termed *maximum-length shift register* sequences or ML sequences (not to be confused with maximum-likelihood (ML) receivers considered in other chapters) or, more simply, m-sequences, the preferred term for this work. Thus the period of an m-sequence is typically $q^\ell - 1$ for some positive integer ℓ, generated by a shift register of length ℓ. Such sequences will have interesting properties that will be discussed.

Define the *characteristic polynomial* of the recursion of the sequence over $\mathbb{F}_q$ of Equation 14.1 as

$$f(x) = x^n - a_1x^{n-1} - a_2x^{n-2} - \cdots - a_n,\ a_i \in \mathbb{F}_q.$$

Consider the question of whether $x_t = \alpha^t, t = 0,1,2,\ldots$ for some element α in an extension field of $\mathbb{F}_q$ is a solution of the recursion by substituting into Equation 14.1, i.e., could it be true that

$$\begin{aligned} &\alpha^t = a_1\alpha^{t-1} + a_2\alpha^{t-2} + \cdots + a_n\alpha^{t-n}, t \geq n \\ \text{or} \quad &\alpha^{t-n}\big(\alpha^n - a_1\alpha^{n-1} - a_2\alpha^{n-2} - \cdots - a_n\big) = 0, \quad t = n, n+1, \ldots \end{aligned}$$

or equivalently $f(\alpha) = 0$, i.e., α must be a root of the recursion characteristic polynomial in order for $x_t = \alpha^t, t = 0,1,2,\ldots$ to be a solution. In such a case it follows that $x_t = \lambda\alpha^t$, for λ a constant also in the extension field, is also a solution of the recursion since the effect is to multiply all terms of the characteristic equation by the same constant. Since α is an element in an extension field of $\mathbb{F}_q$, the sequence would then also be over the extension field. Our interest will be in sequences over $\mathbb{F}_q$. Suppose further that α is an element of order N in the extension field. It follows immediately the sequence it generates is of period N.

By the same reasoning if $\alpha_1, \alpha_2, \ldots, \alpha_n$ are distinct roots of the characteristic polynomial $f(x)$ in some extension field of $\mathbb{F}_q$, say $\mathbb{F}_{q^n}$, then the sequence

$$x_t = \lambda_1\alpha_1^t + \lambda_2\alpha_2^t + \cdots + \lambda_n\alpha_n^t, \quad t \geq n,\ \lambda_i \in \mathbb{F}_{q^n} \tag{14.2}$$

satisfies the linear recurrence Equation 14.1 and produces a sequence over $\mathbb{F}_{q^n}$.

Our (almost exclusive) interest will be in the case where $f(x)$ is a primitive polynomial of degree n over $\mathbb{F}_q$. If α is a zero of $f(x)$, then the other zeros are $\alpha^q, \alpha^{q^2}, \ldots, \alpha^{q^{n-1}}$. From Equation 14.2 a solution to the characteristic Equation 14.1 would be

$$x_t = \lambda_1\alpha^t + \lambda_2\alpha^{tq} + \cdots + \lambda_n\alpha^{tq^{n-1}}$$

for constants $\lambda_i \in \mathbb{F}_{q^n}, i = 1,2,\ldots,n$. The sequence x_t is over $\mathbb{F}_{q^n}$ but notice that if the λ_i are chosen as $\lambda^{q^i}, i = 1,2,\ldots,n$, then

$$x_t = \sum_{i=0}^{n-1} \left(\lambda\alpha^i\right)^{q^i} = \mathrm{Tr}_{q^n|q}(\lambda\alpha^t),\ \lambda \in \mathbb{F}_{q^n}^*$$

is a sequence over $\mathbb{F}_q$ and is a solution of the shift register equation. From the above comments the period of this sequence is the order of the zeros α^{q^i}, which is $q^n - 1$ and these are m-sequences over $\mathbb{F}_q$.

Notice that Equation 14.1 will produce q^n solutions according to the q^n initial conditions of the shift register $\{x_{n-1}, x_{n-2}, \ldots, x_0\}$, including the trivial all-zero solution. The above equation also produces q^n solutions according to the values of $\lambda \in \mathbb{F}_{q^n}$. If these sequences are distinct, then any such sequence that satisfies Equation 14.1 can be so represented. Note the trivial all-zero solution is included in the count. Of course, for a given feedback connection, since each m-sequence cycles through each possible nonzero the $(q^n - 1)$ sequences are simply translates of each other.

It is easy to verify that

$$x_t = \mathrm{Tr}_{q^n|q}(\lambda\alpha^t), \quad \lambda \in \mathbb{F}_{q^n}, \ t \geq 0 \tag{14.3}$$

is a solution of the recursion Equation 14.1 as follows:

$$\begin{aligned} x_t &= a_1 x_{t-1} + a_2 x_{t-2} + \cdots + a_m x_{t-n}, \quad a_i \in \mathbb{F}_q \\ \mathrm{Tr}_{q^n|q}(\lambda\alpha^t) &= a_1 \mathrm{Tr}_{q^n|q}(\lambda\alpha^{t-1}) + \cdots + a_n \mathrm{Tr}_{q^n|q}(\lambda\alpha^{-n}) \\ \text{or} \quad 0 &= \mathrm{Tr}_{q^n|q}\Big(\lambda\big(\alpha^t - a_1\alpha^{t-1} - \cdots - a_n\alpha^{t-n}\big)\Big) \\ \text{or} \quad 0 &= \mathrm{Tr}_{q^n|q}\Big(\lambda\alpha^{t-n}\big(\alpha^n - a_1\alpha^{n-1} - \cdots - a_n\big)\Big) \end{aligned}$$

which follows since α is assumed a zero of the characteristic polynomial. Suppose $\lambda_1 \neq \lambda_2$ and consider if the associated solutions to the recursion are equal, i.e., suppose

$$x_t = \mathrm{Tr}_{q^n|q}(\lambda_1\alpha^t) = x'_t = \mathrm{Tr}_{q^n|q}(\lambda_2\alpha^t), \quad t \geq 0, \quad \lambda_1, \lambda_2 \in \mathbb{F}_{q^n}$$

and so

$$x_t - x'_t = 0 = \mathrm{Tr}_{q^n|q}(\lambda_1\alpha^t) - \mathrm{Tr}_{q^n|q}(\lambda_2\alpha^t) = \mathrm{Tr}_{q^n|q}\big((\lambda_1 - \lambda_2)\alpha^t\big) = 0 \text{ for all } \ t \geq 0.$$

From the onto property of the trace function (Equation 1.5 of Chapter 1) this would imply that $\lambda_1 - \lambda_2 = 0$, a contradiction. Thus every solution over $\mathbb{F}_q$ to the recurrence, corresponding to different initial states of the register, under the above condition, can be represented in the form of Equation 14.3, each field element λ corresponding to an initial condition ([8], theorem 9.2).

The sequence x'_t is called *cyclically equivalent* to the sequence $x_t = \mathrm{Tr}_{q^n|q}\big(\lambda_1\alpha^t\big)$, $\lambda \in \mathbb{F}_{q^n}$, if for some fixed i

$$x'_t = x_{t+i} \quad \text{for all } t \geq 0.$$

Otherwise the two sequences are *cyclically inequivalent* or *cyclically distinct*. Suppose α is an element of order N in $\mathbb{F}_{q^n}$ (hence the sequence is of period N). The two sequences will be cyclically equivalent iff for some i

$$x'_t - x_{t+i} = 0 = \mathrm{Tr}_{q^n|q}\big(\lambda_2\alpha^t\big) - \mathrm{Tr}_{q^n|q}\big(\lambda_1\alpha^{t+i}\big), \text{ or } \mathrm{Tr}_{q^n|q}\big((\lambda_2 - \lambda_1\alpha^i)\alpha^t\big) \text{ for all } t \geq 0.$$

Thus the two sequences are cyclically equivalent iff $\lambda_2 = \lambda_1\alpha^i$, i.e., iff their associated constants must be related by a multiplicative constant of the group $G_n = \{1, \alpha, \alpha^2, \ldots, \alpha^{N-1}\}$, i.e., iff the two constants are in the same coset of this multiplicative subgroup of $\mathbb{F}_{q^n}$.

Interest in these sequences for the remainder of the chapter will be m-sequences, i.e., maximum-length sequences over $\mathbb{F}_q$ where the period is $N = q^n - 1$. From the above observations then α is a primitive element of $\mathbb{F}_{q^n}$ and the sequence characteristic polynomial will be a primitive polynomial over $\mathbb{F}_q$ of degree n. For such a generator, the previous discussion implies **any** two sequences with generator element α will be cyclically equivalent, i.e., for the primitive element $\alpha \in \mathbb{F}_{q^n}$ for any $\lambda_1, \lambda_2 \in \mathbb{F}_{q^n}^*$ the sequences $\text{Tr}_{q^n|q}\left(\lambda_1\alpha^t\right)$ and $\text{Tr}_{q^n|q}\left(\lambda_2\alpha^t\right)$ are cyclically equivalent. As noted, for an m-sequence over $\mathbb{F}_q$, the shift register producing it cycles through each nonzero state in any given cycle.

The sequences corresponding to primitive characteristic polynomials will have interesting properties. Recall that if α is a primitive element of $\mathbb{F}_q$, then α^i will also be primitive iff i and $q^n - 1$ are relatively prime (i.e., $(i, q^n - 1) = 1$). Since the minimal polynomial of any primitive element of $\mathbb{F}_{q^n}$ is of degree n there are exactly $\phi(q^n - 1)/n$ primitive polynomials of degree n over $\mathbb{F}_q$, where it is recalled $\phi(\ell)$ is the Euler ϕ function, the number of integers in $[\ell] = \{1, 2, \ldots, \ell\}$ relatively prime to ℓ. Any of these primitive polynomials could be used to produce an m-sequence. The sequences produced by distinct primitive polynomials are not cyclically equivalent. To see this suppose α is a primitive element of $\mathbb{F}_{q^n}$. Let d be an integer $(d, q^n - 1) = 1$ and $\beta = \alpha^d$. Then β is a distinct primitive element in $\mathbb{F}_{q^n}$. Assume the sequences produced by these elements are cyclically equivalent and so

$$\text{Tr}_{q^n|q}\left(\lambda_1\alpha^t\right) = \text{Tr}_{q^n|q}\left(\lambda_2\alpha^{d(t+i)}\right),\ t \geq 0 \iff \lambda_1\alpha^t = \lambda_2\alpha^{d(t+i)},\ d > 1,\ t \geq 0$$

which is clearly not possible and such sequences cannot be cyclically equivalent. It follows there are $\phi(q^n - 1)/n$ cyclically inequivalent sequences of period $q^n - 1$ over $\mathbb{F}_q$.

For a given m-sequence $x_t = \text{Tr}_{q^n|q}\left(\lambda\alpha^t\right)$, $\alpha \in \mathbb{F}_{q^n}^*$ primitive and $\lambda \in \mathbb{F}_{q^n}^*$, call the sequence by taking every d-th sequence element, so that

$$y_t = x_{dt} \quad \text{for all } t \geq 0$$

the *d-decimated sequence*. It is possible to show that any two cyclically inequivalent m-sequences of the same length over the same field are related by decimation. Suppose x_t and y_t are two m-sequences over $\mathbb{F}_q$. Then there exists a positive integer d, relatively prime to $q^n - 1$, such that

$$y_t = x_{dt} \quad \text{for all } t \geq 0.$$

To see this assume $x_t = \text{Tr}_{q^n|q}(\lambda\alpha^t)$ and the d-decimated sequence be $y_t = x_t^{[d]} = \text{Tr}_{q^n|q}(\lambda\alpha^{dt}) = \text{Tr}_{q^n|q}(\lambda\beta^t)$ where $\beta = \alpha^d$ which, since

$(d, q^n - 1) = 1$, is a primitive element. Thus the number of cyclically inequivalent m-sequences of this period is $\phi(q^n - 1)/n$ and they are generated by one primitive element drawn from each conjugacy class of each primitive polynomial of degree n. It easily follows that any m-sequence can be obtained from any other m-sequence by suitable decimation and translation.

In general if an m-sequence of length $N = q^n - 1$ is d-decimated the period of the resulting sequence will be $N/(N, d)$.

A curious property of m-sequences is that for any translation $\tau \in \mathbb{Z}$ of an m-sequence x_t there exists another unique translation σ such that

$$x_t + x_{t+\tau} = x_{t+\sigma} \quad \text{for all } t \geq 0,$$

often referred to as shift-and-add property [8]. To see this consider a fixed integer τ and note that for some λ

$$\begin{aligned} x_t + x_{t+\tau} &= \mathrm{Tr}_{q^n|q}\left(\lambda\alpha^t\right) + \mathrm{Tr}_{q^n|q}\left(\lambda\alpha^{t+\tau}\right) \\ &= \mathrm{Tr}_{q^n|q}\left(\lambda\alpha^t\left(1 + \alpha^\tau\right)\right) \end{aligned}$$

but since α is primitive in $\mathbb{F}_{q^n}$ there exists a unique σ such that $(1 + \alpha^\tau) = \alpha^\sigma$ and the result follows.

It might be asked if there is regularity as to the number of each finite field element in the period of an m-sequence and the following result is as expected:

$$\mathrm{Tr}_{q^n|q}(\lambda\alpha^t) = \begin{cases} \beta, \ q^{n-1} \text{ times}, \ \beta \in \mathbb{F}_q{}^* \\ 0, \ q^{n-1} - 1 \text{ times}. \end{cases}$$

More generally [6] it is true that for $k \mid n$:

$$\mathrm{Tr}_{q^n|q^k}(\lambda\alpha^t) = \begin{cases} \beta, \ q^{n-k} \text{ times}, \ \beta \in \mathbb{F}_{q^k}^* \\ 0, \ q^{n-k} - 1 \text{ times}. \end{cases}$$

Example 14.1 Consider the two m-sequences over $\mathbb{F}_2$ generated by the circuits in the figure below (the characteristic polynomials are reciprocal polynomials but this property is not important for the development).

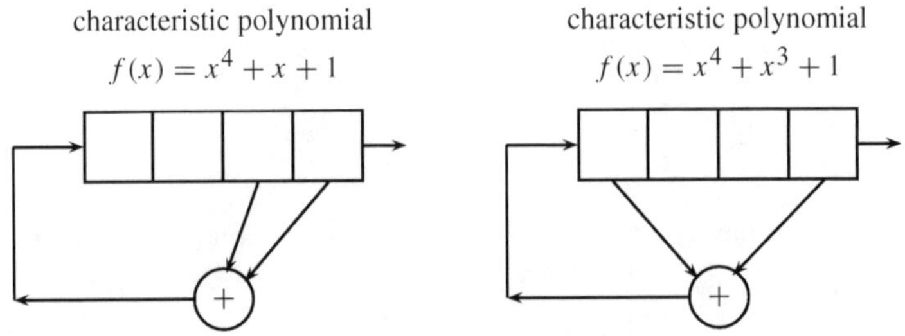

In the tables below, for each circuit, the sequence of states the shift registers go through, the output m-sequences obtained (always the last bit of the state sequence) and the m-sequence produced by the $\mathrm{Tr}_{2^4|2}$ function, with the purpose of showing (for each register) the two m-sequences (from the shift register and from the trace function for that register) are cyclically equivalent but the two shift registers produce cyclically inequivalent m-sequences.

characteristic polynomial $x^4 + x + 1$

register state	output sequence	α^t	$\mathrm{Tr}_{2^4\|2}(\alpha^t)$
0110	0	1	0
1011	0	α	0
0101	1	α^2	0
1010	1	α^3	1
1101	0	α^4	0
1110	1	α^5	0
1111	0	α^6	1
0111	1	α^7	1
0011	1	α^8	0
0001	1	α^9	1
1000	1	α^{10}	0
0100	0	α^{11}	1
0010	0	α^{12}	1
1001	0	α^{13}	1
1100	1	α^{14}	1
– – –	– –	– –	– –
0110	0	1	0

characteristic polynomial $x^4 + x^3 + 1$

register state	output sequence	α^t	$\mathrm{Tr}_{2^4\|2}(\alpha^t)$
0001	0	1	0
1000	1	β	1
1100	0	β^2	1
1110	0	β^3	1
1111	0	β^4	1
0111	1	β^5	0
1011	1	β^6	1
0101	1	β^7	0
1010	1	β^8	1
1101	0	β^9	1
0110	1	β^{10}	0
0011	0	β^{11}	0
1001	1	β^{12}	1
0100	1	β^{13}	0
0010	0	β^{14}	0
– – –	– –	– –	– –
0001	0	1	0

There are numerous other properties of m-sequences but the remainder of the chapter will focus on their correlation properties, of crucial importance to their application.

14.2 Correlation of Sequences and the Welch Bound

Interest in m-sequences lies mainly in their cycle structure, pseudorandomness properties and application to such areas as radar, ranging and spread spectrum where their auto- and cross-correlation properties are of importance. The representation of Equation 14.3 is useful in these studies. A brief overview of their properties and applications is noted here.

While much of the remainder of the chapter will restrict sequences to be binary (either $\{0,1\}$ for developing the theory or $\{\pm 1\}$ for applications) many of their interesting auto- and cross-correlation functions properties do not depend on this and a few of their general properties are discussed before further restricting the alphabet. Let $\boldsymbol{x},\boldsymbol{y}$ be complex sequences with period N, i.e., $x_i = x_{i+N}, y_i = y_{i+N}$ for all i. Define their periodic auto- and cross-correlation functions as

$$\theta_{\boldsymbol{x}}(\tau) = \sum_{i=0}^{N-1} x_i \bar{x}_{i+\tau} \quad \text{and} \quad \theta_{\boldsymbol{x},\boldsymbol{y}}(\tau) = \sum_{i=0}^{N-1} x_i \bar{y}_{i+\tau},\ \tau \in \mathbb{Z},\ x_i, y_j \in \mathbb{C} \tag{14.4}$$

where the bar indicates complex conjugation. Numerous properties of these functions follow from their definition, e.g., [10]

$$\sum_{\ell=0}^{N-1} \mid \theta_{\boldsymbol{x},\boldsymbol{y}}(\ell) \mid^2 \leq \left(\sum_{\ell=0}^{N-1} \mid \theta_{\boldsymbol{x}}(\ell) \mid^2 \right)^{1/2} \left(\sum_{\ell=0}^{N-1} \mid \theta_{\boldsymbol{y}}(\ell) \mid^2 \right)^{1/2}$$

$$\sum_{\ell=0}^{N-1} \mid \theta_{\boldsymbol{x},\boldsymbol{y}}(\ell) \mid^2 \in \left(\theta_{\boldsymbol{x}}(0)\theta_{\boldsymbol{y}}(0) \pm \left(\sum_{\ell=1}^{N-1} \mid \theta_{\boldsymbol{x}}(\ell) \mid^2 \right)^{1/2} \left(\sum_{\ell=1}^{N-1} \mid \bar{\theta}_{\boldsymbol{y}}(\ell) \mid^2 \right)^{1/2} \right).$$

The aperiodic case can also be of interest for some applications. For such a case sequences can be formulated from one period of each sequence with the sequences containing zeros outside of the one period and correlations formed between one sequence and a translated version of the other sequence. Such will not be considered here. Interest will be in sets of sequences with low maximum periodic cross-correlation, with questions such as how large a set can be found with a given maximum cross-correlation? The following definitions are of interest. Assume the set C contains M complex sequences of length N.

Define the peak *out-of-phase* autocorrelation for a sequence $\boldsymbol{x} \in \mathbb{C}^N$ as

$$\theta_a = \max\{\mid \theta_{\boldsymbol{x}}(\ell) \mid,\ 1 \leq \ell \leq N-1\}$$

and the peak cross-correlation as

$$\theta_c = \max_{\substack{x \neq y \\ x,y \in C}} \{ | \theta_{x,y}(\ell) | \, , \, 0 \leq \ell \leq N-1 \}$$

and

$$\theta_{\max} \triangleq \max(\theta_a, \theta_c).$$

Then [9] for the set of sequences C

$$\left(\frac{\theta_c^2}{N}\right) + \frac{N-1}{N(M-1)}\left(\frac{\theta_a^2}{n}\right) \geq 1$$

and it follows that

$$\theta_{\max} \geq N\left[\frac{M-1}{NM-1}\right]^{1/2}$$

which is a special case of the Welch bound (with the parameter $k = 1$) shown below where the sequences have norm unity.

For the remainder of the discussion attention will be restricted to the binary primitive case, i.e., m-sequences although there is a considerable literature on nonbinary sequences and their properties. The length of the m-sequence is $N = 2^n - 1$ and as one considers sliding a window of length n along the sequence each possible $(2^n - 1)$ nonzero n-tuple will appear as a state of the register – only the all-zeros state is missing. As noted above, each period of length N of the sequence contains 2^{n-1} 1's and $(2^{(n-1)} - 1)$ 0's.

For applications, interest is in binary $\{\pm 1\}$ sequences rather than binary $\{0,1\}$ and the usual transition is made via $y_t = (-1)^{x_t}$. The (periodic) autocorrelation function of a sequence of length $N = 2^n - 1$ for the sequence represented by $y_t = (-1)^{x_t}$ where $x_t = \mathrm{Tr}_{q^n|q}(\lambda\alpha^t)$, α a primitive element of $\mathbb{F}_{2^n}$, is

$$\begin{aligned}
\theta_{\mathbf{y}}(\tau) &= \sum_{i=0}^{N-1} y_t y_{t+\tau} \\
&= \sum_{i=0}^{N-1} (-1)^{x_t}(-1)^{x_{t+\tau}} = \sum_{t=0}^{N-1} (-1)^{x_t + x_{t+\tau}} \\
&= \sum_{t=0}^{N-1} (-1)^{\mathrm{Tr}_{q^n|q}(\lambda\alpha^t) + \mathrm{Tr}_{q^n|q}(\lambda\alpha^{t+\tau})} \\
&= \sum_{t=0}^{N-1} (-1)^{\mathrm{Tr}_{q^n|q}(\lambda\alpha^t(1+\alpha^\tau))}.
\end{aligned}$$

As t ranges over 0 to $N-1 = 2^n - 2$, the trace function gives $1, 2^{n-1}$ times and $0, (2^{n-1}-1)$ times yielding a value of -1 for the sum. Thus the autocorrelation of an m-sequence is two-valued, namely

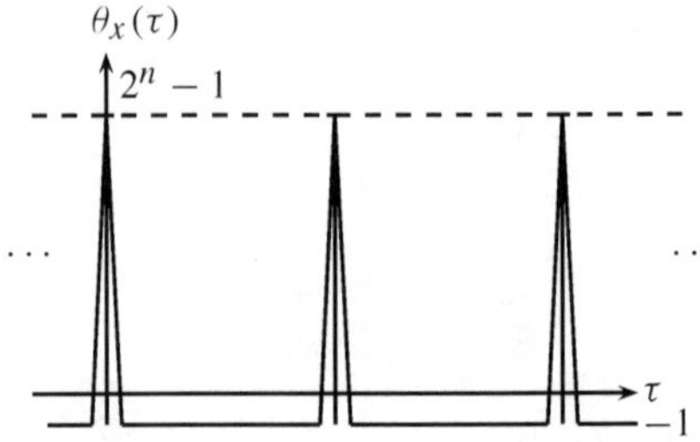

Figure 14.2 Ideal correlation function of m-sequence of period $2^m - 1$

$$\theta_y(\tau) = \begin{cases} N = 2^n - 1, & \text{if } \tau \equiv 0 \mod (2^n - 1) \\ -1, & \text{if } \tau \not\equiv 0 \mod (2^n - 1). \end{cases}$$

It is this peaked correlation function that is the prime characteristic of interest for the application of such sequences. If the ± 1 values modulated pulses are of height $+1$ and width one time unit, the "classic" picture of such an autocorrelation function is shown in Figure 14.2. It is what makes them of importance in applications.

As noted, m-sequences are also referred to as pseudonoise or pn sequences. To explain this terminology, *Golomb's sequence randomness postulates* are noted: A sequence is referred to as a pn-sequence if

(i) the number of 1's in a period differs by at most 1 from the number of 0's;
(ii) in a complete period, the number of runs (of either 0's or 1's), half have length 1, one-fourth have length 2, one-eighth of length 3, etc. (although there will be only one run of 1's of length n and none of 0's of length n and one run of 0's of length $n - 1$ and none of 1's of this length);
(iii) the autocorrelation function of the sequence is two-valued (say k for off-peak and N for peak value where for the m-sequences we have seen $k = -1$).

It is argued that these are properties one might expect of a purely random binary sequence and since the m-sequences, generated deterministically, have such properties, they are often referred to as *pseudorandom* or pn-sequences. The notation m-sequences will be retained here.

For the multiuser application introduced below there will be interest not only in autocorrelations of m-sequences noted above but also cross-correlations of distinct m-sequences (of the same length). The above result indicates that for such sequences it is sufficient to consider decimations of sequences. Suppose x_t and y_t are $\{0, 1\}$ m-sequences of length $N = 2^n - 1$. The periodic cross-correlation of the corresponding $\{\pm 1\}$ sequences will be defined as

$$\theta_{\boldsymbol{x},\boldsymbol{y}}(\tau) = \sum_{t=0}^{N-1} (-1)^{x_t}(-1)^{y_{t+\tau}}$$

and each sequence has the same norm.

Specifically it will be of interest to generate large sets of sequences with the property that for any two sequences in the set, $\boldsymbol{x},\boldsymbol{y},\boldsymbol{x} \neq \boldsymbol{y}$ the largest cross-correlation (since for an m-sequence $\theta_a = -1$)

$$c_{\max} = \max_{\boldsymbol{x} \neq \boldsymbol{y}} |\theta_{\boldsymbol{x},\boldsymbol{y}}(\tau)| \ , \quad \tau = 0, 1, \ldots, N-1$$

is as small as possible. A natural set to consider is the set of all m-sequences of a given length. This turns out not to be useful as the results are poor. A table of such cross-correlations is given in [10] and, e.g., for binary $\{\pm 1\}$ sequences of length $2^7 - 1$, the number of such sequences is $\phi(2^7 - 2)/7 = 18$ and for this set $c_{\max} = 41$.

To examine the trade-off possible between $c_{\max}$, the number of sequences and their length, an influential bound due to Welch [11] is useful. It is developed here, using his original argument with slightly changed notation.

Let C be a set of M sequences of length N over the complex numbers of norm unity. Notice that this includes the alphabet $\left\{ \pm \frac{1}{\sqrt{N}} \right\}$.

$$C = \{\boldsymbol{x}^\nu, \boldsymbol{x}^\lambda, \ldots\}, \ \ |C| = M, \ \boldsymbol{x}^\nu = \left(x_1^\nu, \ldots, x_N^\nu\right) \in \mathbb{C}^N,$$
$$\nu = 1, 2, \ldots, M, \ \textstyle\sum_{i=1}^N x_i^\nu \bar{x}_i^\nu = \sum_{i=1}^N |x_i|^2 = 1,$$

where the bar indicates complex conjugation. The sequences are complex with the norm of each sequence unity. Although this approach seems restrictive for the kinds of binary sequences of interest, it will in fact yield useful bounds. Developing bounds on inner products of these sequences will yield bounds on auto- and cross-correlations of interest.

For the set C of M complex sequences of length N and norm unity, the inner product between two sequences $\boldsymbol{x}^\nu$ and $\boldsymbol{x}^\lambda$ is

$$c_{\nu,\lambda} = (\boldsymbol{x}^\nu, \bar{\boldsymbol{x}}^\lambda) = \sum_{i=1}^N x_i^\nu \bar{x}_i^\lambda, \quad \boldsymbol{x}^\nu, \boldsymbol{x}^\lambda \in C$$

and the maximum of all distinct inner products as

$$c_{\max} = \max_{\nu \neq \lambda} |c_{\nu,\lambda}| \tag{14.5}$$

The following theorem, the Welch bound, that relates $c_{\max}$ to the sequence parameters is of interest:

Theorem 14.2 ([11]) *Let k be a positive integer. Then for any set of M complex sequences of length N and norm unity,*

$$(c_{\max})^{2k} \geq \frac{1}{M-1}\left[\frac{M}{\binom{N+k-1}{k}} - 1\right].$$

Proof: Consider an $M \times N$ matrix $\boldsymbol{M}$ whose rows are the M complex sequences and the $M \times M$ matrix $\boldsymbol{M}\bar{\boldsymbol{M}}^t = (m_{\nu,\lambda})$ where $m_{\nu,\lambda} = c_{\nu,\lambda}$. Denote by B_k the sum of all $2k$ powers of the magnitude of elements of this matrix. Thus

$$B_k = \sum_{\nu,\lambda} |c_{\nu,\lambda}|^{2k} \leq M(M-1)c_{\max}^{2k} + M.$$

An estimate for B_k will be derived that yields a relationship (bound) between $c_{\max}$ and M and N. Clearly

$$\begin{aligned} B_k &= \sum_{\nu,\lambda=1,2,\ldots,M} (c_{\nu,\lambda})^k (\bar{c}_{\nu,\lambda})^k \\ &= \sum_{\nu,\lambda=1,2,\ldots,M} \left(x_1^\nu \bar{x}_1^\lambda + \cdots + x_N^\nu \bar{x}_N^\lambda\right) \cdots \left(x_1^\nu \bar{x}_1^\lambda + \cdots + x_N^\nu \bar{x}_N^\lambda\right) \cdots \\ &\qquad \left(\bar{x}_1^\nu x_1^\lambda + \cdots + \bar{x}_N^\nu x_N^\lambda\right) \cdots \left(\bar{x}_1^\nu x_1^\lambda + \cdots + \bar{x}_N^\nu x_N^\lambda\right), \end{aligned}$$

a summation over the product of $2k$ brackets, each bracket a sum of N terms. In expanding the brackets one term is chosen from each bracket to yield:

$$B_k = \sum_{\nu,\lambda=1,2,\ldots,M} \sum_{\substack{s_1,\ldots,s_k=1 \\ t_1,\ldots,t_k=1}}^{N} \prod_{i=1}^{k} x_{s_i}^\nu \bar{x}_{t_i}^\nu \bar{x}_{s_i}^\lambda x_{t_i}^\lambda$$

and by interchanging the order of the summations it is verified that

$$B_k = \sum_{\substack{s_1,\ldots,s_k=1 \\ t_1,\ldots,t_k=1}}^{N} \left| \sum_{\nu=1}^{M} \prod_{i=1}^{k} x_{s_i}^\nu \bar{x}_{t_i}^\nu \right|^2.$$

As with the usual binomial expansion, many terms in the sum will be the same. The extension to this multinomial case is straightforward. Let

$$\boldsymbol{b} = (b_1, b_2, \ldots, b_n),\ b_i \in \mathbb{Z}_{\geq 0},\ \sum_{i=1}^{n} b_i = k$$

$$\boldsymbol{d} = (d_1, d_2, \ldots, d_n),\ b_i \in \mathbb{Z}_{\geq 0},\ \sum_{i=1}^{n} d_i = k$$

$$\binom{k}{\boldsymbol{b}} = \frac{k}{\prod_i (b_i!)} \quad \text{and} \quad \binom{k}{\boldsymbol{d}} = \frac{k}{\prod_i (d_i!)}$$

and gathering like terms in the previous expression for B_k yields

$$B_k = \sum_{\boldsymbol{b},\boldsymbol{d}} \binom{k}{\boldsymbol{b}} \binom{k}{\boldsymbol{d}} \left| \sum_{\nu=1}^{M} \prod_{i=1}^{N} (x_i^{\nu})^{b_i} (\bar{x}_{t_i}^{\nu})^{d_i} \right|^2 .$$

An inequality is created by dropping (all terms positive) terms for which $\boldsymbol{b} \neq \boldsymbol{d}$ to yield

$$B_k \geq \sum_{\boldsymbol{b}} \left[\binom{k}{\boldsymbol{b}} \sum_{\nu=1}^{M} \prod_{i=1}^{N} |x_i^{\nu}|^{2b_i} \right]^2 .$$

Recall the Cauchy–Schwarz inequality over real numbers as

$$\left(\sum_i v_i w_i \right)^2 \leq \left(\sum_i v_i^2 \right) \left(\sum_i w_i^2 \right)$$

and identifying v_i with 1 and w_i as a term in the sum over the $\boldsymbol{b}$ summation yields

$$B_k \geq \left[\sum_{\boldsymbol{b}} \binom{k}{\boldsymbol{b}} \sum_{\nu=1}^{M} \prod_{i=1}^{N} |x_i^{\nu}|^{2b_i} \right]^2 \Big/ \sum_{\boldsymbol{b}} 1^2 = \left[\sum_{\boldsymbol{b}} \binom{k}{\boldsymbol{b}} \sum_{\nu=1}^{M} \prod_{i=1}^{N} |x_i^{\nu}|^{2b_i} \right]^2 \Big/ \binom{N+k-1}{k}$$

where $\binom{N+k-1}{k}$ is the number of possible vectors $\boldsymbol{b}$, i.e., the number of partitions of the integer N into k parts, each part between 0 and k (Equation 1.10 of Chapter 1), and interchanging the order of the summations and recalling all sequences have a norm of unity gives

$$B_k \geq \frac{\left(\sum_{\nu=1}^{M} \left(\sum_{i=1}^{N} |x_i^{\nu}|^2 \right)^k \right)^2}{\binom{N+k-1}{k}} = \frac{M^2}{\binom{N+k-1}{k}}$$

and from the first equation of the proof

$$M(M-1)c_{\max}^{2k} + M \geq B_k \geq \frac{M^2}{\binom{N+k-1}{k}}$$

from which the theorem statement follows. ■

The above bound refers to correlations of a set of M unit-length complex sequences of length N. To examine the off-peak correlations and cross-correlations we can add all cyclic shifts of all sequences to the list and use the same bound by replacing M by MN. Thus for this case the bound reads (without changing the $c_{\max}$)

$$(c_{\max})^{2k} \geq \frac{1}{MN-1} \left[\frac{MN}{\binom{N+k-1}{k}} - 1 \right]. \tag{14.6}$$

For a set of M distinct m-sequences of length $N = 2^n - 1$, $c_{\max}$ is the max-min cross-correlation. Then the above expression with $k = 1$ gives a lower bound on this value as

$$c_{\max} \geq \sqrt{\frac{M-1}{MN-1}}$$

which for large M and N will be approximately $1/\sqrt{N}$. Thus the smallest maximum cross-correlation any set of M m-sequences of length $N = 2^n - 1$ can have is approximately $\sqrt{N} \approx 2^{-n/2}$.

Numerous properties of m-sequences and their correlations are developed in the seminal paper [10] and a few of these are noted. For $\boldsymbol{x}, \boldsymbol{y}$ two distinct ± 1 m-sequences of length $2^n - 1$, the cross-correlation $\theta_{\boldsymbol{x},\boldsymbol{y}}(\ell)$ is always an odd integer and indeed $\theta_{\boldsymbol{x},\boldsymbol{y}}(\ell) + 1$ is divisible by 8 unless the sequences are generated by reciprocal polynomials. Note that these sequences have length N and the Welch bound Equation 14.6 must be adjusted for this case. Also it is true that

$$\sum_{\ell=0}^{2^n-2} \theta_{\boldsymbol{x},\boldsymbol{y}}(\ell) = +1$$

which implies that the average value of the cross-correlation over a period is nearly 0 and that the cross-correlations take on negative as well as positive values. In addition

$$\begin{aligned}\sum_{\ell=0}^{2^n-2} \mid \theta_{\boldsymbol{x},\boldsymbol{y}}(\ell) \mid^2 &= (2^n - 1)^2 + (2^n - 1) + 1 \\ &= 2^{2n} - 2^n + 1.\end{aligned}$$

As noted previously, the peaked autocorrelation function of an m-sequence as in Figure 14.2 is often described as an "ideal" correlation function. The view of this figure is that of one period of an m-sequence correlating with infinite translations of a version of the same m-sequence. For cross-correlation of distinct m-sequences, the picture is more complex. In particular, if $\boldsymbol{x}$ and $\boldsymbol{y}$ are m-sequences generated by distinct primitive polynomials of degree m, then the cross-correlation function $\theta_{\boldsymbol{x},\boldsymbol{y}}(\cdot)$ must take on at least three values (shown by Golomb). Furthermore, as shown in the previous section, any two m-sequences are related through a d-decimation, i.e., one is the d-decimation of the other. Then the cross-correlation function of this pair of m-sequences is the same as for any other pair of m-sequences related by d-decimation.

Certain sets of m-sequences with good correlation properties have been proposed in the literature and the following is of interest.

Theorem 14.3 ([10], theorem 1) *Let* $\mathbf{x}$ *and* $\mathbf{y}$ *denote two m-sequences of length* $2^n - 1$ *with* $\mathbf{y}$ *the d-decimated sequence of* $\mathbf{x}$ *where* $d = 2^k + 1$ *or* $d =$

$2^{2k} - 2^k + 1$. *If* $e = (n,k)$ *is such that* n/e *is odd, then the cross-correlation* $\theta_{\mathbf{x},\mathbf{y}}$ *is three-valued with the values with frequencies of values:*

$$\begin{array}{ll} -1 + 2^{(n+e)/2} & \textit{occurs } 2^{n-e-1} + 2^{(n-e-2)/2} \textit{ times} \\ -1 & \textit{occurs } 2^n - 2^{n-e} - 1 \textit{ times} \\ -1 - 2^{(n+e)/2} & \textit{occurs } 2^{n-e-1} - 2^{(n-e-2)/2} \textit{ times.} \end{array}$$

As a consequence of this result, it is useful to define the function

$$t(n) = 1 + 2^{\lfloor (n+2)/2 \rfloor}. \tag{14.7}$$

Then it can be shown that for $n \not\equiv 0 \pmod 4$ there exists pairs of m-sequences with three (nontrivial) cross-correlation values $-1, -t(n)$ and $t(n) - 2$. Any such pair with such a three-valued cross-correlation is called a *preferred pair of m-sequences* and the cross-correlation function a *preferred three-valued cross-correlation function.*

It was noted earlier in many applications of m-sequences it is of interest to have large sets of them with as low a cross-correlation between any two sequences as possible. It was also noted that subsets of the set of all $\phi(2^n-1)/n$ distinct m-sequences of length $2^n - 1$ have relatively poor cross-correlation properties. The literature on construction techniques for such sequence sets is large. The constructions due to Gold [1, 2] and Kasami [7], now viewed as the "classical" constructions, are discussed in the next section using the treatment in [10] and the original works. The description of such sequences tends to be complex and only an overview is given.

14.3 Gold and Kasami Sequences

To construct sets of sequences with small peak cross-correlations it is natural to consider cyclic codes since all cyclic shifts of the sequences are in the code and the correlation properties are easily derived from the distance structure of the code. Only an outline of the code constructions is given. Thus if $\boldsymbol{a}, \boldsymbol{b} \in \{0,1\}^N$ and $\boldsymbol{x}, \boldsymbol{y}$ are the equivalent vectors in $\{\pm 1\}^n$ via $x_i = (-1)^{a_i}$, then

$$(\boldsymbol{x}, \boldsymbol{y}) = N - 2w(\boldsymbol{a} \oplus \boldsymbol{b})$$

where $w(\cdot)$ is Hamming weight. Thus from the code distance structure the set of cyclically distinct sequences can be determined as well as the maximum cross-correlation of them.

The following construction appears to have been independently discovered by Gold [1, 2] and Kasami [7]. They constructed a three nonzero weight binary cyclic code of length $N = 2^n - 1$ and dimension $2n$ with generator polynomial

$f_1(x)f_\ell(x)$ where $f_1(x)$ is a primitive polynomial of degree n over $\mathbb{F}_2$ with primitive root $\alpha \in \mathbb{F}_{2^n}$ and $f_\ell(x)$ is an irreducible polynomial with root $\alpha^{2^\ell+1}$, which is primitive provided m is odd and $(\ell, n) = 1$. The resulting binary cyclic code has [2] the following weight distribution:

codeword weight w	number of codewords A_w
2^{n-1}	$(2^n - 1)(2^{n-1} + 1)$
$2^{n-1} + 2^{(n-1)/2}$	$(2^n - 1)(2^{n-2} - 2^{(n-3)/2})$
$2^{n-1} - 2^{(n-1)/2}$	$(2^n - 1)(2^{n-2} + 2^{(n-3)/2})$

The factor $(2^n - 1)$ in the number of codewords refers to the $2^n - 1$ possible cyclic shifts of each distinct codeword. Equivalently [2] the three possible values of the cross-correlations from these codeword weights are

$$\theta_{\boldsymbol{x},\boldsymbol{y}}(\ell) = \begin{cases} 2^n - 1 - 2 \cdot 2^{n-1} = -1 \\ 2^n - 1 - 2 \cdot \left(2^{n-1} \pm 2^{(n-1)/2}\right) \\ -\left(2^{(n+1)/2} + 1\right) \quad \text{or} \quad + \left(2^{(n+1)/2} - 1\right). \end{cases}$$

Taking one sequence from each cyclic equivalency class then gives an alternative way of describing the set of these $2^n + 1$ Gold sequences [2, 6, 10] as:

$$x_t = \mathrm{Tr}_{q^n|q}\left\{a\alpha^t + \alpha^{dt}\right\} \cup \left\{\mathrm{Tr}_{q^n|q}(\alpha^t)\right\}, \; d = 2^\ell + 1, \quad a \in \mathbb{F}_{2^n}. \qquad (14.8)$$

A useful alternative way of describing these Gold sequences [10] is by using the shift operator on vectors (sequences). Thus for $\boldsymbol{x} = (x_0, x_1, \ldots, x_{n-1})$ the operator T is such that $T\boldsymbol{x} = (x_1, x_2, \ldots, x_{n-1}, x_0)$, one cyclic shift to the left. Using this notation, a slightly different definition of Gold sequences is given in [10]. Given the influence of that work, that approach is outlined.

In general let $f(x) = h_1(x)h_2(x)$ be binary polynomials (primitive) with no factors in common generating sequences of periods $2^n - 1$. The set of all sequences with characteristic polynomial $f(x)$ is the set of all binary sequences length $2^n - 1$ of the form

$$\boldsymbol{x} \oplus \boldsymbol{y}$$

where $\boldsymbol{x}$ is generated by $h_1(x)$ and $\boldsymbol{y}$ by $h_2(x)$. Denote by $G(\boldsymbol{x},\boldsymbol{y})$ the set of $2^n + 1$ sequences of length $2^n - 1$

$$G(\boldsymbol{x},\boldsymbol{y}) \triangleq \left\{\boldsymbol{x}, \boldsymbol{y}, \boldsymbol{x} \oplus \boldsymbol{y}, u \oplus T\boldsymbol{y}, \boldsymbol{x} \oplus T^2\boldsymbol{y}, \ldots, \boldsymbol{x} \oplus T^{2^n-2}\boldsymbol{y}\right\}.$$

The above discussion leads to the below definition:

Definition 14.4 Let $\boldsymbol{x},\boldsymbol{y}$ be a preferred pair of m-sequences of period $2^n - 1$ generated by the primitive polynomials $h_1(x)$ and $h_2(x)$. Then the set $G(\boldsymbol{x},\boldsymbol{y})$

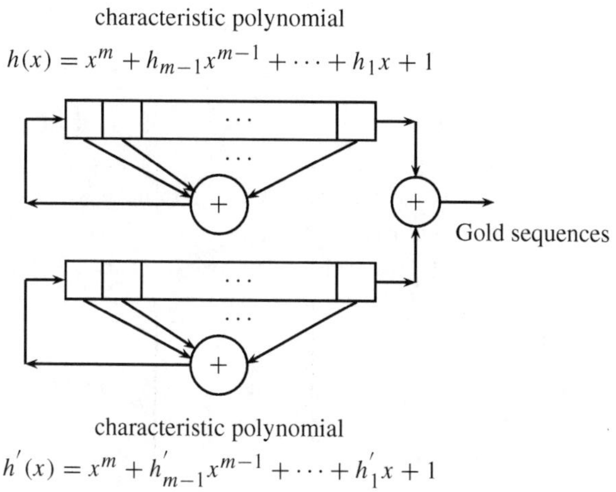

Figure 14.3 Gold sequence set generator, $(h(x), h'(x))$ a preferred pair of m-sequences

of $2^n + 1$ sequences is a set of Gold sequences. For any two distinct sequences $\boldsymbol{u}, \boldsymbol{v} \in G(\boldsymbol{x}, \boldsymbol{y})$ the cross-correlation value is in the set $\{-1, -t(n), t(n) - 2\}$ where $t(n)$ is as given in Equation 14.7. It is noted the nonmaximal-length sequences in the set can be obtained by adding together the outputs of shift registers with characteristic polynomials $h_1(x)$ and $h_2(x)$, respectively. Note also that since each of the two registers could be in the all-zero state, the individual m-sequences generated by $h_1(x)$ and $h_2(x)$ are also in the set.

A generator for Gold sequences is shown in Figure 14.3. To generate the full set, (exactly) one of the generators may be in the all-zero state.

The work of Kasami [7] presented two classes of sequences with low auto- and cross-correlation, the small and large classes of Kasami sequences. Brief descriptions of these classes are given. Let n be an even integer and note that $2^n - 1 = (2^{n/2} - 1)(2^{n/2} + 1)$. Let α be a primitive element of $\mathbb{F}_{2^n}$, a root of the primitive polynomial $h(x)$ of degree n over $\mathbb{F}_2$.

Let $\boldsymbol{u}$ be the m-sequence associated with $h(x)$ and $\boldsymbol{v}$ the decimated sequence $\boldsymbol{v} = \boldsymbol{u}[d]$ where $d = 2^{n/2} + 1$. Thus the sequence $\boldsymbol{v}$ is of length $2^n - 1$ and period $2^{n/2} - 1$.

Let $s(n) = 2^{n/2} + 1 = d$ and $h'(x)$ the polynomial of degree $n/2$ whose roots are the $s(n) = d$ powers of the roots of $h(x)$. Thus the roots of $h'(x)$ are

$$\alpha^d, \alpha^{2d}, \ldots, \alpha^{\alpha^{(2^{n/2}-1)d}} \quad \text{since} \quad \alpha^{2^n} = \alpha$$

and note that

$$\alpha^{2^{n/2}d} = \alpha^{2^{n/2}(2^{n/2}+1)} = \alpha^{2^n+2^{n/2}} = \alpha^{1+2^{n/2}} = \alpha^d$$

and the set of elements in the previous equation is in fact a conjugacy class – the roots of the minimal polynomial over $\mathbb{F}_2$ of α^d.

As noted, the sequence $\boldsymbol{v}$ is a sequence of period $2^{n/2}-1$ and length 2^n-1. Consider sequences generated by the polynomial $h(x)h^{'}(x)$ of degree $3n/2$. Such sequences are of the form [10]

$$K_s \triangleq \left\{\boldsymbol{u}, \boldsymbol{u}\oplus \boldsymbol{v}, \boldsymbol{u}\oplus T\boldsymbol{v}, \ldots, \boldsymbol{u}\oplus T^{2^{n/2}-2}\right\}, \quad |K_s| = 2^{n/2} \tag{14.9}$$

which is the small set of Kasami sequences [7], a set of $2^{n/2}$ sequences of length and period 2^n-1. The correlation functions of these sequences can be shown to belong to the set

$$\{-1, -s(n), s(n)-2\} \quad \text{hence} \quad \theta_{\max} = s(n) = 1+2^{n/2}.$$

Recall the Welch bound was derived for sets of complex sequences of norm unity. Noting that $\theta_{\max}$ is odd, the Welch bound can be improved for binary sequences [10] into

$$\theta_{\max} \geq 1+2^{n/2}$$

and hence the small set of Kasami sequences is an optimal set with respect to this bound.

As before the alternative description of the small set of Kasami sequences [6] can be, for α a primitive element of $\mathbb{F}_{2^n}$, taken as

$$\left\{\mathrm{Tr}_{2^n|2}(\alpha^t) + \mathrm{Tr}_{2^{n/2}|2}(b\alpha^{dt}) \mid d = 2^{n/2}+1,\ b\in \mathbb{F}_{2^{n/2}}\right\}.$$

The definition of the large set of Kasami sequences is more involved to describe although uses concepts already introduced. Briefly, let $h(x)$ be a primitive polynomial of degree n that generates the m-sequence $\boldsymbol{u}$ of length and period 2^n-1. Let $\boldsymbol{v}$ be the decimated sequence $\boldsymbol{v} = \boldsymbol{u}[s(n)]$ where, as before $s(n) = 1+2^{n/2}$, a sequence of period $2^{n/2}-1$ (an m-sequence) by the polynomial $h^{'}(x)$, a polynomial of degree $n/2$, and $\boldsymbol{w}$ the decimated sequence $\boldsymbol{w} = \boldsymbol{u}[t(n)]$ where $t(n) = 1+2^{\lfloor (n+1)/2\rfloor}$, generated by the polynomial $h^{''}(x)$ of degree n, a sequence of period 2^n-1. The large set of Kasami sequences then is generated by the polynomial $h(x)h^{'}(x)h^{''}(x)$ and is a set of sequences of size

$$\begin{array}{ll} 2^{3n/2}+2^{n/2} & \text{if} \quad n = 2 \pmod 4 \\ 2^{3n/2}+2^{n/2}-1 & \text{if} \quad n = 0 \pmod 4 \end{array}$$

and the sequences take correlations in the set $\{-1, -t(n), t(n)-2, -s(n), s(n)-2\}$ with $\Theta_{\max} = t(n)$.

Numerous other sets of sequences appear in the literature including the basic references [6, 10]. The volume [4] is an excellent volume on this topic as well as the references therein.

Comments

The founder of this area of sequence design was Solomon Golomb of USC, author of the influential volume [3] which generations of communications researchers were familiar with. The follow-on volume [4] has also proved a valued contribution as is the more recent [5]. The beautiful little volume [8] has an elegant and useful approach to the description and properties of m-sequences (as well as finite field theory) and was a valuable reference for this chapter. There are numerous survey articles on the general topic of sequence design for low correlation. The one by Tor Helleseth [6] is a readable and comprehensive treatment. The research article [10] contains a wealth of information on such sequences not found elsewhere.

References

[1] Gold, R. 1967. Optimal binary sequences for spread spectrum multiplexing (Corresp.). *IEEE Trans. Inform. Theory*, **13**(4), 619–621.

[2] Gold, R. 1968. Maximal recursive sequences with 3-valued recursive cross-correlation functions (Corresp.). *IEEE Trans. Inform. Theory*, **14**(1), 154–156.

[3] Golomb, S.W. 1967. *Shift register sequences*. Holden-Day, San Francisco, CA.

[4] Golomb, S.W., and Gong, G. 2005. *Signal design for good correlation*. Cambridge University Press, Cambridge.

[5] Goresky, M., and Klapper, A. 2012. *Algebraic shift register sequences*. Cambridge University Press, Cambridge.

[6] Helleseth, T., and Kumar, P.V. 1998. Sequences with low correlation. Pages 1765–1853 of: *Handbook of coding theory, Vol. I, II*. North-Holland, Amsterdam.

[7] Kasami, T. 1966. Weight distribution formula for some class of cyclic codes. *University of Illinois, CSL Technical Report R 285*.

[8] McEliece, R.J. 1987. *Finite fields for computer scientists and engineers*. The Kluwer International Series in Engineering and Computer Science, vol. 23. Kluwer Academic Publishers, Boston, MA.

[9] Sarwate, D. 1979. Bounds on cross-correlation and autocorrelation of sequences (Corresp.). *IEEE Trans. Inform. Theory*, **25**(6), 720–724.

[10] Sarwate, D.V., and Pursley, M.B. 1980. Cross-correlation properties of pseudo-random and related sequences. *Proc. IEEE*, **68**(5), 593–619.

[11] Welch, L. 1974. Lower bounds on the maximum cross-correlation of signals (Corresp.). *IEEE Trans. Inform. Theory*, **20**(3), 397–399.

15
Postquantum Cryptography

Interest in this chapter is in giving some background on the history of classical public-key cryptography and the influence the advent of quantum computing has had on it, leading to the present situation of postquantum public-key cryptography [16]. In particular an attempt is made to understand how the quantum computation model is able to reduce certain problems with exponential complexity on a classical computer to polynomial complexity on a quantum computer. Although it does not involve coding, it does set the stage for the next chapter which overviews the notions of quantum error-correcting codes.

Public-key cryptography arose in the 1970s with seminal papers of Rivest, Shamir and Adelman [26] and Diffie and Hellman [6] based on the assumed difficulty of factoring integers and of finding logarithms in certain finite fields, respectively. Importantly, it led to the notion of asymmetric-key cryptography, cryptosystems with different enciphering and deciphering keys. Such systems led to the notions of secure exchanges of keys over a public network and of digital signatures. With a digital signature, a document can be "signed" by attaching several bytes which give assurance that only the assumed sender could have sent the message.

It will be useful to note the role of the National Security Agency (NSA) and the National Institute of Standards and Technology (NIST) in this work. The NSA, whose budget and workforce size are officially secrets (although estimates of both are found on the Internet), has a wide mandate in intelligence and counterintelligence and security and in the protection of US communication networks and information systems. It was noted on a Wikipedia page the NSA is likely the world's largest employer of mathematicians. NIST has developed a suite of protocols which have proved invaluable to corporate users and individuals involved with communications security around the world. These are referred to as Federal Information Processing Standards (FIPS) and cover several areas (not only security). The standards for a particular cryptographic function are often initiated by an open call to academic and corporate cryptographic groups around the world to propose algorithms which are gathered on the NIST website. A period of evaluation and discussion follows by which finalists are decided upon. The NSA lends their expertise, which is considerable, to this process and is involved in the final selection and suggested parameter selection.

Lurking in the background of public-key cryptography are two associated algorithms, those of symmetric-key encryption algorithms and of hash functions which are used by virtually all such algorithms. A symmetric-key algorithm uses a key, usually several bytes long, and operates on blocks of data of a standard size to produce encrypted blocks of data, usually of the same size. These are very efficient algorithms capable of high data speeds. They are decrypted at a receiver with the same key. The symmetric-key system in almost universal usage is the *Advanced Encryption Standard* (AES) which uses key sizes and data block sizes of between 128 and 256 bits. The FIPS Pub 197, 2001 concerning AES gives a detailed description of the algorithm as well as a set of configurations or modes in which the algorithm can be used. Typically public-key systems are computationally expensive so are used only to establish a common key or signature while a symmetric-key encryption algorithm such as AES is used for the actual encryption once the common key is established.

Another workhorse of the cryptographic world is the hash function, the *Secure Hash Algorithm-2* (SHA-2), which supports a variety of block sizes. The notion of a hash function is to produce a digest of a long message. The digest might be a word of 256 bits, such that if any bit or bits of the message it operates on – which may be several gigabytes – are changed, many bits of the digest (hash) will be changed. When attached to a message, a receiver may produce a hash value from the received message and compare it to the received hash to justify the message was not altered in transmission. They also play a central role in digital signatures. A variety of word and block sizes for the hash function standard are specified in FIPS Pub 180-2. There is also SHA-3.

The reader is reminded that confidence in the security of most cryptographic systems is only established after a large number of capable cryptanalysts with enough resources have tried breaking the systems for a sufficient length of time. Even then there is always the possibility some researcher with a clever new direction of attack may render the system useless.

The advent of quantum computers has had a profound impact on cryptographic practice. Eminent physicists in the 1980s considered that quantum computers might offer computation models that render them capable of providing results not achievable by a classical computer. However, it was the work of Shor [27] who devised algorithms that could run on a quantum computer that reduced classical cryptographic algorithms useless. The capabilities of such a computer were assumed, based on physical models, since no such computer actually existed at that time. This led the NSA to almost immediately recommend all future development effort be on algorithms that resist such a possibility which has led to postquantum cryptography. The term refers to cryptographic algorithms that appear not to run significantly faster on a quantum computer than on a classical one. A suite of such postquantum algorithms has evolved over the past few years and is discussed in Section 15.3. As with classical algorithms there seems no guarantee that a quantum algorithm could not be found for such postquantum or quantum-resistant algorithms to render them susceptible, given enough effort and time.

The next section gives a brief overview of the classical public-key algorithms that were rendered suspect by Shor's algorithms, serving as a reminder of the cryptographic protocols of interest. Section 15.2 attempts to describe the quantum computation model that allowed Shor to devise quantum algorithms for the problems of integer factoring and discrete logarithms in a finite field. These quantum algorithms reduced the complexity of these problems from subexponential on a classical computer to polynomial on a quantum computer. The final Section 15.3 describes a few of the algorithms that are currently viewed as quantum-resistant, i.e., no quantum algorithms are known for them that would allow them to run faster on a quantum computer than on a classical one. They are regarded as postquantum algorithms.

15.1 Classical Public-Key Cryptography

The section recalls basic elements of the role of discrete logarithms and integer factorization in certain public-key cryptographic protocols. This serves as a precursor to the pivotal work of Shor in showing how their complexity under quantum computation is reduced from subexponential to polynomial – and hence useless from a cryptographic point of view. In general, public-key cryptography provides a means for key exchange, digital signatures and encryption, although the number of ingenious other cryptographic protocols that have arisen out of these ideas is almost without limit.

Consider first the basic Diffie–Hellman key exchange algorithm [6] containing perhaps the first idea of what was possible for communicating in secret over a public network. Let $\mathbb{F}_p$ be a prime field, integers modulo a large prime p, and let α be a generator of a large cyclic subgroup $G = \langle \alpha \rangle \subset \mathbb{F}_p^*$ of order q for some prime q. Thus $q \mid (p-1)$. Let $\beta = \alpha^a$, for some positive integer a, a quantity that is efficient to compute using, say, a square and multiply technique. The discrete logarithm of β to the base α then is $\log_\alpha(\beta) = a$. Given p, α and β then, the problem is to determine the integer a, the so-called discrete logarithm problem (DLP). The thesis is that finding such discrete logarithms is much harder than exponentiation, as will be commented on later.

Assuming the DLP is hard, what can this be used for? Diffie and Hellman [6] suggested the following groundbreaking idea. Suppose Alice and Bob wish to communicate in secret over a public network with no prior communications. They publicly share a modulus (prime) p and large subgroup generator α and its order q. Alice chooses a secret exponent a and sends α^a to Bob. Bob chooses a secret exponent b and sends α^b to Alice. They each compute α^{ab} from which the common key is derived. The common key might then be used for a symmetric-key algorithm such as AES. The security of the scheme rests on the presumed difficulty of the noted DLP which will be commented on shortly. Thus an eavesdropper, seeing α^a and α^b in the transmissions, is unable to compute either a, b or α^{ab} (modulo p).

For $n \geq 2$ an arbitrary positive integer, $\mathbb{Z}_n^*$ is the set of integers in $[n]$ relatively prime to n and $\phi(n) = |\mathbb{Z}_n^*|$, $\phi(n)$ the Euler's totient function. The set $\mathbb{Z}_n^*$ is a multiplicative group mod (n). The RSA cryptographic system relies on Euler's theorem which states that for $a \in \mathbb{Z}_n^*$

$$a^{\phi(n)} \equiv 1 \pmod{n}.$$

To introduce the RSA public-key cryptosystem (standing for its inventors Rivest, Shamir and Adleman) and the notion of a digital signature, let $n = pq$ be the product of two large primes (chosen carefully with certain properties), and $\phi(n) = (p-1)(q-1)$. Notice that knowing n is a product of two primes and the value of $\phi(n)$ allows one to factor n. An encryption exponent e is chosen so that $(e, \phi(n)) = 1$, i.e., relatively prime to $\phi(n)$, which allows a unique decryption exponent d to be chosen so that $ed \equiv 1 \mod \phi(n)$.

RSA encryption is as follows. For user A, the integers $n_A = p_A q_A$, the product of two primes, and e_A are chosen by user A and made public (for each system user). Here e_A is the user A public key, while the corresponding decryption exponent, d_A, computed by user A, knowing the factorization of n_A, is kept secret. For Bob to send an encrypted message to Alice he retrieves her public exponent, say n_A and e_A and for the message m, assumed to be an integer in the range $(0, n_A - 1)$ sends Alice $r = m^{e_A} \mod n_A$. On receiving this encryption, Alice computes $m \equiv r^{d_A} \equiv m^{e_A d_A} \equiv m \mod n_A$ thus retrieving the message.

Note that if the integer n_A can be factored, then it is possible to determine $\phi(n_A)$ and hence determine d_A from e_A and the system is insecure. Thus the security of the system rests on the difficulty of factoring large integers.

To recall the notion of a digital signature for a message m, assume that $m \in [1, n-1]$. In practice one would either block the message into suitable subblocks or employ some publicly known redundancy function with the appropriate properties. For Bob to sign the message m to Alice he uses his secret key d_B and computes the signature

$$s \equiv m^{d_B} \mod n_B$$

and sends the pair (m, s) to Alice. On reception Alice retrieves Bob's public key e_B, n_B and computes

$$s^{e_B} \equiv m^{d_B e_B} \equiv m \mod n_B$$

which is compared with the received message m and accepted if they match. In practice of course numerous details are to be accounted for but the essence of the scheme is as above. Details of the system are to be found in the encyclopedic reference [19]. The essential notion of a digital signature is that only Bob could have produced the signature s for the message m, since only he knows his secret key d_B. However, any user, given the message m, s, n_B and the public key e_B can verify it.

It is clear that the DLP can be implemented in any finite group, and of course, in an additive or multiplicative group. A particularly successful group in terms of its complexity characteristics has been the additive group of an elliptic curve. Although somewhat complex to discuss, a brief introduction to this aspect of cryptography is given. An elliptic curve (EC) over a field of characteristic 2, $q = 2^n$, has a representative form [4]

$$E(\mathbb{F}_q):\ y^2 + xy = x^3 + a_2x^2 + a_6, \quad a_2 \in \{0, \gamma\},\ a_6 \in \mathbb{F}_q^*,\ q = 2^n,$$

where γ is a fixed element of trace unity over $\mathbb{F}_2$, i.e., $\mathrm{Tr}_{q|2}(\gamma) = 1$. A point $P = (x, y)$ is on the curve if the coordinates satisfy the equation. The set of all points on the curve $E(\mathbb{F}_q)$ form an additive (and Abelian) group of size

$$|E(\mathbb{F}_q)| = q + 1 - t$$

where t is the *Frobenius trace* of the curve which satisfies $|t| \leq 2\sqrt{q}$. If $P_1 = (x_1, y_1)$ and $P_2 = (x_2, y_2)$ are on the curve, then from the form of the equation, by "drawing a straight line" through the points will intersect the curve in a unique third point $P_3 = (x_3, y_3)$. In a field of characteristic 2 the negative of a point $P_1(x_1, y_1)$ is $-P_1 = (x_1, x_1 + y_1)$ and the sum of a point and its negative gives "point at infinity," the identity of the group. Intuitive addition formulae can be given for the addition of two points on the curve [4] but are omitted.

Each point on the curve has an additive finite order. The DLP on elliptic curves (ECDLP) is then: given $P \in E(\mathbb{F}_q)$ and $Q = aP = P + P +$

$\cdots P$ $\{a$ times$\}$, find a. The surprising fact is that the ECDLP problem is a much harder problem on an elliptic curve than the DLP in a finite field of comparable size and this led to elliptic curve being widely adopted during the 1990s, especially for small devices. Elliptic curve algorithms for encryption, signatures and key exchange are available. The digital signature algorithm (DSA) in both prime fields and in fields of characteristic 2, as well as the elliptic curve DSA (ECDSA), is described in the digital signature standard (DSS) FIPS Pub 186-4 (2013).

To discuss the complexity of the above problems define the function [4, 19]

$$L_n(\alpha, c) = \exp\left(c(\log n)^{\alpha}(\log\log n)^{(1-\alpha)}\right)$$

and note that when $\alpha = 1$ the function is exponential in $\log n$ and when it is zero it is polynomial. For in-between values it is described as *subexponential* for intuitive reasons.

The fastest algorithms for both integer factorization and finding discrete logarithms are versions of the number field sieve. In a prime field they have a complexity of the form $L_n(1/3, c)$ for a small number c, typically less than 2, for n the size (number of bits) of the integer or prime. Recent work on the finite field DLP, however, found weaknesses (lower complexity) for certain cases. For the discrete logarithm a technique referred to as index calculus was found and explored extensively.

The ECDLP, however, had a complexity that is square root of the field of definition of the curve. The best-known algorithm for achieving this complexity is the somewhat naive so-called baby-step-giant-step algorithm (BSGS). Briefly, in this algorithm, for a base point P of order n on an elliptic curve, a table of size $\beta = \sqrt{n}$ is created containing $iP, i = 1, 2, \ldots, \beta$. For a given point Q whose log to the base P is required, the point $R = \beta P$ is successively added until it is a member of the table at which point its log can be calculated. Such an algorithm has square root complexity, i.e., if the order of the finite field of the elliptic curve is $q = 2^n$, the complexity of the BSGS algorithm is $O(2^{n/2})$.

Comparing the complexities for an RSA modulus (product of two primes) with 4096 bits, an ECDLP over a field with 313 bits gave approximately the same level of security, a consideration which led to the popularity of the elliptic curve cryptography for certain applications, although arithmetic on the elliptic curve is more complex than for integers.

As will be discussed in the next section, the discovery of a quantum algorithm for integer factorization and for discrete logarithms (which were shown to include elliptic curve versions) by Shor [27] led the NSA/NIST in 2015 to promote the use of postquantum algorithms (also referred to as quantum-resistant). It resulted in the relegation of the above "classical" algorithms to limited usage. In essence, the possibility of a quantum computer, led to the demise of the above classical public-key systems.

15.2 Quantum Computation

In order to appreciate the need for postquantum cryptography, some knowledge of the techniques of quantum computation is needed, that is, an appreciation of how quantum computers acquire their tremendous speedup for certain algorithms. It is difficult to delve into the intricacies of quantum mechanics to give a detailed description but perhaps an intuitive understanding of the situation might be possible. The next few pages attempt such an overview.

Before beginning it should be mentioned that the construction of quantum computers that achieve the theoretical gains predicted by the algorithms mentioned has been a daunting task. Many authors expressed skepticism that such computers with sufficient memory and sufficient stability to yield useful results on the problems posed will be possible. However, the expected gains are so tempting that numerous large corporations have invested huge resources to make such computers a reality and sufficient progress has been made that it now appears likely that such computers may well appear in the not too distant future.

A word on quantum algorithms is in order. It is not clear which computational problems can be formulated to derive an effective quantum algorithm. Indeed it is questioned in [29] why there are so few good quantum algorithms. In general it seems a difficult problem to pose such problems and generate suitable algorithms for quantum computation.

However, there have been some notable successes and perhaps first and foremost among these are the quantum integer factorization and quantum DLPs of Shor [27, 28]. These reduced their complexity from subexponential on a classical computer to polynomial (on input lengths) on a quantum computer and prompted their demise as trusted public-key algorithms. The quantum algorithms changed the practice of cryptography and data security techniques forever. It was also shown that even the elliptic curve DLP was included in this speedup.

The integer factorization problem used two subalgorithms: (i) that of finding the order of an element modulo a large integer N (i.e., for $x \in [N]$ find the least integer r such that $x^r \equiv 1 \pmod{N}$), and (ii) a quantum Fourier transform (QFT). While the entire integer factorization method of Shor has detail that requires lengthy development, the QFT is sufficiently simple with an impressive performance that an intuitive overview of it will be used for illustrative purposes here.

Among the other notable quantum algorithms that achieve speedup are the Simon algorithm, Grover's algorithm and the hidden subgroup problem. These problems are briefly described.

For Simon's problem we are given a function $f(\boldsymbol{x})\colon \{0,1\}^n \rightarrow \{0,1\}^n$ as a black box where given $\boldsymbol{x} \in \{0,1\}^n$ the box produces $f(\boldsymbol{x})$. We are also given the assurance that for all $\boldsymbol{x},\boldsymbol{y} \in \{0,1\}^n$ there exists an $i^* \in \{0,1\}^n$ such that **either** $f(\boldsymbol{x}) = f(\boldsymbol{y})$ or $f(\boldsymbol{x}) = f(\boldsymbol{y} \oplus i^*)$. The goal is to determine the string i^* by making queries to the black box. This problem is exponential in n on a classical computer and polynomial ($O(n^3)$ or less [11]) on a quantum computer.

Grover's algorithm is a search algorithm that is often cast as a database search algorithm. This is equivalent [30] to the formulation of being given a function $f(\boldsymbol{x})\colon \{0,1\}^n \rightarrow \{0,1\}$ and the assurance that for exactly one $\boldsymbol{x}_0$

$$f(\boldsymbol{x}) = \begin{cases} 1, & \text{if } \boldsymbol{x} = \boldsymbol{x}_0 \\ 0, & \text{if } \boldsymbol{x} \neq \boldsymbol{x}_0. \end{cases}$$

The problem is to determine $\boldsymbol{x}_0$. This problem is $O(2^n)$ on a classical computer and $O(\sqrt{2^n})$ on a quantum computer, a quadratic speedup.

The hidden subgroup problem [11] is: given a group G, a set X and subgroup $H \leq G$ and a function $f\colon G \rightarrow X$ that is constant on cosets of the subgroup H – but has different values on different cosets, determine the subgroup H or a set of generators for it. The quantum algorithm for the case the group G is Abelian is polynomial in $\log \mid G \mid$. A quantum algorithm for non-Abelian groups does not appear to be known. It is interesting [11, 23] that this problem has strong ties to numerous other standard problems of interest, including Shor's algorithms for discrete logarithms and integer factorization.

The remainder of the section contains a discussion of quantum computing and quantum gates with a view to obtaining an intuitive discussion of how they lead to the impressive speedups they do for certain algorithms. The QFT is given as an example of a speedup from exponential complexity on a classical computer to polynomial on a quantum computer in the log of the size of the transform.

In classical computing the fundamental unit is the bit that takes on value of either 0 or 1. In quantum computing the corresponding unit is the *qubit* which is viewed as a superposition of a bit, a vector in two-dimensional complex space of the form

$$|\phi\rangle = a|0\rangle + b|1\rangle, \quad |a|^2 + |b|^2 = 1, \quad a,b \in \mathbb{C}.$$

The notation $|\cdot\rangle$ is referred to as "ket." Notice that the notion of qubit, as a projective quantity, is invariant under scalar (over $\mathbb{C}$) multiplication. The quantities $|0\rangle$ and $|1\rangle$ are viewed as orthogonal two-dimensional (with dimensions labeled with $|0\rangle$ and $|1\rangle$) column vectors in the space, which form a basis of the space:

$$|0\rangle \equiv \begin{bmatrix}1\\0\end{bmatrix} \quad |1\rangle \equiv \begin{bmatrix}0\\1\end{bmatrix} \quad |\phi\rangle \equiv \begin{bmatrix}a\\b\end{bmatrix}.$$

If one measures the qubit the outcome is not deterministic but rather random with a probability of $|a|^2$ of measuring the result $|0\rangle$ and $|b|^2$ of measuring the result $|1\rangle$.

For a two-qubit system, the nature of quantum mechanics dictates that the joint quantum state system is the tensor product of the individual quantum state systems. Thus the orthonormal basis vectors (over $\mathbb{C}$) of the two-qubit system can be expressed as

$$\begin{aligned} |00\rangle &= |0\rangle \otimes |0\rangle = |0\rangle|0\rangle, \quad |01\rangle = |0\rangle \otimes |1\rangle = |0\rangle|1\rangle, \\ |10\rangle &= |1\rangle \otimes |0\rangle = |1\rangle|0\rangle, \quad |11\rangle = |1\rangle \otimes |1\rangle = |1\rangle|1\rangle \end{aligned}$$

with these four basis states corresponding to unit vectors in $\mathbb{C}^4$ as

$$|00\rangle = |0\rangle \otimes |0\rangle \equiv \begin{bmatrix}1\\0\end{bmatrix} \otimes \begin{bmatrix}1\\0\end{bmatrix} = \begin{bmatrix}1\\0\\0\\0\end{bmatrix}, \ |01\rangle \equiv \begin{bmatrix}0\\1\\0\\0\end{bmatrix}, \ |10\rangle \equiv \begin{bmatrix}0\\0\\1\\0\end{bmatrix}, \ |11\rangle \equiv \begin{bmatrix}0\\0\\0\\1\end{bmatrix}.$$

A general element in the quantum state space of the two-qubit system (four-dimensional over $\mathbb{C}$) can be expressed as

$$a_1|00\rangle + a_2|01\rangle + a_3|10\rangle + a_4|11\rangle, \quad a_i \in \mathbb{C}, \quad \sum_i |a_i|^2 = 1$$

and, as with the single-qubit system, measuring this state will produce, e.g., state $|10\rangle$ with probability $|a_3|^2$. Such states can be viewed as four-dimensional column vectors over $\mathbb{C}$ as above. Operations in a two-qubit system correspond to 4×4 unitary matrices with rows and columns labeled with the states $|00\rangle$, $|01\rangle$, $|10\rangle$ and $|11\rangle$, respectively. Thus $|10\rangle \equiv (0,0,1,0)^t$.

In general an n-qubit system will have a quantum state space with 2^n basis states. One of these can be expressed, e.g., as

$$|1\rangle \otimes |1\rangle \otimes |0\rangle \cdots \otimes |1\rangle \stackrel{\Delta}{=} |110\cdots 1\rangle$$

and to describe a general element in the quantum state space, identify the integer $x, 0 \leq x \leq 2^n - 1$ with its binary expansion in $\mathcal{S}_n \stackrel{\Delta}{=} \{0,1\}^n$, and

$$\sum_{x=0}^{2^n-1} a_x|x\rangle = \sum_{x \in \mathcal{S}_n} a_x|x\rangle, \quad \sum_{x=0}^{2^n-1} |a_x|^2 = 1, \quad a_x \in \mathbb{C}$$

and measuring the state of the system will yield the state $|x\rangle$ with probability $|a_x|^2$. Such states can be viewed as column vectors over $\mathbb{C}$ of dimension 2^n. We use position in the binary n-tuple to distinguish the n qubits. The elements $|x\rangle, x \in \mathcal{S}_n$ are viewed as orthonormal column vectors of dimension 2^n over $\mathbb{C}$. The notation $\langle x|, x \in \mathcal{S}_n$ (called "bra") will be viewed as the dual vector to $|x\rangle$, a row vector of the complex conjugates of the column vector x. Further the notation $\langle a \mid b\rangle, a, b \in \mathbb{C}_{2^n}$ for two superimposed states will indicate an inner product, i.e.,

$$a = \sum_{x \in \mathcal{S}_n} a_x |x\rangle,\ b = \sum_{y \in \mathcal{S}_n} b_y |y\rangle \quad \Rightarrow \quad \langle a \mid b\rangle = \sum_{x \in \mathcal{S}_n} a_x^* b_x \in \mathbb{C}$$

where the states $|x\rangle$ and $|y\rangle, x \neq y$ are orthogonal and their inner product $\langle x|y\rangle = 0$. There is also the notion of an *outer product* operator expressed as $|w\rangle\langle v|$, with v, w in the inner product spaces V, W, respectively. Define the outer product operator as:

$$\begin{aligned} |w\rangle\langle v|;\, V &\longrightarrow W \\ |v'\rangle &\mapsto \langle v|v'\rangle\,|w\rangle. \end{aligned}$$

A state $|\phi\rangle$ that cannot be written as a tensor product $|a\rangle|b\rangle$ of two single-qubit states is referred to as an *entangled state*. It can be shown [23] that

$$\frac{1}{\sqrt{2}}\big(|00\rangle + |11\rangle\big)$$

is an entangled state.

As noted it requires $2^n - 1$ complex numbers to specify a state in the projective quantum state space and the state space is $\mathbb{C}^{2^n}$ (actually the unit ball in this space). The exponential dimension of the quantum state space with number of qubits contributes to the computational power of quantum computing. The quantum state space is taken to be a Hilbert space of dimension 2^n, i.e., a complex vector space of dimension 2^n with an inner product that is complete – that every Cauchy sequence of vectors converges. For this discussion it is enough to assume a complex vector space with the usual inner product.

The laws of quantum mechanics allow only computational operations or transformations of the quantum state space that can be modeled as unitary matrices over the complex numbers. A unitary matrix (also referred to as Hermitian adjoint) $\boldsymbol{U}$ is unitary iff its inverse is equal to its conjugate transpose, i.e.,

$$\boldsymbol{U}^{-1} = \boldsymbol{U}^{*t} \triangleq \boldsymbol{U}^{\dagger}$$

where for compatibility with much of the quantum literature $*$ indicates complex conjugate and t indicates transpose and $\dagger$ conjugate transpose. Such matrices map points on the unit ball in $\mathbb{C}^{2^n}$ to points on the unit ball. In addition their rows and columns form orthonormal basis sets as do the set of their eigenvectors. For reference recall a Hermitian matrix $\boldsymbol{H}$ is one where $\boldsymbol{H} = \boldsymbol{H}^\dagger$.

Just as for classical logic gates, a quantum logic gate can be expressed by its truth table. However, [28] not all quantum logic truth tables correspond to realizable quantum gates as many do not correspond to unitary matrices. A quantum logic gate on a single qubit corresponds to a 2×2 unitary matrix (with rows and columns labeled with the basis states $|0\rangle$ and $|1\rangle$) and one on two qubits to a 4×4 unitary matrix (with rows and columns labeled with the basis states $|00\rangle$, $|01\rangle$, $|10\rangle$ and $|11\rangle$). A few examples will illustrate. The unitary matrices of some basic single-input quantum logic gates including the Pauli gates and the Hadamard gate are shown in Equation 15.1.

$$X = \begin{bmatrix} 0 & 1 \\ 1 & 0 \end{bmatrix}, \quad Y = \begin{bmatrix} 0 & -i \\ i & 0 \end{bmatrix}, \quad Z = \begin{bmatrix} 1 & 0 \\ 0 & -1 \end{bmatrix}, \quad H = \frac{1}{\sqrt{2}} \begin{bmatrix} 1 & 1 \\ 1 & -1 \end{bmatrix} \tag{15.1}$$

Pauli-X gate Pauli-Y gate Pauli-Z gate Hadamard gate

These matrices will play a prominent role in the next chapter on quantum error-correcting codes, where they will denote error operators. Specifically in an n-qubit state $|a_1 a_2 \cdots a_n\rangle$ the operators are n-fold tensor products of these matrices. For example, a bit flip in the i-th qubit is designated as X_i and a phase flip by Z_i. Thus

$$X_1 \otimes I_2 \otimes Z_3 |111\rangle = X_1 Z_3 |111\rangle = -|011\rangle.$$

Further properties of the set of operators (or 2×2 matrices) are developed in the next chapter. It is the nature of these operators that requires a different approach to the construction and decoding algorithms for quantum error-correcting codes from classical coding.

The notion of a *controlled gate* where inputs include control bits and target bits, where certain actions are taken on the target bits iff certain conditions hold on the control bits. An important example of this is the simple 2-input 2-output CNOT (controlled NOT) gate whose unitary matrix is shown in Equation 15.2 – the target bit is flipped iff the control bit is 1 – rows and columns labeled with $|00\rangle$, $|01\rangle$, $|10\rangle$ and $|11\rangle$. The quantum logic operation for the gate would be $U_{CNOT} : |c\rangle|t\rangle \longrightarrow |c\rangle|t \oplus c\rangle$:

$$U_{CNOT} = \begin{bmatrix} 1 & 0 & 0 & 0 \\ 0 & 1 & 0 & 0 \\ 0 & 0 & 0 & 1 \\ 0 & 0 & 1 & 0 \end{bmatrix}. \tag{15.2}$$

Thus if the control bit c is 1, the target bit t is flipped to $t \oplus 1$.

An important example of a three-qubit quantum logic gate is the Toffoli gate whose corresponding unitary matrix is given in Equation 15.3 (rows and columns labeled with $|000\rangle, |001\rangle, |010\rangle, |011\rangle, \ldots, |110\rangle, |111\rangle$). There are two control bits c_1 and c_2 and one target bit t and the quantum logic operation is $U_{toffoli}: |c_1\rangle|c_2\rangle|t\rangle \longrightarrow |c_1\rangle|c_2\rangle|t \oplus (c_1 \wedge c_2)\rangle$ – thus the target bit is flipped iff both control bits are set 1: thus the 8×8 matrix

$$U_{TOFFOLI} = \begin{bmatrix} 1&0&0&0&0&0&0&0 \\ 0&1&0&0&0&0&0&0 \\ 0&0&1&0&0&0&0&0 \\ 0&0&0&1&0&0&0&0 \\ 0&0&0&0&1&0&0&0 \\ 0&0&0&0&0&1&0&0 \\ 0&0&0&0&0&0&0&1 \\ 0&0&0&0&0&0&1&0 \end{bmatrix}. \tag{15.3}$$

It is well known in classical logic that a logic function $f: \mathcal{S}_n \rightarrow \{0,1\}^M$ can be realized using only AND, OR and NOT gates. Such a set is referred to as a *universal set* of gates. Equivalently [11] the set NAND and FANOUT is a universal set for such functions.

In a similar fashion there exists universal quantum gate sets in the sense [23] the set will be universal if any unitary operation can be approximated to an arbitrary level of accuracy using only gates in the universal set. Thus the set of all two-level unitary quantum logic gates is universal for quantum computation. Equivalently any unitary matrix can be decomposed into a product of 2×2 unitary matrices. Similarly the set of all single-qubit logic gates and CNOT gates is universal. Also, the set of Toffoli logic gates by itself is universal as is the set of Fredkin gates, not considered here.

From the above comments a quantum algorithm can be thought of conceptually as a set of horizontal 2^n lines representing the states of an n-qubit system. As time proceeds one inserts unitary operations on one-qubit (a 2×2 unitary matrix operating on two of the lines) or a two-qubit operation (a 4×4 unitary matrix operating on four of the lines). For example, suppose $\boldsymbol{A}^{i,j}$ is a 4×4 unitary matrix representing a quantum logic operation operating on two qubits i and j of the n-qubit state. In a two-qubit system the rows and columns would be indexed with the states $|00\rangle$, $|01\rangle$, $|10\rangle$ and $|11\rangle$. In the n-qubit system the rows and columns of the 4×4 matrix $A^{i,j}$ would be indexed with the states

$$\begin{array}{l} |b_1b_2\ldots b_{i-1}0b_{i+1}\ldots b_{j-1}0b_{j+1}\ldots b_n\rangle \\ |b_1b_2\ldots b_{i-1}0b_{i+1}\ldots b_{j-1}1b_{j+1}\ldots b_n\rangle \\ |b_1b_2\ldots b_{i-1}1b_{i+1}\ldots b_{j-1}0b_{j+1}\ldots b_n\rangle \\ |b_1b_2\ldots b_{i-1}1b_{i+1}\ldots b_{j-1}1b_{j+1}\ldots b_n\rangle \end{array}$$

and the remaining $2^n - 4$ of the lines remain unchanged. The logic operation on these two qubits would then be realized by a 4×4 unitary matrix operating on these four lines. Finding such sequences of operations to effect a particular computation/algorithm is the challenge of quantum computing.

It is somewhat surprising that it is not possible to clone an unknown quantum state. The following argument [23] is instructive. Suppose the initial state is $|\,\phi\rangle \otimes |\,s\rangle$ where s is one of the base orthonormal states and ϕ is an unknown quantum state. It is desired to find an operation (unitary) U such that

$$U : |\,\phi\rangle \otimes |\,s\rangle \longrightarrow |\,\phi\rangle \otimes |\,\phi\rangle$$

and assume this can be done for two such states $|\,\phi\rangle$ and $|\,\psi\rangle$ so

$$U\big(|\,\phi\rangle \otimes |\,s\rangle\big) = |\,\phi\rangle \otimes |\,\phi\rangle$$
$$U\big(|\,\psi\rangle \otimes |\,s\rangle\big) = |\,\psi\rangle \otimes |\,\psi\rangle.$$

Taking inner products of the LH sides and RH sides of these equations yields

$$\langle\phi \mid \psi\rangle = \big(\langle\phi \mid \psi\rangle\big)^2 \quad \Rightarrow \quad \langle\phi \mid \psi\rangle = 0 \text{ or } 1$$

which implies either $|\,\phi\rangle = |\,\psi\rangle$ or the states are orthogonal. Thus only states which are orthogonal can be cloned and a general cloning operation is not possible. Thus [30] it is not possible to make a copy of an unknown quantum state without destroying the original, i.e., *one can cut and paste but not copy and paste*.

The QFT is discussed as an example where dramatic gains with quantum computing over classical computing can be realized. Recall the classical discrete Fourier transform (DFT). Only the case of $N = 2^n$ is considered – the extension to arbitrary N is straightforward. The DFT of the vector $\boldsymbol{x} = (x_0, x_1, \ldots, x_{N-1}) \in \mathbb{C}^N$ is $\boldsymbol{y} = (y_0, \ldots, y_{N-1})$ where

$$y_k = \frac{1}{\sqrt{N}} \sum_{j=0}^{N-1} x_j \omega_N^{jk}, \quad k = 0, 1, \ldots, N-1, \quad \omega_N = e^{2\pi i/N}$$

and ω_N an N-th root of unity in $\mathbb{C}$. The classical complexity of taking DFTs is $O(N \log N) = O(n2^n)$.

Let $|\,x\rangle, x = 0, 1, 2, \ldots, N-1$ be the standard orthonormal basis of the quantum state space of an n-qubit system, indexed by the elements of $\{0, 1\}^n$. The QFT then is the map, considered as the transformed basis $|\,\phi_k\rangle$, $k = 0, 1, 2, \ldots, N-1 = 2^n - 1$ map, where

$$\text{QFT} : |\,j\rangle \longrightarrow |\,\phi_j\rangle = \frac{1}{\sqrt{N}} \sum_{k=0}^{N-1} \omega_N^{jk} |\,k\rangle, \; j \in \{0, 1\}^n.$$

Equivalently [23] the QFT is the map on arbitrary superpositioned states

$$\sum_{j=0}^{N-1} x_j | j\rangle \longrightarrow \sum_{k=0}^{N-1} y_k | k\rangle, \quad y_k = \frac{1}{\sqrt{N}} \sum_{\ell=0}^{N-1} x_\ell \omega_N^{k\ell}$$

where $\{x_j\}$ and $\{y_k\}$ are DFT pairs. Note that the QFT matrix, denoted $\boldsymbol{F}_N = (f_{ij}),\ f_{ij} = \frac{1}{\sqrt{N}} \omega_N^{ij}$ is unitary.

In order to implement a QFT on n qubits (quantum state space of dimension $N = 2^n$) it can be shown that only two types of operations are needed [23] which are:

(i) Gate R_j which operates on the j-th qubit of the quantum computer:

$$R_j = \frac{1}{\sqrt{2}} \begin{bmatrix} 1 & 1 \\ 1 & -1 \end{bmatrix}$$

(rows and columns indexed by $| 0\rangle$ and $| 1\rangle$) which operates on the j-th qubit; and

(ii) Gate S_{jk} which operates on qubits i and $j < k$

$$S_{jk} = \begin{bmatrix} 1 & 0 & 0 & 0 \\ 0 & 1 & 0 & 0 \\ 0 & 0 & 1 & 0 \\ 0 & 0 & 0 & e^{i\theta_{k-j}} \end{bmatrix}$$

where $\theta_{k-j} = \pi/2^{k-j}$ and rows and columns indexed with $| 00\rangle, | 01\rangle, | 10\rangle, | 11\rangle$, respectively.

The QFT is then implemented as the sequence of these two operations:

$$R_{n-1} S_{n-2,n-1} R_{n-2} S_{n-3,n-1} S_{n-3,n-2} R_{n-3} \cdots R_1 S_{0,n-1} S_{0,n-2} \cdots S_{0,1} R_0$$

the R gates in decreasing order and the S_{jk} gates for $k > j$ only. There are $n(n-1)/2$ S gates and n R gates and since each such quantum logic gate has a constant complexity the overall complexity of the QFT is $O(n^2)$. This is compared to the DFT complexity mentioned of $O(n2^n)$, a dramatic exponential order difference. It is hoped the manner by which this computational speedup for the quantum case was achieved with the unitary matrices acting on the quantum states gives some indication of the difference between the classical and quantum computational models. The interested reader should consult [23] and [28] for a much more detailed and persuasive discussion than this sketch.

As with the classical DFT, the QFT is a fundamental workhorse for quantum algorithms, including the quantum integer factorization algorithm of Shor [27, 28]. It is suggested that the above discussion of the QFT is representative of the huge complexity advantage of certain quantum algorithms over their classical counterparts. To date only certain problems, such as integer factorization and the DLP, have been found to take advantage of this speedup.

15.3 Postquantum Cryptography

Having discussed the impact of quantum computing on cryptographic practice, it remains to describe the current state of cryptography and in particular those algorithms that have been recommended for use to implement public-key cryptography in this postquantum era. These are commonly referred to as *postquantum* algorithms or *quantum-resistant* algorithms. It should be mentioned that these algorithms are typically determined by extensive research efforts to find quantum algorithms for their solution rather than by observing any particular feature of them. It seems possible that such a postquantum algorithm could migrate into the class of quantum-weak algorithms should some clever researcher discover the right technique.

Below, four commonly accepted postquantum algorithms are described. There are others – such as isogenies of super-singular elliptic curves but the background required for a reasonable discussion of them is excessive.

Many symmetric-key encryption algorithms – such as AES with a sufficiently large key/block size, and hash functions – such as SHA-2 (SHA-256) are regarded as quantum-resistant since no quantum algorithms to "break" these have yet been formulated.

Code-Based Public-Key Cryptosystems

McEliece [17] devised a simple and effective public-key encryption system that has withstood the test of intensive scrutiny and remains an interesting option for quantum-resistant cryptography. It appears that Goppa codes are particularly suitable for this application. The construction and basic properties of Goppa codes are recalled, e.g., ([15, 17]). Let $g(x) \in \mathbb{F}_{q^m}[x]$ of degree t and let $L = \{\alpha_1, \ldots, \alpha_n\} \subset \mathbb{F}_{q^m}$ disjoint from any zeros of $g(x)$ in $\mathbb{F}_{q^m}$. Define a $t \times n$ parity-check matrix

$$\boldsymbol{H} = \begin{bmatrix} g(\alpha_1)^{-1} & \cdots & g(\alpha_n)^{-1} \\ \alpha_1 g(\alpha_1)^{-1} & \cdots & \alpha_n g(\alpha_n)^{-1} \\ \vdots & \vdots & \vdots \\ \alpha_1^{t-1} g(\alpha_1)^{-1} & \cdots & \alpha_n^{t-1} g(\alpha_n)^{-1} \end{bmatrix}.$$

This is a parity-check matrix for an $(n, k \geq n - mt, d \geq 2t + 1)_q$ t error-correcting Goppa code, often denoted $\mathcal{G} = \Gamma(L, g)$. For consistency with previous systems chose $q = 2$.

With the above parameters n, k, t and associated generator matrix $\boldsymbol{G}$ fixed, user Alice A chooses a $k \times k$ binary nonsingular matrix $\boldsymbol{S}_A$ and an $n \times n$ permutation matrix $\boldsymbol{P}_A$. The $k \times n$ generator matrix $\boldsymbol{G}_A^* = \boldsymbol{S}_A \boldsymbol{G} \boldsymbol{P}_A$ is computed.

Alice's public key then is $(\boldsymbol{G}_A^*, t)$ and secret key $(\boldsymbol{S}_A, \boldsymbol{G}, \boldsymbol{P}_A)$. For Bob to send a message $\boldsymbol{m}$ to Alice, assumed to be binary of length k, an error vector $\boldsymbol{e}$ of length n and weight t is chosen and the encrypted message $\boldsymbol{c} = \boldsymbol{m}G^* + \boldsymbol{e}$ is sent. On receiving $\boldsymbol{c}$ Alice computes $\boldsymbol{c}^* = \boldsymbol{c}\boldsymbol{P}_A^{-1}$ and decodes this binary n-tuple with the decoding algorithm for the Goppa code with generator matrix $\boldsymbol{G}$ to give $\boldsymbol{m}^*$. The original message is then computed to be $\boldsymbol{m} = \boldsymbol{m}^*\boldsymbol{S}_A^{-1}$.

Various attacks on this system have been proposed and none have proved successful. An obvious drawback to the system is the rather large number of bits in the public key. The parameters suggested in [18] are $n = 1024, t = 50$ – hence $m = 10$ and $k \geq 1024 - 500 = 524$. Thus the public key size is approximately the number of bits in the generator matrix or 1024×524, a large number. Thus while the encryption operations are efficient, the size of typical public keys is a drawback.

Niederreiter [22] proposed an interesting variant of this scheme. Let $\mathscr{C} = (n, k, d \geq 2t + 1)_q$ be a t-error correcting over $\mathbb{F}_q$, e.g., the Goppa code noted above. Again we could choose $q = 2$ for consistency but the argument is the same for any q. Let H be a full-rank $(n - k) \times n$ parity-check matrix for the code and note that, by the error-correcting properties of the code for any two distinct messages $\boldsymbol{m}_1 \neq \boldsymbol{m}_2 \in \mathbb{F}_2^n$, each of weight $\leq t$, $\boldsymbol{H}\boldsymbol{m}_1 \neq \boldsymbol{H}\boldsymbol{m}_2$. This is the basis of the Niederreiter public-key encryption system, namely: choose an $(n - k) \times (n - k)$ nonsingular matrix $\boldsymbol{R}$ and an $n \times n$ permutation matrix $\boldsymbol{P}$ and let $\boldsymbol{H}^*$ be the $(n - k) \times n$ matrix $\boldsymbol{H}^* = \boldsymbol{R}\boldsymbol{H}\boldsymbol{P}$. The public key for the system is $(\boldsymbol{H}^*, t)$ and the secret key is $(\boldsymbol{R}, \boldsymbol{P})$. All matrices will be subscripted by a user ID.

It is assumed all original messages $\boldsymbol{m}$ are of weight $\leq t$ which can be arranged by a variety of techniques. For Bob to send Alice the message $\boldsymbol{m}, w_{\boldsymbol{H}}(\boldsymbol{m}) \leq t$ he retrieves Alice's public key $\boldsymbol{H}_A^*$ and sends the encryption $\boldsymbol{H}_A^*\boldsymbol{m}^t = \boldsymbol{y}^t$. To decrypt this encryption Alice computes $\boldsymbol{R}_A^{-1}\boldsymbol{H}_A^*\boldsymbol{m}^t = \boldsymbol{H}_A\boldsymbol{P}_A\boldsymbol{m}^t = \boldsymbol{y}^t$. Since $\boldsymbol{P}_A\boldsymbol{m}^t$ is of weight $\leq t$, it can be decoded to z and the original message is recovered by $z(\boldsymbol{P}_A^t)^{-1}$.

Other aspects of quantum-resistant code-based public-key systems can be found in [14], [8] and the chapter by Overbeck et al. in [3].

Lattice-Based Public-Key Cryptosystems

Let $\boldsymbol{v}_1, \boldsymbol{v}_2, \ldots, \boldsymbol{v}_n$ be linearly independent vectors in the n-dimensional real vector space $\mathbb{R}^n$. A lattice will then be a $\boldsymbol{Z}$-linear combination of these vectors, namely

$$L = \left\{a_1\boldsymbol{v}_1 + a_2\boldsymbol{v}_2 + \cdots + a_n\boldsymbol{v}_n, a_1, a_2, \ldots, a_n \in \mathbb{Z}\right\} \subset \mathbb{R}^n.$$

Any set of n linearly independent vectors in the lattice generates L and all such sets have n vectors in them, the dimension of the lattice. A lattice is a discrete subgroup of $\mathbb{R}^n$. Any two bases of the lattice [10] are related by an integer matrix with determinant ± 1.

The *fundamental domain* of a lattice with basis $\{\boldsymbol{v}_1, \ldots, \boldsymbol{v}_n\}$ is the set of points

$$F = \left\{ f_1\boldsymbol{v}_1 + \cdots + f_n\boldsymbol{v}_n,\ 0 \le f_i < 1 \right\}$$

and it is easy to see that any point in $\boldsymbol{z} \in \mathbb{R}^n$ can be expressed as $\boldsymbol{z} = \boldsymbol{w} + \boldsymbol{f}$ for $\boldsymbol{w} \in L$ and $\boldsymbol{f} \in F$, i.e.,

$$\mathbb{R}^n = \bigcup_{\boldsymbol{w} \in L} (\boldsymbol{w} + F)$$

and the translations of the fundamental domain by lattice points cover (tile) the space $\mathbb{R}^n$.

Let $\boldsymbol{V}$ be the $n \times n$ matrix with $\boldsymbol{v}_i$ as the i-th column. Then the volume of the fundamental domain F, vol(F) is given by

$$\mathrm{vol}(F) = |\det(\boldsymbol{V})|$$

and it can be shown that

$$\mathrm{vol}(F) \le |\boldsymbol{v}_1| \cdots |\boldsymbol{v}_n|, \qquad |\boldsymbol{v}_i| = \left(\sum_{j=1}^{n} v_{ij}^2 \right)^{1/2}.$$

Indeed ([10], corollary 6.22), every fundamental domain of a lattice has the same volume and hence det$(\boldsymbol{V})$ is an invariant of the lattice, independent of the choice of basis of the lattice.

There are numerous interesting and important problems associated with lattices and an excellent introductory treatment of them is [10]. The use of lattices in cryptography centers largely around two fundamental lattice problems which, for the right parameters will be presumed hard. Given a lattice L, they are ([10], section 6.5.1):

Definition 15.1 The shortest vector problem (SVP) in a lattice L is to find a shortest nonzero vector in the lattice.

Definition 15.2 The closest vector problem (CVP) is, given a vector $\boldsymbol{w} \in \mathbb{R}^n, \boldsymbol{w} \notin L$, to find a vector $\boldsymbol{v} \in L$ closest to $\boldsymbol{w}$.

There are other interesting problems associated with lattices – e.g., one might ask only for approximate solutions to the above problems, solutions that fall within some prescribed bounds.

Notice in each case the solutions may not be unique. The difficulty of solving these problems for a given lattice and basis, is largely a function of properties of the basis used to represent lattice points. Generally the more nearly orthogonal the basis vectors are, the more amenable the problems are to solution.

While our interest will be only in the public-key encryption system of [9], referred to as GGH (for its authors Goldreich, Goldwasser and Halevi) it should be noted there are many deep and fundamental theoretical results related to lattices. For example, Ajtai and Dwork [1], devised a scheme whose security could be shown to be provably secure unless an equivalent worst case problem could be solved in polynomial time, showing essentially a worst-case/average case connection and a celebrated result of complexity theory.

To discuss the GGH system let $\{\boldsymbol{v}_1, \ldots, \boldsymbol{v}_n\}$ be a basis for the lattice L and the $\boldsymbol{V}$, the $n \times n$ matrix whose i-th column is $\boldsymbol{v}_i$. The following definition will be useful for the discussion:

Definition 15.3 For an $n \times n$ nonsingular matrix $\boldsymbol{V}$ define the orthogonality defect as

$$\text{orth-defect} \triangleq \frac{\prod_{i=1}^{n} |\boldsymbol{v}_i|}{|\det(\boldsymbol{V})|}$$

for $|\boldsymbol{v}_i|$ the norm of the i-th column of $\boldsymbol{V}$.

It is immediate that orth-defect($\boldsymbol{V}$) $= 1$ iff the columns of $\boldsymbol{V}$ are orthogonal and > 1 otherwise.

There are two important results on lattice basis reduction that should be noted. The first is, given a basis for a lattice, the LLL (for its authors Lenstra, Lenstra and Lovász) [13] basis reduction algorithm operates on the basis to produce another basis that has relatively short and quasi-orthogonal basis vectors. This celebrated result has found critical applications to a variety of important lattice complexity problems and is relevant to the problems of interest here. The reader is referred to the original work for a detailed look at the algorithm and to obtain a better appreciation for the terms "relatively short" and "quasi-orthogonal."

Another fundamental and useful work is that of Babai [2] who gave two simple algorithms for solving the CVP problem – up to a multiplicative constant – for lattices with relatively short and quasi-orthogonal bases, *rounding off* and *nearest plane* algorithms. In particular the algorithms tend to be efficient for lattices with an LLL reduced basis (in particular quasi-orthogonal bases). Only the rounding off algorithm will be noted here. Suppose $\{\boldsymbol{v}_1, \ldots, \boldsymbol{v}_n\}$ is a basis for the lattice L and let $\boldsymbol{x} = \sum_{i=1}^{n} a_i \boldsymbol{v}_i \in \mathbb{R}^n$ be an arbitrary element in $\mathbb{R}^n$. Let $\lceil a_i \rfloor$ be the coefficient rounded off to the nearest integer. Then the point $\boldsymbol{w} = \sum_{i=1}^{n} \lceil a_i \rfloor \boldsymbol{v}_i$ is a nearest lattice point up to a multiplicative constant that depends on the dimension n.

The impact of these results is that lattices that have a relatively short and nearly orthogonal basis tend to have efficient solutions to the CVP problem in dimensions up to several hundred. There are other intuitive reasons (apart from Babai's result) and experimental results that indicate the CVP problem and approximate versions [9, 10] of them have a difficulty that increases with the orth-defect. Assuming this, a description of the public-key encryption system of [9] is of interest.

Suppose a lattice L is generated by the basis $\boldsymbol{v}_1, \ldots, \boldsymbol{v}_n$ that has a low orth-defect – the basis vectors are quasi-orthogonal and let $\boldsymbol{V}$ be the matrix whose i-th column is $\boldsymbol{v}_i$. Choose an integer matrix $\boldsymbol{U}$ with determinant ± 1 and let $\boldsymbol{w}_i$ be the i-th column of $\boldsymbol{W} = \boldsymbol{V}\boldsymbol{U}$. The basis $\{\boldsymbol{w}_1, \boldsymbol{w}_2, \ldots, \boldsymbol{w}_n\}$ generates the same lattice. It is assumed this basis has a large orth-defect for which solving CVP is computationally infeasible.

Suppose user Alice generates the matrices $\boldsymbol{V}_A$ and W_A with "good" basis $\boldsymbol{V}_A$ kept secret and makes "bad" basis $\boldsymbol{W}_A$ public. Suppose Bob wishes to send the (column) message $\boldsymbol{m}$ with small components. He generates the lattice point $\boldsymbol{W}_A\boldsymbol{m}$ and sends the perturbed (column) vector z

$$z = \boldsymbol{W}_A \boldsymbol{m} + \boldsymbol{r} = \sum_{i=1}^{n} m_i \boldsymbol{w}_{Ai} + \boldsymbol{r}$$

for some vector $\boldsymbol{r}$ with small components. The vector z is, by definition, close to a lattice point. Alice then uses the Babai rounding off algorithm described above and her private key good basis $\boldsymbol{V}_A$ to find the closest lattice vector (in the bad basis) $z' = \boldsymbol{W}_A \boldsymbol{m}$ and recovers the message by

$$\boldsymbol{W}_A^{-1} z' = \boldsymbol{m}.$$

This is a simple and efficient public-key encryption algorithm. However, the previous comments apply to it, namely that for any high orth-defect basis one could apply the LLL algorithm to determine a "good" basis and hence break the GGH algorithm. However, applying the LLL algorithm becomes more difficult as the dimension of the lattice increases. Thus for a sufficiently high dimension the GGH algorithm may be secure. On the other hand very high dimensions are not attractive due to inefficiency concerns.

In spite of these considerations, various lattice problems remain of interest to postquantum cryptography. No quantum algorithms are known that are better than classical ones for solving lattice problems and the area remains of research interest.

Hash-Based Public-Key Cryptosystems

Most of the hash-based systems in the literature correspond to signature schemes. Some have the disadvantage they can only be used for a small number of signatures. However, they tend to be efficient and there is some confidence as to their quantum-resistance.

A hash function $H(\boldsymbol{x})$ maps $\{0,1\}^*$ a binary input of arbitrary length, to $\{0,1\}^n$ to an output of some fixed length n. Normally it is the case the input to the hash function is (much) longer than the output and the function provides *compression*. The result $\boldsymbol{y} = H(\boldsymbol{x})$ is often referred to as digest of the input $\boldsymbol{x}$. It should have some important properties [19]:

(i) It should be easy/efficient to compute;
(ii) *Preimage resistance*: given $\boldsymbol{y} = H(\boldsymbol{x})$ it should be computationally infeasible to compute $\boldsymbol{x}$;
(iii) *Second preimage resistance*: given $\boldsymbol{x}_1$ and $\boldsymbol{y} = H(\boldsymbol{x}_1)$ it should be computationally infeasible to compute $\boldsymbol{x}_2 \neq \boldsymbol{x}_1$ such that $H(\boldsymbol{x}_1) = H(\boldsymbol{x}_2)$;
(iv) *Collision resistance*: it should be computationally infeasible to compute any two $\boldsymbol{x}_1 \neq \boldsymbol{x}_2$ such that $H(\boldsymbol{x}_1) = H(\boldsymbol{x}_2)$.

Naturally there are relationships between these concepts. The hash function SHA-2 = SHA-256 is a common – and widely used – example of such a function. A related notion is that of a *one-way function* (OWF): An OWF $f(\cdot)$ is such that for any $\boldsymbol{x}$ in the domain of f, $f(\boldsymbol{x})$ is easy to compute while for essentially all $\boldsymbol{y}$ in the range of f it is computationally infeasible to determine $\boldsymbol{x}$ such that $\boldsymbol{y} = f(\boldsymbol{x})$.

Compared to hash functions, an OWF typically has a fixed-size domain and range and need not provide compression. See [19] for an interesting discussion.

There are numerous examples of hash-based one-time digital signature schemes in the literature. Three examples are discussed here.

The Lamport one-time signature scheme [3, 12] requires an OWF $f\colon \{0,1\}^n \to \{0,1\}^n$ and a hash function H that produces digests of n bits. Denote by $\{0,1\}^{(n,m)}$ a set of m binary n-tuples. The signature and verification keys of the Lamport scheme will be

$$\begin{aligned}\text{signature keys: } X &= \left\{\boldsymbol{x}_{0,0},\boldsymbol{x}_{0,1},\boldsymbol{x}_{1,0},\boldsymbol{x}_{1,1},\ldots,\boldsymbol{x}_{n-1,0},\boldsymbol{x}_{n-1,1}\right\} \in \{0.1\}^{(n,2n)}\\ \text{verification keys: } Y &= \left\{\boldsymbol{y}_{0,0},\boldsymbol{y}_{0,1},\boldsymbol{y}_{1,0},\boldsymbol{y}_{1,1},\ldots,\boldsymbol{y}_{n-1,0},\boldsymbol{y}_{n-1,1}\right\} \in \{0.1\}^{(n,2n)}\end{aligned}$$

where the signature keys are chosen uniformly at random, and the verification keys are generated from the signature keys by

$$\boldsymbol{y}_{i,j} = f(\boldsymbol{x}_{i,j}),\ i = 0,1,\ldots,n-1,\ j = 0,1.$$

The signature generation for a message $\boldsymbol{m} \in \{0,1\}^*$ is as follows: Let $\boldsymbol{h} = (h_0, h_1, \ldots, h_{n-1}) = H(\boldsymbol{m})$ be the hash of the message. The signature for $\boldsymbol{m}$ then is

$$\boldsymbol{S_m} = (\boldsymbol{x}_{0,h_0}, \boldsymbol{x}_{1,h_1}, \ldots, \boldsymbol{x}_{n_1,h_n-1}) \in \{0,1\}^{(n,n)} = (\boldsymbol{s}_0, \boldsymbol{s}_1, \ldots, \boldsymbol{s}_{n-1}).$$

It is clear why it is a one-time signature scheme since reusing signature/verification key pairs can be used for forgery. The pair $(\boldsymbol{m}, \boldsymbol{S_m})$ is transmitted.

The signature verification process is straightforward. On receiving $(\boldsymbol{m}, \boldsymbol{\sigma_m})$, $\boldsymbol{\sigma_m} = (\sigma_0, \sigma_1, \ldots, \sigma_{n-1})$, the receiver determines the hash value of the message: $\boldsymbol{h} = (h_0, h_1, \ldots, h_{n-1}) = H(\boldsymbol{m})$ and uses the public OWF f to verify that

$$(\boldsymbol{y}_{0,h_0}, \boldsymbol{y}_{1,h_1}, \ldots, \boldsymbol{y}_{n-1,h_{n-1}}) = (f(\sigma_0), f(\sigma_1), \ldots, f(\sigma_{n-1}))$$

and verifies that $f(\sigma_i)$ is y_{i,m_i}.

While the above one-time signature scheme is quite efficient in terms of the computations required, the signature size can be large. Winternitz [20] proposed an improvement by grouping bits together for signature (rather than single bits) – see [7] and [3] (chapter by Buchmann et al.) for details of this system.

One-time signature schemes are inefficient in that a new set of keys is required to be generated for each signature. Merkle [20] showed how an *authentication tree* can be used to authenticate the verification keys for any one-time signature scheme thereby allowing the one-time signature scheme to be used to sign a predetermined number of messages.

Multivariable Public-Key Cryptosystems

Multivariable public-key cryptosystems typically employ a number of polynomials of small degree over small fields in a large number of variables. The technique includes the *hidden field equations* (HFE) systems. Unfortunately a complete description of such a system requires a level of detail and complexity that it is not pursued here. Readers are referred to [5, 24] and [3] (chapter by Bernstein) [25] and [21] (section 16.3).

The above has been a very brief tour through the main contenders for postquantum public-key cryptosystems. No doubt others will arise and some will fail as further research to improve their security and efficiency continues.

Comments

The section attempted to give an overview of the state of public-key algorithms that are quantum-resistant and the path that led to the current situation. The existence of a quantum computer with a sufficient size, memory and stability seems not yet available to actually render the classical algorithms useless. However, given the enormous power and advantages such machines offer and the enormous efforts being expended by well-funded and determined interests, it seems only a matter of time before their promise is realized. This chapter attempted to give an appreciation of the developments.

References

[1] Ajtai, M., and Dwork, C. 1999. A public-key cryptosystem with worst-case/average-case equivalence. Pages 284–293 of: *STOC '97 El Paso, TX*. ACM, New York.

[2] Babai, L. 1986. On Lovás lattice reduction and the nearest lattice point problem. *Combinatorica*, **6**(1), 1–13.

[3] Bernstein, D.J., Buchmann, J., and Dahmen, E. (eds.). 2009. *Post-quantum cryptography*. Springer-Verlag, Berlin.

[4] Blake, I.F., Seroussi, G., and Smart, N. 1999. *Elliptic curves in cryptography*. Cambridge University Press, Cambridge.

[5] Courtois, N.T. 2001. The security of hidden field equations (HFE). Pages 266–281 of: *Topics in cryptology – CT-RSA 2001 (San Francisco, CA)*. Lecture Notes in Computer Science, vol. 2020. Springer, Berlin.

[6] Diffie, W., and Hellman, M. 1976. New directions in cryptography. *IEEE Trans. Inform. Theory*, **22**(6), 644–654.

[7] Dods, C., Smart, N.P., and Stam, M. 2005. Hash based digital signature schemes. Pages 96–115 of: Smart, Nigel P. (ed.), *Cryptography and Coding, 10th IMA International Conference, Cirencester, UK, December 19–21, 2005, Proceedings*. Lecture Notes in Computer Science, vol. 3796. Springer.

[8] Esser, A., and Bellini, E. 2022. Syndrome decoding estimator. Pages 112–141 of: Hanaoka, G., Shikata, J., and Watanabe, Y. (eds.), *Public-key cryptography – PKC 2022*. Springer, Cham.

[9] Goldreich, O., Goldwasser, S., and Halevi, S. 1997. Public-key cryptosystems from lattice reduction problems. Pages 112–131 of: *Advances in cryptology – CRYPTO '97 (Santa Barbara, CA, 1997)*. Lecture Notes in Computer Science, vol. 1294. Springer, Berlin.

[10] Hoffstein, J., Pipher, J., and Silverman, J.H. 2014. *An introduction to mathematical cryptography*, 2nd ed. Undergraduate Texts in Mathematics. Springer, New York.

[11] Kaye, P., Laflamme, R., and Mosca, M. 2007. *An introduction to quantum computing*. Oxford University Press, Oxford.

[12] Lamport, L. 1979. Constructing digital signatures from a one way function. In: *Report*. SRI INternational.

[13] Lenstra, A.K., Lenstra, Jr., H.W., and Lovász, L. 1982. Factoring polynomials with rational coefficients. *Math. Ann.*, **261**(4), 515–534.

[14] Li, Y.X., Deng, R.H., and Wang, X.M. 1994. On the equivalence of McEliece's and Niederreiter's public-key cryptosystems. *IEEE Trans. Inform. Theory*, **40**(1), 271–273.

[15] Ling, S., and Xing, C. 2004. *Coding theory*. Cambridge University Press, Cambridge.

[16] Luby, M.G. 2002. LT codes. Pages 271 – 280 of: *Proceedings of the 43rd Annual IEEE Symposium on Foundations of Computer Science*.

[17] McEliece, R.J. 1977. *The theory of information and coding*. Addison-Wesley, Reading, MA.

[18] McEliece, R.J. 1978. A public key cryptosystem based on algebraic coding theory. In: *NASA DSN Progress Report*.

[19] Menezes, A.J., van Oorschot, P.C., and Vanstone, S.A. 1997. *Handbook of applied cryptography*. CRC Press Series on Discrete Mathematics and Its Applications. CRC Press, Boca Raton, FL.

[20] Merkle, R.C. 1989. A certified digital signature. Pages 218–238 of: Brassard, G. (ed.), *Advances in cryptology – CRYPTO '89, 9th Annual International Cryptology Conference, Santa Barbara, California, USA, August 20–24, 1989, Proceedings*. Lecture Notes in Computer Science, vol. 435. Springer.

[21] Mullen, G.L. (ed.). 2013. *Handbook of finite fields*. Discrete Mathematics and Its Applications (Boca Raton). CRC Press, Boca Raton, FL.

[22] Niederreiter, H. 1986. Knapsack-type cryptosystems and algebraic coding theory. *Problems Control Inform. Theory/Problemy Upravlen. Teor. Inform.*, **15**(2), 159–166.

[23] Nielsen, M.A., and Chuang, I.L. 2000. *Quantum computation and quantum information*. Cambridge University Press, Cambridge.

[24] Patarin, J. 1996. Hidden fields equations (HFE) and isomorphisms of polynomials (IP): two new families of asymmetric algorithms. Pages 33–48 of: Maurer, U.M. (ed.), *Advances in Cryptology – EUROCRYPT '96, International Conference on the Theory and Application of Cryptographic Techniques, Saragossa, Spain, May 12–16, 1996, Proceeding*. Lecture Notes in Computer Science, vol. 1070. Springer.

[25] Patarin, J., Goubin, L., and Courtois, N.T. 1998. Improved algorithms for isomorphisms of polynomials. Pages 184–200 of: Nyberg, Kaisa (ed.), *Advances in Cryptology – EUROCRYPT '98, International Conference on the Theory and Application of Cryptographic Techniques, Espoo, Finland, May 31 – June 4, 1998, Proceeding*. Lecture Notes in Computer Science, vol. 1403. Springer.

[26] Rivest, R.L., Shamir, A., and Adleman, L. 1978. A method for obtaining digital signatures and public-key cryptosystems. *Commun. ACM*, **21**(2), 120–126.

[27] Shor, P.W. 1994. Algorithms for quantum computation: discrete logarithms and factoring. Pages 124–134 of: *35th Annual Symposium on Foundations of Computer Science (Santa Fe, NM, 1994)*. IEEE Computer Society Press, Los Alamitos, CA.

[28] Shor, P.W. 1997. Polynomial-time algorithms for prime factorization and discrete logarithms on a quantum computer. *SIAM J. Comput.*, **26**(5), 1484–1509.

[29] Shor, P.W. 2003. Why haven't more quantum algorithms been found? *J. ACM*, **50**(1), 87–90.

[30] Yanofsky, N.S., and Mannucci, M.A. 2008. *Quantum computing for computer scientists*. Cambridge University Press, Cambridge.

16
Quantum Error-Correcting Codes

Error-correcting codes play a central and even a critical role in quantum computing in contrast to classical computing where they are not a factor. There are several reasons for this. Quantum computing takes place in a Hilbert space over the (continuous) complex numbers and the codewords are quantum states which are subject to environmental interactions which can alter the state leading to errors or *decoherence*. Thus the physical qubits correspond to some atomic-scale phenomena such as atoms, photons or trapped ions and can become entangled with other states with the potential to destroy the desired computation. The role of error correction in this scenario is thus crucial to give sufficient stability to the system to allow successful completion of the computation.

Quantum codes differ from classical coding in several ways. In classical error correction the error process is often modeled by a discrete memoryless channel as noted in previous chapters and these can generally be thought of as bit flips – $0 \leftrightarrow 1$. In quantum computing errors can involve not only bit flips but phase changes which can be extreme and result in states that evolve continuously from the original state. The state of a quantum computation can thus become entangled with the states of other qubits which make error correction critical.

There are two approaches to quantum error correction, one viewing codewords as elements/vectors of a quantum state space and the other a stabilizer approach where codewords are sets of states fixed by a subgroup of transformation operations. In the first approach one finds a subspace containing the received state and error correction involves determining the transmitted vector in that subspace. In the second approach one finds a coset containing the received element and error correction involves finding the group element in the coset. The situation is reminiscent of syndrome decoding in classical coding where one finds the coset containing the received codeword by computing the syndrome and decodes to find the transmitted word. Both approaches are discussed here, beginning with the state space approach which has notions more closely allied with classical coding.

In addition to the above problems, measurement in quantum systems poses a significant challenge since measuring a qubit will alter its state as well as its correlations with other qubits. Thus care must be taken to arrange such measurements be taken only at the end of computations or in a manner that will not destroy the required end result. This is often done by the introduction of *ancilla states*, extra states where operations and output may be effected without disturbing the main computation.

The next section introduces general properties of quantum error-correcting codes by considering the error operators which are taken as the Pauli operators introduced in the previous chapter. While somewhat abstract, the properties noted are fundamental to the quantum error-correction process. The remainder of the chapter takes a concrete approach to error correction to establish some of the approaches while limiting discussion of such important issues as input/output and effect of measurements of the quantum system.

Section 16.2 introduces the notion of standard simple quantum error-correcting codes, the three-, five- and nine-bit codes that bear resemblance to repetition codes in classical error correction. A seven-bit code will be discussed in the final section. These codes are able to correct bit flips ($| 0\rangle$ to $| 1\rangle$ and vice versa) and phase flips ($| 1\rangle$ to $-| 1\rangle$ and vice versa). The differences of this type of error correction from classical error correction are noted. Many of the quantum error-correction codes rely on notions and constructions from classical coding although with significant differences. Section 16.3 gives an indication of other important types of codes, notably the CSS (for the authors Calderbank, Shor and Steane) codes, the stabilizer codes and codes constructed over $GF(4) = \mathbb{F}_4$.

16.1 General Properties of Quantum Error-Correcting Codes

Given the importance of quantum error-correcting codes for the successful implementation of quantum computers and the huge possibilities for such computers, there has been considerable effort to construct effective and practical error-correcting codes in the quantum domain. This section briefly considers properties that any quantum error-correcting code should have.

Recall from the previous chapter the Pauli matrices:

$$X = \begin{bmatrix} 0 & 1 \\ 1 & 0 \end{bmatrix}, \quad Y = \begin{bmatrix} 0 & -i \\ i & 0 \end{bmatrix}, \quad Z = \begin{bmatrix} 1 & 0 \\ 0 & -1 \end{bmatrix},$$

and recall their action on n-qubit states – e.g., $X_3 Z_5 |\, 11101\rangle = -|\, 11001\rangle$ where the operator subscript indicates the qubit being operated on.

As a trivial example of error correction consider a code with two codewords, $|\,\phi_1\rangle = |\,000\rangle$ and $|\,\phi_2\rangle = |\,111\rangle$ and suppose $|\,\rho\rangle = |\,101\rangle$ is received. Note that

$$Z_1 Z_2 |\,\rho\rangle = -|\,\rho\rangle \quad \text{and} \quad Z_1 Z_3 |\,\rho\rangle = |\,\rho\rangle.$$

The first relation means that qubit one and two of the received word (state) are different while the second means that qubits one and three are the same. The received word can only be $|\,010\rangle$ or $|\,101\rangle$ and it can be established that the most likely transmitted codeword (fewest errors) is $|\,\phi_2\rangle = |\,111\rangle$.

All operators, both error formation and correction, will be assumed to be Pauli operators. Such operators have order 2 and either commute or anticommute with each other. For two operators A and B, define the bracket operations:

$$[A, B] \triangleq AB - BA \quad \text{and} \quad \{A, B\} \triangleq AB + BA.$$

Then A and B are said to commute if $[A, B] = 0$ and anticommute if $\{A, B\} = 0$ – equivalently if $AB = BA$ and $AB = -BA$, respectively.

Notice in particular that all the matrices have order 2 and anticommute:

$$\begin{aligned} XY &= iZ = -YX, \ \{X, Y\} = 0 \\ XZ &= iY = -ZX, \ \{X, Z\} = 0 \\ YZ &= iX = -ZY, \ \{Y, Z\} = 0 \end{aligned}$$

and the identity matrix commutes with all matrices.

The group of tensor products of the Pauli matrices on n qubits is of the form

$$\mathcal{G}_n = \Big\{ \pm i^{\lambda} u_1 \otimes u_2 \otimes \cdots \otimes u_n \Big\}, \ \lambda \in \{0, 1\}, \ u_i \in \{I, X, Y, Z\}.$$

It is clear that $\mathcal{G}_n$ is a group on n generators and that it has order

$$|\mathcal{G}_n| = 2^{2n+2}.$$

Notice that

$$\mathcal{G}_1 = \big\{ \pm I, \pm iI, \pm X, \pm iX, \pm Y, \pm iY, \pm Z, \pm iZ \big\}.$$

This group $\mathcal{G}_n$ describes the set of possible errors on a system with n qubits. It is also the set of operators that may be used on received codewords for error correction. Typically one decides on a subset $\mathscr{E} \subset \mathcal{G}_n$ as the set of errors a code will correct. The *weight* [5] of an operator in $\mathcal{G}_n$ is the minimum number of qubits it acts on which it differs from the identity. A t-error-correcting code will then be able to correct all operators in $\mathcal{G}_n$ of weight $\leq t$. A code to encode k qubits into n qubits will have 2^k basis codewords. If the code can correct errors introduced by operators $E_a, E_b \in \mathcal{G}_n$, it can correct a linear (over $\mathbb{C}$) sum of the errors and it is sufficient to ensure error correction on the basis codewords. The code will be a subspace of dimension 2^k in the Hilbert space of dimension 2^n. A quantum error-correcting code $\mathscr{C}$ coding k qubits to n qubits will be designated as an

$$[\![n,K,d]\!] \quad \text{or} \quad [\![n,k,d)]\!],\ k = \log_2(K)$$

code of length n, dimension $K = 2^k$ and minimum distance d where d is the lowest weight operator in $\mathcal{G}_n$ that can transform one codeword into another codeword. Such a code ([8], chapter of Ezerman) can correct all errors of weight $\leq \lfloor \frac{d-1}{2} \rfloor$.

For a code to be capable of distinguishing an error $E_a \in \mathcal{G}_n$ acting on codeword (state) ϕ_i from error E_b acting on codeword ϕ_j, the two results must be orthogonal, i.e.,

$$\langle \phi_i | E_a^\dagger E_b | \phi_j \rangle = 0$$

as otherwise there will be a nonzero probability of confusion. It is argued in [2, 9] that for all distinct basis codewords $|\phi_i\rangle$, $|\phi_j\rangle \in \mathscr{C}$ and two error operators E_a and E_b in the error set of interest $\mathscr{E} \subset \mathcal{G}_n$ we must have

$$\langle \phi_i | E_a^\dagger E_b | \phi_i \rangle = \langle \phi_j | E_a^\dagger E_b | \phi_j \rangle$$

or else information on the error could be gleaned from computing such quantities. The above relations can be combined into the necessary and sufficient conditions for the code $\mathscr{C}$ to be able to correct all errors in $\mathscr{E}$

$$\langle \phi_i | E_a^\dagger E_b | \phi_j \rangle = C_{ab}\delta_{ij} \tag{16.1}$$

for all possible distinct codewords $|\phi_i\rangle, |\phi_j\rangle \in \mathscr{C}$ and error operators E_a, $E_b \in \mathscr{E}$. The matrix C_{ab} is Hermitian (i.e., view a, b as fixed as the codewords vary to obtain a matrix for each such error pair E_a, E_b). Thus [9] the code $\mathscr{C}$ will correct all errors in a set $\mathscr{E} \subset \mathcal{G}_n$ iff the above condition Equation 16.1 is satisfied for all $E_a, E_b \in \mathscr{E}$ and all codewords in $\mathscr{C}$.

Gottesman [5] defines a code for which there is a pair E_a, E_b such that the matrix C_{ab} is singular as a *degenerate* code and a code for which no such pair exists as a *nondegenerate* code.

In Equation 16.1 for $E = E_a^\dagger E_b$, the weight of the smallest weight $E \in \mathcal{G}_n$ for which the relationship does not hold for $|\phi_i\rangle, |\phi_j\rangle$ in the code under consideration is the distance of the code. As with classical coding, in order for a quantum error-correcting code to correct t errors it must have a distance of at least $2t+1$. Notice that if E is an operator of weight equal to the code minimum distance, then it is possible $E|\phi_i\rangle$ is a codeword and the above relationship would be violated, much as for classical coding.

For the remainder of the chapter several classes of quantum error-correcting codes are discussed, the next section giving three specific codes that have played an important part in the development and Section 16.3 more general classes of codes.

16.2 The Standard Three-, Five- and Nine-Qubit Codes

Three specific codes are discussed in this section. While simple in structure they illustrate the encoding and decoding procedures. The next section considers important and more general construction techniques.

The Three-Qubit Code

A three-qubit code is first considered. The two codewords are given by:

$$|0_L\rangle \mapsto |000\rangle \quad \text{and} \quad |1_L\rangle \mapsto |111\rangle,$$

denoting logical 0 and 1, similar to the repetition code of classical coding. Notice [1] the mapping represents an embedding of the two-dimensional space of the single qubit into a two-dimensional subspace of the eight dimensions of the three-qubit space rather than a cloning.

The code maps the state $a|0\rangle + b|1\rangle$ into

$$a|0\rangle + b|1\rangle \mapsto \phi = a|000\rangle + b|111\rangle.$$

To decode a single bit flip consider the following projection operators [12]:

$$\begin{aligned}
P_0 &= |000\rangle\langle 000| + |111\rangle\langle 111| \\
P_1 &= |100\rangle\langle 100| + |011\rangle\langle 011| \\
P_2 &= |010\rangle\langle 010| + |101\rangle\langle 101| \\
P_3 &= |001\rangle\langle 001| + |110\rangle\langle 110|.
\end{aligned}$$

These are recognized as projection operators as the following example will illustrate.

Suppose a single bit flip has occurred in the state $|\,\phi\rangle$ in position 2 of the codeword above then it is easily computed that

$$\begin{aligned}\langle\phi|\,P_0|\,\phi\rangle &= 0\\ \langle\phi|\,P_1|\,\phi\rangle &= 0\\ \langle\phi|\,P_2|\,\phi\rangle &= 1\\ \langle\phi|\,P_3|\,\phi\rangle &= 0\end{aligned}$$

indicating the bit flip in the second position. To illustrate a computation consider (mapping the quantum states to eight-dimensional vectors):

$$\begin{aligned}\langle\phi|P_2|\,\phi\rangle &= \Big(a\langle 010| + b\langle 101|\Big)\Big(|\,010\rangle\langle 010| + |\,101\rangle\langle 101|\Big)\Big(a|\,010\rangle + b|\,101\rangle\Big)\\ &= \Big(a\langle 010| + b\langle 101|\Big)\Big(a|\,010\rangle + b.0 + a.0 + b|\,101\rangle\Big)\\ &= a^2 + b^2 = 1\end{aligned}$$

using the inner product formula of the previous chapter. Thus if any bit flip occurs it can be corrected in this manner.

An interesting alternative to this method of correcting bit flips is the following [12]. Consider the projection operator on the three qubits:

$$\Big(|\,00\rangle\langle 00| + |\,11\rangle\langle 11|\Big) \otimes I_2 - (|\,01\rangle\langle 01| + |\,10\rangle\langle 10|) \otimes I_2$$

and note the results of the operator acting on the states $|\,00\rangle, |\,01\rangle, |\,10\rangle$ and $|\,11\rangle$ are, respectively, 1, -1, -1, and 1. This can be expressed as $Z_1 Z_2$ which can be interpreted as giving $+1$ if the first two qubits are the same and -1 if they are different. This is a convenient way of expressing the relation

$$Z_1 \otimes Z_2 \otimes I_2$$

where Z_i is the Pauli operator previously defined on the i-th qubit and this matrix acting on the set of states gives

$$\begin{bmatrix} 1 & 0 & 0 & 0 & 0 & 0 & 0 & 0\\ 0 & 1 & 0 & 0 & -1 & 0 & 0 & 0\\ 0 & 0 & -1 & 0 & 0 & 0 & 0 & 0\\ 0 & 0 & 0 & -1 & 0 & 0 & 0 & 0\\ 0 & 0 & 0 & 0 & -1 & 0 & 0 & 0\\ 0 & 0 & 0 & 0 & 0 & -1 & 0 & 0\\ 0 & 0 & 0 & 0 & 0 & 0 & 1 & 0\\ 0 & 0 & 0 & 0 & 0 & 0 & 0 & 1 \end{bmatrix} \cdot \begin{bmatrix} 000\\ 001\\ 010\\ 011\\ 100\\ 101\\ 110\\ 111 \end{bmatrix} = \begin{bmatrix} 000\\ 001\\ -010\\ -011\\ -100\\ -101\\ 110\\ 111 \end{bmatrix}$$

from which the view that $Z_1 Z_2$, as an operator, i.e. as above, is $+1$ for states for which the first two state qubits are the same and -1 if they are different. Similarly we have

$$I_1 \otimes Z_2 \otimes Z_3 = \begin{bmatrix} 1 & 0 & 0 & 0 & 0 & 0 & 0 & 0 \\ 0 & -1 & 0 & 0 & -1 & 0 & 0 & 0 \\ 0 & 0 & -1 & 0 & 0 & 0 & 0 & 0 \\ 0 & 0 & 0 & -1 & 0 & 0 & 0 & 0 \\ 0 & 0 & 0 & 0 & 1 & 0 & 0 & 0 \\ 0 & 0 & 0 & 0 & 0 & -1 & 0 & 0 \\ 0 & 0 & 0 & 0 & 0 & 0 & -1 & 0 \\ 0 & 0 & 0 & 0 & 0 & 0 & 0 & 1 \end{bmatrix} \cdot \begin{bmatrix} 000 \\ 001 \\ 010 \\ 011 \\ 100 \\ 101 \\ 110 \\ 111 \end{bmatrix} = \begin{bmatrix} 000 \\ -001 \\ -010 \\ 011 \\ 100 \\ -101 \\ -110 \\ 111 \end{bmatrix}$$

from which the shorthand observation that Z_2Z_3 operating on the states is $+1$ if the second and third qubits are the same and -1 otherwise.

These quantities Z_1, Z_2 and Z_3 can be viewed as syndromes in classical coding in the sense that, e.g., if $Z_1Z_2 = -1$ and $Z_2Z_3 = +1$ then the first two qubits are different and qubits 2 and 3 are the same, and the first qubit is most likely to have been in error while if $Z_1Z_2 = -1$ and $Z_2Z_3 = -1$ the second qubit is most likely an error, etc. Quantum gates representing the unitary operators Z_i are simple to implement.

Suppose now that a phase flip has occurred in a state $|\,1\rangle \mapsto -|\,1\rangle$ in any of the three positions of the three-bit code. To see how this can be corrected consider rotating the axis of the basis to states

$$|\,p\rangle = (|\,0\rangle + |\,1\rangle)/\sqrt{2} \quad \text{and} \quad |\,m\rangle = (|\,0\rangle - |\,1\rangle)/\sqrt{2}$$

where the p and m are "plus" and "minus" rotations. This is equivalent to operating on the data with a Hadamard operator. The above three-bit code in the original basis now maps to the codewords

$$|\,p_L\rangle = |\,ppp\rangle \quad \text{and} \quad |\,m_L\rangle = |\,mmm\rangle$$

where the subscript L denotes logical as before.

In this rotated basis a phase shift (the Pauli Z operator) $1 \leftrightarrow -1$ acts as a bit flip operator on the $|\,p\rangle \leftrightarrow |\,m\rangle$ states. Thus a phase flip in the $|\,0\rangle$, $|\,1\rangle$ basis corresponds to a bit flip in the $|\,p\rangle$, $|\,m\rangle$ basis. The rotated basis is easily derived from the original basis using unitary transformations corresponding to CNOT gates and Hadamard transforms.

Thus to correct phase flips in the original basis one can use the previous bit flip procedure on the rotated basis. It can be shown that the previous projection operators, in a rotated basis resolve to

$$P_j' \longrightarrow H^{\otimes 3} P_j H^{\otimes 3}$$

and, similarly, the syndrome measurements resolve to

$$H^{\otimes 3} Z_1 Z_2 H^{\otimes 3} = X_1 X_2 \quad \text{and} \quad H^{\otimes 3} Z_2 Z_3 H^{\otimes 3} = X_2 X_3.$$

It can be shown that $X_1 X_2$ will be $+1$ if the first two qubits in the rotated basis are the same – either $|\,p\rangle|\,p\rangle \otimes (\cdot)$ or $|\,m\rangle|\,m\rangle \otimes (\cdot)$ otherwise. Then, from these syndromes, if a bit flip is detected in the first qubit of the rotated basis ($|\,p\rangle \leftrightarrow |\,m\rangle$) this corresponds to a phase flip in the original basis ($|\,1\rangle \leftrightarrow |-1\rangle$) and is corrected by applying the operator $HX_1H = Z_1$.

Thus the three-bit code can correct a single bit flip in the original basis or a single phase flip in the rotated basis. The nine-qubit Shor code will correct both a bit flip and a phase flip (not necessarily on the same qubit) using many of the same ideas. For a more detailed discussion of this interesting approach see [5, 12].

The Five- Qubit Code

A five-qubit quantum single error-correcting code was found [2, 10]. The two logical codewords (codewords that correspond to the transmission of a data 0 and a 1) are sums of 16 states as follows:

$$\begin{aligned} |\,0_L\rangle = {} & |\,00000\rangle + |\,10010\rangle + |\,01001\rangle + |\,10100\rangle \\ & + |\,01010\rangle - |\,11011\rangle - |\,00110\rangle - |\,11000\rangle \\ & - |\,11101\rangle - |\,00011\rangle - |\,11110\rangle - |\,01111\rangle \\ & - |\,10001\rangle - |\,01100\rangle - |\,10111\rangle + |\,00101\rangle \end{aligned}$$

and

$$\begin{aligned} |\,1_L\rangle = {} & |\,11111\rangle + |\,01101\rangle + |\,10110\rangle + |\,01011\rangle \\ & + |\,10101\rangle - |\,00100\rangle - |\,11001\rangle - |\,00111\rangle \\ & - |\,00010\rangle - |\,11100\rangle - |\,00001\rangle - |\,10000\rangle \\ & - |\,01110\rangle - |\,10011\rangle - |\,01000\rangle + |\,11010\rangle \end{aligned}$$

The codeword $|\,1_L\rangle$ is the complement of $|\,0_L\rangle$ (16 terms, each the complement of the above statements with the same sign). The codeword $|\,0_L\rangle$ contains all terms with an odd number of 0's and $|\,1_L\rangle$ with an even number. That these two codewords can correct an arbitrary (bit flip, phase flip) error the reader is referred to the arguments in [7, 12]. This code will be discussed further in the next section from a stabilizer construction point of view. In particular, as it is a $[\![5,1,3]\!]$ code its parameters will be shown to meet the quantum Hamming bound for single error-correcting codes (see Equation 16.3 for the general Hamming bound)

$$(1+3n)2^k \leq 2^n$$

and hence is a perfect code (in analogy with classical coding), one of many single error-correcting perfect quantum codes. A more detailed discussion of the structure of this code is given in [11].

There is an interesting seven-qubit error-correcting code which is a CSS code, also referred to as the Steane code, which is discussed in the next section.

The Shor Nine-Qubit Code

Using the notation above the two codewords of the Shor nine-qubit code are

$$\begin{aligned} |\,0_L\rangle &\mapsto \frac{1}{2\sqrt{2}}\big(|\,000\rangle + |\,111\rangle\big)\big(|\,000\rangle + |\,111\rangle\big)\big(|\,000\rangle + |\,111\rangle\big) \\ |\,1_L\rangle &\mapsto \frac{1}{2\sqrt{2}}\big(|\,000\rangle - |\,111\rangle\big)\big(|\,000\rangle - |\,111\rangle\big)\big(|\,000\rangle - |\,111\rangle\big). \end{aligned}$$

Consider the operator $X_1X_2X_3X_4X_5X_6$ acting on the state

$$\big(|\,000\rangle + |\,111\rangle\big)\big(|\,000\rangle \pm |\,111\rangle\big).$$

The result will be

$$\begin{aligned} X_1X_2X_3X_4X_5X_6&\big(|\,111\rangle + |\,000\rangle\big)\big(|\,111\rangle + |\,000\rangle\big) \\ &= \big(|\,000\rangle + |\,111\rangle\big)\big(|\,000\rangle + |\,111\rangle\big) \\ X_1X_2X_3X_4X_5X_6&\big(|\,111\rangle + |\,000\rangle\big)\big(|\,000\rangle - |\,111\rangle\big) \\ &= \big(|\,000\rangle + |\,111\rangle\big)\big(|\,111\rangle - |\,000\rangle\big) \\ &= -\big(|\,000\rangle + |\,111\rangle\big)\big(|\,000\rangle - |\,111\rangle\big) \end{aligned}$$

Thus the operator $X_1X_2X_3X_4X_5X_6$ takes on the value $+1$ if the signs in the two brackets are the same and -1 if they are different. Similarly for the syndrome $X_4X_5X_6X_7X_8X_9$ for the signs in the last two brackets. A complete set of syndromes then for the Shor nine-qubit code is

$$\begin{aligned} &S_1 = Z_1Z_2,\ S_2 = Z_2Z_3,\ S_3 = Z_4Z_5,\ S_4 = Z_5Z_6, \\ &S_5 = Z_7Z_8,\ S_6 = Z_8Z_9,\ S_7 = X_1X_2X_3X_4X_5X_6,\ S_8 = X_4X_5X_6X_7X_8X_9. \end{aligned}$$

The first six syndromes determine operations on the bits within a bracket and the last two the signs between brackets.

As an example of the code correcting one bit flip and one phase flip in the same qubit, suppose the second codeword above is transmitted and suffers both a bit flip and phase flip in the fifth qubit. Thus the received word is

$$\boldsymbol{r} = \big(|\,000\rangle - |\,111\rangle\big)\big(|\,010\rangle + |\,101\rangle\big)\big(|\,000\rangle - |\,111\rangle\big).$$

The relevant syndromes are $S_1 = +1, S_2 = +1, S_3 = -1, S_4 = -1, S_5 = +1, S_6 = +1, S_7 = -1, S_8 = -1$. To correct bit flips first, the syndromes $S_3 = -1,\ S_4 = -1$ indicate a bit flip in the fifth bit. It could also have indicated two bit flips in the fourth and sixth qubits but the former is more likely. It is corrected to

$$\boldsymbol{c} = \big(|\,000\rangle - |\,111\rangle\big)\big(|\,000\rangle + |\,111\rangle\big)\big(|\,000\rangle - |\,111\rangle\big).$$

The syndromes $S_7 = -1$, $S_8 = -1$ indicate a phase flip among the one qubits of the middle bracket – it could have been in any of them. These syndromes could also have resulted from phase flips in the first and third bracket but the former is more likely. It is corrected to

$$\boldsymbol{c} = \big(|\,000\rangle - |\,111\rangle\big)\big(|\,000\rangle - |\,111\rangle\big)\big(|\,000\rangle - |\,111\rangle\big).$$

16.3 CSS, Stabilizer and $\mathbb{F}_4$ Codes

The CSS codes form an interesting class of quantum error-correcting codes, derived from classical error-correcting codes that are able to use the classical error-correcting properties and algorithms to correct quantum errors in the qubits. They are based on the work of Calderbank and Shor in [3] and Steane in [13] and discussed extensively in [5, 12].

Let $C_1 = (n,k_1)_2$ and $C_2 = (n,k_2)_2$, $C_2 \subset C_1$ be two (classical) linear codes over $\mathbb{F}_2$ such that $k_1 > k_2$ and C_1 and $C_2^{\perp}$ are able to correct t errors, i.e., the minimum distance of the respective codes is at least $2t+1$ in the classical code/channel case. It is shown [12] how to derive a quantum n-qubit error-correcting code, designated $CSS(C_1,C_2)$ capable of correcting t-qubit errors (bit flips and phase flips – a total of t) using the classical code error-correction algorithms. The code will contain $2^{k_1-k_2}$ codewords.

The quantum states will be on n qubits, designated $|\,\boldsymbol{x}\rangle, \boldsymbol{x} \in C_1$ giving a space of dimension 2^n and the states are orthogonal. Denote the superimposed quantum state

$$|\,\boldsymbol{x}+C_2\rangle = \frac{1}{\sqrt{|C_2|}}\sum_{\boldsymbol{y}\in C_2}|\,\boldsymbol{x}\oplus\boldsymbol{y}\rangle = \frac{1}{2^{k_2}/2}\sum_{\boldsymbol{y}\in C_2}|\,\boldsymbol{x}\oplus\boldsymbol{y}\rangle, \quad \boldsymbol{x}\in C_1.$$

Denote the set of cosets of C_2 in C_1 as $\mathscr{C}_{C_1|C_2}$. The set can be viewed as a $2^{k_1-k_2} \times 2^{k_2}$ array of codewords of C_1. The rows of the array (cosets) are disjoint. Label the codewords in the first column of this array as $\boldsymbol{x}_j, j = 0,1,\ldots,2^{k_1-k_2}-1$ called coset leaders (a standard term in classical coding where they are usually chosen as minimal weight within the coset). Any codeword in a coset can act as a coset leader as the cosets are closed under addition of codewords of C_2. One can define addition on the cosets in an obvious manner and the set of cosets is closed under this addition.

For $\boldsymbol{x},\boldsymbol{x}'$ in the same coset, $|\,\boldsymbol{x}+C_2\rangle$ is equal to $|\,\boldsymbol{x}'+C_2\rangle$. For $\boldsymbol{x},\boldsymbol{x}'$ not in the same coset, the states $|\,\boldsymbol{x}+C_2\rangle$ and $|\,\boldsymbol{x}'+C_2\rangle$ are orthogonal, being the disjoint sums of orthonormal states. The CSS code is then the set of states

$$CSS(C_1,C_2) = \Big\{|\,\boldsymbol{x}_j + C_2\rangle,\ j = 0,1,2,\ldots,2^{k_1-k_2}\Big\}.$$

Consider first the correction of bit flips in a state. Consider a corrupted codeword $|\boldsymbol{x}+C_2\rangle, \boldsymbol{x} \in C_1$ with at most t bit flips among the n qubits of the code. This can be modeled as the state

$$\frac{1}{2^{k_2/2}} \sum_{y \in C_2} |\boldsymbol{x} \oplus \boldsymbol{y} \oplus \boldsymbol{e}_1\rangle, \ \boldsymbol{x}+\boldsymbol{y} \in C_1$$

where $\boldsymbol{e}_1 \in \mathbb{F}_2^n$ contains ones where the bit flips occur – of weight at most t. Let an $(n-k_1)\times n$ parity-check matrix for C_1 be H_1. By definition $H_1(\boldsymbol{x}+\boldsymbol{y}+\boldsymbol{e}_1) = H_1\boldsymbol{e}_1$. Add a sufficient number of ancilla (extra, initially empty) states to the system state to allow the syndrome $H_1\boldsymbol{e}_1$ to be derived, i.e., by applying the unitary operators to the states $|\boldsymbol{x} \oplus \boldsymbol{y} \oplus \boldsymbol{e}_1\rangle$ to implement the effect of H_1 on the system state which in effect is computing the syndrome of the "received word" $\boldsymbol{x} \oplus \boldsymbol{y} \oplus \boldsymbol{e}_1$. The result $|H_1\boldsymbol{e}_1\rangle$ is stored in the ancilla states and the entire system plus ancilla states give the state:

$$|\boldsymbol{x}+\boldsymbol{y}+\boldsymbol{e}_1\rangle |H_1\boldsymbol{e}_1\rangle.$$

The quantity $H_1\boldsymbol{e}_1$ is the syndrome of the error pattern and this can be read from the ancilla states which are then discarded. From this syndrome the classical error-correcting algorithm can be used to determine the error (bit flip) positions which are then corrected in the system quantum state.

Consider [12] the case of t phase flips errors and the corrupted codeword represented as

$$\frac{1}{\sqrt{|C_2|}} \sum_{y \in C_2} (-1)^{(\boldsymbol{x} \oplus \boldsymbol{y}, \boldsymbol{e}_2)} |\boldsymbol{x} \oplus \boldsymbol{y}\rangle \tag{16.2}$$

where the phase flip errors occur where the vector $\boldsymbol{e}_2$ is one. The result of the phase flips on the codeword component $|\boldsymbol{x} \oplus \boldsymbol{y}\rangle$ is to yield $-|\boldsymbol{x} \oplus \boldsymbol{y}\rangle$ if the number of positions where the word $\boldsymbol{x} \oplus \boldsymbol{y}$ and where the phase flips occur is odd, i.e., if the inner product $(\boldsymbol{x} \oplus \boldsymbol{y}, \boldsymbol{e}_2)$ is odd.

By applying Hadamard gates to each of the n qubits it can be shown [12] the problem reduces to one similar to the bit flip case, an argument that was used previously. This operation wields 2^n terms for each of the n-qubit states $|\boldsymbol{x} \oplus \boldsymbol{y}\rangle$ in the above equation. For example, applying Hadamard gates to each of the three qubits in the state $|001\rangle$ yields the set of states:

$$\frac{1}{2\sqrt{2}}\Big((|0\rangle + |1\rangle)(|0\rangle - |1\rangle)(|0\rangle - |1\rangle)\Big)$$

and expanding these out yields eight terms:

$$\frac{1}{2\sqrt{2}}\Big(|000\rangle - |001\rangle - |010\rangle + |011\rangle + |100\rangle - |101\rangle - |110\rangle + |111\rangle\Big).$$

In general, all 2^n possible terms are present and the sign of the terms is to be determined. The following shows the computation.

Applying this to Equation 16.2 can be shown [12] to yield

$$\frac{1}{\sqrt{|C_2|\,2^n}} \sum_{z\in\{0,1\}^n} \sum_{y\in C_2} (-1)^{(\boldsymbol{x}\oplus\boldsymbol{y},\boldsymbol{e}_2\oplus\boldsymbol{z})}|\boldsymbol{z}\rangle.$$

Changing the variable to $\boldsymbol{w}=\boldsymbol{z}\oplus\boldsymbol{e}_2$ this can be expressed as

$$\frac{1}{\sqrt{|C_2|\,2^n}} \sum_{w\in\{0,1\}^n} \sum_{y\in C_2} (-1)^{(\boldsymbol{x}\oplus\boldsymbol{y},\boldsymbol{w})}|\boldsymbol{w}\oplus\boldsymbol{e}_2\rangle.$$

Now for any binary linear code C, if $\boldsymbol{x}\in C^{\perp}$ then $\sum_{y\in C}(-1)^{(x,y)}=|C|$ while if $\boldsymbol{x}\notin C$ then $\sum_{y\in C}(-1)^{(x,y)}=0$. This follows since the number of codewords $\boldsymbol{y}\in C$ such that $(\boldsymbol{x},\boldsymbol{y})=0, \boldsymbol{x}\notin C$ is the same as the number for which the inner product is 1 and hence the result. (The set for which the inner product is zero is a subspace.) This argument reduces the above equation to

$$\frac{1}{\sqrt{2^n/\sqrt{|C_2|}}} \sum_{w\in C_2^{\perp}} (-1)^{(\boldsymbol{x},\boldsymbol{w})}|\boldsymbol{w}\oplus\boldsymbol{e}_2\rangle.$$

The form of this equation is precisely that used to correct bit flips (the $(-1)^{(\boldsymbol{x},\boldsymbol{w})}$ term does not affect the argument). Thus, returning to the original equation up to t phase flips can be corrected using the classical decoding algorithm to determine the positions of the phase flips.

The above development treated the cases of bit flips separate to phase flips. Clearly the analysis [12] could have considered the combined expression

$$\frac{1}{\sqrt{|C_2|}} \sum_{y\in C_2} (-1)^{(x\oplus y,e_2)}|\boldsymbol{x}\oplus\boldsymbol{y}+\boldsymbol{e}_1\rangle$$

for a total of t errors although the analysis of the two cases remains the same.

Steane [13] independently arrived at an equivalent construction and the code contained in the following example is usually referred to as the Steane seven-qubit quantum code ([3, 7, 12]):

The Steane Seven-Qubit Code

Let C_1 denote the Hamming $(7,4,3)_2$ classical Hamming code and $C_2=C_1^{\perp}$, a $(7,3,4)_2$ code consisting of those codewords of C_1 of even weight. The Steane code is the $CSS(C_1,C_2)$ code and a codeword of this seven-qubit code is

$$| 0_L \rangle = \frac{1}{\sqrt{8}} \Big\{ | 0000000\rangle + | 1010101\rangle + | 0110011\rangle + | 1100110\rangle$$
$$+ | 0001111\rangle + | 1011010\rangle + | 0111100\rangle + | 1101001\rangle \Big\}$$

and the codeword $| 1_L \rangle$ takes the complements of these states. Again, this is a $[[7, 1, 3]]$ quantum error-correcting code capable of correcting an arbitrary error in a single qubit.

To elaborate on its construction of this code C_1, let a generator and a parity-check matrix for the $(7, 4, 3)_2$ Hamming code be denoted

$$G = \begin{bmatrix} 1 & 0 & 0 & 0 & 0 & 1 & 1 \\ 0 & 1 & 0 & 0 & 1 & 0 & 1 \\ 0 & 0 & 1 & 0 & 1 & 1 & 0 \\ 0 & 0 & 0 & 1 & 1 & 1 & 1 \end{bmatrix} \quad \text{and} \quad H = \begin{bmatrix} 0 & 1 & 1 & 1 & 1 & 0 & 0 \\ 1 & 0 & 1 & 1 & 0 & 1 & 0 \\ 1 & 1 & 0 & 1 & 0 & 0 & 1 \end{bmatrix},$$

respectively. The codewords of $C_2 = C_1^{\perp}$ are

0000000 0111100 1011010 1101001
1100110 1010101 0110011 0001111

which gives the codeword $| 0_L \rangle$ above. The coset of C_2 in C_1 is given by adding a binary 7-tuple not in C_2 to the words of C_2, say 1111111 which gives the complements of the components of $| 0_L \rangle$ for $| 1_L \rangle$. The code is considered further later from a stabilizer/generator point of view.

Many of the quantum codes developed are related to classical codes and, in some manner, decode using a variant of the classical decoding algorithm. As noted earlier, many works denote a binary linear classical code as $(n, k, d)_2$ and a quantum code on n qubits as $[[n, k, d]]$ with double brackets, the convention which is followed here. Such a code is of dimension 2^k.

Given the relationship to classical coding notions, it is natural to consider quantum code bounds analogous to the classical code bounds. Corresponding to the three Pauli matrices, there are three types of quantum errors a given qubit can suffer and hence a quantum code that can encode k qubits into n qubits and correct t quantum errors must satisfy the relation

$$\left(\sum_{i=0}^{t} 3^i \binom{n}{i} \right) 2^k \leq 2^n, \qquad \text{quantum Hamming bound} \quad (16.3)$$

which is the analog of the classical coding Hamming bound – referred to as the quantum Hamming bound. As with classical codes, if equality is achieved the code is *perfect*. In particular for $t = 1$ the bound reduces to

$$(1 + 3n)2^k \leq 2^n$$

and a code meeting this bound will be an $[\![n,k,3]\!]$ code and a perfect single error-correcting code, an example of which is the five-qubit code discussed above.

In a similar manner, using arguments as in the classical coding situation, it can be shown that if the following condition is satisfied

$$\left(\sum_{i=0}^{d-1} 3^i \binom{n}{i}\right) 2^k \leq 2^n, \qquad \text{quantum Varshamov–Gilbert bound}$$

then an $[\![n,k,d]\!]$ quantum error-correcting code will exist. Using standard approximations to the binomial coefficients these two relationships can be expressed as ([3], [6], part of theorem 4):

Theorem 16.1 *For large n, $R = k/n$ and $p = d/2n$ fixed, the best nondegenerate quantum codes satisfy*

$$1 - 2p\log_2 3 - H_2(2p) \leq R \leq 1 - p\log_2 3 - H_2(p)$$

where $H_2(\cdot)$ is the binary entropy function.

Similarly the quantum bound analogous to the classical Singleton bound for an $[\![n,k,d]\!]$ quantum code can be shown as [9]

$$n - k \geq 2d - 2. \qquad \text{quantum Singleton bound}$$

To discuss the stabilizer approach to quantum codes it will be helpful to recall a few notions from group theory. For an arbitrary finite group $\mathcal{G}$ the *center* of $\mathcal{G}$, usually denoted $Z(\mathcal{G})$, is defined as the set of elements that commute with all elements of $\mathcal{G}$, i.e.,

$$Z(\mathcal{G}) = \{a \in \mathcal{G} \mid ag = ga \ \forall g \in \mathcal{G}\}.$$

Also a subgroup $N \subset \mathcal{G}$ is a *normal* subgroup, denoted $N \lhd \mathcal{G}$ if for all $n \in N$ and $g \in \mathcal{G}$, $gng^{-1} \in N$.

More generally, for S a subset (not necessarily a subgroup) of $\mathcal{G}$ the *centralizer* of S in $\mathcal{G}$ is

$$Z_{\mathcal{G}}(S) = \{g \in \mathcal{G} \mid gs = sg \ \forall s \in S\},$$

i.e., the set of group elements that commute with all elements of S. Similarly the *normalizer* of S in $\mathcal{G}$ is

$$N_{\mathcal{G}}(S) = \{g \in \mathcal{G} \mid gS = Sg\}.$$

Clearly the centralizer is a more stringent requirement and $Z_{\mathcal{G}}(S) \subseteq N_{\mathcal{G}}(S)$ and both are subgroups.

As a trivial introduction to stabilizer codes consider the two three-qubit codewords of C given by

$$|\phi_0\rangle = \frac{1}{\sqrt{2}}\Big(|000\rangle + |111\rangle\Big) \quad |\phi_1\rangle = \frac{1}{\sqrt{2}}\Big(|000\rangle - |111\rangle\Big).$$

Consider the error operator $Z_1Z_2 \in \mathcal{G}_3$ acting on these codewords giving

$$Z_1Z_2|\phi_0\rangle = |\phi_0\rangle \quad \text{and} \quad Z_1Z_2|\phi_2\rangle = |\phi_2\rangle$$

which can be described as the operator Z_1Z_2 *stabilizing* the codewords of C. In a similar argument the codewords are stabilized by Z_1Z_3 and Z_2Z_3 and the set of operators

$$\{I, Z_1Z_2, Z_1Z_3, Z_2Z_3\}$$

is a subgroup of $\mathcal{G}_3$. Note that while $X_1X_2X_3$ stabilizes $|\phi_0\rangle$ it does not stabilize $|\phi_1\rangle$.

The plan, which appears in Gottesman [5] in his Caltech thesis and independently in [4], is that by considering the stabilizer of a code one can determine a set of error operators that it is able to correct. An outline of this innovative work is described.

Specifically suppose C is a quantum error-correcting code that encodes k qubits into n qubits, and S a subset of $\mathcal{G}_n$ that fixes (stabilizes) the codewords. Typically each codeword will contain a sum of states and the operators of S might permute the terms of a codeword while fixing the codeword itself. If $g_1, g_2 \in S$, then for $|\phi\rangle \in C$, $g_1g_2|\phi\rangle = g_1|\phi\rangle = |\phi\rangle$ and $g_1g_2 \in S$ and S is a subgroup of $\mathcal{G}_n$. Furthermore $g_1g_2|\phi\rangle = g_2g_1|\phi\rangle$ and S is Abelian. Alternatively one could define a code by specifying a subgroup $S \subseteq \mathcal{G}_n$ and setting the code as

$$C = \{|\phi\rangle \mid g|\phi\rangle = |\phi\rangle \ \forall g \in S\}.$$

To encode k qubits into n the code C must have dimension 2^k and thus [5, 12] S is a subgroup with 2^{n-k} elements. Such a group can be generated by $n-k$ *independent* elements, i.e.,

$$S = \langle g_1, g_2, \ldots, g_{n-k}\rangle, \ g_i \in sG_n$$

where the angle brackets indicate the set of operators generated by the enclosed elements and where by independent is meant deleting any one element from the set will result in the generation of a strict subset of S.

Notice that $-I_n \notin S$ as otherwise we would have $-I_n|\phi\rangle = |\phi\rangle$ implying $|\phi\rangle = 0$ and the code would be trivial. It is interesting to observe [12] that if $S = \langle g_1, g_2, \ldots, g_{n-k}\rangle$, then $-I_n \notin S$ iff $g_j^2 = I_n$ for all j and $g_j \neq -I_n$ for all j and $g^2 = I_n$ for all $g \in S$ and that $g^\dagger = g$.

Similarly if $g_1, g_2 \in S$, as products of Pauli matrices they either commute or anticommute. Suppose they anticommute and for each state $|\phi\rangle$ in C, $|\phi\rangle = g_1g_2|\phi\rangle = -g_2g_1|\phi\rangle = -|\phi\rangle$, again a contradiction.

Recall that the elements of $\mathcal{G}_n$ are all unitary (hence $g^\dagger = g^{-1} \forall g \in \mathcal{G}_n$) and either Hermitian ($g^\dagger = g$) or anti-Hermitian ($g^\dagger = -g$) (the operators of $g \in G_n$ can be viewed as matrices). In addition recall the elements of $\mathcal{G}_n$ either commute or anticommute.

It has been noted $Z_{\mathcal{G}_n} \subseteq N_{\mathcal{G}_n}$ and indeed [5]:

$$Z_{\mathcal{G}_n}(S) = N_{\mathcal{G}_n}(S)$$

and to see this let $g \in \mathcal{G}_n$, $s \in S$ and $g \in N_{\mathcal{G}_n}(S)$ and consider $g^\dagger s g$. Since $s, g \in \mathcal{G}_n$ they either commute or anticommute so

$$g^\dagger s g = \pm g^\dagger g s = \pm s.$$

However, $-s \notin S$ and so $g^\dagger s g = s$, i.e., $g \in Z_{\mathcal{G}_n}(S)$. Similarly it is shown [5] that $S \lhd N_{\mathcal{G}_n}(S)$ and that $N_{\mathcal{G}_n}(S)$ has $4 \cdot 2^{n+k}$ elements.

The conclusion of these observations ([5], section 3.2, [12], theorem 10.8) is that the code C generated by the stabilizer group $S \subseteq \mathcal{G}_n$ will correct a set of errors $\mathcal{E} \subseteq \mathcal{G}_n$ if $E_a E_b \in S \cup \mathcal{G}_n \backslash N_{\mathcal{G}_n}(S))$ for all $E_a, E_b \in \mathcal{E}$.

Thus the stabilizer approach is able to both provide a means for the efficient description of a code and an estimate of the set of errors it is able to correct, an impressive achievement of the approach.

Several examples of the stabilizer approach are given in the following, using the codes already described.

The Five-Qubit Code

The codewords of this code were previously given as:

$$\begin{aligned} |0_L\rangle = {} & |00000\rangle + |10010\rangle + |01001\rangle + |10100\rangle \\ & +|01010\rangle - |11011\rangle - |00110\rangle - |11000\rangle \\ & -|11101\rangle - |00011\rangle - |11110\rangle - |01111\rangle \\ & -|10001\rangle - |01100\rangle - |10111\rangle + |00101\rangle \end{aligned}$$

and the complements for $|1_L\rangle$. It can be shown [7, 12] that the stabilizer of this code is

$$S = \langle g_1, g_2, g_3, g_4 \rangle$$
where $g_1 = X_1Z_2Z_3X_4,\ g_2 = X_2Z_3Z_4X_5,\ g_3 = X_1X_3Z_4Z_5,\ g_4 = Z_1, X_2, X_4Z_5$
and
$$S = \{I, g_1, g_2, g_3, g_4, g_1g_2, g_1g_3, g_1g_4, g_2g_3, g_2g_4, g_3g_4, g_1g_2g_3, g_1g_2g_4, g_1g_3g_4, g_2g_3g_4, g_1g_2g_3g_4\}.$$

As an example of the computations in terms of operators from $\mathcal{G}_5$, consider e.g.,

$$g_2g_4 = X_2Z_3Z_4X_5Z_1X_2X_4Z_5 = Z_1X_2^2Z_3Z_4X_4X_5Z_5.$$

Operators with different subscripts commute and from the form of the Pauli matrices

$$ZX = iY \quad \text{and} \quad XZ = -iY \quad \text{where } Y = \begin{bmatrix} 0 & 1 \\ -1 & 0 \end{bmatrix}.$$

and so

$$g_2g_4 = Z_1Z_3(iY_4)(-iY_5) = Z_1Z_3Y_4Y_5.$$

For the record we have:

$$\begin{array}{llll} 1 = I_5 & g_1g_2 = X_1Y_2Y_4X_5 & g_3g_4 &= Y_1X_2Y_3Y_4 \\ g_1 = X_1Z_2Z_3X_4 & g_1g_3 = Z_2Y_3Y_4Z_5 & g_1g_2g_3 &= Y_2X_3X_4Y_5 \\ g_2 = X_2Z_3Z_4X_5 & g_1g_4 = Y_1Y_2Z_3Z_5 & g_1g_2g_4 &= Y_1Z_2Y_3X_4 \\ g_3 = X_1X_3Z_4Z_5 & g_2g_3 = X_1X_2Y_3Y_5 & g_1g_3g_4 &= Z_1Y_2Y_3X_4 \\ g_4 = Z_1X_2X_4Z_5 & g_2g_4 = Z_1Z_3Y_4Y_5 & g_2g_3g_4 &= Y_1Y_3X_4X_5 \\ & & g_1g_2g_3g_4 &= Z_1Z_2X_3X_5 \end{array}$$

and hence the codewords are generated by the action of this subgroup on $|\,00000\rangle$ and $|\,11111\rangle$:

$$\begin{aligned} |\,0_L\rangle &= S|\,00000\rangle \\ &= |\,00000\rangle + |\,10010\rangle + |\,01001\rangle + |\,10100\rangle \\ &\quad +|\,01010\rangle - |\,11011\rangle - |\,00110\rangle - |\,11000\rangle \\ &\quad -|\,11101\rangle - |\,00011\rangle - |\,11110\rangle - |\,01111\rangle \\ &\quad -|\,10001\rangle - |\,01100\rangle - |\,10111\rangle + |\,00101\rangle \end{aligned}$$

with similar computations for $|\,1_L\rangle$. Notice the cyclic nature of the generating set. It was previously noted this five-qubit code is perfect in that it meets the Hamming bound (Equation 16.3) for $t = 1$, $n = 5$, $k = 1$:

$$(1 + 3n)2^k \le 2^n \quad \text{or} \quad (1 + 15) \cdot 2 = 2^5 = 32.$$

It is noted [5] that for n of the form $(2^{2j} - 1)/3$ for positive integers $j > 1$ then $(1 + 3n)$ will be a power of 2 (e.g., $n = 5, 21, 85 \ldots$) and hence there is a possibility of perfect codes for such lengths: i.e., $[\![(2^{2j} - 1)/3, (2^{2j} - 1)/3 - 2j, 3]\!]$ and such perfect single error-correcting quantum codes do exist for such parameters.

The Steane Seven-Qubit Code

Recall the codewords of the Steane code (an example of a CSS code) discussed earlier are

$$
\begin{aligned}
|\,0_L\rangle = \frac{1}{\sqrt{8}}\Big\{&|\,0000000\rangle + |\,1010101\rangle + |\,0110011\rangle + |\,1100110\rangle \\
&+|\,0001111\rangle + |\,1011010\rangle + |\,0111100\rangle + |\,1101001\rangle\Big\}
\end{aligned}
$$

and its complements giving $|\,1_L\rangle$. From earlier comments [12] the generators can be derived from the parity-check matrix of the dual of the $(7,4,3)_2$ Hamming code as

$$
\begin{aligned}
S &= \langle g_1 = X_4X_5X_6X_7,\ g_2 = X_2X_3X_6X_7,\ g_3 = X_1X_3X_5X_7\rangle \\
&= \Big\{I_7,\ X_4X_5X_6X_7,\ X_2X_3X_6X_7,\ X_1X_3X_5X_7,\ X_2X_3X_4X_5, \\
&\qquad X_1X_3X_4X_6,\ X_1X_2X_5X_6,\ X_1X_2X_4X_7\Big\}.
\end{aligned}
$$

The terms in the sum of states for each codeword are then obtained, as above, by the operators of the stabilizer group S acting on an initial state, e.g.,

$$
|\,0_L\rangle = S|\,0000000\rangle \quad \text{and} \quad |\,1_L\rangle = S|\,1111111\rangle.
$$

The Shor Nine-Qubit Code

The codewords of the Shor nine-qubit code were reported earlier as:

$$
\begin{aligned}
|\,0\rangle &\mapsto \frac{1}{2\sqrt{2}}\big(|\,000\rangle + |\,111\rangle\big)\big(|\,000\rangle + |\,111\rangle\big)\big(|\,000\rangle + |\,111\rangle\big) \\
|\,1\rangle &\mapsto \frac{1}{2\sqrt{2}}\big(|\,000\rangle - |\,111\rangle\big)\big(|\,000\rangle - |\,111\rangle\big)\big(|\,000\rangle - |\,111\rangle\big).
\end{aligned}
$$

The stabilizer description of this code has a stabilizer group

$$
\begin{aligned}
S = \Big\{&g_1 = Z_1Z_2,\ g_2 = Z_2Z_3,\ g_3 = Z_4Z_5,\ g_4 = Z_5Z_6,\ g_5 = Z_7Z_8, \\
&g_6 = Z_8Z_9,\ g_7 = X_1X_2X_3X_4X_5X_6,\ g_8 = X_4X_5X_6X_7X_8X_9\Big\}.
\end{aligned}
$$

The code generation from the stabilizer group is not pursued here.

The following gives an overview of the interesting work of [3] that uses classical error-correcting codes over $\mathbb{F}_4$ to construct large classes of quantum codes. Denote by $\bar{E}$ the quotient group

$$
\bar{E} \stackrel{\Delta}{=} \mathcal{G}_n/\{\pm I,\ \pm iI\}
$$

which is an elementary Abelian group of order 2^{2n} and hence is isomorphic to a binary vector space of dimension $2n$. Elements of $\bar{E}$ are written as $(a \mid b)$ (catenation), a, b binary n-tuples and the space is furnished with the inner product

$$
\big((a_1 \mid b_1), (a_2 \mid b_2)\big) = (a_1, b_2) + (a_2, b_1) \in \mathbb{F}_2
$$

where $(\cdot, \cdot)$ is the usual inner product. This is a symplectic inner product since $\big((a \mid b), (a \mid b)\big) = 0$. The weight of element $(a \mid b) \in \bar{E}$ is the number of the n coordinates i such that at least one of a_i, b_i is 1. The following theorem (theorem 1 of [3]) sets the stage for a further theorem that derives quantum codes from classical error-correcting codes.

Theorem 16.2 *Let $\bar{S}$ be an $(n-k)$-dimensional linear subspace of $\bar{E}$ which is contained in $\bar{S}^{\perp}$ (by the above symplectic inner product) such that there are no vectors of weight $< d$ in $\bar{S}^{\perp} \backslash \bar{S}$. Then there is a quantum error-correcting code mapping k qubits to n qubits which can correct $\lfloor (d-1)/2 \rfloor$ errors.*

Codes obtained from this theorem are termed *additive codes*.

Let ω be a zero of the polynomial $x^2 + x + 1$, irreducible over $\mathbb{F}_2$ and note that $\mathbb{F}_4 = \{0, 1, \omega, \bar{\omega} = \omega^2 = \omega + 1\}$. Associate to $v = (a \mid b) \in \bar{E}, a, b$ binary n-tuples, the $\mathbb{F}_4$ n-tuple $\phi(v) = \omega a + \bar{\omega} b \in \mathbb{F}_4^n$ and note that

$$\begin{aligned} \bar{E}\text{-weight of } v(a \mid b) \in \bar{E} &= \text{usual Hamming weight of } \phi(v) \in \mathbb{F}_4^n \\ \bar{E}\text{-distance of } v_1 = (a_1 \mid b_1),\ v_2 = (a_2 \mid b_2) \in \bar{E} & \\ &= \text{Hamming distance } d_H(\phi(v_1), \phi(v_2)). \end{aligned}$$

A key theorem of the work then is:

Theorem 16.3 *Suppose C is an additive self-orthogonal subcode of $\mathbb{F}_4^n$ containing 2^{n-k} vectors, such that there are no vectors of weight $< d$ in $C^{\perp} \backslash C$. Then any eigenspace of $\phi^{-1}(C)$ is an additive quantum error-correcting code with parameters $[\![n, k, d]\!]$.*

The theorem applied to classical codes leads to a wide array to quantum error-correcting codes. In essence the elements of the field $\mathbb{F}_4$ are associated with the four Pauli error matrices.

Comments

The chapter has introduced a few of the ideas related to the construction and performance of quantum error-correcting codes although many important aspects of the subject, such as the implementation of the codes to obtain reliable quantum computation and the measurement of computation outputs, were not addressed. The literature on these codes is substantial and goes far beyond the ideas mentioned here. The importance of quantum error correction and its potential impact on the implementation of quantum computers will ensure the importance of research on such codes.

References

[1] Bennett, C.H., and Shor, P.W. 1998. Quantum information theory. vol. 44. IEEE.

[2] Bennett, C.H., DiVincenzo, D.P., Smolin, J.A., and Wootters, W.K. 1996. Mixed-state entanglement and quantum error correction. *Phys. Rev. A*, **54**, 3824–3851.

[3] Calderbank, A.R., and Shor, P.W. 1996. Good quantum error-correcting codes exist. *Phys. Rev. A*, **54**(Aug), 1098–1105.

[4] Calderbank, A.R., Rains, E.M., Shor, P.W., and Sloane, N.J.A. 1997. Quantum error correction and orthogonal geometry. *Phys. Rev. Lett.*, **78**, 405–408.

[5] Gottesman, D. 1997. Stabilizer codes and quantum error correction. arXiv:quant-ph/9705052.

[6] Gottesman, D. 2002. An introduction to quantum error correction. Pages 221–235 of: *Quantum computation: a grand mathematical challenge for the twenty-first century and the millennium (Washington, DC, 2000)*. Proceedings of Symposia in Applied Mathematics, vol. 58. American Mathematical Society, Providence, RI.

[7] Huang, L., and Wu, X. 2021. New construction of nine-qubit error-correcting code. arXiv::quant-ph2110.05130v4.

[8] Huffman, W.C., Kim, J.L., and Solé, P. (eds.). 2021. *A concise encyclopedia of coding theory*. CRC Press, Boca Raton, FL.

[9] Knill, E., and Laflamme, R. 1997. A theory of quantum error correcting codes. *Phys. Rev. A*, **55**, 900–911.

[10] Laflamme, R., Miquel, C., Paz, J.P, and Zurek, W.H. 1996. Perfect quantum error correcting code. *Phys. Rev. Lett.*, **77**, 198–201.

[11] Mermin, N.D. 2007. *Quantum computer science: an introduction*. Cambridge University Press, Cambridge.

[12] Nielsen, M.A., and Chuang, I.L. 2000. *Quantum computation and quantum information*. Cambridge University Press, Cambridge.

[13] Steane, A.M. 1998. Introduction to quantum error correction. *R. Soc. Lond. Philos. Trans. Ser. A Math. Phys. Eng. Sci.*, **356**(1743), 1739–1758.

17
Other Types of Coding

From the contents of the previous chapters it is clear the term "coding" covers a wide variety of applications and techniques. The literature contains a much more impressive array of topics. It seems that as each new technological development arrives, a form of coding arises to improve its performance. The purpose of this chapter is to give brief and shallow descriptions of a few of them, ranging from a paragraph to a few pages for each topic to illustrate, hopefully extending slightly the range of coding considered in this volume. The list remains far from exhaustive. The range of mathematics beyond that already encountered and the ingenuity of techniques used to effect solutions to intriguing problems is interesting.

17.1 Snake-in-the-Box, Balanced and WOM Codes

Snake-in-the-Box Codes and Gray Codes

Let C_n denote the binary cube in n dimensions, i.e., a graph with 2^n vertices where each vertex has n neighbors. It is convenient to label the vertices with binary n-tuples in such a way that two vertices are adjacent iff their labels have Hamming distance of one. A Gray code then will be an ordered list of binary n-tuples (vertex labels) such that two consecutive codewords have distance one. Clearly such a code corresponds to a Hamiltonian path where every vertex is visited once. The added constraint that the list of codewords represents a cycle in the graph where the last codeword is distance one from the first, is referred to as a Hamiltonian cycle. That a cube always has a Hamiltonian cycle is well known ([22], theorem 10.1.1). One application for such a code might be in labeling the levels of a quantizer with the codewords. If a single bit error is made, the level will be received as an adjacent level with limited impact (except for the first and last words).

Perhaps of more interest is the notion of a *snake-in-the-box* code introduced in [29]. This also is represented by a cycle in C_n (hence an ordered list of binary n-tuples) except that a codeword is adjacent (distance one) to the two codewords preceding it and succeeding it in the list and to no other codeword in the list. In graph terms it represents a cycle in C_n without chords where a *chord* is an edge of the graph C_n which joins two vertices of the cycle (which are then adjacent – distance one) but is not an edge of the cycle. Thus a snake-in-the-box code is a simple cycle of C_n without chords. Said another way, such a code has each word adjacent to two other words of the cycle (the ones preceding it and succeeding it) but to no other word of the code. One could define such a code as a path in the cube rather than a cycle (first and last words distance one apart) but the literature tends to consider cycles which is of interest here.

Notice that if a single error is made in a codeword, it will result in either an adjacent word or a noncodeword. In this sense the code will detect a single error to a nonadjacent word. The original work on such codes is [29].

Example 17.1 Consider a code for $n = 5$ ([29]):

$$\begin{array}{ccccccccccccc} 00000 & \rightarrow & 00001 & \rightarrow & 00011 & \rightarrow & 00111 & \rightarrow & 01111 & \rightarrow & 11111 & \rightarrow & 11101 \\ \uparrow & & & & & & & & & & & & \downarrow \\ 00100 & \leftarrow & 10100 & \leftarrow & 10110 & \leftarrow & 10010 & \leftarrow & 11010 & \leftarrow & 11000 & \leftarrow & 11001 \end{array}$$

Of interest in such codes is the length of the longest snake for each dimension, a problem addressed in numerous works. Let $S(n)$ be the length of the longest snake in the n-dimensional binary cube C_n. $S(n)$ is given in [29] for small values of n. Since $S(5) = 14$ the above example is maximal. It is known [1, 51] that

$$2^{n-1}\left(1 - \frac{1}{89n^{1/2}} + O\left(\frac{1}{n}\right)\right) \leq S(n).$$

Balanced Codes

For some applications in memory and transmission systems it assists system performance to transmit balanced codewords, words that have an equal (or approximately equal) number of ones and zeroes (or $+1, -1$'s), e.g., if the system suffers from DC wander. The encoding of arbitrary bit strings into such balanced words has received attention in the literature [3, 5, 30, 37, 45]. A brief introduction to such coding using the results of Knuth [30] is discussed.

It is desired to encode binary information words of length n,$\boldsymbol{x} \in \mathbb{F}_2^n$, into balanced codewords by operating on $\boldsymbol{x}$ and adding a prefix $\boldsymbol{u} \in \mathbb{F}_2^m$ of m bits to allow for decoding of the concatenated codeword word to restore $\boldsymbol{x}$.

The transmitted codewords will have weight $(n+m)/2$. To construct such codewords, let $\boldsymbol{x}^{(k)}$ denote the information n-tuple $\boldsymbol{x}$ with the first (say from the left) $k, 0 \leq k \leq n$, bits complemented. Suppose $\boldsymbol{x}$ has weight ℓ. As k increases from 0 to n the weight of $\boldsymbol{x}^{(k)}$ goes from ℓ to $n-\ell$. As k varies the weight of $\boldsymbol{x}^{(k)}$ can vary outside this range but certainly all weights in the interval $[\ell, n-\ell]$ will be obtained for any initial word $\boldsymbol{x}$. Notice that $n/2$ in particular is in this interval. Thus as k increases from 0, the weight $w(\boldsymbol{x}^{(k)})$ increases or decreases by one and there will be a minimum value for k such that the weight $w(\boldsymbol{x}^{(k)})$ achieves any given value in the range $[\ell, n-\ell]$. In particular there will be a minimum value for k such that for n even, $w(\boldsymbol{x}^{(k)}) = n/2$, a balanced word. One might consider encoding this minimum value of k that achieves this balance into the prefix m-tuple $\boldsymbol{u}$ and transmit the concatenated codeword $\boldsymbol{u}\boldsymbol{x}^{(k)}$. In general this word will not be balanced as $\boldsymbol{u}$ will not in general be balanced. There will be a simple fix for this. Before considering this it is asked how large m must be to accommodate this type of encoding?

The number of balanced binary words of length $(n+m)$ (assume balance means words of weight $\lfloor (m+n)/2 \rfloor$) will have to be at least as large as 2^n to allow for the encoding of all binary information n-tuples – thus it is required that

$$\binom{n+m}{\lfloor (n+m)/2 \rfloor} \geq 2^n. \tag{17.1}$$

By using Stirling's approximation good approximations can be obtained to show that one requires the number of prefix bits to be at least

$$m \gtrapprox \frac{1}{2}\log_2(n) + \frac{1}{2}\log_2\left(\frac{\pi}{2}\right) \approx \frac{1}{2}\log_2(n) + 0.326. \tag{17.2}$$

The work of Knuth [30] contains several encoding methods, including the following technique. Suppose $w(\boldsymbol{x}) = \ell$ and assume $n = 2^m$, $\boldsymbol{x} \in \mathbb{F}_2^n$ (so the above inequality is satisfied). As noted, it will always be possible to find an integer k that $w(x^{(k)}) \in [\ell, n-\ell]$. Assume for convenience n is even. For each $\ell \in [0, n]$, the weight of $w(\boldsymbol{x})$, it is desired an m-tuple $\boldsymbol{u}_\ell$ such that

$$w(\boldsymbol{u}_\ell) + w\left(\boldsymbol{x}^{(k)}\right) = (n+m)/2$$

hence giving balanced codewords $\boldsymbol{u}\boldsymbol{x}^{(k)}$ where we are free to choose k (as long as $w(\boldsymbol{x}^{(k)})$ is in the interval $[\ell, n-\ell]$) and $\boldsymbol{u}_\ell$ so that this is true. Assuming this can always be done, the decoding is as follows. On receiving a codeword, the prefix $\boldsymbol{u}_\ell$ is used by table look-up to give the weight of the original information word $\boldsymbol{x}$ and the bits of the rest of the received codeword $\boldsymbol{x}^{(k)}$ are complemented successively from the left until that weight is reached.

The following example illustrates the technique.

Example 17.2 Let $\boldsymbol{x} \in \mathbb{F}_2^{16}$ be a binary information word of weight ℓ and length 16 and let $\sigma_k(x) = s_\ell$ be the weight of $\boldsymbol{x}^{(k)}$ where $s_\ell \in [\ell, n-\ell]$ and k the smallest integer that achieves this weight. For each possible weight ℓ a value of s_ℓ is chosen in the interval and associated with a binary 4-tuple $\boldsymbol{u}_\ell$ so that the codeword $\boldsymbol{u}_\ell \boldsymbol{x}^{(k)}$ of length 20 has weight 10.

ℓ	$[\ell,n-\ell]$	s_ℓ	u_ℓ	ℓ	$[\ell,n-\ell]$	s_ℓ	u_ℓ
0	[0, 16]	6	1111	8	[8, 8]	8	0110
1	[1, 15]	7	1110	9	[7, 9]	8	0101
2	[2, 14]	7	1101	10	[[6, 10]	8	0011
3	[3, 13]	7	1011	11	[5, 11]	9	1000
4	[4, 12]	7	0111	12	[4, 12]	9	0100
5	[5, 11]	8	1100	13	[3, 13]	9	0010
6	[6, 10]	8	1010	14	[2, 14]	9	0001
7	[7, 9]	8	1001	15	[1, 15]	10	0000

On receiving the codeword $\boldsymbol{u}_\ell \boldsymbol{x}^{(k)}$, the weight of the information word $\boldsymbol{x}$ is determined by table lookup using the prefix $\boldsymbol{u}_\ell$. The bits of the received n-tuple $\boldsymbol{x}^{(k)}$ are then complemented (from the left) until this weight is achieved to retrieve $\boldsymbol{x}$ (the smallest number of complementations to achieve this weight).

It can be established [30] that this procedure will work for all positive integers m for data words of length $n = 2^m$ and m parity bits satisfying the above bound. The important point is to verify it is always possible to arrange the association of prefix m-tuples with the modified information words $\boldsymbol{x}^{(k)}$ such that their total weight is $(n+m)/2$, to yield the correct weight. The problem was further examined in [37] who managed to reduce the redundancy to about $\log_2(n/2+1)$. As noted earlier numerous researchers have contributed further to the problem including [3, 5, 45].

Write-Once Memory Codes

Certain storage media such as paper tape, punch cards and more recently optical discs and flash memories have the feature that writing to them is irreversible. Thus initially the medium, considered as n cells (often called *wits* in the literature), may be viewed as being in the "all-zero" state. Writing a 1 in a cell is irreversible in the sense the cell cannot be returned to the 0-state. In a surprising development it was shown [36] how such *write-once memories* (WOMs) can be reused by using coding techniques to write information to its cells several times, each time of course avoiding cells that are already in the one state. An indication of how this can be accomplished is discussed. The elegant demonstration of this in the seminal work [36] writes two bits to three cells twice – thus writing four bits to three cells.

The storage ability, capacity, of WOMs was considered from an information-theoretic point of view in [47]. Consider a WOM with n cells. Information is written on the device in t *stages* or *generations*. The maximum amount of information (number of bits) that can be written and retrieved from the device in the t stages, the capacity, is denoted by $C_{t,n}$.

As a matter of terminology denote by $\mathrm{supp}(\boldsymbol{x}) \subseteq [n]$, $\boldsymbol{x} \in \mathbb{F}_2^n$ the set of coordinate positions containing 1's and denote by $\boldsymbol{y} \geqslant \boldsymbol{x}$ if $\boldsymbol{y}$ covers $\boldsymbol{x}$, i.e., $\mathrm{supp}(\boldsymbol{y}) \supseteq \mathrm{supp}(\boldsymbol{x})$.

Consider the following informal definition (following the notation of [50]).

An $\{n,t;M_1,\ldots,M_t\}$ t-write WOM code C is a coding scheme with n cells and t pairs of encoding and decoding maps E_i and $D_i, i = 1,2,\ldots,t$ such that the first write operation maps one of M_1 possible information symbols m_1 (hence $\log_2(M_1)$ bits) into a binary n-tuple say $\boldsymbol{c}_1$. At the second write one of M_2 possible information symbols, m_2, is encoded into a binary n-tuple $\boldsymbol{c}_2$ such that the support of $\boldsymbol{c}_2$ is disjoint to that of $\boldsymbol{c}_1$ (i.e., $\mathrm{supp}(\boldsymbol{c}_2) \cap \mathrm{supp}(\boldsymbol{c}_1) = \phi$) and so on for the t-th write, at each stage of writing the support of the codeword used is disjoint to that of the previous stages.

For each stage of decoding there is a decoding function:

$$\begin{aligned} D_i : \{0,1\}^n &\longrightarrow \{1,2,\ldots,M_i\} \\ \boldsymbol{c}_i &\mapsto m_i. \end{aligned}$$

The decoding operation of the i-th write may depend on knowing the state of the system (the contents of the cells) at the previous stage.

The *sum-rate* of this t-write WOM code is defined as

$$\mathcal{R} = \frac{\sum_{i=1}^{t} \log_2 M_i}{n}$$

and, as above, the maximum achievable sum-rate is the capacity of the n-cell t-write WOM $C_{t,n}$. To consider the capacity, there are four cases to be considered, depending on whether the encoder (resp. decoder) needs to know the state of the decoder at the previous state in order to perform its operations. In the case where neither the encoder nor the decoder knows the previous state the capacity is shown to be

$$C_{t,n} < \pi^2 n / 5 \ln 2 \approx 2.37n. \quad \text{(previous encoder/decoder state unknown)}$$

In the other three cases the capacity is

$$C_{t,n} = n \log(t+1).$$

It seems interesting that the case of multiple writes increases the capacity of the memory over the 1-write case by a factor of $\log(t+1)$.

Some constructions of WOM codes use linear codes and covering codes, codes for which the placing of spheres of a certain radius around each codeword results in the entire space being covered (see [20]). Other approaches to the problem include [26, 28, 39, 40, 48, 50]. As noted earlier, the original seminal paper of the subject [36] contains a simple, elegant and compelling example of WOMs, giving a 2-write WOM writing two bits in three cells twice – thus a total of four bits. It is viewed in [20, 49] as a coset coding example.

17.2 Codes for the Gaussian Channel and Permutation Codes

Codes for the Gaussian Channel

An undergraduate course on digital communications often begins by defining a digital communication system which transmits one of M possible continuous signals of length T seconds, every T seconds and to which white Gaussian noise is added to it in transmission. This is the *additive white Gaussian noise* (AWGN) channel as discussed in Chapter 3. Even for channels where it is not a good model its tractability may give useful intuition. Such a system can be shown to be reducible to a discrete-time model in Euclidean n-space (for some n) where the code is a set of points in the n-space (either within a sphere of some radius – the power-constrained case – or on the surface of a sphere – the equal power case). In either case the problem can be described as a packing problem – either packing spheres of a given radius within a larger sphere or packing spherical caps on the surface of a sphere. When a given point (signal) is transmitted channel noise (Gaussian with mean 0 and variance $N_0/2$ as discussed in Chapter 3) perturbs the point off the transmitted point and the decoding problem is to determine the closest code point to the received point. The related coding problem then is reduced to finding sets of points in Euclidean n-space $\mathbb{R}_n$, as far apart as possible under a given constraint. Such a strategy can be shown to optimize system performance.

The approaches are summarized as:

(i) equal power constraint which corresponds to the mathematical problem of finding M points on a sphere in Euclidean n-space of some radius; or
(ii) maximum power constraint, which corresponds to the mathematical problem of finding M points as far apart as possible within a sphere of a given radius.

Both of these signal constraints have been extensively studied in the mathematical and engineering literature and many volumes on them are available. The discussion here gives a brief indication of two of the approaches.

An interesting approach to the equal power problem appeared in [42], using group representations in terms of orthogonal matrices acting on an initial vector resulting in points on the unit sphere in $\mathbb{E}_n$. An approach to the signal design problem with a maximum power constraint uses the theory of lattices very successfully.

The studies of group representations and lattices are well-established mathematical disciplines and their use for coding often translates many of their results to a different domain, while introducing new aspects to the problem.

A group representation of a finite group G degree n is a homomorphism from G into the set of $n \times n$ orthogonal matrices over $\mathbb{R}$:

$$\begin{aligned} \rho : G &\longrightarrow O_{n \times n} \\ g &\longmapsto \rho(g). \end{aligned}$$

For the current interest, the representations are simply a method to generate a set of orthogonal matrices. Representation theory is one more mathematical tool to investigate the structure of the group. A group code is then formed by choosing an *initial vector* $\mathbf{x}$ on the unit sphere in $\mathbb{R}_n$ and using the orthogonal matrices of the above representation to generate n points (hopefully distinct) on the unit sphere. The code then is

$$C = \{\rho(g)\boldsymbol{x} \mid g \in G, \ |\boldsymbol{x}| = 1\}.$$

Assuming the generator vectors are distinct, define the minimum distance of the code with initial vector $\boldsymbol{x} \in \mathbb{R}^n$ as

$$d_n(\boldsymbol{x}) = \min_{g \in G, g \neq 1} |\rho(g)\boldsymbol{x} - \boldsymbol{x}| \, .$$

The initial vector problem is to find the initial vector $\boldsymbol{x} \in \mathbb{R}^n$, $|\boldsymbol{x}| = 1$, i.e., find $\boldsymbol{x}$ such that

$$d_n(\boldsymbol{x}) = \sup_{\boldsymbol{y} \in \mathbb{R}^n, |\boldsymbol{y}| = |\boldsymbol{x}| = 1} d_n(\boldsymbol{y})$$

is maximized. This approach was initiated by the work [42] with further contributions in [12, 23] and [25, 35]. The approach is mathematically challenging and has produced results of limited interest.

A particularly interesting class of equal energy codes – and of the above group representation approach – for the Gaussian channel is that of permutation codes [41]. One version is to choose the initial vector as balanced (around 0) and components equally spaced – thus

$$x = \left(\tfrac{n-1}{2}\alpha, \tfrac{n-3}{2}\alpha, \ldots, -\alpha, 0, \alpha, \ldots, -\tfrac{n-3}{2}\alpha, -\tfrac{n-1}{2}\alpha,\right), \quad n \text{ odd}$$
$$x = \left(\tfrac{n-1}{2}\alpha, \ldots, \tfrac{\alpha}{2}, -\tfrac{\alpha}{2}, \ldots, -\tfrac{n-1}{2}\alpha,\right), \quad n \text{ even}$$

where α is chosen to make $|x|= 1$, i.e. in either case

$$\alpha = \left[\frac{12}{(n-1)n(n+1)}\right]^{1/2}.$$

The code is the set of all vectors obtained by permuting the initial vector in all possible ways. The maximum distance of this natural representation operating on the above initial vector then is

$$d_{\min} = \sqrt{2}\alpha. \tag{17.3}$$

The use of lattices to generate good codes in Euclidean n-space, and more generally to obtain good sphere packing in Euclidean space, has achieved exceptional results. The notion of lattices was considered in the previous chapter for their use in postquantum cryptosystems. For convenience some of the notions introduced there are repeated.

The fundamental reference on lattices is the encyclopedic and authoritative volume [21]. Only real lattices are considered, i.e., additive subgroups of points in $\mathbb{R}^n$. There is a slight repetition of the material here with that introduced in the discussion of quantum-resistant cryptography of Chapter 15.

Let $\mathcal{X} = \{u_1, u_2, \ldots, u_n\}$ be vectors in $\mathbb{R}^m$ and define a *lattice* as the set of points

$$\Lambda_n = \left\{ y = \sum_{i=1}^{n} \alpha_i u_i,\ \alpha_i \in \mathbb{Z} \right\}. \tag{17.4}$$

The vectors $\mathcal{X}$ will be referred to as a basis of the lattice which will often (although not necessarily) be assumed to be linearly independent over $\mathbb{R}$. If $u_i = (u_{i1}, u_{i2}, \ldots, u_{im})$, then a generator matrix of the lattice is

$$M = \begin{bmatrix} u_{11} & u_{12} & \cdots & u_{1m} \\ u_{21} & u_{22} & \cdots & u_{2m} \\ \vdots & \vdots & \vdots & \vdots \\ u_{n1} & u_{n2} & \cdots & u_{nm} \end{bmatrix}.$$

A mathematical objective for lattices is to choose the basis vectors in such a way that the lattice points are as far apart as possible which often yields lattices with interesting mathematical properties. This is the same objective for constructing lattices for the AWGN channel, since the further apart the points, the greater the noise immunity of the system to the additive noise. For the power-constrained channel all lattice points within a sphere of a given radius might be used. For equal energy signals all lattice points on a sphere of a given radius would be used.

The *fundamental domain* of the lattice is the set of points

$$\alpha_1 \boldsymbol{u}_1 + \alpha_2 \boldsymbol{u}_2 + \cdots + \alpha_n \boldsymbol{u}_n, \quad 0 \leq \alpha_i < 1$$

and it is readily seen that each lattice point has such a region associated with it and the set of all such regions exhausts the space. The *determinant* of Λ for the case of linearly independent basis vectors is given by

$$\det \Lambda = (\det \boldsymbol{M})^2 \quad \left(\text{or } \det (\boldsymbol{M}\boldsymbol{M}^T) \text{ for } \boldsymbol{M} \text{ nonsquare}\right)$$

and the volume of a fundamental domain in this case is given $(\det \Lambda)^{1/2}$.

A question of fundamental interest – mathematical and practical – is the packing of spheres of equal radius in $\mathbb{R}^n$ and the study of lattices has been used extensively in this study. Thus surrounding each lattice point with nonintersecting spheres of radius ρ, a question of interest is what fraction of the space in $\mathbb{R}^n$ is taken up by the spheres?

Define a *Voronoi region* of a lattice point of $\boldsymbol{x} \in \Lambda$ as the set of all points of $\mathbb{R}^m$ closest to $\boldsymbol{x}$ than to any other lattice point. This notion is of more relevance to the coding problem than that of the fundamental domain. Clearly the set of all Voronoi regions around the set of lattice points exhausts the space as does the set of translated fundamental domains as noted previously. Denote by ρ the radius of the maximum radius possible for nonintersecting spheres of a given lattice. Thus such a sphere of radius ρ around each lattice point fits, by definition, within the Voronoi region for that lattice point. Denote by Δ the ratio of the volumes of all the spheres to the total volume of the space and, recalling the volume of an n-dimensional sphere of radius ρ in $\mathbb{R}^m$ is

$$V_m(\rho) = \frac{\pi^{m/2}}{\Gamma\left(\frac{m}{2}+1\right)} \rho^m, \quad \Gamma \text{ the Gamma function, } \Gamma(x+1) = x\Gamma(x),$$

note that

$$\begin{aligned} \Delta &= \frac{\text{volume of a sphere of radius } \rho}{\text{volume of fundamental region}} \\ &= \frac{\pi^{m/2}\rho^m}{\Gamma\left(\frac{m}{2}+1\right)} \Big/ (\det \Delta)^{1/2}. \end{aligned}$$

Perusal of the volumes [46] and [21] is a rewarding exercise.

Permutation Codes for DMCs

The first notion of the use of permutation groups for coding appears to be the work of Slepian [41] who constructed codes for the Gaussian channel in Euclidean n-space $\mathbb{R}_n$ using, naturally, the Euclidean metric. He termed these code "permutation modulation," as discussed above. The notion of permutation codes for discrete channels with the Hamming metric appears to have been initiated in the work [14]. A few of the results of that work and [41] are outlined here. The coding notions are a direct consequence of properties of the Hamming metric and the properties of permutation groups.

Consider the set $N = [n] = \{1, 2, \ldots, n\}$. Let G be a permutation group of degree n and order $|G|$ whose elements act on the set N. Thus a codeword is formed from $\sigma \in G$ by $\sigma(N) = (\sigma(1), \sigma(2), \ldots, \sigma(n))$. A permutation code C on N is a set

$$C = \{\sigma(N), \sigma \in G\}.$$

An (n, M, d) permutation code C of size M is such that the Hamming distance between any two distinct codewords is at least d. For an arbitrary permutation group it may be quite difficult to determine the minimum distance of the code. For the transitive and sharply transitive groups, however, the minimum distance is immediate.

Definition 17.3 A permutation group of degree n is *k-transitive* if for any two ordered k sets of distinct elements of K, $\{i_1, i_2, \ldots, i_k\}$ and $\{j_1, j_2, \ldots, j_k\}$ if there exists an element $\sigma \in G$ such that $\sigma(i_\ell) = j_\ell, \ell = 1, 2, \ldots, k$. If there is exactly one such element the group is called *sharply* k-transitive.

In a sharply k-transitive, every nonidentity element can move at least $n - k + 1$ elements. It is immediate that the order of such a group is

$$|G| = \prod_{i=0}^{k-1}(n - i)$$

and such groups give rise to $(n, \prod_{i=0}^{k-1}(n - i), n - k + 1)$ code with an alphabet size of n.

In a sense the codes corresponding to sharply transitive groups are optimum for their parameters. The structure of all sharply k-transitive groups for $k \geq 2$ is known [34] and the corresponding codes are summarized in [12]. Bounds on permutation codes (arrays) are given [14].

There is a sizeable literature on these and related arrays of which [14] is an introduction and survey up to its date. The problem gives interesting slants on standard combinatorial and group-theoretic problems.

17.3 IPP, Frameproof and Constrained Codes

Codes with the Identifiable Parent Property

Consider an alphabet $\boldsymbol{Q}$ with q letters, $|\boldsymbol{Q}|= q$, and words $\boldsymbol{a} = (a_1,a_2,\ldots,a_n)$, $\boldsymbol{b} = (b_1,b_2,\ldots,b_n) \in \boldsymbol{Q}^n$ and define the set of words in $\boldsymbol{Q}^n$, referred to as *descendants* of $\boldsymbol{a}$ and $\boldsymbol{b}$ as

$$\text{desc}(\boldsymbol{a},\boldsymbol{b}) = \Big\{\boldsymbol{x} = (x_1,x_2,\ldots,x_n) \in \boldsymbol{Q}^n \mid x_i \in \{a_i,b_i\}\Big\}. \tag{17.5}$$

In this case the words $\boldsymbol{a},\boldsymbol{b}$ are called *parents* of the descendants $\text{desc}(\boldsymbol{a},\boldsymbol{b})$. Thus a descendant of a pair of words is a word that can be formed by successively choosing a coordinate from each of the words, referred to as *parents* in all possible ways. Thus for binary words of length 4

$$\text{desc}(0100,1110) = \{0100,1110,0110,1100\}.$$

More generally let $\boldsymbol{A} = \big\{\boldsymbol{a}_1,\boldsymbol{a}_2,\ldots,\boldsymbol{a}_t\big\}$, $\boldsymbol{a}_j \in \boldsymbol{Q}^n, \boldsymbol{a}_j = (a_{j1},a_{j2},\ldots,a_{jn})$, $a_{ji} \in \boldsymbol{Q}$ be a set of t words in $\boldsymbol{Q}^n$. As for the $t = 2$ case above define the set of descendants as

$$\text{desc}(\boldsymbol{A}) = \Big\{\boldsymbol{y} \in \boldsymbol{Q}^n \mid y_i \in \{a_{1i},a_{2i},\ldots,a_{ti}\},\ i = 1,2,\ldots,n\Big\}. \tag{17.6}$$

Definition 17.4 Let $C \subseteq \boldsymbol{Q}^n$ be a code. The code C is said to have the *identifiable parent property of order t* (t-IPP) [4] if for any word $\boldsymbol{d}$ of $\boldsymbol{Q}^n$, either $\boldsymbol{d}$ is not a descendant of any set of t codewords of C or, for any set of t codewords $\boldsymbol{P} \subset C$ of size t such that $\boldsymbol{d} \in \text{desc}(\boldsymbol{P})$ at least one of the parents (elements of $\boldsymbol{P}$) can be identified.

These codes were first discussed in [27] for the case $t = 2$ although the idea was present in the work of Chor et al. [18] on traitor tracing to be discussed next with which they have a natural association.

A small example may be illustrative (essentially theorem 3 of [27]).

Example 17.5 Let C be an equidistant code over $\mathbb{F}_q$ with odd distance d. Then C is 2-IPP. To see this consider two distinct codewords $\boldsymbol{x},\boldsymbol{y} \in C \subset \mathbb{F}_q^n$ which without loss of generality can be viewed as

$$\begin{aligned} \boldsymbol{x} &= x_1,x_2,\ldots x_d,z_1,z_2,\ldots z_{n-d} \\ \boldsymbol{y} &= y_1,y_2,\ldots y_d,z_1,z_2,\ldots z_{n-d}, \quad x_i \neq y_i, \quad i = 1,2,\ldots,d. \end{aligned}$$

To form a descendant $\boldsymbol{u}$ of these two words, one can choose $i \leq d$ coordinates from among the first d of $\boldsymbol{x}$ and $d - i$ from $\boldsymbol{y}$. The last $n - d$ coordinates of $\boldsymbol{u}$ will be $z_1,\ldots,z_{n-d}$. Thus the descendant $\boldsymbol{u}$ will be distance $< d/2$ to one of the codewords $\boldsymbol{x}$ or $\boldsymbol{y}$ and a greater distance to all other codewords and hence a parent is identified. It is noted [27] that if d is even and $n < 3d/2$ the code is also 2-IPP.

With similar logic one can show ([9], proposition 1.2) that if C is an $(n,k,d)_q$ code over $\mathbb{F}_q$, then it is t-IPP if

$$d > (1 - 1/t^2)n.$$

For example, if C is the RS code over $\mathbb{F}_{2^k}$ $(2^k, 2^{k-2}, 2^k - 2^{k-2} + 1)_{2^k}, k > 2$, then it is 2-IPP.

Numerous properties of such t-IPP codes have been considered including the following:

Lemma 17.6 ([27], lemma 1) *A code $C \subset \mathbf{Q}^n$ is 2-IPP iff:*
(i) $\mathbf{a}, \mathbf{b}, \mathbf{c}$ *distinct in* $C \Rightarrow a_i, b_i, c_i$ *distinct for some* i;
(ii) $\mathbf{a}, \mathbf{b}, \mathbf{c}, \mathbf{d} \in C$ *with* $\{\mathbf{a}, \mathbf{b}\} \cap \{\mathbf{c}, \mathbf{d}\} = \phi \Rightarrow \{a_i, b_i\} \cap \{c_i, d_i\} = \phi$ *for some* i.

The first of these properties of *trifference* was studied in [31] as of independent interest.

Problems of interest for these codes include construction of such codes and the maximum size of such codes. Define

$$\boldsymbol{F}_t(n,q) = \max \left\{ |C| \;\middle|\; \exists C \subseteq \boldsymbol{Q}^n \text{ such that } C \text{ is } t\text{-IPP} \right\}.$$

Numerous works on this function as well as t-IPP constructions are found in the literature including [15] ($t = 1$), [6] ($t = 2, n = 4$), [9] (shows for arbitrary $t \leq q - 1$ there exist sequences of codes with nonvanishing rate), [7] (relationship of t-IPP to certain notions of hashing), [4] (establishes general bounds on $\boldsymbol{F}_t(n,q)$ with particular attention to constructions and bounds for $t = 3$ and $n = 5, 6$), [11] (introduces the notion of a *prolific* t-IPP code). These works are suggested as representative of the large literature on the subject.

Frameproof Codes and Traitor Tracing Schemes

The notion of a *traitor tracing scheme* arose in the context of distributing content (say music or pay-per-view video) to a set of authorized users in such a way that if a subset of authorized users (traitors) collude to sell the access mechanism to an unauthorized user, the distributor, on discovering this user's access, will be able to determine at least one of the traitors. Similarly the notion of a *frameproof code* consists of assigning authorizations to content in such a way that a subset of authorized users of less than a certain size is unable to pool their access information in such a way to derive access information for an unauthorized user (i.e., the subset cannot *frame* another user). Typically these concepts are realized through the use of key distribution techniques.

The concepts arose in the work of Chor et al. [18, 19] (for traitor tracing) and Fiat et al. [24] for broadcast encryption for a version of frameproof codes. The notion of a frameproof code is also inherent in the work of Boneh et al. [16] for watermarking content in such a manner that collusions to provide access to an illegal copy can be traced. Our interest here is merely to define the ideas in an informal and limited manner and give simple examples of them. There are slight variations in the literature in the concept definitions. The approaches of [44] and [10] are of particular interest.

The definition of a c-frameproof code [10] is straightforward.

Definition 17.7 Consider a set of m users $\mathcal{U} = \{U_1, U_2, \dots, U_m\}$ and F an alphabet of q symbols and C a set of codewords over F of length n, $C \subseteq F^n$. Assume each codeword is assigned to a user. The code C is a c-frameproof code if, for all subsets $P \subseteq C$, $|P| \leq c$, $\mathrm{desc}\{P\} \cap C = P$ (see Equation 17.5 for the definition of desc).

The definition implies that with the code C, if any subset of c users collude in trying to construct another codeword from using only the coordinates of its own sets, they will be unable to do so – hence they will be unable to frame another user.

The same work [10] contains an interesting nontrivial example of such codes.

Example 17.8 Let the alphabet $F = \mathbb{F}_q$ and $\alpha_1, \alpha_2, \dots, \alpha_n$ be distinct elements of $\mathbb{F}_q$ and let C be

$$C = \Big\{(f(\alpha_1), f(\alpha_2), \dots, f(\alpha_n)),\ \deg(f) < \lceil (n/c \rceil \Big\}.$$

To show that C is c-frameproof, suppose $P \subseteq C$, $|C| \leq c$ and let $\boldsymbol{x} \in \mathrm{desc} P \cap C$. Each coordinate of $\boldsymbol{x}$ agrees with at least one of the codewords of P and hence there must be a codeword that agrees with $\boldsymbol{x}$ in at least $\lceil n/c \rceil$ positions. But by construction this means $\boldsymbol{x}$ is equal to that codeword since it uniquely specifies the codeword polynomial. Hence the code is c-frameproof.

Other interesting examples of binary c-frameproof codes are found in [44].

Consider now the notion of a *traceability scheme* [18, 19, 44] for which the following model is introduced. A large file is to be transmitted to a set of m legitimate users $\mathcal{U} = \{U_1, U_2, \dots, U_m\}$, $m = 2^r$. The content provider generates a set of v keys, T, and gives each user a subset of r keys, referred to as the user's personal keys and denoted $P(U_i)$ for user $U_i \in \mathcal{U}$, $i = 1, 2, \dots, m$. The large file is broken up into N segments and each segment is transmitted as an encrypted *content block*, $CB_i, i = 1, 2, \dots, N$, block CB_i encrypted with symmetric algorithm key i^* (say AES). Along with each encrypted block is an *enabling block*, $EB_i, i = 1, 2, \ldots, N$. For each user, the information in the enabling block, along with the users set of personal keys, $P(U_i)$, allows the user to derive the symmetric key i^* for each block (the key could be different for each block). Several interesting traceability schemes based on this model are given in [19]. The simplest one is described in the following example:

Example 17.9 Following the system description above suppose there are $m = 2^r$ authorized users and users are identified with binary r-tuples, user $\boldsymbol{u}_j$ assigned an ID of $(u_{j1}, u_{j2}, \ldots, u_{jr}) \in \mathbb{F}_2^r$. The encrypted block EB_i is encrypted with the key $\boldsymbol{s} \in \mathbb{F}_2^\ell$ which is implemented as the XOR of r keys, i.e., $\boldsymbol{s} = \oplus_{j=1}^r \boldsymbol{s}_j$, $\boldsymbol{s}_j \in \mathbb{F}_2^\ell$. The content provider generates r pairs of keys $\mathcal{A} = \{(\boldsymbol{a}_1^{(0)}, \boldsymbol{a}_1^{(1)}), (\boldsymbol{a}_2^{(0)}, \boldsymbol{a}_2^{(1)}), \ldots, (\boldsymbol{a}_r^{(0)}, \boldsymbol{a}_r^{(1)}), \text{ all in } \mathbb{F}_2^\ell\}$. User $\boldsymbol{u}_j$ is assigned the set of keys $P(U_j) = \{\boldsymbol{a}_1^{(u_{j1})}, \boldsymbol{a}_2^{(u_{j2})}, \ldots, \boldsymbol{a}_r^{(u_{jr})}\}, j = 1, 2, \ldots, m$. The content provider includes in the enabling block for each segment of the content the encryption of each key $\boldsymbol{s}_j$ by both of the keys $\boldsymbol{a}_j^{(0)}$ and $\boldsymbol{a}_j^{(1)}$, $j = 1, 2, \ldots, r$, a set of $2r$ encryptions.

Clearly each legitimate user is able, from the information in the enabling block to retrieve each of the subkeys $\boldsymbol{s}_j$ and hence $\boldsymbol{s}$. Furthermore if a pirate content decoder box is discovered the content provider will be able to look at the keys inside $\boldsymbol{a}_1^{(u_{p1})}, \boldsymbol{a}_2^{(u_{p2})}, \ldots, \boldsymbol{a}_r^{(u_{pr})}$ which will identify the user who provided them. Hence the scheme is 1-traceable.

More generally interest is in protecting content against largest sets of colluders. Such schemes are of great interest but tend to be more intricate than space here allows. Suppose a set of users $\Pi \subset \mathcal{U}$ (traitors) conspire to form a set of r keys given to a *pirate decoder* Φ, $\Delta \subset T$, $|T| = v$ such that $\Delta \subseteq \cap_{U \in \Pi} P(U)$. To detect at least one of the traitors one could form the set $|\Delta \cap P(U)|$ for all users $U \in \mathcal{U}$. If $|\Delta \cap P(U)| \geq |\Phi \cap V|$ for all users $V \neq U$, then user U is identified as an *exposed user* in that they have contributed more keys to the pirate decoder than other users. In this case [44]:

Definition 17.10 Suppose an exposed user U is a member of the coalition Π for a pirate decoder Φ produced by Π, $|\Pi| \leq c$. Then the traitor tracing scheme is referred to as a c-traceability scheme.

Interesting c-traceability schemes are constructed in [44] from t-designs and other combinatorial structures. The relationships between frameproof codes and traceability schemes are discussed in numerous works, including [43, 44].

Constrained Coding and Run-Length Limited Codes

For many types of storage and communication systems it is necessary for the transmitted system to possess certain properties for their correct operation. For example, certain disk storage systems had timing and synchronization circuits derived from the data sequences themselves. If the data had too many consecutive zeros in a sequence, the dc value tended to wander, making bit detection inaccurate, while if it had too few the synchronization circuits tended to lose bit synchronization. These resulted in the (d,k)-sequences where the sequences must have the property of having at least d and at most k zeros between any two ones [8, 33, 38]. Such sequences were used in certain IBM storage systems.

Such a constraint can be modeled by the finite-state machine (FSM) as shown in Figure 17.1 for $(d,k) = (2,4)$. Similarly, one could consider a constraint of sequences that have no runs of either zeros or ones of length greater than k, referred to as (n,k)-sequences [13], or simply as k-sequences.

Any path around the FSM for a given constraint will satisfy the constraint and the issue will be to construct such codes as well as encoding and decoding algorithms. In practice such codes might have a finite length n and one might add the constraint that the concatenation of any two codewords also satisfies the constraint. Such issues have been considered in the literature and many have been implemented in practice.

The only issue of interest in this work is the notion of the capacity of the sequences, i.e., what is the maximum amount of information that can be carried by a coded bit. This question was answered by Shannon in his foundational paper [38]. Let $\boldsymbol{D}$ be the directed transition matrix of the FSM representing the constraint, i.e., d_{ij} is 1 if there is a transition from state i to state j and 0 otherwise (it is assumed there is not more than one edge between any two states – although it is only necessary for the symbols on multiple outgoing branches to be distinct). $\boldsymbol{D}$ is a $K \times K$ matrix for a FSM with K states. Shannon showed that the capacity of the sequence is

$$C = \log_2 \lambda \quad \text{bits per code symbol} \tag{17.7}$$

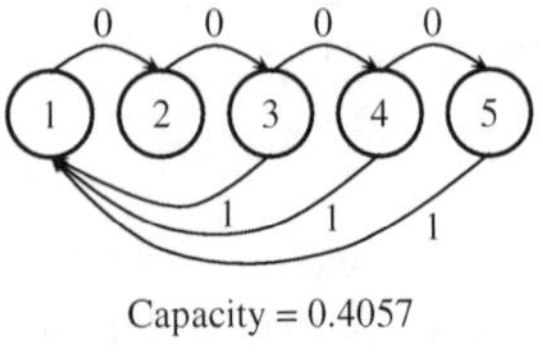

Figure 17.1 Constrained code (2, 4)

where λ is the largest eigenvalue of the transition matrix $\boldsymbol{D}$ representing the constraint. Thus a long codeword of length n satisfying the constraint would represent λn information bits. Equivalently the capacity of the constrained sequences is the largest real root of the polynomial expressed by

$$\det(\boldsymbol{D}z^{-1} - \boldsymbol{I}) = 0. \tag{17.8}$$

Given the transition matrix $\boldsymbol{D}$ representing the constraint, this equation leads to a polynomial equation. For example, the (d,k) constraint the above matrix equation can be reduced [8, 33, 38] to the polynomial equation

$$z^{k+2} - z^{k+1} - z^{k-d+1} + 1 = 0.$$

Thus the capacity of the constraint is $\log_2 \lambda$ where λ is the largest real root of the polynomial equation. Similarly the polynomial equation for the k-sequences can be derived as

$$z^{-2k}\left\{z^{2k} - z^2((1-z^k)/(1-z))^2\right\} = 0$$

and again the capacity of the constraint is the logarithm to the base 2 of the largest root of this equation. The capacity of the example displayed in Figure 17.1 is shown there. An extensive table of the capacity of the (d,k)-sequences is given in [33].

Of the literature on this subject the paper [2] is seminal and the book [32] is an excellent source.

Comments

The purpose of this chapter has been to note further types of coding beyond the types of the preceding chapters by providing (relatively) brief descriptions, simply to expand the number of coding systems of interest. Readers might well be aware of numerous other types of coding not mentioned. These might have included arithmetic coding or source coding topics such as Lempel–Ziv codes and enumerative source coding among numerous other types. Another type of coding is designed to reduce adjacent channel intersymbol interference or crosstalk (see [17] and the references therein). Similar ideas have already found commercial application to improve chip-to-chip transmission speeds significantly. Indeed many of the types of coding considered in this volume have been of value in system implementations.

References

[1] Abbott, H.L., and Katchalski, M. 1988. On the snake in the box problem. *J. Combin. Theory Ser. B*, **45**(1), 13–24.

[2] Adler, R., Coppersmith, D., and Hassner, M. 1983. Algorithms for sliding block codes – an application of symbolic dynamics to information theory. *IEEE Trans. Inform. Theory*, **29**(1), 5–22.

[3] Al-Bassam, S., and Bose, B. 1990. On balanced codes. *IEEE Trans. Inform. Theory*, **36**(2), 406–408.

[4] Alon, N., and Stav, U. 2004. New bounds on parent-identifying codes: the case of multiple parents. *Combin. Probab. Comput.*, **13**(6), 795–807.

[5] Alon, N., Bergmann, E.E., Coppersmith, D., and Odlyzko, A.E. 1988. Balancing sets of vectors. *IEEE Trans. Inform. Theory*, **34**(1), 128–130.

[6] Alon, N., Fischer, E., and Szegedy, M. 2001. Parent-identifying codes. *J. Combin. Theory Ser. A*, **95**(2), 349–359.

[7] Alon, N., Cohen, G., Krivelevich, M., and Litsyn, S. 2003. Generalized hashing and parent-identifying codes. *J. Combin. Theory Ser. A*, **104**(1), 207–215.

[8] Ashley, J., and Siegel, P. 1987. A note on the Shannon capacity of run-length-limited codes. *IEEE Trans. Inform. Theory*, **33**(4), 601–605.

[9] Barg, A., Cohen, G., Encheva, S., Kabatiansky, G., and Zémor, G. 2001. A hypergraph approach to the identifying parent property: the case of multiple parents. *SIAM J. Discrete Math.*, **14**(3), 423–431.

[10] Blackburn, S.R. 2003. Frameproof codes. *SIAM J. Discrete Math.*, **16**(3), 499–510 (electronic).

[11] Blackburn, S.R., Etzion, T., and Ng, S.-L. 2008. Prolific codes with the identifiable parent property. *SIAM J. Discrete Math.*, **22**(4), 1393–1410.

[12] Blake, I.F. 1974. Configuration matrices of group codes. *IEEE Trans. Inform. Theory*, **IT-20**, 95–100.

[13] Blake, I.F. 1982. The enumeration of certain run length sequences. *Inform. and Control*, **55**(1–3), 222–237.

[14] Blake, I.F., Cohen, G., and Deza, M. 1979. Coding with permutations. *Inform. and Control*, **43**(1), 1–19.

[15] Blass, U., Honkala, I., and Litsyn, S. 1999. On the size of identifying codes. In: *In AAECC-13, LNCS No. 1719.*

[16] Boneh, D., and Shaw, J. 1998. Collusion-secure fingerprinting for digital data. *IEEE Trans. Inform. Theory*, **44**(5), 1897–1905.

[17] Chee, Y.M., Colbourn, C.J., Ling, A., Zhang, H., and Zhang, X. 2015. Optimal low-power coding for error correction and crosstalk avoidance in on-chip data buses. *Des. Codes Cryptogr.*, **77**(2–3), 479–491.

[18] Chor, B., Fiat, A., and Naor, M. 1994. Tracing traitors. Pages 257–270 of: *Advances in cryptology – CRYPTO '94*. Lecture Notes in Computer Science, vol. 839. Springer, Heidelberg.

[19] Chor, B., Naor, M., and Pinkas, B. 2000. Tracing traitors. *IEEE Trans. Inform. Theory*, **46**(3), 893–910.

[20] Cohen, G., Godlewski, P., and Merkx, F. 1986. Linear binary code for write-once memories (Corresp.). *IEEE Trans. Inform. Theory*, **32**(5), 697–700.

[21] Conway, J.H., and Sloane, N.J.A. 1999. *Sphere packings, lattices and groups*, 3rd ed. Grundlehren der Mathematischen Wissenschaften [Fundamental Principles of Mathematical Sciences], vol. 290. Springer-Verlag, New York. With additional contributions by E. Bannai, R. E. Borcherds, J. Leech, S. P. Norton, A. M. Odlyzko, R. A. Parker, L. Queen and B. B. Venkov.

[22] Diestel, R. 2018. *Graph theory*, 5th ed. Graduate Texts in Mathematics, vol. 173. Springer, Berlin.

[23] Djoković, D.Ž., and Blake, I.F. 1972. An optimization problem for unitary and orthogonal representations of finite groups. *Trans. Amer. Math. Soc.*, **164**, 267–274.

[24] Fiat, A., and Naor, M. 1994. Broadcast encryption. *Adv. Cryptol.*, **773**, 480–491.

[25] Fossorier, M.P.C., Nation, J.B., and Peterson, W.W. 2007. A note on the optimality of variant-I permutation modulation codes. *IEEE Trans. Inform. Theory*, **53**(8), 2878–2879.

[26] Fu, F.-W., and Han Vinck, A.J. 1999. On the capacity of generalized write-once memory with state transitions described by an arbitrary directed acyclic graph. *IEEE Trans. Inform. Theory*, **45**(1), 308–313.

[27] Hollmann, H.D.L., van Lint, J.H., Linnartz, J.-P., and Tolhuizen, L. M. 1998. On codes with the identifiable parent property. *J. Combin. Theory Ser. A*, **82**(2), 121–133.

[28] Horovitz, M., and Yaakobi, E. 2017. On the capacity of write-once memories. *IEEE Trans. Inform. Theory*, **63**(8), 5124–5137.

[29] Kautz, W.H. 1958. Unit-distance error-checking codes. *IRE Trans. Electron. Comput.*, **EC-7**(2), 179–180.

[30] Knuth, D.E. 1986. Efficient balanced codes. *IEEE Trans. Inform. Theory*, **32**(1), 51–53.

[31] Körner, J., and Simonyi, G. 1995. Trifference. *Studia Sci. Math. Hungar.*, **30**(1–2), 95–103.

[32] Lind, D., and Marcus, B. 1995. *An introduction to symbolic dynamics and coding*. Cambridge University Press, Cambridge.

[33] Norris, K., and Bloomberg, D. 1981. Channel capacity of charge-constrained run-length limited codes. *IEEE Trans. Magnet.*, **17**(6), 3452–3455.

[34] Passman, D. 1968. *Permutation groups*. W. A. Benjamin, New York/Amsterdam.

[35] Peterson, W.W., Nation, J.B., and Fossorier, M.P. 2010. Reflection group codes and their decoding. *IEEE Trans. Inform. Theory*, **56**(12), 6273–6293.

[36] Rivest, R.L., and Shamir, A. 1982. How to reuse a "write-once" memory. *Inform. and Control*, **55**(1–3), 1–19.

[37] Schouhamer Immink, K.A., and Weber, J.H. 2010. Knuth's balanced codes revisited. *IEEE Trans. Inform. Theory*, **56**(4), 1673–1679.

[38] Shannon, C.E. 1948. A mathematical theory of communication. *Bell Syst. Tech. J.*, **27**, 379–423.

[39] Shpilka, A. 2013. New constructions of WOM codes using the Wozencraft ensemble. *IEEE Trans. Inform. Theory*, **59**(7), 4520–4529.

[40] Shpilka, A. 2014. Capacity-achieving multiwrite WOM codes. *IEEE Trans. Inform. Theory*, **60**(3), 1481–1487.

[41] Slepian, D. 1965. Permutation modulation. *Proc. IEEE*, **53**, 228–236.

[42] Slepian, D. 1968. Group codes for the Gaussian channel. *Bell System Tech. J.*, **47**, 575–602.

[43] Staddon, J.N., Stinson, D.R., and Wei, R. 2001. Combinatorial properties of frameproof and traceability codes. *IEEE Trans. Inform. Theory*, **47**(3), 1042–1049.

[44] Stinson, D.R., and Wei, R. 1998. Combinatorial properties and constructions of traceability schemes and frameproof codes. *SIAM J. Discrete Math.*, **11**(1), 41–53 (electronic).

[45] Tallini, L.G., Capocelli, R.M., and Bose, B. 1996. Design of some new efficient balanced codes. *IEEE Trans. Inform. Theory*, **42**(3), 790–802.

[46] Thompson, T.M. 1983. *From error-correcting codes through sphere packings to simple groups*. Carus Mathematical Monographs, vol. 21. Mathematical Association of America, Washington, DC.

[47] Wolf, J.K., Wyner, A.D., Ziv, J., and Krner, J. 1984. Coding for a write-once memory. *AT&T Bell Lab. Tech. J.*, **63**(6), 1089–1112.

[48] Wu, Yunnan. 2010. Low complexity codes for writing a write-once memory twice. Pages 1928–1932 of: *2010 IEEE International Symposium on Information Theory*.

[49] Yaakobi, E., Kayser, S., Siegel, P.H., Vardy, A., and Wolf, J.K. 2010. Efficient two-write WOM-codes. Pages 1–5 of: *2010 IEEE Information Theory Workshop*.

[50] Yaakobi, E., Kayser, S., Siegel, P.H., Vardy, A., and Wolf, J.K. 2012. Codes for write-once memories. *IEEE Trans. Inform. Theory*, **58**(9), 5985–5999.

[51] Zémor, G. 1997. An upper bound on the size of the snake-in-the-box. *Combinatorica*, **17**(2), 287–298.

Appendix A Finite Geometries, Linearized Polynomials and Gaussian Coefficients

This appendix presents a few standard facts on geometries, certain enumeration problems and related polynomials drawn mainly from the excellent references [1, 6, 8]. The results are relevant largely to the chapters on rank-metric codes and network codes.

Finite Geometries

In general terms, a finite geometry is a set of finite points along with a set of axioms as to how the lines, (a subset of points) and points, interact. The Euclidean geometry $EG(n,q)$ is often introduced through the Projective geometry $PG(n,q)$ of dimension n over the finite field $\mathbb{F}_q$, which is briefly introduced as follows. Denote by $V_{n+1}(q)$ the vector space of $(n+1)$-tuples over $\mathbb{F}_q$. For the points of $PG(n,q)$ identify scalar multiples (over $\mathbb{F}_q$) in $V_{n+1}(q)$ to yield a total of

$$\frac{q^{n+1}-1}{q-1} = 1 + q + q^2 + \cdots + q^n = |PG(n,q)|$$

points. For two points $\boldsymbol{p}_1 = (a_1, a_2, \ldots, a_{n+1})$ and $\boldsymbol{p}_2 = (b_1, b_2, \ldots b_{n+1})$ (not scalar multiples) the line through these points is the equivalence class of

$$\alpha \boldsymbol{p}_1 + \beta \boldsymbol{p}_2 = (\alpha a_1 + \beta b_1, \alpha a_2 + \beta b_2, \ldots, \alpha a_n + \beta b_n)$$

and hence there are $(q^2-1)/(q-1) = q+1$ points on a line in $PG(n,q)$. Any two lines in $PG(n,q)$ intersect. Subgeometries of $PG(n,q)$ of dimension m are referred to as m-flats, $PG(m,q)$, which can be defined in the obvious manner. The number of such m-flats in $PG(n,q)$, $m < n$, is given by

$$\prod_{i=0}^{m} \frac{(q^{n+1-i}-1)}{(q^{m+1-i}-1)}.$$

The formal manner of defining a Euclidean geometry is often to start with $PG(n,q)$ and delete a hyperplane or $(n-1)$-flat resulting in

$$|EG(n,q)| = q^n.$$

The point/line behavior of this derived geometry is inherited from the projective geometry. For example, the number of ℓ-flats in $EG(n,q)$ is calculated as

$$q^{n-\ell} \prod_{k=1}^{\ell} \frac{(q^{n+1-k}-1)}{(q^{\ell+1-k}-1)}.$$

Notions of parallel lines and flats are restored in such a geometry. These can be viewed as cosets of a vector subspace $V_n(q)$. The remainder of this appendix focus will be on a vector space over $\mathbb{F}_q$, $V_n(q)$, and its (vector) subspaces.

Subspace Enumeration and Gaussian Coefficients

Definition A.1 The *Gaussian coefficients* (also referred to as *q-binomial coefficients*) $\begin{bmatrix} n \\ k \end{bmatrix}_q$ are defined as the number of k-dimensional subspaces of $V_n(q)$, i.e.,

$$\begin{aligned}\begin{bmatrix} n \\ k \end{bmatrix}_q &= \frac{(q^n-1)(q^n-q)\cdots(q^n-q^{n-(k-1)})}{(q^k-1)(q^k-q)\cdots(q^k-q^{k-1})} \\ &\stackrel{\Delta}{=} \frac{(q^n-1)(q^{n-1}-1)\cdots(q^{n-(k-1)}-1)}{(q^k-1)(q^{k-1}-1)\cdots(q-1)}.\end{aligned}$$

The numerator of the first equation in the definition is the number of ordered ways of choosing a k basis vectors in $V_n(q)$ and its denominator is the number of ordered ways to choose a basis of a k-dimensional subspace, the ratio being the number of distinct subspaces of dimension k.

These coefficients share a number of interesting properties similar to those of the ordinary binomial coefficients. These properties are of independent interest and a number of them will be stated without proof. All can be found in the references [1, 8].

It is interesting to note that if one views the Gaussian coefficient as a rational function of the variable q, it is in fact a polynomial of degree $k(n-k)$ as a consequence of it taking on an integral value for an infinite number of integral values of q. The coefficients of this polynomial have combinatorial significance ([8], theorem 24.2), i.e.,

$$\begin{bmatrix} n \\ k \end{bmatrix}_q = \sum_{i=0}^{k(n-k)} a_i q^i.$$

Example A.2

$$\begin{bmatrix} 5 \\ 2 \end{bmatrix}_q = \frac{(q^5-1)(q^4-1)}{(q^2-1)(q-1)} = (q^4+q^3+q^2+q+1)(q^2+1).$$

Let $\mathcal{P}_q(n)$ denote the set of all subspaces of the vector space $\mathbb{F}_q^n$ and $\mathcal{G}_q(n,k)$ denote the set of all k-dimensional subspaces of the $\mathbb{F}_q^n$, referred to as the *Grassmannian of order k* of $\mathbb{F}_q^n$. By definition the cardinality of this Grassmannian is

$$|\mathcal{G}_q(n,k)| = \begin{bmatrix} n \\ k \end{bmatrix}_q. \tag{A.1}$$

The Gaussian coefficients play a significant role in the subject of subspace coding and the following simple bound can be useful [3] and ([4], Kschischang chapter):

$$q^{k(n-k)} \leq \begin{bmatrix} n \\ k \end{bmatrix}_q \leq K_q^{-1} q^{k(n-k)} \tag{A.2}$$

where $K_q = \prod_{i=1}^{\infty}(1 - q^{-i})$. The constant K_q^{-1} is monotonically decreasing with q with a maximum of 3.4594 at $q = 2$ and a value 1.0711 at $q = 16$.

The relations

$$\begin{bmatrix} n \\ 0 \end{bmatrix}_q = \begin{bmatrix} n \\ n \end{bmatrix}_q = 1, \text{ and } \begin{bmatrix} n \\ m \end{bmatrix}_q = \begin{bmatrix} n \\ n-m \end{bmatrix}_q, \quad 0 \leq m \leq n,$$

are verified by direct computation. Similar to the binomial relation

$$\binom{n+1}{k} = \binom{n}{k} + \binom{n}{k-1}$$

there is the Gaussian coefficient relation:

$$\begin{bmatrix} n+1 \\ k \end{bmatrix}_q = \begin{bmatrix} n \\ k \end{bmatrix}_q + q^{n+1-k} \begin{bmatrix} n \\ k-1 \end{bmatrix}_q \tag{A.3}$$

and, additionally [2]:

$$\begin{bmatrix} n \\ m \end{bmatrix}_q \begin{bmatrix} m \\ p \end{bmatrix}_q = \begin{bmatrix} n \\ p \end{bmatrix}_q \begin{bmatrix} n-p \\ n-m \end{bmatrix}_q, \quad 0 \leq p \leq m \leq n.$$

These relations can be verified by direct computation.

It is also noted [1] that as a consequence of applying l'Hopital's rule

$$\lim_{q \to 1} \frac{q^x - 1}{q^y - 1} = \lim_{q \to 1} \frac{x q^{x-1}}{y q^{y-1}} = \frac{x}{y}$$

and hence

$$\lim_{q \to 1} \begin{bmatrix} n \\ k \end{bmatrix}_q = \lim_{q \to 1} = \frac{n(n-1)\cdots(n-k+1)}{k(k-1)\cdots 1} = \binom{n}{k}.$$

A version [1] of the usual binomial theorem is

$$(1+t)^r = \sum_{i=0}^{r} \binom{r}{k} t^k$$

valid for any real number r where

$$\binom{r}{k} = \frac{r(r-1)\cdots(r-k+1)}{k!}.$$

An analog of this for q-binomial coefficients is, for $n \geq 1$

$$\prod_{i=0}^{n-1}(1 + q^i t) = \sum_{k=0}^{n} q^{k(k-1)/2} \begin{bmatrix} n \\ k \end{bmatrix}_q t^k$$

or, similarly

$$\prod_{i=0}^{n-1} \frac{1}{(1-q^i t)} = \sum_{i=o}^{\infty} \begin{bmatrix} n-i+1 \\ i \end{bmatrix}_q t^i.$$

Another binomial relation is

$$\binom{n+m}{k} = \sum_{i=0}^{k} \binom{m}{i} \binom{n}{k-i}$$

which is easily argued by the number of ways of choosing k items from a set of $n+m$ items and considering the number of such sets obtained by choosing i from a subset of m of the items and $k-i$ from the complement set. For the q-binomial analog the argument might be as follows. Let U be a vector space over $\mathbb{F}_q$ of dimension $n+m$ and V a subspace of dimension m. Let W be a subspace of V of dimension i – there are $\begin{bmatrix} m \\ i \end{bmatrix}_q$ such subspaces. We ask how many ways can W be completed with vectors entirely outside of V to form a subspace of U of dimension k. A sum of such quantities over i will yield $\begin{bmatrix} n+m \\ k \end{bmatrix}_q$, the number of subspaces of U of dimension k. The quantity sought is:

$$\begin{aligned}
&(q^{n+m}-q^m)(q^{n+m}-q^{m+1})\cdots(q^{n+m}-q^{m+(k-i)-1})\Big/ \\
&\qquad (q^k-q^i)(q^k-q^{i+1})\cdots(q^k-q^{k-1}) \\
&= q^{m(k-i)+1+2+\cdots+(k-i)-1}(q^n-1)(q^{n-1}-1)\cdots(q^{n-(k-i)+1}-1)\Big/ \\
&\qquad q^{(k-i)i+1+2+\cdots+(k-i)-1}(q^{k-i}-1)(q^{k-i-1}-1)\cdots(q-1) \\
&= q^{(k-i)(m-i)} \begin{bmatrix} n \\ k-i \end{bmatrix}_q .
\end{aligned}$$

This leads to the relation [8]

$$\begin{bmatrix} n+m \\ k \end{bmatrix}_q = \sum_{i=0}^{k} q^{(k-i)(m-i)} \begin{bmatrix} m \\ i \end{bmatrix}_q \begin{bmatrix} n \\ k-i \end{bmatrix}_q .$$

In addition there is an expansion relation:

$$\sum_{j=0}^{n} (-1)^j \begin{bmatrix} n \\ i \end{bmatrix}_q z^j q^{j(j-1)/2} = (-1)^n (z-1)(zq-1)\cdots(zq^{n-1}-1)$$

and a type of inversion relation:

$$\text{If } a_m = \sum_{j=0}^{m} \begin{bmatrix} d+m \\ d+j \end{bmatrix}_q b_j, \text{ then } b_m = \sum_{j=0}^{m} (-1)^{m+j} \begin{bmatrix} d+m \\ d+j \end{bmatrix}_q q^{(m-j)(m-j-1)/2} a_j .$$

This last relation was used in [2] for the derivation of the rank distribution of a linear MRD rank-matrix code.

Numerous other analogs between binomial and Gaussian coefficients exist.

Linearized Polynomials

Assume that q is a fixed prime power.

Definition A.3 A polynomial of the form

$$L(x) = \sum_{i=0}^{n} a_i x^{q^i}, \quad a_i \in \mathbb{F}_{q^m}$$

is called a linearized polynomial over $\mathbb{F}_{q^m}$. It is also referred to as a q-polynomial, the term which will be used here. For $L(x)$ a q-polynomial over $\mathbb{F}_{q^m}$, a polynomial of the form

$$A(x) = L(x) - \alpha, \quad \alpha \in \mathbb{F}_{q^m}$$

is called an affine q-polynomial.

The standard reference for the properties of such polynomials is the indispensable volume [6]. Some of their properties are discussed in abbreviated form here. Let $\mathbb{F}_{q^m}$ be an extension of the field $\mathbb{F}_q$. One property that makes such polynomials of interest is the fact the set of roots of such a polynomial form a vector space over $\mathbb{F}_q$, a fact that follows from the properties that for $L(x)$ a q-polynomial over $\mathbb{F}_{q^m}$ we have

$$\begin{aligned} L(\alpha + \beta) &= L(\alpha) + L(\beta), \quad \forall \alpha, \beta \in \mathbb{F}_q \\ L(c\beta) &= cL(\beta), \quad \forall c \in \mathbb{F}_q, \text{ and } \forall \beta \in \mathbb{F}_q. \end{aligned}$$

It is shown ([6], theorem 3.50) that for $\mathbb{F}_{q^s}$, an extension of $\mathbb{F}_{q^m}$, that contains all the roots of $L(x)$, then each root of $L(x)$ has the same multiplicity which is either 1 or a power of q and the set of such roots forms a vector subspace of $\mathbb{F}_{q^s}$ viewed as a vector space over $\mathbb{F}_q$. A converse of this statement ([6], theorem 3.52) is that if U is a vector subspace of $\mathbb{F}_{q^m}$ over $\mathbb{F}_q$, then for any nonnegative integer k the polynomial

$$L(x) = \prod_{\beta \in U} (x - \beta)^{q^k}$$

is a q-polynomial over $\mathbb{F}_{q^m}$.

Similar statements hold for affine q-polynomials with subspace in the above replaced by affine space.

For the material on rank-metric codes it has already been noted that the following matrix ([6], lemma 3.51) for elements $\beta_1, \beta_2, \ldots, \beta_n \in \mathbb{F}_{q^m}$:

$$\mathscr{B} = \begin{vmatrix} \beta_1 & \beta_1^q & \beta_1^{q^2} & \cdots & \beta_1^{q^{n-1}} \\ \beta_2 & \beta_2^q & \beta_2^{q^2} & \cdots & \beta_2^{q^{n-1}} \\ \vdots & \vdots & \vdots & \vdots & \vdots \\ \beta_n & \beta_n^q & \beta_n^{q^2} & \cdots & \beta_n^{q^{n-1}} \end{vmatrix} \qquad \text{(A.4)}$$

is nonsingular iff $\beta_1, \beta_2, \ldots, \beta_n$ are linearly independent over $\mathbb{F}_q$. Indeed the formula for the determinant [6] is given as

$$\det(\mathscr{B}) = \beta_1 \prod_{j=1}^{n-1} \prod_{c_1, \ldots, c_j \in \mathbb{F}_q} \left(\beta_{j+1} - \sum_{k=1}^{j} c_k \beta_k \right).$$

The matrix is sometimes referred to as a *Moore matrix*.

Clearly the q-polynomials have interesting properties and it is natural to ask, for a given q polynomial $L(x) = \sum_{i=0}^{n} a_i x^{q^i}$ how its properties relate to the properties of the associated polynomial $\ell(x) = \sum_{i=0}^{n} a_i x^i$. The two polynomials will be referred to simply as q-associates of each other, the direction being clear from the context. Interesting properties of these polynomials are found in [6].

Random Matrices over Finite Fields

In connection with the random packet coding scheme noted in Section 2.3 it is of interest to determine the rank properties of random matrices over finite fields. This is a well-investigated problem in the literature and only elementary properties are considered here, specifically the probability that a random binary $k \times (k+m)$ matrix over $\mathbb{F}_q$ is of full rank. The number of $s \times t$ matrices of rank r over a finite field $\mathbb{F}_q$, denoted by $\Phi_q(s,t,r)$, is first enumerated (although interest here is almost exclusively in the binary $q = 2$ case). This quantity satisfies the recursion [5]:

$$\Phi_q(s,t,r) = q^r \Phi_q(s-1,t,r) + (q^t - q^{r-1})\Phi_q(s-1,t,r-1) \tag{A.5}$$

with initial conditions

$$\Phi_q(s,t,0) = 1, \quad \Phi_q(s,t,r) = 0 \quad \text{for } r \geq \min(s,t)$$

which is the number of ways of augmenting an $(s-1) \times t$ matrix of rank r to an $s \times t$ matrix of rank r plus the number of ways of augmenting an $(s-1) \times t$ matrix of rank $r-1$ to an $s \times t$ matrix of rank r. With the stated initial conditions the solution can be determined as

$$\Phi_q(s,t,r) = \prod_{i=0}^{r-1} (q^t - q^i)\frac{(q^s - q^i)}{(q^{i+1} - 1)}, \quad r \leq s \leq t. \tag{A.6}$$

It must also be the case that the total number of matrices is

$$\sum_{r=0}^{s} \Phi_q(s,t,r) = q^{st} \quad \text{for } s \leq t.$$

It follows that for the equally distributed case the probability a random $s \times t$ matrix over $\mathbb{F}_q$ has rank r is

$$p_q(s,t,r) = \Phi_q(s,t,r)/q^{st}.$$

The probability a $k \times (k+m)$ binary matrix (i.e., over $\mathbb{F}_2$) is of full rank (which would allow Gaussian elimination to be successful) is given by the expression in Equation A.7:

$$\Phi_2(k,k+m,k)/2^{k(k+m)} = Q_{k,m} = \prod_{\ell=m+1}^{k+m} \left(1 - \frac{1}{2^\ell}\right) \approx \prod_{\ell=m+1}^{\infty} \left(1 - \frac{1}{2^\ell}\right). \tag{A.7}$$

This quantity is approximated by

$$1 - \sum_{j=m+1}^{k+m} \frac{1}{2^j} < Q_{k,m} < 1 - \sum_{j=m+1}^{k+m} \frac{1}{2^j} + \sum_{i,j=m+1}^{k+m} \frac{1}{2^{i+j}}$$

and it is straightforward to establish that

$$1 - \frac{1}{2^m}\left(1 - \frac{1}{2^k}\right) < Q_{k,m} < 1 - \frac{1}{2^m}\left(1 - \frac{1}{2^k}\right) + \frac{1}{2^{2m}}.$$

The equation implies that for a sufficiently large k, the probability the matrix is of full rank tends to unity with m fast, independently of k. For example, for k large the probability a $k \times (k + 10)$ matrix is of full rank is in the range

$$0.999 < Q_{k,k+10} < 0.999 + \epsilon, \quad \epsilon < 10^{-6}. \tag{A.8}$$

An interesting conjecture in [7] is that the average number of extra columns m for a $k \times (k + m)$ matrix over $\mathbb{F}_q$ to achieve full rank is only $m = 1.606695$, a conjecture supported by a significant number of simulations. It would be of interest to have a theoretical result to back this assertion up. These results are of interest to the random fountain coding scheme of Section 2.3.

References

[1] Cameron, P.J. 1994. *Combinatorics: topics, techniques, algorithms*. Cambridge University Press, Cambridge.

[2] Gabidulin, È.M. 1985. Theory of rank codes with minimum rank distance. *Prob. Peredachi Inform.*, **21**(1), 1–12.

[3] Gadouleau, M., and Yan, Z. 2008. On the decoder error probability of bounded rank-distance decoders for maximum rank distance codes. *IEEE Trans. Inform. Theory*, **54**(7), 3202–3206.

[4] Greferath, M., Pavčević, M.O., Silberstein, N., and Vázquez-Castro, M.A. (eds.). 2018. *Network coding and subspace designs*. Signals and Communication Technology. Springer, Cham.

[5] Landsberg, G. 1893. Ueber eine Anzahlbestimmung und eine damit zusammenhngende Reihe. *J. fr die Reine und Angew. Math.*, **111**, 87–88.

[6] Lidl, R., and Niederreiter, H. 1997. *Finite fields*, 2nd ed. Encyclopedia of Mathematics and Its Applications, vol. 20. Cambridge University Press, Cambridge.

[7] Studholme, C., and Blake, I.F. 2010. Random matrices and codes for the erasure channel. *Algorithmica*, **56**(4), 605–620.

[8] van Lint, J.H., and Wilson, R.M. 1992. *A course in combinatorics*. Cambridge University Press, Cambridge.

Appendix B Hasse Derivatives and Zeros of Multivariate Polynomials

The basic reference for this subject is [4] with interesting discussions in [5, 6] and [8] and further information in [1, 3]. Such derivatives are considered to have more interesting and useful properties for polynomials over a finite field than the usual derivative for fields of characteristic 0. The results of this section are mostly related to the discussion of multiplicity codes in Chapter 8.

Consider first the case of univariate functions. The n-th Hasse derivative of a univariate function $f(x)$ will be denoted $f^{[n]}(x)$ and for a polynomial x^m it is defined, for characteristic of the field either 0 or greater than n, as

$$(x^m)^{[n]} = \begin{cases} \binom{m}{n} x^{m-n}, & m \geq n \\ 0, & \text{otherwise,} \end{cases}$$

and so for a univariate polynomial $f(x) = \sum_{i=0}^{m} f_i x^i$

$$f^{[n]}(x) = \sum_{i=n}^{m} \binom{i}{n} f_i x^{i-n} \quad \text{and} \quad f^{[n]}(x-a) = \sum_{i=n}^{m} \binom{i}{n} f_i (x-a)^{i-n}.$$

It follows from the definition that if the ordinary (over fields of characteristic 0) derivative of f is denoted $f^{(n)}(x) = d^n f(x)/dx^n$ then

$$f^{[n]}(x) = \frac{1}{n!} f^{(n)}(x).$$

For a bivariate polynomial $f(x_1, x_2) = \sum_{i_1, i_2} f_{i_1, i_2} x_1^{i_1} x_2^{i_2}$ the ordinary partial derivative is given by

$$\frac{\partial^{(j_1+j_2)} f(x_1, x_2)}{\partial x_1^{j_1} \partial x_2^{j_2}} = \sum_{i_1, i_2} j_1! \binom{i_1}{j_1} j_2! \binom{i_2}{j_2} f_{i_1, i_2} x_1^{i_1 - j_1} x_2^{i_2 - j_2}, \quad i_1 \geq j_1,\ i_2 \geq j_2$$

while the equivalent Hasse derivative is

$$\frac{\partial^{[j_1+j_2]} f(x_1, x_2)}{\partial x_1^{[j_1]} \partial x_2^{[j_2]}} = \sum_{i_1, i_2} \binom{i_1}{j_1} \binom{i_2}{j_2} f_{i_1, i_2} x_1^{i_1 - j_1} x_2^{i_2 - j_2}, \quad i_1 \geq j_1,\ i_2 \geq j_2.$$

Notice ([4], p. 148) that for indeterminates u, v

$$f(x_1 + u, x_2 + v) = \sum_{j_1, j_2 \geq 0} \frac{\partial^{[j_1+j_2]} f(x_1, x_2)}{\partial x_1^{[j_1]} \partial x_2^{[j_2]}} u^{j_1} v^{j_2}.$$

The m-variate case is now straightforward to consider ([2, 3, 4, 5, 8]). Recall the notation of $\boldsymbol{x} = \{x_1, x_2, \ldots, x_m\}$ for the set of m variables and $f(\boldsymbol{x}) = f(x_1, x_2, \ldots, x_m) \in \mathbb{F}_q[\boldsymbol{x}] = \mathbb{F}_q[x_1, \ldots, x_m]$ and $[m] = \{1, 2, \ldots, m\}$. For nonnegative integer m-tuples $\boldsymbol{i} = \{i_1, i_2, \ldots, i_m\}$ and $\boldsymbol{j} = \{j_1, j_2, \ldots, j_m\}$ denote $\boldsymbol{i} \geq \boldsymbol{j}$ if $i_k \geq j_k$, $k = 1, 2, \ldots, m$ and $\boldsymbol{i} > \boldsymbol{j}$ if $\boldsymbol{i} \geq \boldsymbol{j}$ and $i_k > j_k$ for some $k \in [m]$. Denote the total weight $wt(\boldsymbol{i}) = \sum_{j=1}^{m} i_j$ the weight of the m-tuple $\boldsymbol{i} \in \mathbb{Z}_{\geq 0}^m$. As above, a term of the form

$$\boldsymbol{x}^{\boldsymbol{i}} = x_1^{i_1} x_2^{i_2} \cdots x_m^{i_m} \quad \text{for} \quad \boldsymbol{i} = \{i_1, i_2 \ldots, i_m\} \in \mathbb{Z}_{\geq 0}^m$$

is referred to as a *monomial* and is of degree $wt(\boldsymbol{i}) = \sum_{j=1}^{m} i_j$. A multivariate polynomial in m variables is a sum of such monomials and its degree is the maximum of the degrees of its monomials:

$$f(x_1, x_2, \ldots, x_m) = \sum_{i_1, \ldots, i_m} f_{i_1, i_2, \ldots, i_m} x_1^{i_1} x_2^{i_2} \cdots x_m^{i_m} = f(\boldsymbol{x}).$$

Further, the m-variate polynomial $f \in \mathbb{F}_q[\boldsymbol{x}]$ is *homogeneous* of weight d if it is the sum of monomials each of weight d. For m-tuples $\boldsymbol{i}$ and $\boldsymbol{j} \leq \boldsymbol{i}$ denote

$$\binom{\boldsymbol{i}}{\boldsymbol{j}} = \prod_{\ell=1}^{m} \binom{i_\ell}{j_\ell}.$$

The following definition is equivalent to the previous one:

Definition B.1 For m-variates $\boldsymbol{x}, \boldsymbol{z}$ and an m-variate polynomial $f(\boldsymbol{x}) = \sum_{\boldsymbol{k} \in \mathbb{Z}_{\geq 0}^m} f_{\boldsymbol{k}} \boldsymbol{x}^{\boldsymbol{k}} \in \mathbb{F}_q[\boldsymbol{x}]$ and a nonnegative integer m-tuple $\boldsymbol{i} \in \mathbb{Z}_{\geq 0}^m$ the Hasse derivative $f^{[\boldsymbol{i}]}(\boldsymbol{x})$ is defined as the coefficient of $\boldsymbol{z}^{\boldsymbol{i}}$ in the expansion

$$f(\boldsymbol{x} + \boldsymbol{z}) \triangleq \sum_{\boldsymbol{k} \in \mathbb{Z}_{\geq 0}^m} f^{[\boldsymbol{k}]}(\boldsymbol{x}) \boldsymbol{z}^{\boldsymbol{k}} \text{ where } f^{[\boldsymbol{k}]} = \sum_{\boldsymbol{j} \geq \boldsymbol{k}} f_{\boldsymbol{j}} \binom{\boldsymbol{j}}{\boldsymbol{k}} \boldsymbol{x}^{\boldsymbol{j}-\boldsymbol{k}}.$$

Several useful properties follow from such a definition. It is noted [3] that in fields of finite characteristic this definition of derivative is more interesting and in particular allows a form of Taylor series expansion to be given. The following properties [3, 4, 5, 8] follow readily from the definition and are stated without proof. For two m-variable functions $f, g \in \mathbb{F}_q[\boldsymbol{x}]$ and nonnegative integer vectors $\boldsymbol{i}, \boldsymbol{j}$

(i) $(\lambda f)^{[\boldsymbol{i}]}(\boldsymbol{x}) = \lambda f^{[\boldsymbol{i}]}(\boldsymbol{x})$,

(ii) $f^{[\boldsymbol{i}]}(\boldsymbol{x}) + g^{[\boldsymbol{i}]}(\boldsymbol{x}) = (f + g)^{[\boldsymbol{i}]}(\boldsymbol{x})$,

(iii) $(f \cdot g)^{[\boldsymbol{i}]}(\boldsymbol{x}) = \sum_{\boldsymbol{j}} f^{[\boldsymbol{j}]}(\boldsymbol{x}) g^{[\boldsymbol{i}-\boldsymbol{j}]}(\boldsymbol{x})$,

(iv) for a homogeneous polynomial of weight d,

$f^{[\boldsymbol{i}]}(\boldsymbol{x})$ is homogeneous of degree $d - wt(\boldsymbol{i})$,

(iv) $(f^{[\boldsymbol{i}]}(\boldsymbol{x}))^{[\boldsymbol{j}]} = \binom{\boldsymbol{i}+\boldsymbol{j}}{\boldsymbol{i}} f^{[\boldsymbol{i}+\boldsymbol{j}]}(\boldsymbol{x})$.

The notion of multiplicities of zeros of multivariable polynomials is of interest in the definition of multiplicity codes of Chapter 8. The notion for univariate polynomials may be expressed that a univariate polynomial $f(x) \in \mathbb{F}_q[x]$ has a zero of order $s > 0$ at $a \in \mathbb{F}_q$ if $(x - a)^s \mid f(x)$. For multivariate polynomials the concept is slightly more complicated.

Definition B.2 The multiplicity of a zero of $f(\boldsymbol{x}) \in \mathbb{F}_q[\boldsymbol{x}]$ at a point $\boldsymbol{a} \in \mathbb{F}_q^m$, denoted mult$(f,\boldsymbol{a})$, is the largest integer s such that for every $\boldsymbol{i} \in \mathbb{Z}_{\geq 0}^m$, $wt(\boldsymbol{i}) < s$ it is true that $f^{[\boldsymbol{i}]}(\boldsymbol{a}) = 0$. Alternatively $f(\boldsymbol{x})$ has a zero of multiplicity s at $\boldsymbol{a} \in \mathbb{F}_q^n$ if $f(\boldsymbol{x}+\boldsymbol{a})$ has no monomials of degree less than s.

Note that the number of monomials of degree less than s, by Equation 1.11 (with $d = s-1$) is $\binom{m+s-1}{m}$, which will be used in the discussion of multiplicity codes. The number of zeros of multivariable polynomials is of interest for the definition of such codes. It is first noted ([7], theorem 6.13) that the number of zeros of the m-variate polynomial $f(\boldsymbol{x}) \in \mathbb{F}_q[\boldsymbol{x}]$ of degree d is

$$\leq d \cdot q^{m-1}. \tag{B.1}$$

The proof is straightforward using a double induction on the number of variables m and d and is omitted. A useful consequence of these observations is that two distinct m-variate polynomials over $\mathbb{F}_q$ of total degree d cannot agree on more than dq^{m-1} of points of $\mathbb{F}_q^m$ since otherwise their difference, a polynomial of total degree at most d, would have more than this number of zeros and the two polynomials are equal. A refinement of the above result [3, 5, 6] is that:

Lemma B.3 *For $f \in \mathbb{F}_q[\mathbf{x}]$ an m-variable polynomial of total degree at most d, for any finite subset $S \subseteq \mathbb{F}_q$*

$$\sum_{a \in S^m} mult(f,\mathbf{a}) \leq d \cdot |S|^{m-1}.$$

In particular, for any integer $s > 0$

$$Pr_{\mathbf{a} \in S^m}(mult(f,\mathbf{a}) \geq s) \leq \frac{d}{s\,|S|}.$$

It follows that for any m-variate polynomial over $\mathbb{F}_q$ of degree at most d

$$\sum_{\boldsymbol{a} \in \mathbb{F}_q^m} mult(f,\boldsymbol{a}) \leq dq^{m-1}. \tag{B.2}$$

An interesting property of the multiplicity of m-variate polynomials is:

Lemma B.4 ([3], lemma 5) *If $f(\mathbf{x}) \in \mathbb{F}_q[\mathbf{x}]$ and $\mathbf{a} \in \mathbb{F}_q^m$ are such that $mult(f,\mathbf{a}) = m$, then $mult(f^{[\mathbf{i}]},\mathbf{a}) \geq m - wt(i)$.*

Notice that for the univariate case $m = 1$ the notion of multiplicity coincides with the usual notion, i.e., a root a of the univariate polynomial $f(x_1)$ of multiplicity m implies that $(x_1 - a)^m \mid f(x_1)$ or $x_1^m \mid f(x_1 + a)$.

The following result ([3], proposition 10) is also of interest:

Proposition B.5 *Given a set $K \subseteq \mathbb{F}_q^m$ and nonnegative integers m, d such that*

$$\binom{m+n-1}{n} \cdot |K| < \binom{d+n}{n}$$

there exists a nonzero polynomial $p \in \mathbb{F}_q[\mathbf{x}]$ of total degree at most d such that $mult(p,\mathbf{a}) \geq m \ \forall \mathbf{a} \in K$.

References

[1] Asi, H., and Yaakobi, E. 2019. Nearly optimal constructions of PIR and batch codes. *IEEE Trans. Inform. Theory*, **65**(2), 947–964.

[2] Augot, D., Levy-dit Vehel, F., and Shikfa, A. 2014. A storage-efficient and robust private information retrieval scheme allowing few servers. Pages 222–239 of: *Cryptology and network security*. Lecture Notes in Computer Science, vol. 8813. Springer, Cham.

[3] Dvir, Z., Kopparty, S., Saraf, S., and Sudan, M. 2009. Extensions to the method of multiplicities, with applications to Kakeya sets and mergers. Pages 181–190 of: *2009 50th Annual IEEE Symposium on Foundations of Computer Science – FOCS '09*. IEEE Computer Society, Los Alamitos, CA.

[4] Hirschfeld, J.W.P., Korchmáros, G., and Torres, F. 2008. *Algebraic curves over a finite field*. Princeton Series in Applied Mathematics. Princeton University Press, Princeton, NJ.

[5] Kopparty, S., Saraf, S., and Yekhanin, S. 2011. High-rate codes with sublinear-time decoding. Pages 167–176 of: *Proceedings of the Forty-Third Annual ACM Symposium on Theory of Computing*. STOC '11. ACM, New York.

[6] Kopparty, S., Saraf, S., and Yekhanin, S. 2014. High-rate codes with sublinear-time decoding. *J. ACM*, **61**(5), Art. 28, 20.

[7] Lidl, R., and Niederreiter, H. 1997. *Finite fields*, 2nd ed. Encyclopedia of Mathematics and Its Applications, vol. 20. Cambridge University Press, Cambridge.

[8] Yekhanin, S. 2010. Locally decodable codes. *Found. Trends Theor. Comput. Sci.*, **6**(3), front matter, 139–255 (2012).

Index

Printed in the United States
by Baker & Taylor Publisher Services